James Emerson

Treatise Relative to the Testing of Water-Wheels and Machinery

also of inventions, studies, and experiments, with suggestions from a life's

experience

James Emerson

Treatise Relative to the Testing of Water-Wheels and Machinery
also of inventions, studies, and experiments, with suggestions from a life's experience

ISBN/EAN: 9783337094782

Printed in Europe, USA, Canada, Australia, Japan

Cover: Foto ©berggeist007 / pixelio.de

More available books at **www.hansebooks.com**

TREATISE

RELATIVE TO THE

Testing of Water-Wheels

AND

MACHINERY, ·

ALSO OF

Inventions, Studies, and Experiments, with Suggestions
from a Life's Experience.

By JAMES EMERSON,

WILLIMANSETT, MASS., U. S.

FOURTH EDITION.

PRICE, $1.00. POSTAGE, 10 CENTS—STAMPS.

1892.

INTRODUCTION.

Some ten years since, through the invention of an instrument for weighing the power required to drive machinery, I became interested in the testing of turbine water wheels. Previously such tests had only been possible for the wealthy. The apparatus used for the purpose, though expensive, was crude, clumsy and unreliable, while the formulas for computing water used were tedious for the initiated and impossible of application by the multitude, consequently few of those using wheels were able to demonstrate the absurdity of the fabulous claims made by the most of the turbine builders, and for years confusion had reigned, alike injurious to the manufacturer and honest builder. Years of experience that will be mentioned in the last part of this work made me fully aware of the task it would be to bring order out of such confusion; still the attempt was made, and has since been continued without a thought of abandonment. Those who have only witnessed the test of turbines at the Holyoke flume have little idea of the operation as conducted by engineers of the past; barrels of oil and a small army of assistants were required, so that the cost run up into the thousands. The average cost per wheel in 1869 was $2,500. The superintendent of the Niles Iron Works of Cincinnati, O., came to me at Lowell, in order to make arrangements for the test of a Kindleberger wheel; he offered $600, but under the then existing conditions it could not be done. Weeks and even months were consumed in the test of a single wheel. The experience that year convinced me that such expenditures of time and money were entirely unnecessary, and plans were soon completed for demonstrating that fact. Many ideas then prevalent had to be considered. In the first place, a testing flume with suitable apparatus was an expensive affair, while my means were very limited; then again it was supposed by all, that wheels of the same make were all of the same proportional efficiency, so that each builder would only need to have one wheel tested. consequently the patronage would be very irregular, while the expense would be constant, as experienced help would be required, and such help could only be retained by constant employment, or at least constant pay; the latter difficulty was surmounted by doing all of the most difficult and hardest work myself, simply employing a laborer for each test, while my daughter *timed;* kept the records of gauges during the trials, gave me the power every two minutes, in order to enable me to change the weight correctly, then made the computations and copied the results. This continued for a year or more; then Miss Charla A. Adams. "Charla" succeeded my daughter, and such success as I have had in aiding the improvement of turbines, by enabling builders of small means to ascertain the exact value of their numerous plans, and establishing the testing system, is due in a great measure to her unwearied patience, care and attention. She has had the entire mathematical part of the work to do. not only of the tests, but that necessary for the preparation of a large portion of the tables published in this work; she has kept records of all tests, and prepared numerous copies of the same for public institutions and for turbine builders; in all, she has proved her fitness for the purpose, and not only her fitness, but woman's adaptability for such work. The practice of testing turbines has caused many changes and exploded many theories; of course this has not been done without destroying the hopes of many builders, at the same time it has been the means of bringing the best wheels prominently before the public. The tests have at all times been open to the public; builders have been desired to bring engineers to assist, and such have ever been welcome. It is a difficult matter to make purchasers realize that wheels made from the same patterns vary exceedingly in efficiency, yet there are few manufacturers ignorant of the fact that a wheel of any make doing well in a mill gives no assurance that another of the same make will give equal satisfaction. Ninety per cent. wheels are much sought for, but there are plenty of 80 per cent. wheels that will do far better than many that have given higher results. Ninety per cent. is only obtained under the most favorable conditions, and such can not be continued long in practical use.

Illustrations published in the first edition of this work have been found very convenient for reference in law and other cases, consequently a greater variety has been published in this edition.

That I know but little about the exact lines necessary for the production of a good turbine is not, perhaps, a legitimate excuse for the absence in this work of directions for turbine building, because the most minute formulas to be found upon the subject have undoubtedly been published by those who knew still less about it than myself, but such formulas seem to have hindered rather than to have aided turbine improvements, for it is very certain that the best turbine builders have given little heed to such formulas, hence I have not attempted to do what I could not do well.

Some of my Annual Reports were electrotyped, and various items from those have been used in this work, and where such reports of tests have been used, the numerous changes of weights are given in full; while in others only the best test at whole gate is given; and it may be well here to state that there is a certain speed at which any turbine does its best, and to find that point it is necessary to try many changes of weights. Wheels made from the same patterns seldom do their best at the same speed, and this variation is the cause of considerable loss of power through incorrect gearing for speed.

It is also necessary to state that there is always a leakage into the measuring pit during a test, which is to be deducted from the quantity flowing over the weir; this leakage may not be given in some of the reports, but if the depth on the weir is given, the difference between the quantity as found per tables for that depth, and the cubic feet given in the report of test will give the leakage, that is, if the length of weir is given. The omissions are owing to the use of only a part of the electrotyped reports.

JAMES EMERSON.

WILLIMANSETT, MASS., October 1, 1878.

INTRODUCTION

TO THE FOURTH EDITION.

THE success of my previous publications warrants a continuance of my previous methods.

As a rule, technical works only interest such as are engaged in the study of the subject treated ; my practice has been to give the best possible information relative to the subject treated upon, then to mix in other matter of varied interest, causing others to look the work through, thus making, perhaps in itself, a dull technical matter generally familiar. In this edition the mixture has been carried to a greater extent and quite likely may be objectionable to some, but to others that may be the very cream of the whole, and I hope to give each his or her money's worth. And here it may be stated that the price and cost of this book have little relation ; the contents epitomize a life's work and study. It is difficult to-day to decide what will be popular a half century hence, and it is a matter of far greater interest to me to be popular then than now. A comparison of this edition of my book with any and all works treating upon hydrodynamics printed previous to 1870 I think will prove that I have as hydrodynamic engineer done more in the past quarter of a century to establish a practical knowledge of milling hydrodynamics, and in the production of instruments to render that knowledge evident, than has been done by all other engineers for a century past.

Still further evidence may be found in a comparison of the plans proposed in the book and the costly and obsolete methods continued at Lowell, Lawrence, Holyoke, Cohoes, Minneapolis, and all other water powers under the charge of engineers graduated from our colleges and technical institutions, where text books a century behind our time continue to be used. But the real difficulty consists in the fact that the instructors and instructed too often are entirely unfitted by nature for engineers ; *expedients for emergencies* and good judgment are absolute requirements for an engineer; mathematics as tools are excellent, but a poor substitute for brains. Mechanical ability of the highest standard will enable man or woman to reach a high plane as an engineer, but without such ability one had better turn to some other calling where less skill is required.

"The Eminent Hydraulic Engineer," page 41, is intended as the description of a class much too numerous, and has no reference to any individual.

JAMES EMERSON.

July 10, 1892.

INDEX.

Tests of Water Wheels and Machinery

DESIGNED TO AID ALL INTERESTED IN HYDRAULICS,
Particularly Turbine Builders, Manufacturers, Owners
of Water Power, and Counsel Managing
Cases in Litigation.

———— ◆ ◆ ◆ ————

TESTIMONIALS.

———

OFFICE OF THE

PROPRIETOR OF THE LOCKS AND CANALS ON THE MERRIMAC RIVER,
LOWELL, MASS., February 5, 1879.

JAMES EMERSON, Willimansett, Mass.

Dear Sir: Your work on water wheels and machinery was left here yesterday by Mr. Swain.

My father (James B. Francis) is at present in Europe, and probably will not return before next August. I take the liberty to thank you for him, and to assure you that your book contains a fund of information of the kind we want. How to utilize water power to the best advantage is one of the great problems of the day, and I am sure you have contributed much information on the subject.

Very truly yours,

JAMES FRANCIS, *Ass't Engineer.*

———

NORTH CHELMSFORD, MASS., February 7, 1879.

JAMES EMERSON, Willimansett, Mass.

Dear Sir: I have examined your book, "Treatise on Tests of Water Wheels and Machinery," and find in it a very large amount of valuable information, in a simple form, not obtainable in any other work. I have copies, in full, of all your tests, dating back nearly ten years, prepared at (to me) great expense, and of course I have faith in their reliability. (At all events, twenty times their cost would not buy them, were no other copy available.) The publication of your experience, in so convenient a form, must prove to be of very great benefit to manufacturers, turbine builders, and such of the legal fraternity as have hydraulic cases to manage. Every millwright ought certainly to possess a copy. The tables of "Velocities due Head," alone, will save him in time every year several times the cost of the book.

Very truly yours,

A. M. SWAIN.

———

BOND BROS. & BOTTUM LAW OFFICE,
NORTHAMPTON, February 7, 1879.

JAMES EMERSON, Holyoke, Mass.

Dear Sir: I have examined your work on "The Testing of Water Wheels and Machinery," *with matters pertaining to Hydraulics* (2d Ed. 1878), and find that it contains in a very convenient form a large amount of information which every lawyer must obtain from some source before he can safely advise a client or properly try a case concerning water power or the power of water wheels. Accept my thanks for your treatise; it will be of great use to me in my professional work. Yours truly,

D. W. BOND, *District Attorney.*

LAW OFFICE,
NEW HARTFORD, CONN., February 8, 1879.

JAMES EMERSON:

Dear Sir: I have examined your book, "Treatise on Tests of Water Wheels, &c.," and find it to be really *multum in parvo*. It contains in simple form much information needed by members of the legal profession who are engaged in suits involving hydraulics, power, flow of water, and kindred subjects.

I am very truly yours,

JARED B. FOSTER.

LAW OFFICE OF J. P. BUCKLAND, COUNSELOR IN PATENT CAUSES,
SPRINGFIELD, MASS., February 8, 1879.

JAMES EMERSON:

Dear Sir: I assure you that my examination of your new work, entitled "Tests of Water Wheels and Machinery," has given me a great deal of pleasure. Many a time have I searched for hours to find some of the many data with which the book is crowded.

In the preparation and trial of cases involving questions of water power, mill-rights, leases of power and the performance of machines built under contract, and kindred matters which are constantly coming before courts and arbitrators for settlement, your work will be a valuable aid to lawyers and parties. I know of no single book which has within its covers so many practical data for use in the above line of cases, and I am bound to say that I think the legal profession is under much obligation to you for the preparation of it.

Yours truly,

J. P. BUCKLAND.

SENATE CHAMBER, WASHINGTON, D. C., February 9, 1879.

JAMES EMERSON:

My Dear Sir: Permit me to thank you for a copy of your book, "Tests of Water Wheels and Machinery." I have read it with as much care as I could find time from my official duties here to do, and have no hesitancy in saying that it must prove a very valuable work, as well to lawyers conducting litigations, as to mill owners seeking to avoid them.

Thanking you again, I am, truly yours

H. L. DAWES.

OFFICE OF
AMERICAN PRINT WORKS, FALL RIVER, MASS., March 3, 1879.

JAMES EMERSON:

DEAR SIR: Please accept my thanks for the copy of your book "Tests of Water Wheels and Machinery, received a day or two since. It contains much valuable information relating to water power, is arranged in a very concise form and will prove a very convenient source of reference."

Very truly yours,

THOS. J. BORDEN, *Treas. Watuppa Reservoir Co.*

BOOTT COTTON MILLS, LOWELL, MASS., March 11, 1879.

JAMES EMERSON:

DEAR SIR: Permit me to thank you for a copy of your "Tests of Water Wheels and Machinery." I have examined the same with great care and must say that you have given a fund of information to Manufacturers, Turbine Builders, Owners of Water Power, and the Legal Profession who have suits involving Hydraulics, in a most simple and concise form. There is no subject to-day connected with manufacturing that there is so much ignorance about as the economical use of water power, best wheels, and appurtenances to utilize it. Your book cannot fail to have a large sale, as you have contributed so much information on the subject.   Yours truly,   A. G. CUMNOCK, *Agent.*

NEW HAVEN, CONN., April 29, 1879.
MR. JAMES EMERSON:

DEAR SIR,—Allow me to take this opportunity to thank you for the copy of your Treatise relative to the Testing of Water Wheels and Machinery, which you did me the favor to send me some time since. I have found it to be a mine of valuable information, and have often had occasion to consult it. It certainly effectively supplies a great desideratum. The manufacturing interests of this country owe you a debt of gratitude, to speak of nothing more, for the valuable work you have done in testing water wheels and machinery.

Yours truly,

W. A. NORTON,

Prof. of Civil Engineering in Sheffield Scientific School.

LEBANON, N. II., March 7, 1879.

JAMES EMERSON: I will do all I can to recommend your book. I think it the best book I have seen for a millwright.

Yours truly, WILLIAM DUNCAN, *Engineer and Millwright.*

UNITED STATES SENATE CHAMBER, WASHINGTON, Feb. 1, 1879.
JAMES EMERSON:

DEAR SIR: Yours of the 30th ult. has just been received, with your book, for which I heartily thank you. Of course, I have not yet had time to examine it, but will do so and write you about it. I think from a hasty glance I cast over it, that its contents will be of value, and I shall preserve it as a memento of one of the intelligent patentees of the country, who have done so much for its interests. Yours truly,

BAINBRIDGE WADLEIGH.

My correspondence with Mr. Wadleigh had reference to a change in the Patent Law advocated by him, a change which to me seems unjust to the inventor and unlikely to benefit the public. The defects of the present law are mainly due to its loose administration. To make a great show of income, patents are daily granted for frivolous devices unworthy of such distinction; worse still, patents are constantly being granted to several different applicants for substantially the same device, a practice alike injurious to the real inventor and the public. A single amendment of the present law, with a return to the conservative practice of granting patents followed thirty years since, would do away with the real objections to such grants. A large proportion of the patents that are granted are not only worthless but they stand in the way of real improvement. Amend the present law so that provisional protection shall be granted for three years upon application as at present, *for real inventions*, anytime previous to the expiration of that period, upon the payment of one hundred dollars; issue a patent for twenty years from the date of application; under no circumstances allow an extension of such grant. The provisional term for the sum of twenty dollars will enable the applicant to judge of the validity, value and advisability of its continuance. If it is deemed advisable to protect the plans of geniuses who saw a barrel in halves, thus making two tubs of it; or point one end of a friction match stick, so that it may be used to light a pipe or pick the teeth, do so; but do it for a shorter period and under some designation equivalent to that of "Trade Mark." It discourages the real inventor and debases the term, Patent, to put such devices upon the same plane as those that have required years of study to perfect.

THAYER SCHOOL OF CIVIL ENGINEERING.
Dartmouth College,
HANOVER, N. II., August 22, 1879.

JAMES EMERSON.

Dear Sir: Permit me to say that your book on Tests of Water Wheels and Machinery has greatly interested me. To students of hydraulics it has a special value in affording so many ractical results of American construction and opera-

Holyoke Testing Flume.

My connection with the above closed in 1880, yet parties often write to me for information about tests made there which cannot be furnished because the tests now made there are private, a practice which, according to my experience, makes the tests of little value. While the flume was in my charge it was a common matter for testers to request and at times insist that certain results should be suppressed.

One man asked me to give him such reports as I gave another whose name he mentioned and draw upon him at sight for one thousand dollars. Another said he would give me four thousand dollars in greenbacks if I would comply with a similar request. Two well known builders first tried to buy me and then threatened vengeance if correct reports were given. William M. Mills took the whole gate tests of one wheel and the part gates of another and published them as the test of one wheel.

Stilwell & Bierce used certificates obtained from Clemens Herschel of 88 per cent. tests from wheels that I had twice tested and been unable to get 80 per cent. These wheels, two of them at least, were made from the same patterns as the 92 per cent. wheel (see page 342) used to make the belt, gear, and draft tube experiments which gave such remarkable results that it was tested and retested by delegations from Yale, Boston School of Technology, and other places. It was taken from the flume, reset and retested, but invariably repeated results within one-fourth of one per cent. The Holyoke Machine Company bought up 150 copies of my book, all that remained of the second edition, and used them for fuel because the low as well as high results of their tests were published. In short, to be of general use, all tests should be public. There are monomaniacs rated as engineers that pretend that the so-called Boyden wheel is of a superior character, while in fact it was never in any particular equal to the Tyler scroll wheel.

Frank Sicman, now in charge of the Holyoke flume, I believe to be capable and honest, but he has too many gauge hands (See difference of results, pages 36 and 37, tests of the same wheel with same gauge hands.)

It is not likely that any one will again have the chance I did to make such tests, and economy compelled me to do most of the work myself, so that it was brought to the most perfect exactitude and simplicity. The three tests reported on pages 352, 353, 354, were taken offhand, without a thought of extra care, yet I do not believe it would be possible to make such tests again.

Much has been learned in testing since the system commenced with the Swain wheel in 1869. That wheel reached about 80 per cent. Had it been properly set and tested it should easily have exceeded 85 per cent. Wheel builders often request me to reply to articles slurring the test, which may be done when an erroneous report can be named. In 1879 four parties objected to my having charge of the free public tests, promising to have wheels tested if some other engineer was employed, but failed to send wheels. One of the four, I have since ascertained, furnished the article for publication (page 204), "A man of courage." Another, my friend I. X. L. Humphrey, an amusing writer of fiction relative to turbine of his own make and hydrodynamics in general, being interested as defendant in an expensive hydraulic suit, accidentally met the plaintiff waiting at the depot for a train to Willimansett. Excusing himself for a moment, he put his I. X. L. into practical operation and made better time on foot across the bridge than I had ever been able to get from his wheel, and retained me for his side just as the plaintiff arrived at the door.

JAMES EMERSON.

July 16, 1892.

BROWN'S FACING MILL, POWER, ETC.

Willimansett, Mass.

In this case the ordinary question cannot be raised as to whether steam or water power is cheapest, because the great expense necessary for land, dam, turbine, etc., etc., has already been incurred, consequently the loss of the water power would necessitate the expense almost of a double plant without any corresponding gain, for the present plant is located close to the depot, upon the road to Springfield, in proximity to Holyoke, and probably in as convenient and desirable a place as could be found in the New England States.

The whole plant must now stand to the owner at a cost of some seven or eight thousand dollars and completely fitted for business, to be operated by water power.

This power consists of an almost unchanging supply of water falling sixteen feet. Very few water powers can be found so regular in quantity as this. About three hundred cubic feet per minute can be depended upon the year round, though in extreme droughts it may be a little less, and during the spring something more. Three hundred cubic feet of water per minute falling sixteen feet evolves nine horse power; ponding the twenty-four hours' supply and using it in eight hours furnishes twenty-seven horse power, of which eighteen may be utilized or, say, fifteen horse power, ten hours per day, and this without waiting to get up steam as would be the case for every little job with a steam plant.

Mr. Brown's work is not such as to require constant power, hence the advantage of the power that may be called upon to operate the machinery for ten minutes, or an hour, as the case may be, then stopped and remain idle without expense or care until again needed, which could not be the case with steam power. Besides, with steam power to be safe for mill and neighborhood, a qualified engineer would be necessary to take charge of engine and boiler. I do not mean to be understood that engineers are always employed in such cases, but I do mean decidedly that, to obtain the same safety and convenience now enjoyed by Mr. Brown, one thousand dollars per year will not make an equivalent for the value of his water power, which is now in good condition throughout, to the best of my belief.

JAMES EMERSON.

WILLIMANSETT, MASS., Feb. 21, 1892.

The Law Antagonistic to Knowledge and Justice.

A thorough study for a score of years of hydrodynamics makes it evident at least to myself that, except through imperfect deeds, no cause for disputes in milling matters can arise that may not be made so clear as to leave no just cause for litigation. Every effect has a cause, and such cause may readily be ascertained by an intelligent engineer.

Such cases usually have an individuality and each must be considered in itself.

The ordinary surveyor's level between mills is seldom of much account in backwater cases so far as determining the cause of complaint. A number of such cases will be reported in this edition of my work, from which information of my method of ascertaining facts may be obtained. A wide experience of the practice of law in many states in hydraulic cases convinces me that while Massachusetts from its early manufacturing should be one of the most intelligent in such matters it is in fact like its Andover theology and gallows witness stand more iron-clad and backward in its rules than any other. The assessors' valuation of a plaintiff's property, the verdict of a parallel case in the same county, mention of the obsolete character of the property in question, nor in fact anything that would show the utter worthlessness of the whole claim was presented in a recent case for damage. This may have been owing to the incompetency of the attorney, but certainly the most essential evidence necessary to enable the jury to decide intelligently and justly was left out; both attorneys seemed destitute of knowledge in such matters and equally desirous of preventing the jury from obtaining knowledge.

The Willimantic Borough case offered another lesson for litigants. It would hardly be possible to find a case more decisive in character; one any intelligent manufacturer would easily understand.

The case is one for mutual concession by which both could be benefited. There is a trifle less than three feet in the level between the level of the crests of the Borough and Johnson's dams. Johnson claims to be desirous of digging a new tail-race 12 feet wide and 400 feet in length extending from his wheel down to Borough line where the water backs up 2 1-2 feet in depth.

Mr. Johnson desires to send down through the said tail-race 150 cubic feet of water per second, which moving three feet per second would require a depth of four feet, or a foot and a half more than the depth of water at the dividing line of the two properties, consequently the set-back of water could cause no real loss of power to Johnson. Like all such conditions the discharge from the race meeting the standing water of the pond would soon form a bar that would constantly increase in height and reduce the head on Johnson's wheels. By mutual concession such bar could yearly be removed. During the average flow of water the supply gorges the channel below Johnson's land so that there is no backwater at the line at all, and the channel is constantly filling up so that the rock B represented in the illustration of 1890 is now submerged and invisible.

Instead of defending the case upon its merits, the defense was that the river was variable in supply and of little value. There is little encouragement for knowledge if the law is to render such knowledge worthless.

What is the difference in principle, for lawyers to band together in Congress and Legislature and use their influence to perpetuate laws and rules of a barbarous age for personal benefit, and the uniting in mobs to destroy power looms and other improved machinery as was done by the laboring classes?

On another page I have stated that I seldom leave the witness stand without feeling outraged, but since attorneys of reputation confine their cross examination of my testimony to reading extracts from my legal criticisms it rather looks as though they consider their case weak and "get off their head," and so almost unconsciously give out information where it *should do the most good.*

ATWATER MANUFACTURING COMPANY,
PLANTSVILLE, CONNECTICUT.

The annexed illustrations show the dam substantially as it appeared in 1880. The coping stone of the dam had in several places been crushed and carried away by ice, leaving the iron dowels projecting above as shown. The coping stones N and O remained, as was the case with some at the other end of the dam. The necessity for repairs of the dam at that time was so apparent that it was made a matter of my report and record. The crest of dam has since been evened up by timber and cement on a level with the top of the original coping, as shown by the stone N. Soon after a suit was commenced by the mill owners above, I examined the dam and saw that no improper raising had been made, nor was there any indication that the set-back of water from the dam interfered with the mills above. I went to the plaintiffs and proposed to test the power of their mills under existing conditions, then draw down the pond several feet and again test their power.

The land above and below is swampy; on such foundation a stone dam is sure to settle, as well as to be worn away on top by the overflow. The mill upon the east side has three turbines, the Manufacturing Co., one. For years the four wheels had kept the water constantly drawn down, leaving the banks of streams and the marshes uncovered, so as to cause sanitary complaint. At about the time the dam was repaired, the three wheels in the old mill were closed for good, since when the pond has remained full, perhaps causing the belief that the dam had been raised. Had that been the case there would be a belt of dead trees up the banks of the pond and river, for those banks were lined with trees, but no dead ones were to be found.

The Manufacturing Co. purchased the place under the assurance that the fall was eight feet. The dam has been the "overflow" of the pond and much debris had settled below the dam in the tail race at F. The small island marked D, and loose stone and gravel, have been removed and the head is now but seven and a half feet. The pond is quite extensive and extends considerable of the distance towards the mills above, and of course substantially is level from end to end, while the current from the pond to the mills above is quite rapid; at the line marked A, it is one foot per second when the wheels above are in operation. A surface mark was made at L just before noon; as the wheel gates closed in the mills above, the water settled two inches during the noon hour, though it must have risen on the Atwater dam; for the wheel gate was closed there. In the tail race at V, the surface fell six or more inches. Several hundred feet down stream from the plaintiff's mill, a stake is driven, the top of which is five and one-eighth inches above the crest of Atwater dam; with the wheels above in full operation the depth of water on stake was one and three-eighths inches. From the stake up, the current was very rapid, so as to make it hard to row a boat. Under the bridge on Main street, a sewer and the road wash has partially formed a bar, through the middle of which the current has cut its way, carrying the debris down near the line marked X, where it meets the set-back of water from the Atwater dam, and there has formed an extensive bar across the stream that raises the water above and in the vicinity, and this bar, the whole cause of the misunderstanding, had not been found by the several different civil engineers employed in the case, though the action of the water easily made such obstruction apparent.

This bar has doubtless rapidly increased since the stopping of the three wheels in the old Atwater mill, for previous to that the water was so constantly drawn down that the meeting of the waters would have been in pond.

To understand the conditions fully, consideration must be had of the fact that the plaintiff ponds the twenty-four hours' supply of the stream, and then sends the whole down in ten hours or less; consequently as the Atwater wheel can use but a portion of that quantity, much of it must go over the dam, causing the surface there to be higher than if only the natural flow of the stream came down.

Now one hundred dollars properly expended as I proposed, by testing the power, would have made the matter so plain that there could have remained no possible cause for dispute, instead of which the law was invoked, to the great benefit of the lawyers at least.

In court the case must be fitted to the law, not the law to the case. Each witness is sworn to tell the truth, the whole truth, and nothing but the truth, and then every effort is made to suppress all undesirable truth by one side, the other side trying to confuse and make the witness _e if possible.

Representation of Dam in 1880, and plan of surroundings. 1890

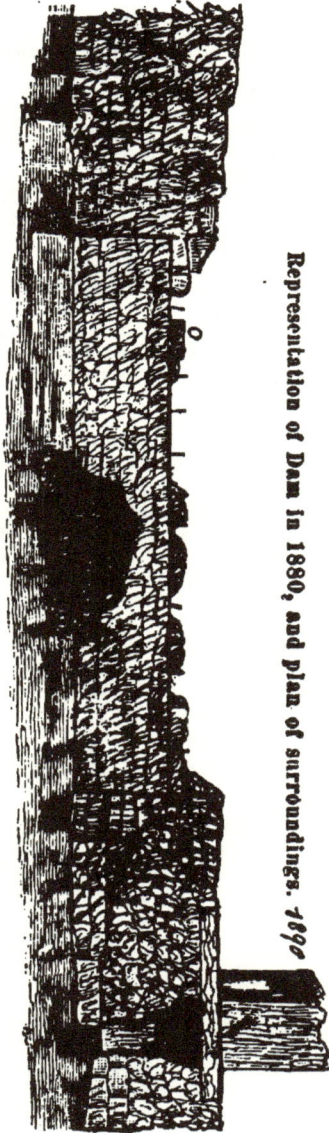

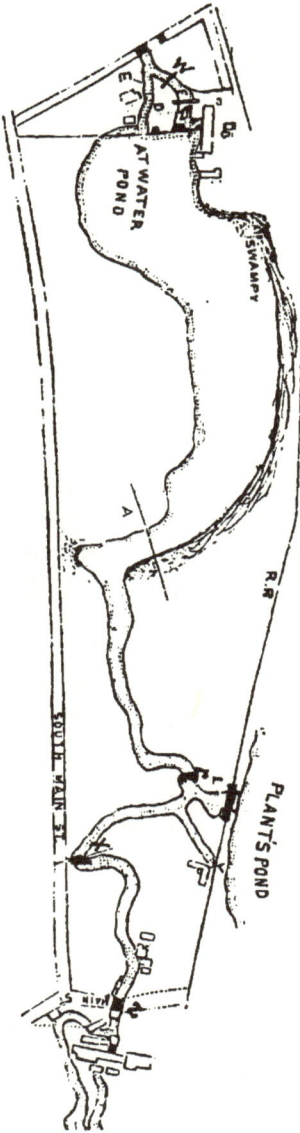

WILLIMANSETT, MASS., July 14, 1890.

To the Board of Water Commissioners of Willimantic, Conn.

GENTLEMEN :—Having made several examinations of your arrangements for supplying the borough with water, I report as follows : My first examination was made when the estimated flow of water over Mr. Johnson's dam was eighteen inches in depth, the next when it was ten, and one last week when there was but a slight flow over it. At the same time the water was drawn down several feet at the pumping station dam, leaving only the flow of the natural stream at Johnson's line. From surface of water there to the level of your dam it was two feet five and a half inches. The lower line in the sketch annexed represents the approximate surface there at that time, and I believe the higher lines would do the same under the conditions named, could your pond be drawn down at such times. Two feet average for the year round of backwater would certainly be more than Mr. Johnson would suffer from your dam.

The sketch of dam also annexed does not represent your dam as it is, but as I would earnestly advise it be made and the dam backed with gravel before the season closes. The cracks in the walls of the station, embankments, and dam show that a rapid process of disintegration is going on. The mortar—it has no claim to be called cement—is poor. With such an abundance of water it is hard to conceive why more than fifteen feet head was ever desired. The present dam, made with an overflow, and waste-gates that can be used in time of need as I have shown, well backed with gravel, should stand.

The wheels selected show the engineer to be ignorant of the improvements made in turbines, and the general mechanical construction displays a lamentable lack of mechanical ability ; while placing the waste-gates where it is impossible to use them when most needed, shows a lack of ability and judgment that seems incredible.

James B. Francis of Lowell, Mass., has had constant care of dams for nearly sixty years. I would seriously advise you to employ him to make an examination of your dam ; have the water drawn down at the time. Stone dams have a fatal tendency to tumble down, and there seems a possibility of yours doing so. Owing to the poor mortar there is little strength in the dam except in its weight.

Very respectfully yours,

JAMES EMERSON.

The above reported works were planned by Mr. J. T. Fanning, and constructed under his supervision. The pumps work at half speed designed but broke the iron frames connecting the pumps to the wheels at that speed soon after starting. The waste-gates are placed about fifty feet from the abutments of the dam, so that it is impossible to get at them unless the water is below crest of dam.

Johnson's Dam.

At rock B there was a rapid current at my first visit.
A, Islet and narrow passage.

SUIT FOR DAMAGES.

PALMER (Mass.) WATER WORKS *vs.* STONE, *Plaintiff*.

Visited the reservoir and streams, also pond, and works of Mr. Stone, in June. Stream below water works dry, as was the pond of plaintiff. His shop was closed.

At the reservoir there were plain indications that the supply was but little more than equal for water company's use.

Early in November made another examination with similar results, except that the plaintiff's pond was full, yet his shop remained idle.

November 14, made a more thorough examination of the streams, reservoirs, and stream supplying reservoirs ; made a crude measurement of the supplying stream, also of capacity of plaintiff's pond.

There was more water the 14th than at the previous examinations, yet evidently but little surplus, as the lower reservoir was not quite full. There had been quite a heavy rain, for a day, between my second and last visit.

The crude measurement of the supplying stream showed about one-half of a cubic foot per second, which, for safety and convenience of computation, I call four gallons per second as the total supply to the reservoirs, though there may be small springs in or near the edges of the reservoirs, yet as there are but about two hundred families supplied, aside from depots, hotels, and wire mills, the supplying stream will seem sufficient.

Four gallons per second will supply 5,760 persons each with sixty gallons of water per day of twenty-four hours.

The plaintiff's pond is irregular in shape and depth, estimated surface ninety by one hundred feet, of which two and one-half feet in depth may be used, or say 22,500 cubic feet. This is gauged to the conduit to wheel through a ten-inch pipe the top of which is three feet below the surface of the water when the pond is full.

Overshot wheel coarsely made, set and supplied with water. Area of gauge, 78.5 square inches, sixty per cent. of which is 47 square inches.

Spurting velocity of water under say an average head of two feet is 11.84 feet per second, equaling a discharge of 3.7 cubic feet per second ; 3.7 cubic feet per second falling sixteen feet evolves 6.7 h.p. propelling force, for which the old wheel could not return a co-efficient of more than fifty per cent., or 3.4 h.p. Two hours' run at that rate would draw all of the water that could be utilized from the pond, then it would require twelve hours to refill the pond so that two and one-half to three hours per day would be all the time during the working hours of the day the machinery could be kept in use.

In the spring and during the melting of the snow, and heavy rains, undoubtedly there is a larger supply of water, but the water courses below or above the reservoirs do not indicate a much larger quantity generally than at the present time. One horse power ten hours per day is as much as could be made available.

JAMES EMERSON.

WILLIMANSETT, MASS., NOV. 16, 1891.

L. L. DEAN & CO., AMSTERDAM, N. Y.

GENTLEMEN:—In accordance with your request, I have at different times made examinations of your mill and its surroundings at Rock City and report as follows:—

On May 28, current year, made my first examination; the water was high, flowing over the dam below your mill 24 inches in depth. The next day, after the water had fallen on the dam to six inches in depth, the depth in tail race back of your mill on the boundary line was about thirty-four inches. I made examinations of the surroundings and found the conditions good so far as they could be ascertained until the water could be drawn from the pond below your mill. I was then informed that a suit had been commenced to compel the lowering of the dam below sufficiently to prevent its backing the water upon the turbine that furnished the power to drive the machinery in your mill, and was requested to ascertain if the dam did back the water upon your turbine, and, if so, to what extent; then to ascertain by the most accurate method the loss so caused, then to ascertain the cost of an equivalent power by steam at the mill.

Stumps in the water on the west side of the pond offer positive evidence that the dam below has been raised within a comparatively recent period.

To ascertain the cost of the equivalent a dynamometer or power scale was placed in the main driving pulley of the mill. The machinery of the mill, except the dynamo for lighting, was then driven by the steam engine during the day, the coal as used being carefully weighed, when it was found that each h. p. required 4.28 pounds of coal per hour or 102.72 pounds of coal each day of twenty-four hours per h. p.

Sept. 24, the water in the pond below your mill was drawn down so that the surface set back exactly to the boundary line back of your mill. The surface at the dam was twenty-four and a half inches below the crest of dam and twenty-seven and a half below the usual water line permanently marked by discoloration on the abutments. This would show a loss from backwater of about eighteen h. p. As neither the turbine nor steam engine can work economically so harnessed together as is usually done to make up for the loss of water power, it would be a moderate estimate to rate the loss at twenty horse power, requiring a ton of coal for each twenty-four hours. Taking the coal at five dollars per ton, the pay of two engineers, and wear and tear of machinery, into consideration, the loss cannot be less than ten dollars per day of twenty-four hours, though the cost of steam power would be less if furnished by a large engine running regularly for furnishing power.

The plaintiff's attorney requested me to proceed and state the facts in my own way which I proceeded to do, when the defendant's attorney, with his long arms wildly flying about his head like those of a frantic windmill let loose, shouted, "I object, your honor, I object!" and *his honor* sustained his objections for no perceptible reasons other than that such positive evidence so easily verified left no chance for the quibbles of the law to continue the case to an indefinite period.   JAMES EMERSON.

AMSTERDAM, N. Y., Sept. 24, 1890.

MASCOMA RIVER IMPROVEMENT CO., PLAINTIFF,

vs.

EMERSON EDGE TOOL CO.

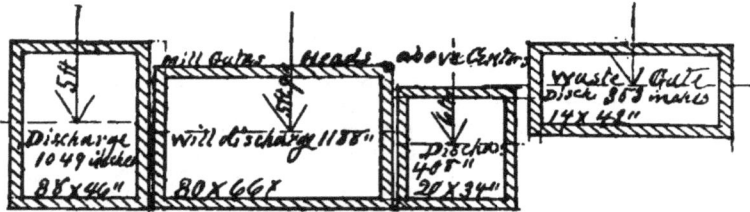

GENTLEMEN, this certifies that on the 27th day of this month I made an examination of the reservoir dam above your works at East Lebanon, N. H. ; that I had the apron planking removed at four several places and that I found the main timbers sound and the whole structure in a condition to render it safe for many years to come. To continue its duration, however, I would advise that annually as convenient the cribbing or frame work of the dam be filled more and more with cobble stone, and gravel be added to that above the dam. By so doing in a few years the whole may be rendered permanent for ages to come.

I would also earnestly advise not only you but all other owners of dams to fix in some unchanging place such as a ledge in the side of the pond at the exact level of the crest of the dam a mark that can never be changed to denote the height of dam and have such mark recorded. The neglect to take this precaution has been the cause of innumerable cases of litigation where dams have gone out, as is so common, or where they have settled or worn away as all dams do in time. This mark need not be near the dam, for the water will give the level when at crest of dam.

The deed conveying the right to this dam, pond, and water power is wildly worded, rendering it necessary for you to observe great care in carrying out its conditions. The reservoir is to be so maintained that at all times sufficient water may be sent down to furnish power for all the mills that now are or may be erected ; a condition very likely soon to become an impossibility from lack of water, hence you must use discretion as to the quantity you allow to pass your gates.

The mill owners have the to-be-unquestioned right to open or close the waste gate in the dam at discretion but in no way to meddle with the old mill gates of yours represented above. To do so must render them liable not only as trespassers, but also for heavy damages that may occur through fires caused by the starting up of machinery when there is no one present to care for it, or by rendering fire pumps useless by the shutting the gates, etc., etc.

Whenever there is a scarcity of water my advice to you is to regulate your gates so that the water in the reservoir decreases but slowly ; then if more is required grant the request only when made in writing by a responsible agent, who shall hold you harmless. Even then you will care for your own rights, and riparian rights below.

The three gates of the old mills represented above, when the reservoir is full, will discharge 345.2 cubic feet of water per second ; less in proportion of course with decrease of head ; 345.2 cubic feet of water falling 14 feet evolves 547.7 h.p.; adding the discharge of the waste gate the center of which has 7 feet 3 inches head and will discharge 53 cubic feet per second, sufficient water is sent down the river to produce a force of 631.7 h. p. where the fall is 14 feet, other falls in proportion. 67 per cent. of such force is the estimated average realized.

There are numerous places below your works before arriving at the Lebanon Mills where dams and mills may be constructed. Such mills will if constructed have the right to the natural flow of the river.

JAMES EMERSON.

WILLIMANSETT, MASS., NOV. 28, 1891.

F. A. Smith, Jr., Treasurer Electric Light and Power Co., Waterville, Maine :

This certifies that from the 9th to the 16th of this month I have made numerous tests at the works of your company, to determine the power required to drive your lighting machinery, also to ascertain the maximum power your two * turbines can furnish, it being understood that the head of water during the year is often less than at present, and that at the present stage it is hardly sufficient to operate the six dynamos, as desired.

Your company have an alternating dynamo connected with a circuit of six hundred incandescent lights, and five other dynamos of a different style for arc lights, all of the same make and capacity, operating five separate circuits, which circuits differ somewhat in conditions and number of lights.

The tests were made by belting from the pulleys on driving shaft to dynamo through an Emerson No. 4 power scale, that carries its load nine feet at each revolution of shaft and scale, the weight being shown in pounds, as upon the ordinary platform scale.

Operation for computation of data: Multiply each revolution by 9, and that product by the weight as shown, then dividing by 33,000 shows the amount of work done in horse power.

Test of alternating dynamo: Revolution of scale per minute, 495; of dynamo, 1,500; weight, 390 pounds.

$495 \times 9 = 4,455 \times 390 = 1,737,450 \div 33,000 = 62.65$ horse power. Test of arc dynamo connected with a circuit of thirty arc lights, 1,200 candle power each: Revolutions of dynamo, 900; of scale, 323 per minute; weight carried, 180 pounds.

$323 \times 9 = 2,907 \times 180 = 523,260 \div 33,000 = 15.85$ horse power. Test second, connected to different circuit: Revolution of scale, 324 per minute; weight, 180 pounds.

$324 \times 9 = 2,916 \times 180 = 524,880 \div 33,000 = 15.90$ horse power. Test of third circuit: Revolution of scale, 325 per minute; weight, 190 pounds.

$325 \times 9 = 2,925 \times 190 = 555,750 \div 33,000 = 16.84$ horse power. Test of fourth circuit: Revolution of scale, 324 per minute; weight, 200 pounds.

$324 \times 9 = 2,916 \times 200 = 583,200 \div 33,000 = 17.84$ horse power. Fifth circuit, same as the first arc circuit tested.

The six dynamos with all lights in use require 150 horse power to be safe.

The two turbines at the best cannot transmit over 140 horse power, oftentimes not that.

---

* Risdon, the most unreliable turbine I ever tested. Owing to some peculiarity, it could never be told until tested whether it would do well or not. One set by a millwright was tested that gave 73; the next after resetting it gave 87 per cent.

JAMES EMERSON.

WILLIMANSETT, MASS., Feb. 18, 1891.

"THE LAW."

The study of untold centuries to enable influential rascals to escape responsibility.
Its only science, the knowledge of its uncertainty.
Is that the law? Never the question, is that the right?
Will it never be understood that the law was made for the benefit of man, and not man for the law?

Equally.

Progressive.

BOND vs. CITY OF SPRINGFIELD.

WILLIMANSETT, Jan. 6, 1892.

To the Water Commissioners of Springfield:—

All through the New England States may be seen relics of old mills, wheelwright shops, etc., located upon streams of little capacity, except in the spring of the year or during heavy rains. These were very useful in early times, but now almost every want is supplied by large manufacturers at a lower price than the raw material would cost at these isolated places, hence few of them continue in operation and such as do bear the marks of a lingering old age going to seed.

Portable saw mills are now moved to timbered lands, and the lumber is sawed, the slabs and refuse wood furnishing the fuel for steam power, at less cost than the timber can be drawn to stationary mills on water powers.

Mr. Bond, of Belchertown, continues the use of one of these ancient mills. The grist mill building was burned a few years since, and its place supplied by a superannuated depot building. The saw mill seems to be dependent upon the most primitive means for taking logs from the pond. The old cobble stone or bowlder dam and rotting surroundings offer evidence that at no distant day extensive and expensive repairs will have to be made from the foundation to retain the pond, and evidence that the future prospects have not warranted the expense of repairs and improvements. The assessed value of mills, houses, blacksmith shop, etc., is $2,150, and were the property well advertised for sale there is no reason to think the rush of purchasers would be so great as to cause the suspicion of undue favoritism on assessed valuation.

Springfield takes about one-half of Mr. Bond's water supply during the summer.

October 3, 1891, the flow into Bond's pond was 72.77 cubic feet per minute, while the flow in Springfield canal was 93.90 cubic feet; but from this quantity must be deducted as a constant a half million gallons daily, or 46.4 cubic feet per minute supplied by springs in the bottom of the canal below the place where the water is diverted from Mr. Bond.

Taking the 46.4 from the 93.90 of course does not leave a quantity equal to the flow to Mr. Bond, but the Springfield supply varies as the mill above is or is not in operation.

November 25, the flow direct to the Bond pond was 252 cubic feet per minute; in the Springfield canal, less 46.4 for percolation, 295.6 cubic feet per minute. But the mill above was then in operation, and so large a flow would at the most continue but for ten of the twenty-four hours, while during the other fourteen there would be much less.

Two millions of gallons daily is a liberal allowance for all of the water that Springfield can draw from Mr. Bond's supply. This quantity, 185.6 cubic feet per minute, falling 13 feet evolves 4.56 h. p. Holyoke Water Power Company furnishes such power per year for $4.33 per h. p. free from all expense to purchaser for maintaining dam, canal, etc. Willimansett brook has a fall and constant water supply for 25 h. p. the year round within a third of a mile of depot and within a mile of the business center of Holyoke, that has run to waste for twenty years, no one considering it worth the expense of fitting it up for utilization. A steam engine that would cost $500 would do more work than all of Mr. Bond's water power much of the year and be far more reliable.

There are two classes of milling men that I often come in contact with that do not seem creditable to the age. The first are shocked and filled with indignation at the mere mention of making examination or measurements in their mills on Sunday, yet in the most bare-faced manner they will steal water for power every other day of the week the year round. The other class seem to consider it a commendable token of smartness to extort ten times the value of a thing from a corporation if possible.

Springfield is able to pay a fair price for what it needs. I would advise a tender of $1,500 to Mr. Bond for the water taken, and, at the utmost, if $2,000 will not satisfy his demands, then decidedly let the courts settle it.

JAMES EMERSON.

THE LAW.

During the past decade complaint has frequently been made of a lack of reverence for the law.

Has the law, with a continuation of its obsolete absurdities and rules requiring cases to be fitted to the law, instead of making the law fit the case in hand, any just claim to respect? Has there been any attempt made by the fraternity to simplify and bring it up to the necessities of the time?

Nearly twenty-four hundred years ago Herodotus wrote of its delays and uncertainties just as is done to-day. At the left of the illustration of court-room is a view of what a century since was thought desirable to put back of the house while the travel was in front. Once in a groove the plan has continued, though a hundred from the car windows are now annoyed by the disgusting sight where one would have been formerly if placed in front. The two views are intended to show the force of habit. The "O yes, O yes," miniature gallows, and bullying of witnesses should be matters of the past.

THE LAW OF THE LAWYERS.

Safeguards for Professional Honesty—The Attorneys' Oaths, 1884.

A frequent charge against members of the bar, made indeed facetiously in most cases, is that of insincerity and lack of veracity. The attorneys' oath of office is in all conscience strict enough, and if there is such a thing as a dishonest lawyer he must be a perjurer as well. Below is given the form at present in use.

THE ATTORNEYS' OATH, 1884.

You solemnly swear that you will do no falsehood, nor consent to the doing of any in court; you will not wittingly or willingly promote or sue any false, groundless, or unlawful suit, nor give aid or consent to the same; you will delay no man for lucre or malice; but you will conduct yourself in the office of an attorney within the courts according to the best of your knowledge and discretion, and with all good fidelity as well to the courts as your clients. So help you God.

1. A lawyer ought to be a gentleman. His function as an attorney gives him no dispensation to disregard the ordinary rules of good manners, and the ordinary principles of decency and honor. He has no right to slander his neighbor, even if his neighbor be the defendant in a cause in which he appears for the plaintiff. He has no right to bully or browbeat a witness in cross-examination, or artfully to entrap that witness into giving false testimony. Whatever the privilege of the court may be, the lawyer who is guilty of such practices in court is no gentleman out of court.

2. A lawyer ought not to lie. He may defend a criminal whom he knows to be guilty, but he may not say to the jury that he believes this criminal to be innocent. It is notorious that some lawyers who would think it scandalous to tell a falsehood out of court in any business transaction lie shamelessly in court in behalf of their clients, and seem to think it part of their professional duty. That bar of justice before which by their professional obligations they are bound to the most stringent truthfulness is the very place where they seem to consider themselves absolved from the common law of veracity. So long as the legal mind is infected with this deadly heresy we need not wonder that our courts of justice often become the instruments of unrighteousness.

3. A lawyer ought not to sell his services for the promotion of injustice and knavery. Swindlers of all types are aided by lawyers in their depredations upon society. It would be more difficult to believe this if its truth were not so often illustrated in the stupendous frauds and piracies of great corporations, all of which are carefully engineered by eminent lawyers. Our modern "buccaneers"—our brave railroad wreckers—are in constant consultation with distinguished lawyers. They undeniably have "the best of legal advice" in planning and executing their bold iniquities.

Bob Ingersoll rails much against a venal priesthood, yet defends Star Route thieves with a gusto that denotes a labor of love. The mote he so dislikes is not small, but he seems to carry a whole lumber yard in his eye without inconvenience.

IS THE LAW ITSELF MORE COMMENDABLE?

Governor Butler said: "Shall I call your attention to the time when no lawyer was allowed to practice?" and he added, "It was a credit to the legal profession that no lawyer had participated in the witchcraft tricks,"—and so it was; but when he said, "No judge presided over them," he simply blundered, for it is well known to every school boy familiar with the history of those times, that it was the notorious Chief Justice Sewell who, in his blind bigotry and desire to serve two masters, both God and man, at the same time, as he thought, condemned twenty-four innocent people to death, and afterward stood up in church in Boston, with bowed-down head and sorrowful countenance, while a paper was read, in which he begged the prayers of the congregation, that the innocent blood which he had erringly shed might not be visited on the country or on him.

"As far as we know," says *Texas Siftings*, "there is not a single instance on record in Texas of a murderer of means having been punished by law, no matter how many homicides he committed." Texas is not an exceptional state where such cases transpire.

The result of the Sellon trial confronts the people of this community with some serious questions. Where and what is the influence which renders the conviction of a man for the taking of human life impossible? How is it that the machinery of the law is wrenched and money poured out like water to convict two men of a crime which a majority of people believe to-day was never committed by anybody, while three men, each with the blood of a fellow being on his hands, walk the streets free men, one of them not even having been indicted for his crime? It has become so in this community that if a murder is committed and the man who does the deed has any influence, political, pecuniary, or social, which can be brought to bear, it is immediately taken for granted that he will not be punished for the crime.

THE KEMMLER REPRIEVE.

The case of the condemned murderer Kemmler certainly offers the most remarkable instance of judicial procrastination on record in this country.

A FATAL FLAW IN THE INDICTMENT.

A highly respected citizen was arraigned before court for shooting and killing a friend. The evidence was direct, and after exhaustive arguments had been made the judge said:—

"It is clearly proven that you are guilty, as charged by the indictment."

"But I protest my innocence," replied the prisoner. "The indictment reads that I did shoot and kill the gentleman with powder and a leaden bullet. This is a mistake. I had no bullets at the time, so I loaded my gun with powder and a horseshoe nail."

"That indeed alters the case," said the judge. "The indictment said bullet, when it should have said nail. You are discharged, sir."

Frank Weiss, the editor of an illustrated German comic paper at Erie, Pa., is on trial for libel, and has succeeded in fighting the law with its own weapons in a very amusing way. The district attorney at the opening of the prosecution claimed the right to "stand aside" jurors under an ancient law of Edward I., never repealed and once sustained by the supreme court of Pennsylvania by some musty decision. In this way, every German or Irish juror was thrown out, the court assenting to the absurd supremacy of this law of 900 years ago. Weiss, who is a small, feeble, melancholy-looking man, then concluded if they were going in for mediæval law he would have some. So he insisted on the trial of the case by ordeal of fire and by combat! He floored the court with his citations of unrepealed law, and at last accounts the suit was still in progress, with more fun in the court than there ever was in the newspaper.

The fallibility of juries has recently had a striking illustration in the case of a man under life sentence for murder in Michigan, having been recently pardoned, after passing twenty-seven years in prison, on the ground that he is innocent of the crime for which he was convicted. He was convicted mainly on the false testimony of a worthless wretch who had a grudge against him, and who afterwards confessed that the evidence given by him at the trial as a lie. It is sad enough to consider the long years of confinement suffered an innocent man, but still more sad to think of his blasted life, and that now he is set free he has no remedy or redress for the suffering and shame endured or the gross injustice of which he has been the victim.

A QUESTION OF PARDON.

I see that the papers notice the "pardon of an innocent man." How can an innocent man be pardoned? What is there to pardon him for? If there is anybody to be pardoned, isn't it the ones who imprisoned him?

Has a people that will allow such a damnable law to continue to exist any claim to be considered civilized? Surely, if the safety of the community requires the punishment of a supposed guilty person, the commonest justice requires the most ample retraction and compensation in case innocence is afterward proved.

A well known lawyer said: "If I had my way, I would abolish all the courts in the state once every ten years. The courts are the masters of the people. Talk about their being the servants of the commonwealth—they are its masters. You can see how it is when anything is attempted at the Legislature which touches any of these courts. If a measure is proposed which would disturb any of them, it is impossible to get it through the Legislature. They have such control over the senators and representatives that nothing can be done. A judge has so much prestige that the representative thinks he is doing just the right thing if he votes as the judge thinks is the best way, and the consequence is that it is absolutely impossible to get any reform through. A judge isn't any better man after he goes on the bench than he was before. Giving him a commission doesn't make any better man of him, or give him any new faculties, or make his opinion any more entitled to respect than it was before."

And if I could have my way the Legislature should meet but once in ten years; then select a few fundamental principles of justice, never exceeding one hundred in number; then repeal all previously existing laws from the beginning of recorded acts, and have all disputes settled by arbitration, allowing no lawyer to be employed.

JURY TRIAL.

What a travesty upon both law and justice; agree or starve! One venal member, by providing beforehand, could easily compel the others to submit

Yet further, it is a well-established point of law that an agreement under duress is illegal

A diligent reader with a good memory may be a successful lawyer without being a statesman or much of a man.

Congress is rotten with lawyers and notoriously lacks statesmanship.

It is a strange condition of society that its laws that all are to live by become so complicated that lawyers at from five to a thousand dollars per day must be employed to explain their meaning.

IN CONCLUSION.

Can any intelligent person accustomed to our courts, witnessing the silly, obsolete forms for opening and closing, its suppression of undesired evidence, its use of private correspondence, its attempts to trick witnesses into contradictory statements, its Jarndyce and Jarndyce procrastinations, its breaking of wills, its pandering to the influential, have any respect therefor or look upon it in any other light than that it is a bondage alike disgraceful to those who practice and those who endure its continuance?

Expurgation and Pretension.

The real value of a book consists in its representation of its time, to expurgate destroys its representative character.

Expurgate the atrocity and obscenity of the Bible and only spiritualism would remain. Expurgate what at this day cannot publicly be read from Shakespeare's works, and the pith is gone. Expurgate the loathsome filthiness from Rabelais' description of the Christianity of his time, and only the covers of his book remain. The delicacy that causes the teacher to send the bare legged boy from school does not prevent her from displaying more than legs at the bathing beach. The age that sentences the poor thief to years of imprisonment for stealing a suit of clothes, pronounces the rich railroad director free from guilt, though, in defiance of law, he has caused the death of passengers by roasting. Talmadge in his church, the clown in his circus, and the self-styled statesman, each worship the Christian's God, and there can be little doubt but that the clown does the most for humanity. The unpretentious farmer that places a watering trough by the wayside for the thirsty man or horse, in my opinion, does more for the elevation of man and glory of God than the rich man who builds a church or endows a college.

Rotten Statutes.

As a people few are more ready than ourselves to censure the tolerance of abuses by others, or more servile in submitting to such of our own. How we smile at the Jay Goulding of a railroad through the chicanery of the law, or even the acquittal of a murderer by the resurrection of a rotten statute that should have been buried by obliteration centuries since. Such successes in any other walk of life would be considered infamous, but in law successful rascality is called smartness. If law is designed to aid justice why is such rascality tolerated by a people claiming to be civilized? Savages would scorn such trickery.

Patents, notes of hand, judgments, etc., etc., are limited in duration. Why not statutes? With nearly fifty independent States each constantly issuing volumes of new statutes, where is it to end? Lawyers produce nothing but strife and their support comes from labor. Will the laborer forever continue to support a class so useless yet so expensive? We claim to be a free people, but can there be freedom with such a mountainous pile of rotten statutes hanging over us? Can anything be more senseless than the common practice of legislators referring matters pertaining to the law to the judiciary committee? Lawyers if no worse certainly are no better than others. Simple laws are not for their interest. The ideal law of the lawyer is of the mattock and spade, mailed shirt and bow and arrow age. If the steel plow, harvester, rifle cannon and repeating rifle are superior to those, then in proportion has the mechanic proved his superiority to the lawyer. Then what excuse is there for suffering the designing or inferior to determine the laws for the superior? Law is for man not man for the law. Get up out of the ruts, Messrs. Legislators! if your heads hit the roof when doing so your brains are safe, that is not their location. Why not to every new statute enacted add, "and all previous enactments inconsistent with this are hereby repealed"? The best governed people are those governed the least. Blot out every statute over twenty years of age, and the occupation of the *smart* lawyer would be gone.

Arbitration.

When a proposition is made by one of a party to leave a case in dispute out to three disinterested persons to decide and the other refuses to do so, we invariably believe the latter to be the one in the wrong. Then why not make such arbitration obligatory, whenever one party demands it? Do away with the so-called law and lawyers. Have fewer officials and those directly amenable to the people. Form a general plan for arbitration and make such decisions final, except in cases of finding new and undeniable evidence, then in serious cases, such as unjust criminal convictions and punishments, have the highest official of the state apologize for the wrong and so far as possible make the fullest restitution for the injustice, instead of as is now done adding outrage to injustice by the mockery of *pardoning* a martyr, what a barbarian would be ashamed to do. In God's name, is there not statesmanship in Massachusetts sufficient to remedy a wrong so glaring?

The Sacredness of an Oath.

In a story about Catiline, a companion says, "Who believes in an oath? Did you ever believe in one, Catiline?" "Well, perhaps so, when a boy," was the reply.

Those accustomed to the usual style of administering the oath, "Hold up your hand. You solemnly yum, yum, yum, s'elp you God," can hardly be much impressed with its sanctity, and the observance of the interested witness with his "I don't remember." and burning face when the question has struck home, will be likely to cause the observer to come to the conclusion that the person who in ordinary conversation embellishes his story merely for self exaltation, will hardly hesitate to lie when under oath if it is for his interest to do so. There are penalties for perjury; why not depend upon those?

Irresponsible Commissions.

If "eternal vigilance is the price of liberty," can it be well to take the power to act direct from the people, and place it in the care of a commission chosen more through political partisanship than personal fitness?

Are three hackneyed politicians more likely to be just to all than those interested for the best good of their homes? Are our schools as effective now in producing practical men and women, as formerly under the old district or local governing system?

Responsibility begets consideration. How quick staid citizens, after enlisting as soldiers during the late war and losing their personal responsibility, became like unruly boys, often worse.

Is it democracy to place the governing power in the hands of a minority of wire pullers? Can a single instance of such a course being found conducive to general good be named since the beginning of history? Then why ignore such ages of experience and abandon the principle of self-government?

SHOW INSTITUTIONS OF THE HILLS,

For the Blind, the Halt, the Idiotic, the Insane, the Pauper,
and Criminal.

Can such an ostentatious display denote a high civilization ?

"EDUCATE THE IDIOTIC."

Can education to such bring happiness ?   If ignorance is bliss
under any condition, it would seem to be so with the idiotic.

Would it not indicate greater intelligence

To seek for the cause and try to stop the production of idiots,
paupers, lunatics, invalids, and criminals ?

Whether entirely satisfactory to the patients,

These institutions are convenient retreats for retiring rival politicians, at the same time producing hot-house culture of "offen-
sive partisanship."

The Testing System.

Having terminated my connection with the business of testing turbines, it may be well to give a brief account of the conception of the business as a system.

Such tests were made in Europe early in the present century; in this country, by Uriah A. Boyden, from 1843 to 1859. I have found it impossible to obtain any authentic record of Mr. Boyden's tests, though there are rumors of fabulous results. Mr. Francis, in the work called "Lowell Hydraulic Experiments," states that data furnished him for computation gave 88 per cent. He does not vouch for the data furnished, nor does it appear that such data was furnished by a disinterested engineer in any case. Mr. Francis followed Mr. Boyden in making such tests, but he, like the former, made them so expensive as to be beyond the reach of any but wealthy corporations, while the manufacturing interest required a definite knowledge of the efficiency of the various kinds of turbine plans then springing into existence.

In 1859–60, the city of Philadelphia gratuitously tested a variety of small wheels for different builders, but the plan for doing it was so defective that the tests had but little influence. In 1867, the Chase Turbine Co., of Orange, Mass., employed me to construct a dynamometer or brake for testing turbines. The friction bands that may be seen on the ship windlass, in another part of this work, gave me the idea of controlling a turbine in that way, for I had brought many a ship to by such bands. The Prony brake had never been heard of by me at that time, nor until my brake was completed.

In 1868, A. M. Swain asked me to get up a suitable brake, and test one of his wheels at Putnam, Ct. Six months' time and $1,700 were expended in preparing the instrument. The company was persuaded to construct a flume at the "overflow" of the Wamesit Power Co., Lowell, Mass. A 42-inch Swain wheel was set, and tested by Mr. Swain and myself. The results were such that the company was urged to employ an engineer with at least a theoretical knowledge of such tests. H. F. Mills, then of Boston, was selected for the purpose. The company then held a meeting and authorized Mr. Swain and myself to make arrangements for a public trial, and the following notice was issued:

IMPORTANT TEST OF TURBINE WATER WHEELS, AT LOWELL, MASS.,
JUNE 16, 1869.

SIR: The Swain Turbine Co. has just completed extensive arrangements for a competitive test of Turbine Water Wheels. A flume and weir of the most approved plan, to supply and measure the water used, has been constructed. Emerson's Dynamometer will be used to test the power of the wheels.

The "pit" is fourteen feet in width; head of water varying from twelve to sixteen feet. Each competitor will select size and finish of wheel to suit himself. The Swain Wheel to be tested was built before the test was thought of, and is in no way superior to the average of wheels furnished by the company. It is forty-two inches in diameter, and will be tested on the 16th day of this month.

The Swain, Leffel, Bodine-Jonval and Bryson Turret wheels were entered. The measuring pit was fourteen feet wide, thirty in length and at first a little over three feet in depth below crest of weir—the wheels, standing inside at the upper end in a quarter turn or iron flume, being about twenty feet from the weir. In this distance there were three separate racks to check the rushing water.

The Swain wheel had thin sheet steel buckets, which made it very light for its diameter; yet, when set, it was barely possible to turn it by the coupling upon the top of its shaft—the coupling being twenty inches in diameter, made that size to connect with brake. Mr. Swain "guessed the wheel would go, only put the water to it."

The Leffels knew better than to lose fifty or a hundred pounds in that way, so, when their wheel was set, it turned about as easy as a child's top. Of course, an engineer of experience would have refused to have tested a wheel running as hard as the Swain did, or to have tested a wheel of that size at all in a pit so small and filled with racks, for a good wheel would have little chance against one of low efficiency. The working surfaces of the brake and band were made of steel and iron. Both being fibrous, little strips tore from each, often checking, and at times bringing the wheel to a sudden stop, so that it was difficult to make steady tests of many minutes' duration. A bell was connected to the wheel-shaft, which struck at each fifty revolutions of the wheel. Instead of making each test with a given weight separate and distinct by itself, observers were placed at the different gauges, with watches set to the same time. As the wheel ran very unsteady at the best—often stopping entirely—it was necessary to reject many of the observations, and it will readily be seen that the difficulty would be in placing the right patches together. That this is not imaginary, the following tables of results are given. The first is a copy of Mr. Mills' report, the second is a record of tests taken by myself, the same gauge hands being employed in each case, and the conditions being precisely the same for both. My tests, however, were taken upon the same plan that I have followed continuously for more than ten years: that is, to make each test for a given weight complete and distinct in itself. Mr. E. A. Thissel made a record of the gauges, as given by each of the hands employed, and as it agreed exactly with the notes I had taken of all, his record is given in the table.

Mode of Conducting the Experiments.

Observers were stationed at various points, as follows:

Mr. J. B. Hale, at the hook-gauge, observed every minute, and a part of the time every thirty seconds, the reading of the hook-gauge, which indicated the depth of water upon the weir.

Mr. R. A. Hale observed the height of the water in the forebay and in the pit, by means of the scale (D) passing from the lower box to the upper every minute.

Mr. E. A. Thissel noted the time of the striking of the bell, which indicated the speed of the wheel, to the nearest quarter second.

Mr. James Emerson, by means of the hand-wheel (M) regulated the friction so that the index (E) should be kept as near to zero as possible, and thus the scale beam be kept level.

Another assistant observed as rapidly as possible the actual position of the index during the experiment.

Another kept the oil cups (T) supplied with oil, and, by a cock attached to each, regulated the amount flowing upon the friction surfaces.

Another attended the gate and kept the racks clear of obstructions.

The writer kept a record of the weights in the scale-pan, the heights of gate, all irregularities in the motion or disturbing causes of any kind that would affect the results of the experiment, and sufficient observations of each class to check the accuracy of all of the notes.

At intervals, during a series of experiments, all of the watches were compared with the standard, and differences noted, that there might be no difficulty in selecting the observations which applied to the time when the conditions for accurate results obtained. Recorded in the following manner:

EXPERIMENTS UPON THE 40-INCH LEFFEL WHEEL, AT LOWELL, MASS.

Date. 1869.	Temp. of Water. Fah.	Opening of Gate. Inches.	TIME. FROM Hour.	Min.	Sec.	TO Hour.	Min.	Sec.	Duration of Experiments. Seconds.	Total No. Revolutions.	No. Rev. of Wheel per second.	Weight in Scale. Pounds.	Height of Water In Flume. Feet.	In Pit. Feet.	Fall acting on Wheel. Feet.	Depth of Water on Weir. Feet.	Quan. of Water passing the Weir. c. ft. per sec.	Rel. Veloc.	Rates of useful effect to the power expended.
Oct. 11.	59°	2.25	12	19	15.0	12	22	57.5	222.5	500	2.2471	80.0	15.575	1.140	14.435	1.099	38.013	.772	.788
"	"	"	12	23	27.0	12	29	46.0	379.0	850	2.2427	80.0	15.563	1.140	14.423	1.098	37.963	.773	.788
"	"	"	12	39	32.0	12	42	10.0	158.0	350	2.2151	82.0	15.560	1.140	14.420	1.0985	37.981	.762	.707
"	"	"	12	46	27.0	12	50	29.0	242.0	500	2.0661	87.0	15.493	1.130	14.363	1.0978	37.943	.712	.793
"	60°	2.10	3	51	39.0	3	53	56.0	137.0	300	2.1898	75.0	15.115	1.090	14.025	1.0410	35.358	.764	.796
"	"	"	3	3	53.0	3	7	42.0	229.0	500	2.1834	75.0	15.190	1.080	14.110	1.0480	35.408	.759	.788
"	"	"	3	59	54.0	4	2	58.0	184.0	400	2.1739	75.0	15.130	1.087	14.043	1.0470	35.358	.757	.789
12.	59¼°	"	3	53	4.0	3	56	7.5	183.5	900	2.1798	76.0	15.245	1.057	14.199	1.0617	35.589	.755	.788
"	"	"	4	00	4.0	4	6	11.0	367.0	900	2.1798	76.0	—	1.053	14.195	1.0519	35.589	.755	.788
"	"	"	3	48	3.0	3	48	45.0	162.0	350	2.1605	76.0	15.217	1.045	14.172	1.0510	35.559	.749	.788
"	"	"	4	14	11.0	3	18	49.0	278.0	600	2.1582	76.5	15.235	1.053	14.182	1.052	35.606	.748	.783
"	"	"	3	19	56.0	3	23	23.5	209.5	450	2.1480	75.0	15.240	1.030	14.210	1.0307	34.536	.744	.786
19.	51°	1.97	4	53	40.0	4	58	49.0	309.0	650		72.0	15.287	1.002	14.285	0.990	32.633	.727	.789
11.	60°	1.96	5	11	30.0	5	15	47.0	257.0	550	2.1030	72.5	15.265	1.048	14.219	1.0048	33.254	.741	.783
"	"	1.84	5	5	16.0	5	11	7.0	351.0	750	2.1367	72.5	15.255	1.042	14.213	1.0050	33.264	.740	.789
12.	51°	"	4	48	37.0	3	50	37.0	120.0	250	2.0633	75.0	15.220	1.060	14.170	1.0060	33.314	.723	.788
"	60°	1.80	3	45	5.5	4	0	34.5	329.0	700	2.1276	72.0	15.299	1.004	14.285	0.9904	32.853	.735	.796
11.	"	"	4	58	19.0	4	1	24.0	190.0	400	2.1053	72.5	15.223	1.007	14.216	0.9983	32.954	.729	.784
12.	60°	1.69	4	31	20.0	4	36	12.5	292.5	600	2.0512	70.0	15.232	0.983	14.249	0.9540	30.779	.710	.783
"	59°	"	12	36	30.5	4	38	14.0	103.5	200	1.9324	75.0	15.237	0.985	14.252	0.9600	31.060	.668	.787
"	"	1.68	12	16	38.5	12	23	1.5	383.0	800	2.0888	67.0	15.048	1.030	14.018	0.9424	30.228	.729	.787
"	"	"	12	26	9.0	12	32	59.0	410.0	850	2.0732	67.5	15.060	1.044	14.016	0.9440	30.299	.723	.794
"	"	"	12	37	40.5	12	44	25.0	404.5	850	2.1013	66.0	15.058	1.058	14.000	0.9415	30.188	.731	.792
Oct. 11.	59°	2.25	11	24	22.0	11	30	29.0	367.0	900	2.4523	60.0	15.203	1.092	14.116	1.070	36.525	.852	.686
"	"	"	11	34	14.0	11	43	22.0	548.0	1300	2.3723	65.0	15.159	1.100	14.059	1.073	36.678	.826	.719
"	"	"	11	46	12.0	11	52	39.0	387.0	900	2.3255	70.0	15.198	1.110	14.088	1.078	36.941	.807	.752
12.	59¼°	1.26	4	33	3.0	4	35	54.5	171.5	400	2.3324	40.0	15.373	0.770	14.603	0.7842	22.994	.797	.668
"	"	"	4	36	56.0	4	39	36.0	160.0	350	2.1900	45.0	15.400	0.780	14.620	0.7946	23.443	.748	.692
"	"	"	4	40	26.0	4	43	53.0	207.0	450	2.1739	46.0	15.413	0.780	14.633	0.7957	23.495	.742	.699
"	"	"	4	49	56.0	4	53	10.0	254.0	550	2.1653	46.5	15.458	0.780	14.678	0.7982	23.583	.738	.699
"	"	"	4	51	13.0	4	53	17.0	124.0	250	2.0161	52.5	15.470	0.795	14.675	0.8083	24.096	.687	.720
"	"	0.84	5	6	46.0	5	10	15.0	209.0	450	2.1531	32.0	15.553	0.615	14.938	0.6256	16.422	.727	.675

Tests on Leffel Wheel.

Date, Oct. 12, 1869. The first 19 tests L of weir was 10.052. Correction, .733.
"   "   13, "   The next 21 "   "   "   "   "   "   "
"   "   1, "   The next 7   "   "   "   "   "   No correction.
"   "   11, "   The next 5   "   "   "   "   "   "

Weight of water used for all these tests was 62.33 lbs. per cubic foot. In the group of "7 tests," the first 3 were made with holes in wheel plates closed, and in the remaining 4 with holes open.

No. Test.	Head.	Weir.	Gate.	Weight.	Rev. per min.	Horse Power.	Cu. feet per sec.	Per Cent.	Rel. Veloc
1	14.28	1.730	4-5	670	138	42.03	32.66	.795	.795
2	14.17	1.727	"	"	135	41.11	32.52	.787	.799
3	14.01	1.671	"	"	127	38.68	29.84	.817	.739
4	"	1.669	"	"	124	37.76	29.75	.799	.722
5	"	1.674	"	675	125	38.35	29.98	.806	.725
6	"	1.673	"	"	"	"	29.94	.807	.725
7	13.98	1.672	"	660	126	37.80	29.89	.798	.733
8	13.86	1.671	"	700	120	38.18	29.84	.815	.701
9	14.48	1.761	7-8	740	130	43.73	34.18	.780	.744
10	14.21	1.761	"	750	128	43 64	34.18	.793	.738
11	14.18	1.762	"	755	126	43.24	34.22	.786	.728
12	14.16	1.780	Full	"	130	44.61	35.11	.792	.753
13	14.17	1.782	"	760	"	44.91	35.21	.789	.747
14	14.20	1.781	"	"	"	"	35.16	.794	.747
15	14.16	1.782	"	765	"	45.20	35 21	.800	.748
16	14.18	1.782	"	"	"	"	35.21	.790	.747
17	14.925	1.378	2-5	400	115	20.91	17.12	.722	.649
18	"	1.364	"	340	125	19.32	16.57	.690	.703
19	"	1.354	"	320	128	18.62	16.18	.680	.720
1	15.21	1.136	1-5	175	111	8.83	7.88	.650	.619
2	"	1.134	"	170	115.2	8.90	7.82	.660	.643
3	15.22	1.132	"	100	"	8.38	7.77	.626	.642
4	15.21	1.129	"	150	122.4	8.35	7.68	.630	.683
5	15.23	1.127	"	140	127.8	8.13	7.63	.617	.713
6	15.24	1.127	"	135	130.2	7.99	7.63	.606	.726
7	"	1.126	"	130	133.2	7.69	7.60	.586	.742
8	"	1.127	"	135	130.2	7.99	7.63	.606	.726
9	14.26	1.526	3-5	500	122.4	27.82	23.26	.740	.705
10	14.21	1.527	"	"	"	"	23.01	.741	.707
11	"	1.535	"	475	124.8	26.95	23.33	.707	.720
12	14.23	1.521	"	460	127.8	26.72	23.05	.719	.737
13	"	1.521	"	465	"	27.01	23.05	.727	.737
14	14.56	1.147	1-5	130	133.2	7.87	8.18	.583	.760
15	14.46	1.149	"	135	133.0	8.10	8.24	.600	.755
16	14.45	1.146	"	140	127.8	8.14	8.15	.610	.732
17	14.16	1.741	Full	725	133.2	43.90	33.20	.822	.770
18	14.13	1.736	"	"	"	"	32.95	.832	.771
19	"	1.742	"	730	127.8	43.41	33.24	.797	.740
20	14.15	1.745	"	735	129.6	43.30	33.39	.809	.750
21	14.02	1.775	"	"	"	42.70	34.87	.771	.743
1	12.22	1.061	None given	700	120	38.18	35.81	.770	.747
2	12.09	1.045	"	725	115.4	38.03	35.01	.793	.721
3	12.01	1.045	"	675	125	38.35	35.01	.804	.784
4	12.73	1.077	"	"	"	"2	36.61	.726	.761
5	12.51	1.077	"	700	120	38.18	36 61	.736	.738
6	12.30	1.061	"	725	115.38	38.02	35.81	.762	.715
7	12.12	1.061	"	750	111	37.84	35.81	.769	.694
1	14.23	1.010	"	725	127.6	42.05	33.29	.783	.737
2	14.08	1.047	"	730	136.6	47.45	35.11	.847	.793
3	14.105	1.076	"	800	127.6	46.40	36.56	.794	.741
4	14.08	1.011	"	750	125	42.61	33.34	.801	.724
5	14.08	1.034	"	600	150	40.91	34.47	.744	.870

I have seen sufficient the past year to convince me that tests made with so many gauge hands are very unreliable.

I would not be understood as vouching for the efficiency of the wheel, as given by Mr. Mills or myself, for my experience since has made me very skeptical about tests made in pits so limited as to require the use of racks to still the water discharged; but, as those tests were made under the same conditions, the discrepancies have made me cautious about using unnecessary formula for mere effect. That much of the formula for testing turbines, published by Mr. Francis, is for effect, it is charitable to believe. The plan is undoubtedly that followed by Mr. Boyden, and it is not creditable to his ability to suppose he believed several pipes, leading from different heads, would fill a tank to the average depth of the whole, yet that is what his perforated pipes around the wheel and across the pit leading to the gauge-tanks mean. With filtered water, plenty of help, abundance of time, and no regard for expense, the plan would not prevent accuracy; but for practical tests under ordinary conditions, with sediment in the water, such pipes are anything but desirable, and under no possible conditions are they necessary. The dash-pot is another source of error. It is absolutely necessary, with such a brake as Mr. Boyden used, also with the best brake that can be made, for some wheels, while there are others that can be tested without it; but the greatest care should be taken to have the plunger work as sensitively as possible. The pipes connecting the gauge-tanks with pit and forebay are matters of great importance. Of course, the smaller they are, the steadier the level of the surface in the tanks. The machine engineer likes small pipe connections, but the practical engineer has them large, that the surface of the water in the tanks may represent the true surface in pit or forebay. The water may rise and fall quick, as it should if it does so in pit or forebay, but it is easy to get the mean of the variations by observing the extremes. Racks, as usually constructed, take up one-half of the cross-section of the pit; a very fine rack more than that, if made of wood, and of course stops the water, causing it to be higher above than below them. This gives accelerated velocity to the water. Following the plan faithfully for two years, it proved to be a perfect trap for catching errors. The tank connections were then enlarged, the pit lengthened and made deeper; the perforated pipes and racks were abandoned, the dash-pot was reduced in size, and the plunger made perfectly free—after which changes, there was no difficulty in making tests that would repeat—a very necessary achievement in a business where suspicious patrons were in the habit of keeping tested wheels months, perhaps years; then, after repainting, return them as new to be retested, as was often done. The bane of engineering has been too much desire for display of mathematical exactitude, without much regard for the mechanical devices used with which to procure data to work from. Look at the coarse brake and scale beam used by Mr. Boyden, also by Mr. Francis, then at proportions as given by the latter in Lowell Hydraulic Experiments:

Prony Brake.

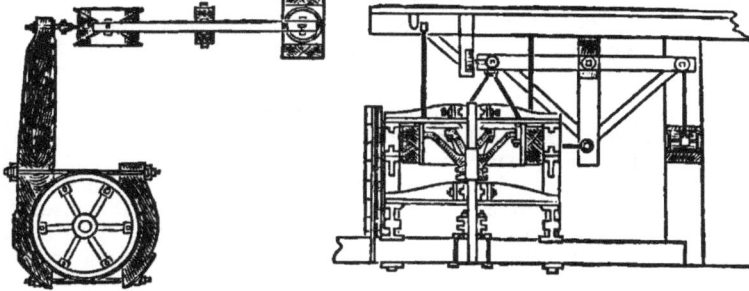

Length of brake was found to be 9.745 feet.
Effective length of vertical arm,   4.500 "
Effective length of horizontal arm, 5.000 "
Consequently, effect in length was $9.745 \times 5 \div 4.5 = 10.827778$ feet.

Why not have made the brake and arms of lengths readily expressed in whole numbers, thus doing away with decimals? Made in any lengths, a coarse oak timber, with an inch and a half round iron bolt through it for a fulcrum, would be a poor substitute for a light iron scale-beam with knife-edge pivots. Weighing what a turbine will pull, means the same as what groceries weigh, and needs the same perfection of weighing apparatus to do it well. The plan, when used by Mr. Boyden, was up to his time, perhaps, but a generation has since passed away, and vast improvements in almost every mechanical device have been made in the time, and practical engineers accept the improvements in turbine testing, as in other matters; but the machine engineer turns back to the oak brake and many decimals as anxiously as a duck takes to water. Turbine building is not a science, nor is it likely to be, until reputable builders, who would willingly test wheels before delivery, are protected from ruinous competition by the ignorant and irresponsible, who promise so readily, caring little about the efficiency of their wheels so long as they sell. To test each wheel before delivery would necessitate its being done quickly and cheaply, which would be impossible with the Boyden-Francis apparatus, nor would it be possible under any conditions with such an apparatus to make such tests as were easily taken to determine the effect of flanged cylinder gate and flaring draft tube, recorded in the report of Hydrodynamic Experiments.

Engineers.

Of the hundreds of young men who yearly graduate from our educational institutions, how few of them are ever likely to reflect credit upon the name, simply because nature never intended them for the business. The term is derived from the word ingenuity; geniuses are not the product of schools, but of birth. No education will ever produce an engineer or mechanic, though it may machines. No mere aptitude for mathematics will make up for lack of fertility in expedients so often demanded. An engineer should have ingenuity, sound judgment, and decision of character for emergencies. Without such characteristics no one will ever make a permanent reputation as an engineer. The calling has received the most of its renown from those who made no pretense of being engineers. Watt, Fulton, Stevenson, and others of the kind, were only considered engineers after their reputation had been made. Our yellow-plush propensity to accept heroes at their own estimate, if they only shout loud enough, has much to do with the continuance of unfounded pretensions. Many will remember the shout that went up at the *debut* of the Monitor. "Form a national society of engineers, and place John Erricsson at the head," was the cry. Had the Monitor encountered a storm on her passage out, as she did when she became the coffin for a hundred men, how different the result. For years previous, Mr. Erricsson had been the laughing stock of the country, and his achievements, before and since, indicate that, though he may- have some original ideas, he lacks the judgment necessary to make them safely useful.

Of our many engineers, we doubtless have those who, if favored with opportunities, would deservedly become noted; but the terrible disasters of the past few years, caused by the destruction of dams and bridges, would hardly indicate that the best have been employed in the most responsible positions.

It is not my purpose to write of engineers in general, but of those who are called, or who call themselves, hydraulic engineers; of this class J. B. Francis has long stood at the head, so far as the calling relates to milling matters. For many years Mr. Francis has had charge of all the property of the Lowell Water Power Co., and general supervision of from twenty to forty large mills. He is thoroughly versed in all of the theories, but it would be absurd to suppose he has had much time to devote to the details that make up the supposed knowledge of a hydrodynamic engineer. The continued use of poor turbines, when those much better could be had at one-half the cost of those used, prove plainly that he knows but little of the common characteristics of the ordinary turbine. The Francis weir formula is excellent, but I have had very disagreeable reasons for doubting whether he, or any of the so-called hydraulic engineers, realize how slight a change in proportion of pit renders the formula worthless.

H. F. Mills of Lawrence, Mass., has experimented much, and, in my opinion, is as good an engineer of the class as can be found: but he travels in a fixed groove. That he measures the water used by the mills there as accurately as may be done by the machine methods, I have no doubt; nor any doubt that it might be done still more accurately by simpler plans, at one-tenth of the cost at which he does it. There are many others that might be named, but they are all of about the same pattern—much formula relating to ancient theories, but with little practical knowledge of the requirements necessary to make manufacturing profitable under the sharp competitive conditions of to-day. Economy seems to be one of the lost arts with the whole class, but the following cross-examination of one of them will speak for itself:

"EMINENT HYDRAULIC ENGINEER."

The announcement may often be seen in the papers that John Smith, the eminent hydraulic engineer, has been called in to examine some prospective water power, mill, reservoir, dam, embankment or some milling matter of interest. Civil engineering seems to cover canal, mill, reservoir and dam building, so it is reasonable to suppose Mr. Smith, as a hydraulic engineer, has been called in to advise about the use of water power or its transmission. And that those interested may banish future anxiety, should Mr. Smith report favorably, we will put him on the stand for examination. If the reader thinks some other engineer more eminent than Mr. Smith, No! well, then, Mr. Smith will you please take the stand.

Mr. Smith, what is your age?

Ans. Fifty-seven years.

What is your occupation or profession?

Ans. Hydraulic engineering.

How long have you followed that business?

Ans. I served seven years apprenticeship and have followed the business thirty years.

You are thoroughly informed in all the minutia of the business?

Ans. (Modestly) I believe I have the credit of being so.

You understand water power and the various means used for its transmission and application to drive machinery?

Ans. I think I do, thoroughly.

You also understand the various methods used for measuring water used to drive machinery?

Ans. I do.

Name the various methods with which you are familiar.

Ans. The weir, aperture, floats and current metre.

You are often called upon by mill owners to measure water?

Ans. Quite often.

Which of the methods named do you consider best?

Ans. Well, where it is convenient, the weir.

Have you ever personally verified measurements made by either methods, so as to be able to vouch for their accuracy?

Ans. W-e-l-l—N-o, not personally.

Suppose the flume leading to a wheel to be so large that the water flows, say, one-half foot per second, would not the slip with a current metre be so great as to leave little chance for accuracy?

Ans. W-e-l-l—it might.

Do you, of your own knowledge, know that accurate measurements of water can be made with a current metre under any conditions?

Ans. No.

In measuring with floats, do you make an allowance for the average instead of apparent velocity? If so, how much?

Ans. I make an allowance of 20 per cent.

Is 20 per cent. fixed upon as a matter of judgment or positive knowledge?

Ans. W-e-l-l—that is the allowance generally made with float measurements.

Then float and current metre measurements have considerable guess work about them?

Ans. W–e–l–l—under favorable conditions they may approximate.

Can you personally vouch for the accuracy of aperture measurements?

Ans. W–e–l–l—N–o.

Do you know the least possible cross section of stream in measuring pit in proportion to the flow on the weir that will give correct measurement?

Ans. I do not.

Suppose the pit to be fourteen feet wide, with vertical sides; place a weir across, with end contractions, depth below the crest four feet, length of weir ten feet; then further down stream have another weir exactly the same, except the depth below the crest to be two feet; let the discharge from the mill flow over both weirs, would the depth on each show the same, supposing the discharge to be fifty feet per second?

Ans. W–e–l–l—r–e–a–l–l–y—I—well, I don't know.

Suppose the end contractions to be removed, what allowance would be necessary to deduct from the width to correct for the friction of the flowing water upon the rough side walls?

Ans. Well, something; I don't know just how much.

You have had experience with all of the water wheels in use from the old undershot to the modern turbine?

Ans. Constant experience for more than thirty years.

You often advise manufacturers as to the best kind for use?

Ans. Very often.

You understand the principle of each?

Ans. I think so, thoroughly.

The undershot is designed for low heads, is it not?

Ans. It is.

Which is the most efficient, undershot or breast wheel?

Ans. Oh, breast wheel, by all means.

Do you mean to say that for one foot head, a breast wheel would do better than an undershot?

Ans. Oh—w–e–l–l—for one foot—well, I don't know.

What is the maximum useful effect of an undershot wheel?

Ans. I don't know.

What is the exact relative velocity for an undesrhot wheel?

Ans. I don't know.

Have you had much to do with breast and overshot wheels?

Ans. Yes, indeed, very much.

Which is best?

Ans. W–e–l–l—some think the breast, others the overshot.

Never mind what others think. What do you know?

Ans. W–e–l–l—I never tested either, but I think—

Don't want to know what you think. Do you know?

Ans. No.

What is the proper velocity for the periphery of either?

Ans. W–e–l–l—some say five feet per second; from five to eight feet per second is probably the—

Don't want any probably. Do you know?

Ans. No.

What is the maximum useful effect a breast wheel will give?

Ans. W–e–l–l—I have read of 75.

Don't care anything about what you have read. Do you know?

Ans. No.

Do you know any better about the overshot?

Ans. No.

Mr. Smith, you are well informed as to turbine wheels?

Ans. Certainly; intimately so.

Which is the best discharge for a turbine—inward, outward or downward?

Ans. W–e–l–l—there are many opinions about that.

Wasn't asking about opinions, but about what you know.

Ans. Well, the Boyden turbine is outward discharge, and I believe that—

Don't want to know about what you believe. Do you know?

Ans. Well, every body knows the Boyden has given the highest useful effect.

Don't care for what every body knows. Do you know?

Ans. Well, I know Mr. Boyden reported—

Did you ● ⤥ test a Boyden wheel?

Ans. No.

Did you ever know of a disinterested engineer testing one who reported remarkably high efficiency?

Ans. W–e–l–l—no.

Did you ever know of a Boyden wheel being used where the water supply was insufficient for over half gate, or half of whole gate discharge, several months of the year, that gave satisfaction?

Ans. W–e–l–l—no—perhaps not.

Have you taken pains to ascertain whether there are other turbines that are better than the Boyden?

Ans. No, for I don't believe there are such.

Please give your reasons for such belief.

Ans. W–e–l–l—I—well—oh, cause I don't believe it.

So you have never taken pains to ascertain the real efficiency of the many other kinds of turbines?

Ans. No.

What is the proper relative velocity of the turbine with the water that drives it?

Ans. I don't know.

How do you know what proportional gears to use to connect turbine with the machinery to be driven?

Ans. Oh, I gear according to the table representing wheel.

What, when you know nothing certainly of the wheel?

Ans. W–e–l–l—yes—there is no other means of doing it.

Are all turbines of the same make of the same efficiency?

Ans. Certainly, or, at least, I suppose so.

You never have been to see such wheels tested in order to learn their peculiarities?

Ans. No, not I.

And why not? Has it not been your duty to do so before advising manufacturers in such matters?   .

Ans. Well, I have no faith in the testing that has been done.

Why not? Have you any real cause for doubt?

Ans. Well, many wheels have been reported as giving better results than is claimed for the Boyden, and—well, I don't believe it at all.

Do you, of your own knowledge, know that there are not fifty kinds of turbines better than the Boyden?

Ans. Oh, of course I know there are not.

Do you solemnly swear that you know there are not?

Ans. Oh, well, perhaps I can not swear that I know, but then you know I—

Please remember you are under oath. Do you mean to be understood that, of your own knowledge, you know anything about the matter?

Ans. Well, perhaps not; but I know what I think.

Quite likely, but that is not important.

Are you aware that the turbine will do considerable work while running at a greater velocity than the water that drives it?

Ans. I have heard so, but do not know it to be so.

Supposing it to be so, can you account for its so doing?

Ans. I can not account for it.

What is the proper shape for a turbine bucket, and in what direction should it project from the center of the wheel?

Ans. Oh, there are many opinions; I don't know.

Please give the exact positions for the chutes to stand.

Ans. Oh, each builder suits himself; I don't know.

Which should have the largest openings, the chutes or buckets?

Ans. Some builders think the chutes, others the buckets; I don't know.

Why is it that two wheels, built exactly alike, placed in the same pit side by side—in one the step burns down every month, in the other never?

Ans. I don't know.

Which is best for buckets, sheet iron, sheet steel, bronze or cast iron.

Ans. I don't know.

In all parts of the country water powers of any size are owned by several parties. Do you know of any means for dividing the water so that each may have his proper share, whether the supply is much or little?

Ans. I do not know of any means for such division.

Does a turbine, having a draft tube for part of the fall, do as well as one set in the tail water?

Ans. I don't know.

Have you taken no pains to ascertain?

Ans. Well—no.

What is the proper diameter for draft tube for a given discharge?

Ans. I don't know.

Suppose a draft tube to lead down stream at an angle of forty-five degrees, or still nearer a horizontal line, what would be the effect?

Ans. I suppose they would do well; I don't know.

Which transmits power with the least loss, belts or gears ?

Ans. Oh, belts, I think, decidedly.

Do you know anything about it positively ?

Ans. No.

Which causes the greatest loss, bevel or spur gears ?

Ans. Oh, bevel, by all means; at least I think so.

Do you know ?

Ans. W–e–l–l—no.

Have you ever taken any pains to ascertain the loss, if there is any, caused by the use of belts, gears, or draft tubes ?

Ans. W–e–l–l—no, not personally.

Mr. Smith, will you be so kind as to state what knowledge about hydrodynamics is actually necessary to entitle a person to be considered an eminent hydraulic engineer ?

Ans. Oh, well—he must know all about water power and mills and things.

Certainly, but please give particulars.

Ans. Oh—well—he must know— why, he must know all about it.

Well, Mr. Smith, that will do for the present.

"OVER-EDUCATION.

" Like over-production, our caption is in some senses a misnomer, for no one can be over-educated in the true development of his best faculties for worthy ends. But there is a great deal of school and college education that is aimless, disproportionate, and cumbersome. There are too many mediocre professional men, lawyers, doctors, ministers, school teachers, writers; few skilled artisans, farmers, gardeners, intelligent laborers technically educated for various spheres that are fundamental to well-ordered society. Society is top-heavy, with too much top and too little bottom. There is too much high-school dabbling that is not thorough enough for mental gymnastics, nor practical enough for the utilitarian necessities of those who must graduate into the hard work of the common and laborious pursuits which ballast society. The great law will assert itself, and all true education must lay its account with it, that by the sweat of the brow we must eat our bread. That is not good American education which would spoil a farmer's boy for the old homestead, or the farmer's girl for housekeeping. There is too large a crowd of unfit female school teachers. There are too many useless, third-rate lawyers hankering after office; too many goodish ministers, unskilled doctors, ignorant apothecaries and engineers. Hence there are multitudes of our boys and girls who are over-educated, in the sense that they are unfitted by an aimless and merely bookish education for any patient and earnest life-work which will utilize them as producers, and develop their individuality into the manly or womanly consummation of a stanch character and a robust and useful life."

Our common school system is at fault for this. What would be thought of the person who should treat everything growing upon his farm with the same care—planting beans, strawberries, cabbages, onions, wheat, weeds, and pumpkins all in the same way ; plowing a little here, digging a little there, going over much surface—none deep ? Would not the results resemble the product of our schools— a smattering of everything, a real knowledge of nothing ? every graduate rushing for the position of major-general—not one willing to accept that of private ? Is it not evident that the system is productive of the idea that honest labor is degrading ? that the proper aim for the young man is office or a profession ; for the young woman, wealthy marriage ? Under its influence, are our Presidents, members of Congress and Legislatures, and officials in general, selected from the first or even second class minds of the country ? Will our officers or teachers, male or female, compare favorably, intellectually, with our native mechanics ? Pay high salaries, and get the best ! is the constant shriek of the office-holder and teacher—which means, get those who will shriek loudest for more pay and less labor. Of all the trashy ideas prevalent, there is none more shallow than the pretense that high salaries insure the best services. High salaries to the few means degradation to the many—really a relic of barbarism—the feudal lord and subjected serf. Salaries so high as to be desirable in themselves are far more likely to be obtained by the unscrupulous pretender than the worthy proficient, as is patent to every one having any knowledge of the way the offices throughout the country are filled. I hope and think the time will come when our school system will limit the studies to the common English branches, and in those, give every child in the country a thorough course, leaving those desiring a higher education to obtain it at their own expense as a luxury—a real luxury—to the proper minds, but unappreciated by the multitude. Even were it possible to give every child a thorough education, gratuitously, in all the studies now merely skimmed over, it would be a matter of very doubtful utility. Possessions are valued somewhat in proportion to their difficulty of attainment ; inherited property is seldom valued like that earned by years of hard labor. It can hardly dignify the high educational system to have the brilliant valedictorian wait idly for a year or two for something grand to turn up and then settle down as keeper of a peanut stand. Limiting the education at the public expense to the branches named will, I believe, produce a higher civilization than the present trashy method—less of the professional, more of the practical ; better mechanics, farmers, engineers, doctors, teachers, fathers, mothers, and wives.

Massachusetts School System.

Nursery for a nation of Roughs and Gamblers.

First dude—Going to the party this evening, dear boy?
Second dude—No, mether isn't very well, and I can't go without a chaperone, don't you know?

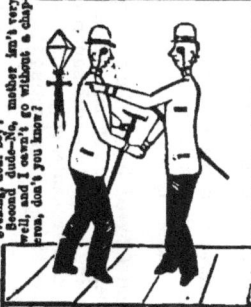

Fitzadolphus — Why, the guvner wants me to help in the shop!
Angelina — Oh cruel! And ma wants me to darn my own stockings.

Prof. Sneezer — Can't spell? What have I to do with spelling? I'm professor of Greek!

God made individuals. I wanted system! Alas and alas, behold the result!

Ma and the guvner.

The Hon. John, or Prof. Sullivan.

Willimansett Spout Experiments,

Made to determine the co-efficient of discharge through such spouts.

These spouts were made one-eighth size of some used in a mill early in the century, at Plattsburgh, N. Y. From the flume downward they pitched four inches to the foot and had vent-holes in their tops just outside of flume. No. 1 was 24 inches in length through center of sides; the increased length was to determine the effect of the extra length; area of opening at lower end, 18 inches, or 4 x 4½; area of opening at upper end, 28 inches, or 4½ x 6¼, which could be increased to 36 inches by withdrawing a wedge of plank.

No. 3 was the proportional standard, 15 inches in length through center of sides, with same openings as No. 1.

No. 4 was of same size as No. 3, placed horizontally.

No. 2 was the same, curved.

No. 5 was one-eighth size of No. 3.

All were tested first with lower end submerged, under three feet head. All, whether long, curved, pitched, or placed horizontally, gave the same discharge and co-efficient.—No. 5 of course proportionally only. The wedge plank was then withdrawn, the opening of the upper end then being 36 inches. With that increase the discharge and co-efficient increased two per cent.

All were then raised, the center of opening of 1, 3, and 4 being three and one-half inches above surface of tail-water, the lower end of No. 2 three inches above the surface. With 2 feet 9 inches of head above lower end of curved spout and center of spouts 1, 3, and 4, the discharge and co-efficient were the same as when tested under three feet head and lower end of spout submerged, showing that the cohesion of water kept the column of discharge solid for about two-thirds of the diameter of the column of discharge below lower end of spouts. The vents had no effect upon the discharge otherwise than to increase it to the extent of the spurt through the vent-hole. The co-efficient of discharge through the standard No. 3 was 91.39 per cent.

The spouts in the old Plattsburgh mill varied considerably in the proportion of openings at their lower and upper ends, as would be likely to be the case with such in other mills, which would affect their co-efficients as shown by increasing the opening in the upper end of No. 1. The inside corner of the planks of all the spouts tested were rounded, producing a flare of the upper end of spout.

At Highgate, Vt., a tub wheel was in use in 1885, and I tested its efficiency by grinding and measuring the water used. The conduit conducting the water to the wheel was an open spout with parallel sides, the ends of planks inside of flume left square, as was the bottom of gate, forming an aperture with square corners which would allow of but sixty per cent. co-efficient. There was 7 feet 3 inches head over center of gate opening, the spout, 8 feet in length, pitched so that the water striking the half depth of wheel gave 11 feet of head acting thereon. Under such conditions it required 5.2 h.p. of water to grind a bushel of ordinary wheat, and 5.9 h.p. to grind a bushel of hard Minnesota wheat.

Willimansett Spout Experiments.

Nos. 2, 3, and 4 are shown in the position of trial, but when tested were placed where No. 1 is shown, the testing pit not extending back of penstock.

THE EMERSON POWER SCALE.

To produce the perfect instrument herewith illustrated has required perhaps a hundred plans and changes, made at a cost of some $30,000, and a quarter of a century in time.

Each size is graduated upon a circle of a given number of feet, and the revolutions per minute must be multiplied by that number in computing the results of trial.

As these scales are all constructed upon the same principle as the ordinary platform scales, and are common in the best mills, it is unnecessary to describe them here.

The illustrations represent the perfected scale, which weighs after connection, let the shaft run either way ; also the register counter.

The ability to weigh when the shaft is running in either direction is made practicable by the use of the double connections 1 1 to the bell crank levers K K, the connections 1 1 being slotted at connecting point as shown in Fig. 3.

The register counter shown in Fig. 1 consists of worm M on shaft, into which works gear N having a hundred teeth, and the head of pendulum B, which forms a shield over nine-tenths of the ratchet gear A back of shield.

The pendulum B raises one-tenth of a circle, the ratchet gear has one hundred teeth, and if the weight was always at the maximum, say 100 pounds, the hook C would rotate the ratchet gear at every ten movements, but as the weight constantly varies, often from zero to the maximum, the shield prevents the hook C from carrying the ratchet gear any more than due the weight at each movement.

As it requires ten operations of the hook C to cause a complete rotation of the ratchet A, supposing the weight to be at its maximum, a cipher must invariably be added to the registered figures shown on the register H, as 976 must read 9760.

To get the real revolutions of the shaft, two ciphers must be added to the registered figures on register I, as the 12035 must read 1203500, which divide by the number of minutes in the run, say for a week of sixty hours, or 3600 minutes, as follows :—

Maximum graduation of quadrant, 100 pounds ; registered figures as shown, 976 ; add cipher, 9760 ; registered figures on register I, 12035 ; divide the figures of register H by those of register I, 9760 ÷ 12035 = .81 as the average weight during the sixty hours' trial.

Now to obtain revolutions per minute take 12035, add two ciphers, 1203500 ÷ 3600 = 334.3 revolutions per minute, multiplied by, say, graduation of No. 3 scale, 6 ft. = 2006 ft. × 81 pounds= 162486 ft. pounds ÷ 33000 = 4.92 h.p.

For information about the scale inquire of the manufacturers,

EMERSON POWER SCALE CO.

FLORENCE, MASS.

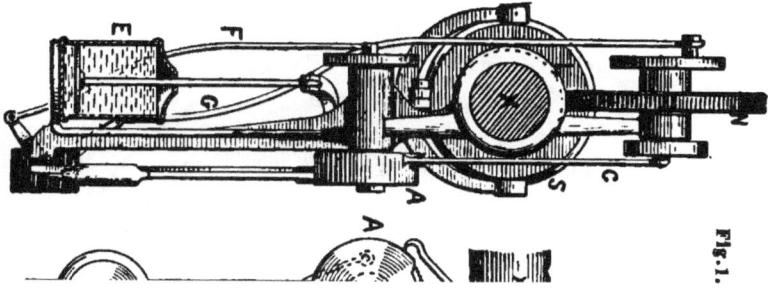

Fig. 1.

Fig. 2.

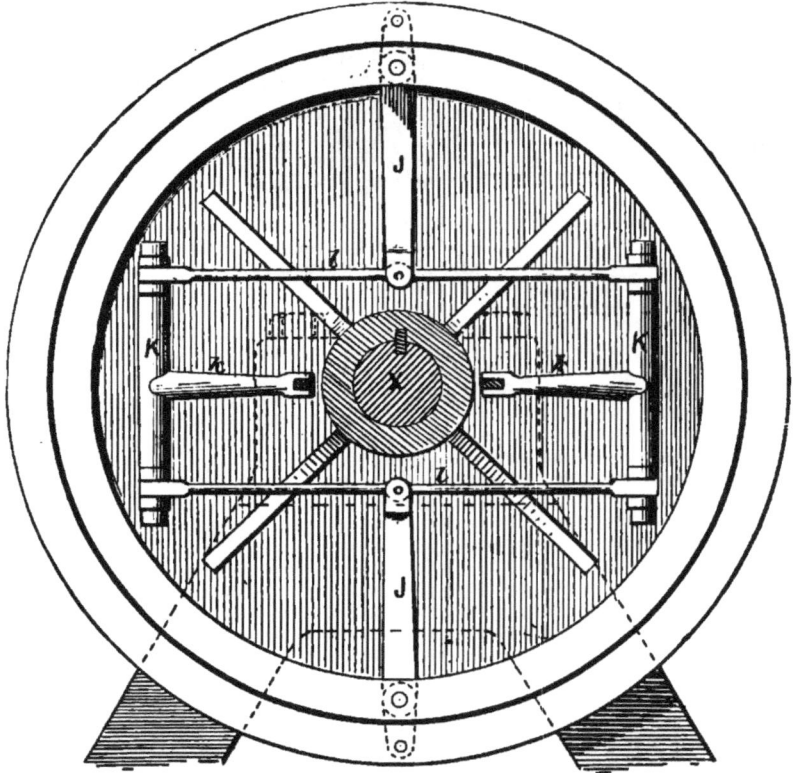

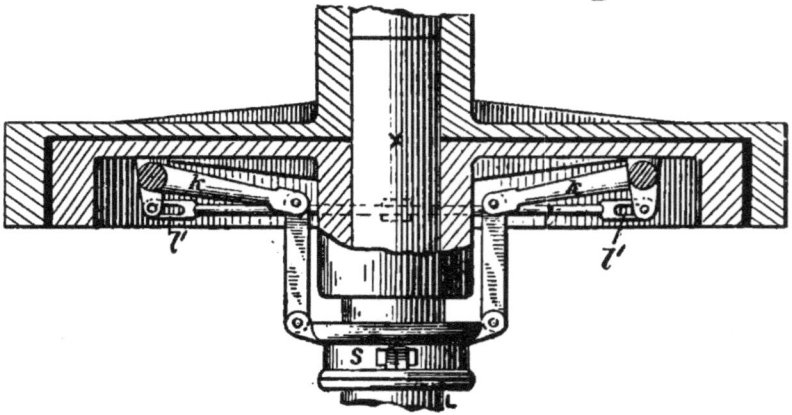

Fig. 3.

THE COTTON MILL SCALE.

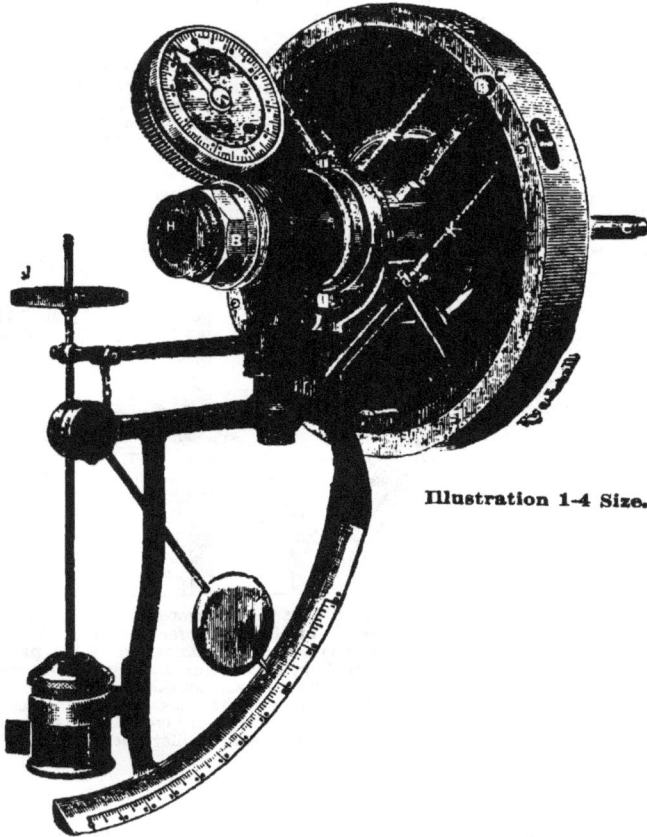

Illustration 1-4 Size.

The above illustration represents scales designed for cotton mills, to be used in testing the power required to drive spinning frames, fly frames, slubbers, and other light running machines, having tight and loose pulley outside of frames.

Graduated upon a two-foot circle.

Water Measurements.

The lack of a practical knowledge of hydraulics a generation since caused a looseness in contracts pertaining to milling matters that has been productive of an immense amount of vexatious and expensive litigation. It is only necessary to glance at the methods adopted by the various Water Power Companies of the country for determining the quantity of water leased, as published on preceding pages, to learn that there has been no generally recognized standard for such measurements even among those claiming to be engineers and experts in such matters; it would seem that the average boy, ten years of age, who has ever played with toy water wheels would be able to provide something more definite than the Oswego plan. One great cause for the looseness in contracts has been the difference between the actual and theoretical discharge of water through an aperture of any size under a given head. The difference is only understood now by a very few. There are turbine builders who suppose that their wheels discharge the full quantity theoretically due their openings, while those calling themselves engineers generally believe the discharge of such wheels to invariably be about 60 per cent. due their openings, when in fact the discharge of turbines varies all the way from 35 to 100 per cent., and in special cases perhaps still more. The discharge through an aperture in the side of a penstock may be made to differ 50 per cent. An aperture one foot square, its center under two feet head, cut with edges at right angles with the face of the planks inside the penstock, leaving perfect sharp corners presented to the water as it issues, (see Fig. 1,) will dis-

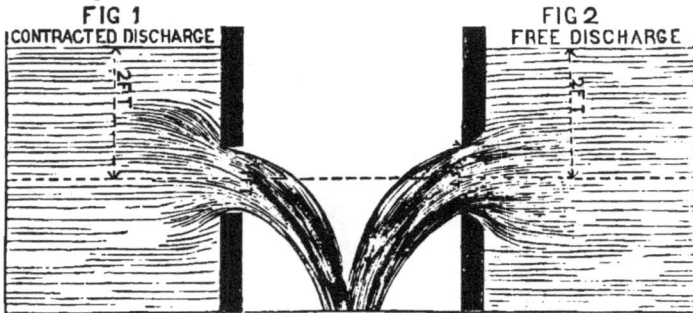

FIG 1
CONTRACTED DISCHARGE

FIG 2
FREE DISCHARGE

charge about 6¾ cubic feet per second; but with a proper flare of the aperture (as in Fig. 2,) the discharge will be about 10½ cubic feet per second, and the same relative percentage for other heads. An examination of the problems demonstrated in Evans' "Work on Milling, Hydraulics, &c.," published as late as 1848, will show that this important difference was not taken into consideration in preparing a work that was to be offered to the public as a guide in such matters. The following extract from the Work, page 96, is given, however, to show that the publisher had an impression that there was a difference.

Article 55.

OF THE FRICTION OF THE APERTURES OF SPOUTING FLUIDS.

The doctrine of this species of friction appears to be as follows :—

1. The ratio of the friction of round apertures, is as their diameters nearly; while the quantity expended is as the squares of their diameter.

2. The friction of an aperture of any regular or irregular figure is as the length of the sum of the circumscribing lines, nearly; the quantities being as the areas of the apertures.* Therefore,

*This will plainly appear, if we consider that the friction does sensibly retard the velocity of the fluid to a certain distance; say half an inch from the side or edge of the aperture, towards its center; and we may reasonably conclude that this distance will be nearly the same in a two and twelve inch aperture; so that in the two inch aperture, a ring on the outside half an inch wide, is sensibly retarded, which is about three-fourths of the whole; while in the twelve inch aperture there is a ring on the outside half an inch wide, retarded about one-sixth of its whole area.

3. The less the head or pressure, and the larger the aperture, the less the ratio of the friction; therefore,

4. This friction need not be much regarded, in the large openings or apertures of undershot mills, where the gates are from 2 to 15 inches in their shortest sides; but it very sensibly affects the small apertures of high overshot or undershot mills, with great heads, where their shortest sides are from five-tenths of an inch to two inches.

This seems to be proved by Smeaton, in his experiments; (see table, Art. 67;) where, when the head was 33 inches, the sluice small, drawn only to the first hole, the velocity was only such as is assigned by theory to a head of 15.85 inches, which he calls virtual head. But when the sluice was larger, drawn to the sixth hole, and head 6 inches, the virtual head was 5.33 inches. But seeing there is no theorem yet discoverd by which we can truly determine the quantity or effect of the friction according to the size of the aperture and height of the head, we cannot, therefore, by the established laws of hydrostatics, determine exactly the velocity or quantity expended through any small aperture; which renders the theory in these cases but little better than conjecture.

ANCIENT AUTHORITIES IN HYDRAULIC CASES IN LITIGATION.

In milling cases on trial, old English or American works are brought in as authority. These a half century since were useful because there was nothing better, but a revolution has taken place in such matters and there is now no difficulty in elucidating any matter pertaining to milling hydrodynamics so as to leave no just cause for dispute.

Oliver Evans has perhaps been considered the best milling authority up to 1860, but he simply copied the most of his ideas from old English works. His ideas of spouting fluids, article 55, show beyond chance for dispute that neither he nor his authorities knew anything about the law governing such spouting or the discharge through apertures.

It is now positively known that all apertures, large or small, round or square, discharge about 60 per cent. of the theoretical quantity due the opening, if the aperture is cut squarely through the plank, leaving sharp corners, as shown fig. 1st, opposite page.

The following note, copied from page 114 of his book, shows how little reliance can be placed in his authorities:—

"After having published the first edition of this work, I have been informed, that, by accurate experiments made at the expense of the British government, it was ascertained that the power produced by 40,000 cubic feet of water descending 1 foot will grind and bolt 1 bushel of wheat. If this be true, then to find the quantity that any stream will grind per hour, multiply the cubic feet of water that it affords per hour, by the virtual descent, (that is, half of the head above the wheel added to the fall after it enters an overshot wheel,) and divide that product by 40,000, and the quotient is the answer in bushels per hour that the stream will grind."

It certainly should do so, for 40,000 cubic feet of water falling one foot evolves 75.5 h. p. Quite likely some essential feature of the experiment is left out so that the statement is worthless, as is invariably the case with their reports.

For, owing to their want of knowledge in such matters, they failed to give the necessary data to make their statements useful; for instance, in mentioning the discharge of water through apertures, they don't describe the form of the apertures, yet, as may be seen by the diagrams opposite, the discharge may be made to vary through the same sized aperture more than fifty per cent.

From personal acquaintance with turbine builders and their ways it has seemed doubtful to me whether any work published previous to the commencement of the testing system in 1869 has, except in a negative way, been of any help towards the improvement of the turbine or knowledge of milling hydrodynamics.

To ascertain whether the opinion was well or ill founded, the following letter was sent to John B. McCormick, who, through personal predilection, perseverance and unequaled opportunity for experimenting, unquestionably stands unrivaled in the knowledge of turbine construction.

Willimansett, Mass., Feb. 27, 1892.
John B. McCormick, Holyoke, Mass.

Dear Sir :—Believing that the continued use of old text books as authority in matters pertaining to hydrodynamics has a tendency to cause the production of an inferior class of engineers, I would ask whether, except to avoid their errors, you have been aided in your turbine improvements by any hydraulic work published previous to the publication of tests in 1869.

Yours truly,
JAMES EMERSON.

REPLY.

Holyoke, Mass., March 1, 1892.
James Emerson, Willimansett, Mass.

Dear Sir :—Yours of the 27th duly received, and in reply will say : The old text books have not been beneficial to the writer, and their teachings were entirely disregarded in the production of the "Hercules" and other wheels which have been produced and perfected since by the undersigned.

Yours truly,
JOHN B. McCORMICK.

But the worst of all is Haswell, who poses as universal instructor for the present time, and presents a *hash* of old theories that have been out of date for a generation past, seriously describing the construction of undershot, overshot, breast, Poncelet, Fourneyron, Boyden, Jouval, and other antique water wheels that have as little chance for future use as has the old stage coach of a half century since.

He gives the possible efficiency of the overshot at 84 per cent., and that of the breast wheel at 93. As actual trial under the same conditions proves that the turbine will do nearly double the work that can be done with the breast wheel, it may safely be stated without fear of successful contradiction that the breast or overshot wheel was never made that could exceed 67 per cent. useful effect.

Mr. Haswell asserts that large turbines give a higher efficiency than small ones, but the testing of twenty years proves the contrary to be the case, as quite likely it would with the breast and overshot.

Mr. Haswell's mind is in an excellent condition to receive instruction in hydrodynamics.

It is the study of such authorities that produce such depositions as the following :—

ETHAN S. REYNOLDS
vs.
INDIANA PAPER CO. et al.

In St. Joseph Circuit Court, }
State of Indiana. }

Complaint No. 2560.

Depositions of Mr. Clemens Herschel, duly sworn, testifies as follows :—

Direct Examination.

Q. You may state your name and residence and occupation?

A. Clemens Herschel, hydraulic engineer, at Holyoke, Mass.

Q. How long have you been hydraulic engineer, located in Holyoke?

A. I have been here since April, 1880.

Q. How long have you been practicing your profession as a hydraulic engineer?

A. Twenty odd years.

Q. What institutions are you a graduate of?

A. I am a graduate of the Lawrence Scientific School, Harvard University, and Polytechnic School of Karlsruhe, Germany.

Q. What position do you occupy in Holyoke with reference to the Holyoke Water Power Co.?

A. I am their hydraulic engineer.

Q. Why is the amount of discharge different under different heads? Will you explain that to us?

A. That is because it is an impossibility that the head, acting on the wheel, shall ever be the same as the head contained in the race, and the allowance for that difference which I made to get the water off and on the wheel, as it is called, is one foot, that being my judgment, and also being a usual measurement, and contained in a great many leases with which I am acquainted.

Q. I understand you to say, as an engineer, that the allowance of one foot is a proper allowance to make, and one that is usually used or allowed?

A. It is both a proper and a usual one. One foot off of six feet is a difference of $16\frac{2}{3}\%$, one foot off of ten feet is only a difference of 10%; that is a reason the quantities I have reached vary from 2074 to 2156, at six and ten feet respectively.

Q. Would measurements of the depth of water at the flume alone indicate the head?

A. It would not.

Cross Examination by Mr. Hubbard, for the Plaintiff.

Q. Mr. Herschel, why is the difference between $16\frac{2}{3}\%$ off for six foot head and 10% off for ten foot head made?

A. Because in any case, this per cent. represents just one foot, and one foot is the usual and customary allowance, and the proper one, in my opinion, and the one that obtains in actual practice.

Q. Would the same percentage be true as to the cubic feet discharged per second or per minute under the same head? That is, six and ten feet off, $16\frac{2}{3}\%$ and 10% respectively?

A. By no means.

Q. Please explain how you arrive at the 16⅔% deduction on account of a difference of one foot between the actua! level of the water in the canal and the tail race, and the actual level between the water immediately above and below the wheel?

A. That percentage is arrived at only in the case of a six foot head being the total, which we in Holyoke call available head. The allowance of one foot is made to get the water to and off the wheel, and one which is customary and proper, as I have explained. One foot being one-sixth of six, it results in reducing the head available, in order to get the head acting on the wheel, by one-sixth, or 16⅔% in this particular case.

Q. Is it not a fact, then, that if the mills were located at, say, ten rods distance from the main canal, and the flumes were too small in proportion to the amount discharged by the wheel to maintain a constant, or nearly so, level in the flume, then the loss of head might be more than one foot?

A. It would be, under those circumstances, more than one foot. I have known it to be one or two feet, and perhaps, in extreme cases, four feet. I arrived at the figure, one foot, from reports made to me by Mr. Smith of the locality, and in the exercise of such judgment as I have in these matters.

Q. This means, then, does it, in short, a deduction for the loss of head in getting the water to the wheel depending upon the distance of the wheel from the canal, and the size of the flume and fore-bay?

A. It depends upon that and other facts. The construction of what is called the rack, in front of the fore-bay, has usually quite an effect on it, the size of the flume, and whether the water turns at right angles or not, and how it turns. The mere length of the flume and tail race has rather a minor influence than some other structures and circumstances that occur in these cases.

Q. Then you include in addition to the items mentioned in my previous questions the loss of head by the means of the tail race?

A. Yes, sir.

Q. And you arrive at this from statements made to you by Mr. Smith of the conditions of the premises of the Indiana Paper Co., in September, 1888, do you not?

A. Partly so, but more largely from my judgment as to the propriety of the allowance of one foot from such loss of the total available head, in order to get the head acting in the wheel, which latter is the head which gives the discharge for the wheel.

Q. You have never seen the premises of the Indiana Paper Co.?

A. Never.

Q. Personally, you know nothing of the actual construction of the head and tail race except as reported by others?

A. I know it only from the report of Mr. Smith and others, and also from my judgment of what such structures look like in the Western states.

CLEMENS HERSCHEL.

HENRY K. HAWES,
Notary Public.

Had I not heard the foregoing deposition read in court, I should have been slow to believe that any one claiming to be an engineer would utter such stuff.

The slightest acquaintance with water powers shows that all vary in head more or less, consequently an allowance is made so that a tenant shall have no cause for action if the head drops somewhat from the usual height. This is done at Holyoke; nineteen feet are deeded, where there usually are twenty. Mr. Herschel has mistaken this practice for safety, as the rule for head when computing the discharge of a wheel.

All he was required to do for the Indiana Paper Co. was to measure the apertures of the several wheels, then give their discharge for given heads, say three, four, five, and six feet.

His success as engineer while at Holyoke hardly warranted his gratuitous fling at Western water powers.

There are many dams built by farmers and mechanics at the West, that such engineers as Mr. Herschel would find it difficult to equal; the one at South Bend, upon which the Indiana Paper Co. is located, is across the St. Joseph River, the bottom of which is so soft that the dam is constantly settling.

It was testified in court at the time Mr. Herschel's deposition was read, that the year before a part of the dam had been raised eighteen inches to restore it to its original height. At Mishawaka, fifteen miles east of South Bend, the dam was built by a farmer and is really a creditable piece of engineering for a professional dam builder, as are many other dams and mill arrangements that may be found West. Their worst feature is that they are nearly all over-worked.

DAMS.

Engineers differ much in opinion as to the proper way to construct dams. Stone dams, as a rule, have not proved so safe as one would naturally expect; yet with proper construction and sufficient material such dams should stand.

That pent up water has mighty force is proved by the vast ravines and notches in mountain ranges wherever such ranges exist.

I have had occasion to admire dams built of the boles of trees, the butts down stream packed closely and bolted one upon another from bottom to the top, then loaded down with rocks and gravel. These structures are often built upon soft mud bottoms or quicksand by men making no claim to be considered engineers, yet their work is perhaps superior to many professional engineering jobs.

There is a stone dam at Windsor, Vt., forty feet in height, that has stood a half century and seems good for the other half. The stones are laid without cement, but planked upon the up-stream side. A stone dam with earth embankment below to me seems a poor arrangement, frost or no frost, while such embankment above or up-stream should be very useful.

THE HERCULES TURBINE.

In March, 1876,—Centennial year—several of these wheels were brought to the Holyoke testing flume to be tested by me. The builders, Messrs. McCormick & Brown, made such extravagant claims in advance that they were laughed at as visionary cranks of the then usual hydrodynamic species.

A week spent in testing and re-testing, changing wheels, again testing, proved the claims of the builders to be well founded. Leading turbine builders were called in to assist in making the tests, for it was evident the wheel marked a new era in hydrodynamics.

After an exhaustive series of testing, a report of the results was made public, and an effort made to have all water wheel builders examine the plans and start anew, and each strive to make the plans still more effective ; but my idea of the matter may be found in report of Holyoke Machine Co. as turbine builders, in the third edition of my work, " Hydrodynamics," 1881. The success of the Hercules led to the production of the Victor and New American, either of which, a few years earlier, would have been considered phenomenal. The Victor in several cases gave high whole gate tests ; the New American high average tests from half to whole gate, so that I recommended their use to parties in Holyoke, and I think that some eight or ten of each were purchased and set in mills there. But, for reasons satisfactory to the purchasers, these, each and every one, I think, have since been replaced by the Hercules, and now, after sixteen years' trial, I can and do say the Hercules turbine *has no equal, it stands alone,* and manufacturers owe much to the Holyoke Machine Co. for its liberality and perseverance in bringing the wheel to its present state of perfection, for it has cost a fortune to do it. Its variations may be realized by an examination of the diagrams showing the efficiency of various wheels tried.

Substantially the same wheel in general appearance is now made by the Holyoke Machine Co. and the Jolly Brothers, both of Holyoke.

Only a test of each wheel can decide which is best. There is an individuality in turbines. No man has yet lived that can build two wheels with absolute certainty that they will give the same results ; a glance at the diagrams and tests will show great variations. The intelligent purchaser has his wheel tested before acceptance, and rejects such as do not reach what he has bargained for. Who, Mr. Purchaser that thinks it won't pay to test, do you suppose gets those *rejected wheels ?*

LITIGATION TO SETTLE QUESTIONS IN DISPUTE.

All who have read Juvenal's Satires will recall the surprise he expresses, that where ropes, daggers, and high buildings render suicide so easy, any man can be fool enough to marry ; so it is equally a matter for surprise that a man having a mill pond large enough to drown himself in should resort to *law* to decide who owns the pond.

A lawyer that takes up a case desires to win, and, as is natural, will do so if he can, right or wrong. Any trickery that can be made to appear legal may be resorted to with approval.

A sucking Blackstone with impudent assurance may browbeat and bully a witness so long as he keeps within the legal ruts, and a very shallow fool can, and often does, ask questions that a wise man cannot answer simply because he is not allowed to explain and show that the question has no application to the case in hand. An annoyance that practical witnesses often have to contend with are works of shallow, conceited aspirants, who desire to shine as that "Eminent Hydraulic Engineer," or as the "Great Doctor Squills." The less such authors know, the more hair-splitting and profound will be their theories,—that is if profundity consists in unintelligibility. Could such frauds be examined by capable members of their calling their pretensions would at once be made apparent, as in the case of the eminent engineer, John Smith.*

A sharp, unscrupulous attorney might, in fact often does, study up such shallow publications, and seemingly confounds an intelligent engineer or physician, simply because either has such contempt for the ignorant stuff presented as science, that, feeling that others should see the palpable absurdity as well as themselves, they treat the whole with contempt. There are few cases in milling matters that cannot readily be explained in a few minutes if the attorney would state the case clearly, then allow the witness to tell what he knows about it in as few words as possible. Certainly such would be much the quickest way to obtain the merits of a case from an intelligent expert; instead of which he is often kept under a shower of questions, for hours, nine-tenths of which have little bearing upon the case in hand, the attorney upon his side treating him like a charge of dynamite, likely to explode unexpectedly, the opposing attorney operating from the start as though he had a criminal to deal with.

For myself I can say with truth that I never took the witness stand with a desire to favor either side, and have seldom left it without feeling *outraged*. The dignity of the law and courts are often lauded, but my experience has not enabled me to see it.

Think of the immense flunkyism there must be latent in human nature to cause the free-born citizen to dress in his granny's old silk gown in order to equip himself for the supreme bench. No wonder the owl, the stupidest of birds, is selected to represent wisdom.

---

*For untold ages it has been found impossible to make laws that human ingenuity cannot evade; then why, like Mrs. Partington, continue to attempt the impossible ? why not obliterate every statute, then re-enact a few broad principles and compel the settlement of all disputes by arbitration in the light of current intelligence ?

The Selection of Turbines

is a matter upon which a manufacturer's success in business often depends, yet in which the least practical knowledge is generally used. The common practice is to guess at the power required, the water at command, the best kind of wheel; finally, at the size of that. That such a system exists is owing to two facts; First, that we have had no really practical milling engineers; Second, to man's desire to get more than he is willing to pay for —to the same disposition that causes him to buy lottery tickets, or to gamble in stocks—and he exclaims: "I do not see why, if one is good, another of the same kind must not be so, too." Suppose he does not see, does he not know of plenty of cases to prove that it is not so? And there are good reasons for its not being so. For a number of years certain turbine builders made expensive efforts to gain high results. So long as the greatest possible care was given to each branch of the business, so long were high results generally obtained; but the moment such care was abandoned, and the business conducted with the ordinary care common in foundry and machine work, the ninety per cent. wheels dropped to eighty or less; then, in a little time, the patterns became warped or worn, or less care was used in setting them exact, as they were being molded, and the wheels made from them would give seventy-four or seventy-five per cent., though wheels made from the same patterns a year before often gave from eighty-five to ninety per cent. Too much time and money have been expended upon such wheels, any way, though in years past it was a matter of less consequence than now, except that it created or encouraged a false idea of the value of such wheels.

The Boyden and Tyler scroll wheels were rivals for a generation—the Boyden being used by large corporations under the most favorable conditions; the Tyler in the backwoods, under conditions in which the Boyden would have been unable to work at all. Many of each have been used twenty years without requiring repairs. If the point could be accurately determined as to the economy in the use of water, there is not a shadow of proof to show that the decision would be favorable to the Boyden; while the cost would be ten and the trouble in keeping the wheels clean and in working condition would be as a hundred to one in favor of the Tyler. Both are now, however, of the past, and out of place where economy is desirable. But, says a manufacturer, "My mill is on the upper level, where the head is always the same, and I buy so many cubic feet per second; so what use is it for me to have a particularly good part gate wheel?"

There are two good reasons for preferring such wheels: First, a good part gate wheel uses water in proportion to the work it has to do, and there are times in all mills when more or less of the work is stopped. Good part gate wheels save water at such times, which benefits all on the same fall; but a more important point is, that during low water in the dry season, when the supply is insufficient to do the work without the aid of steam, the mill having good part gate wheels can utilize whatever there is of water, while those having Boyden, or any of the popular whole gate wheels, can realize but little benefit from a two-thirds and nothing from a half supply.

There is one, and only one, method of securing a valuable turbine without any risk, and that is to ascertain first exactly what is needed, which may readily be done by measuring the water that is to be used and the power the mill requires; then apply to a respectable turbine builder, use ordinary common sense in the matter, and not expect that a wheel of a given capacity can be made in so perfect and durable a manner for four hundred as one that costs four thousand dollars. The idea is equivalent to the quandary of the young man who hesitated as to whether he should give his girl a piano or a pint of peanuts. Pay a fair price, and insist that the wheel shall be thoroughly made in every way, and tested before acceptance; and, unless it gives an average useful effect of 76 per cent. from half to whole gate, refuse to take it. A wheel that will give such an average is good, and will do a third more work with the same water, under the ordinary working conditions, than any Boyden or Victor ever made. There is another and very erroneous plan of fitting up mills: that is, to use wheels much too large for the work with the ordinary head, in order to avoid stoppage during backwater. Such wheels are entirely out of place, for if geared for the ordinary head they run at great loss through waste of water at all times—during the ordinary head, because too large; and, during backwater, because geared for a high speed.

Turbines Running Faster than the Water that Drives Them.

We often hear of destructive collisions when heavy bodies meet, but never when two bodies are moving in the same direction—the forward one the faster; yet the turbine often moves faster than the water that drives it, and does good work. [See, for example, Upham wheel, test 13; weight, 100 pounds; revolutions, 300 per minute.] The wheel was 30 inches in diameter, on what would be the pitch-line of gear of that shape. Any one acquainted with such matters can get the circumference and spurting velocity of water for the head given, and thus verify the statement. Such turbine builders as claim to be scientific have a theory to fit the case, but do not agree well with each other. Will not some of our college professors or students, those engaged in such studies, give it attention? and in so doing take into consideration the fact that the Upham wheel discharges the water obliquely outwards near the periphery of the wheel, where its velocity is greatest, instead of near the center, where the velocity of the wheel is less than the spurting velocity of the water—seemingly a sufficient proof that theories based upon the central discharge idea are incorrect.

Many explanations have been sent to me in relation to the above, none from the colleges or engineers. Judge Waldron of Maine readily accounted for the fact upon the same principle that an ice boat often sails faster than the wind that drives it. Many of the explanations have been lengthy, accompanied with diagrams, but the simplest solution that occurs to me is the wedge that often flies from the frosty log; the wedge to open the cleft one inch may enter three, consequently moves three times as fast as the cleft parts when it flies out.

Backwater under Conditions Difficult of Settlement.

Many cases of backwater for which complaints have been and still are being made, have arisen through the effect of a rapid current produced by a fall in the stream or the discharge of water from a mill located upon the fall—the current having carried the loose sand, mud, gravel, sawdust, bark, or other debris forming the bed of the stream down to a wider or more level place where the velocity was less, and there depositing it, forming a bar across of a greater or less height, as the case might be, raising the water above causing a fall below. In earlier times, when locating a mill upon such a fall, the wheels were seldom placed so low as to receive the full effect of the fall, for, through the abundance of water, the comparatively little power required could be obtained at less expense with a portion of the available head. In time, another mill was erected further down stream, the dam for which flowed the water back upon the bar above, without in any way interfering with the power of the mill above. These conditions continued for years without question. As the country became settled, the supply of water grew less, the power more valuable and better cared for. The upper mill was enlarged, the wheel-pit lowered, the wheels placed at the bottom, and the bar removed. Of course the water from the dam below flowed back into the upper wheel-pit and obstructed the wheels. Under such circumstances, it is apt to cause the owner of the upper mill to insist that the lower dam has gradually been raised above the title thereto. There are plenty of mills yet, the discharges from which are raising such bars, and so gradually as to be overlooked and neglected, which will surely cause trouble in time.

Testing Flume and Turbine Testing.

The testing system, or practice of testing turbines before purchase to determine their value, has become so general that there is no turbine builder of any reputation, who has not found it necessary to submit his wheels to such trial, in order to enable him to sell them; this being the case it is proper that the method by which such tests are determined should be made familiar to all interested. Ten years since the testing of a turbine was a serious matter, and could only be accomplished at a great outlay of time and money, the expense extending into the thousands; while the apparatus used was so crude, and the complications were so numerous, that the matter was understood by but few, and was believed in by less; thousands and tens of thousands of dollars have since been expended in simplifying the process of computation of results obtained, the manner of obtaining them, and in ridding the system of rubbish of no earthly use. In the first place it should be thoroughly understood, that weighing the power of a wheel, or in other words what it will pull while running at a certain speed, is precisely the same in principle as to weigh what a horse or man can pull while traveling at a fixed speed, or as in weighing groceries; consequently an accurate scale beam with knife edges and sealed weights are required as much in the one case as the other; the pounds named in testing a wheel mean precisely the same as in weighing hay or sugar; and if a proper weighing and controlling instrument is used, the wheel will be kept at the same speed so long as a given weight is carried: consequently the gauges remain constant with the same weight on scale, and with the same head of water, so that six different persons taking the gauges add exactly six times to the chances for errors in testing a wheel, and as much more to the cost. Testing with proper apparatus and conveniences is a very simple matter, but it requires experience to make such test reliable; and though an engineer may have the formula committed to memory, he will need considerable experience practically before he will be able to make tests that can be depended upon.

WEIR MEASUREMENTS.

Within the past few years much has been said and written for and against the reliability of measurements of water flowing over weirs; this has arisen through the great diversity of results obtained by different persons, who have used the same formula for computation of data. Turbines of almost every make, tested by their builders, have seemingly given high useful effect; while in actual use few of them have proved economical in the use of water. This has had a tendency to discredit weir measurements, but unjustly so, as may readily be explained, for the matter is one of great simplicity, notwithstanding the complications thrown around it by those who have supposed a long array of decimals denote profundity and accuracy. Any weir under exactly the same conditions will repeat results invariably; but a formula based upon certain conditions, will not give correct results if those conditions are changed. All brooks and rivers vary much in width and depth, yet the same water flows through the narrow as well as the wide places, the velocity, of course, varying with the cross section of the stream. The velocity, however, does not cease immediately upon entering a wider or deeper part, but continues until the momentum is lost, and the general level attained; this of itself would prove the necessity of placing a weir at a considerable distance from the discharge of a higher head. The Francis formula is based upon the natural flow of the water, which for a depth of one foot over a weir is about three feet four inches per second; and it must be evident that such formula is entirely inapplicable where the velocity is four or five feet per second, as it may be if the weir is placed close to the discharge of a poor turbine, where the water leaves the wheel with half the velocity due the head; or where a cross section of pit or stream approaching the weir is but little greater than the capacity of the weir itself. It is plain that under such conditions the velocity will vary according to the useful effect of the wheel, and equally plain that no reliable correction for velocity can be applied. Had this been considered, much trouble and expense might have been saved the past twenty-five years; for it is not likely any builder would have knowingly continued the manufacture of forty per cent. turbines. The cross section of a pit or stream, up stream from a weir, should be at least five times the cross section of the stream flowing over it; and for a discharge of two thousand cubic feet per minute, the weir should be fifty feet from the discharge of the turbine, or opening into pit. Racks should never be used, as they obstruct and raise the water so that it passes through with renewed velocity. If there is a horizontal discharge towards the weir, check the current by zigzag breakwaters. For measuring the flow of a river the weir or dam cannot be too large, but it may be for measuring the discharge from a mill where a governor is used, as the varying discharge, caused by adding or throwing off machinery, may prevent accuracy if too much time is required for the water to find its proper level.

THE SAME WHEELS TESTED IN PITS OF DIFFERENT CAPACITY.

July 24 and 26, two wheels were tested at Holyoke flume; these had previously been tested in another flume, the measuring pit of which was about nine feet in width, two feet in depth below crest of weir, while the weir itself was twenty feet from the wheel. The following results were obtained:

Largest Wheel: Stilwell & Bierce Flume.

Head, 7.64 feet   Discharged, 1178.00 cubic feet.   Percentage, .8785

Holyoke Flume, largest wheel:

Head, 1840   Discharged 2233.55 cubic feet.   Percentage, .7520

Reset and again tested:

Head, 18.07   Discharged 2214.66 cubic feet.   Percentage, .7533

Theoretical discharge for head of 18.40 feet, based upon the Stilwell & Bierce test should be 1828.7 cubic feet.

Smallest Wheel: Stilwell & Bierce Flume.

Head, 7.82 feet   Discharged 761 cubic feet.   Percentage, .8604

Holyoke Flume:

Head, 18.33   Discharged 1387.27 cubic feet.   Percentage, .7777

Taken out, overhauled, then re-tested:

Head, 18.44   Discharged 1400.31 cubic feet.   Percentage, .7753

The head was then reduced, and it was again tested:

Head, 7.85   Discharged 869.34 cubic feet.   Percentage, .7724

Theoretical discharge, based upon Stilwell & Bierce test, for 18.44 feet head, should be 1168.5 cubic feet.

These tests show how little reliance can be placed in measurements made in a pit of insufficient capacity, yet how accurately a proper pit and weir will repeat; at the same time they explain how the high results reported so often by interested parties are obtained.

Illustrations and description of testing flume and apparatus of the present time are herewith given: Fig. 1, represents the dynamometer, or weighing instrument; Fig. 2, an elevation of a testing flume; Fig. 3, a plan view of the same; Fig 4, the hook gauge. Through an opening in the side of fore-bay Fig. 1, may be seen a turbine wheel with its shaft extending upwards, on the upper end of which, above fore-bay, is secured the instrument for weighing the power transmitted from the water discharged. To ascertain the useful effect it is necessary to know the head under which the wheel works, also the quantity of water discharged by it in a given time. The head is the difference in height between the surface level of water in pit and fore-bay *when the wheel is running,* at which time there is generally too much disturbance in the water to allow of accuracy by direct measurement, thus necessitating the use of the tanks A and B; the tank A is connected with water in fore-bay by a short piece of three-fourths inch steam or gas pipe, through which the water flows too slowly to cause ebullition, but fast enough to keep the surface in tank equal in height with that in fore-bay; from the bottom of the tank a rubber pipe extends to the bottom of a glass tube, placed beside the measuring pole at the right. The tank B is connected with the water in pit by a rubber or flexible pipe, that the tank may be raised or lowered, in order to keep the top of the tank nearly even with the surface of tail water in the pit; with this arrangement the point of the hook, which may be seen at the lower end of the measuring pole, will be perceptible the instant it breaks the surface of the water in the tank. This hook and the pole is raised or lowered by a hand nut shown above the tank. The pole is graduated in tenths and hundredths of feet from the point of the hook to the top of the pole, so that after the point of the hook is adjusted to the surface of the water in the tank, the exact head may be found opposite the surface in the glass tube or tank A. The tank C, which is also connected with the water in the pit by a flexible pipe, slides up or down on two parallel rods, and is kept at any height by a counterpoise; above this the hook gauge is firmly fixed to a timber in such a position that the point of the hook will drop in a perpendicular line through the center of the tank, and it will save making corrections for each measurement by placing the point of the hook exactly level with the crest of the weir when the scale of the gauge is standing at zero.

Emerson's Improved Brake.

Manufactured by the Fales & Jenks Machine Co., Pawtucket, R. I.

FIG. 1.

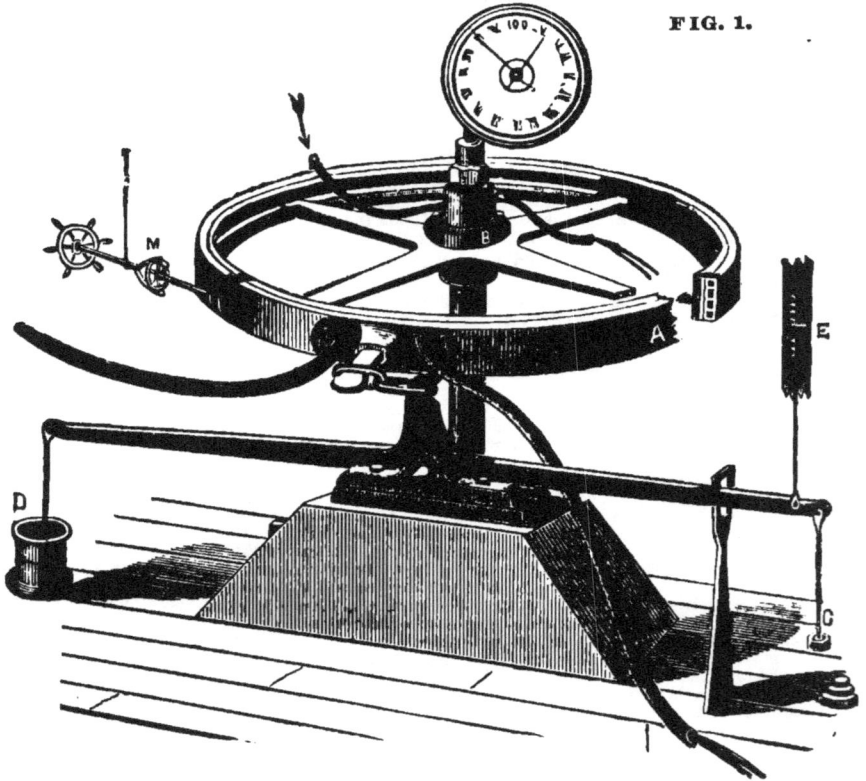

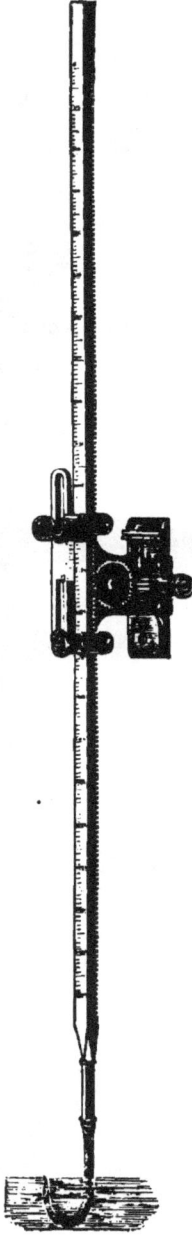

The Emerson Hook Gauges.

Manufactured by Hamilton J. Sawyer, Lowell, Mass.

NO. 1.

NO. 2.

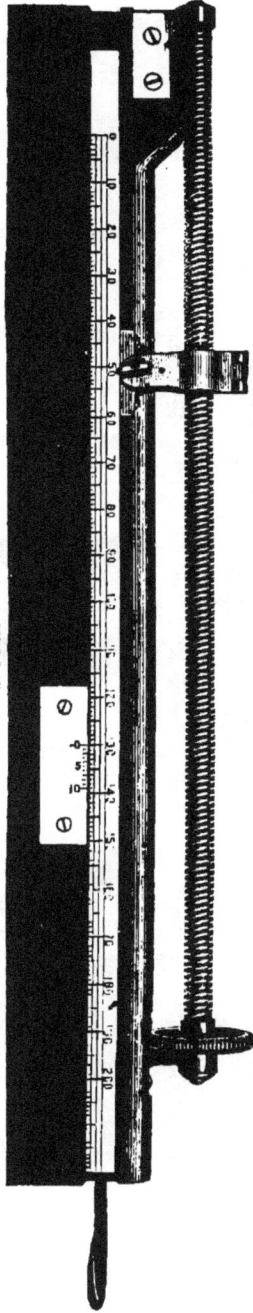

The first hook gauge that I ever saw was some six or seven feet in length, and, when boxed, was a good load to shoulder. It seemed needlessly cumbersome and inconvenient. A short experience caused me to plan the one here illustrated, marked No. 2. that weighed 14 pounds, and was 30 inches in length, which still seemed too heavy for convenience. After much study, No. 1 was produced, that weighed about four pounds, and is so entirely superior in convenience to No. 2, or any other hook gauge that I have ever seen, that I consider it complete.

Those desiring to purchase, should apply direct to Mr. Sawyer, who owns the patterns, and is a very skillful workman.

Plan of the Lowell Testing Flume and Pit of 1869.

FIG.3.

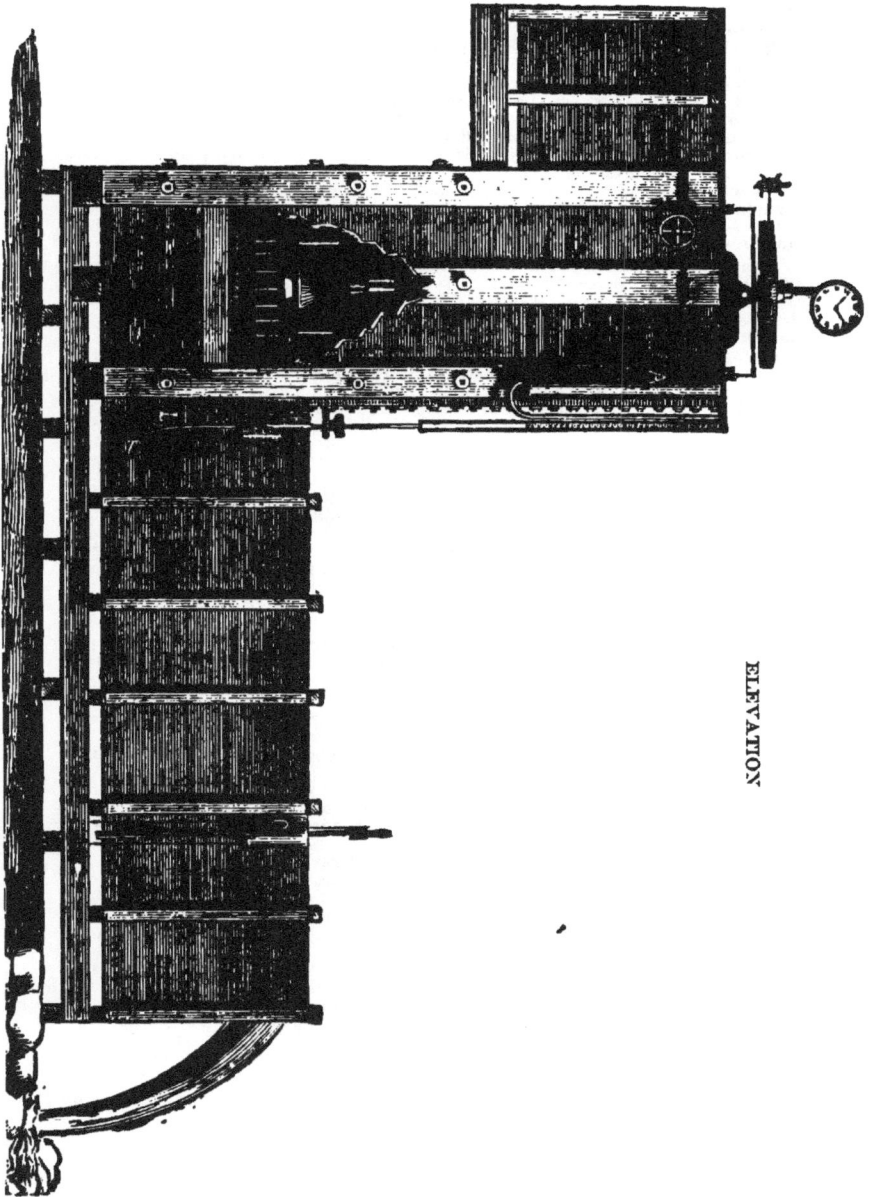

ELEVATION

The proper dimensions for a testing flume are, of course, determined by the size of the wheels to be tested. The fore-bay, in diameter, should at least be twice that of any wheel placed in it, while the width of the pit should equal one and a half times the length of the weir; below the crest of which the depth should equal four times the depth of the stream likely to flow over it. The weir should stand at least twenty feet from the wheel, and at an exact right angle with the flow of the water.

The dynamometer, or instrument used to determine the power transmitted, is simply an improved "prony brake." The wheel B is secured to the shaft of the water-wheel, and its speed is controlled by the friction-band A, which is connected to the scale-beam as shown, the point of connection describing a circle of a given number of feet. The rim of the wheel and the friction-band are hollow, and are kept cool by streams of cold water passing through them; the water in the rim of the wheel being supplied through its hollow arms and the pipe, shown in the engraving. The wheel B, is made of cast iron, the friction-band of "composition" or "gun metal." The hands of the "counter" are so arranged in connection with a worm gear, that they can be made to rotate in the same direction the hands of a clock move, whichever way the wheel being tested may revolve.

The hand wheel for operating the friction-band through the screw M, has a "universal joint" in its shaft, which is arranged with a slide to prevent fraud while testing. The connection of the band with the scale-beam is made by knife-edged links, and the pivot of the beam is also knife-edged. The weights are suspended at one end of the beam as shown at C; at the other end is the "dash-pot" D, (it is better to have "dash-pot" at the same end as the weights,) filled with water to hold the beam steady. The pot is made of cast iron, bored out perfectly true. The plunger on the end of the rod is a thin disk of iron turned to fit the pot loosely, so as to allow it to move perfectly free; it has six three-eighths inch holes through it, stopped with brass thumb screws; one or more of these may be removed at any time to render the beam more sensitive, but the screws must be left lying on the plunger, that the weight may not be changed. To prepare the instrument for testing, the "dash-pot" should be filled with water, the screws removed from the holes in the plunger, but left upon it, the beam leveled with the indicator standing at zero, as shown at E: then place a small weight in the scale-pan, and observe the number of seconds required for the weighted end to settle one-half inch; then change the weight to the other end of the beam, the same distance from the fulcrum, and change the balance weight until the beam is balanced; then return the screws to the holes in the plunger, and connect the beam to the friction-band by the links for that purpose.

When testing, I find that the simplest and surest method of obtaining the correct number of revolutions of the wheel, is to hold the hands of the counter at zero until the "timer" is ready; then to run several minutes, and divide the number run to obtain the revolutions per minute.

The most perfect measurement with the hook gauge can be obtained by keeping the top of the tank C, nearly level with the surface of the water in it, then by looking across it the point of the hook may be seen the moment it breaks the surface.

In testing a wheel I begin with a light weight, say for a 30-inch wheel under fifteen feet head, start with 100 pounds, run two minutes—the man at the wheel keeping the beam level—then change to 125 pounds and repeat. Continue to change 25 pounds every two minutes until the speed of the wheel is reduced below its best point, which is reached, we will say, when it is carrying 250 pounds; then reduce the weight to 235 pounds, and change ten pounds every two minutes until the best point is again passed, which is found, say, when it is carrying 255 pounds; reduce the weight to again, say, 242½ pounds and change the weight five pounds at a time every five minutes. Sometimes, when not in a hurry, I commence with 100 pounds and run to 700, or even 800; then again, I might start on the same wheel (if I knew about the proper weight for it) say with 600 pounds, and not change more than 100 during the whole test. Some parties desire to have their wheels tested with as short a range of weights as can be used and the wheel's best speed be found, for the purpose of showing even results through the whole test; but to the initiated, such results would appear no better than where greater changes were recorded if the weights varied with the speed. Of course, the more the speed of a wheel can be varied without affecting its percentage the better, but that is only determined by using a long range of weights while testing it.

The power transmitted by the wheel is determined as follows: Suppose the scale beam is attached to the friction brake at a point, which, if revolving, would describe a circle of 20 feet, and the wheel running one hundred revolutions per minute, holds the beam at zero when loaded with 500 lbs., $20 \times 100 = 2000 \times 500 = 100000 \div 33000$ gives 30.30 horse-power; divide the transmitted power, by the power of the water used, to ascertain the useful effect of the wheel

An example is here given of finding the useful effect, after testing a turbine, as followed in 1869; and when it is understood that a hundred different weights might be tried in testing a wheel, and that during the trial some six or seven different observers were taking notes every thirty seconds, and that all of these observations had to be made to agree it will readily be seen that there were wide openings for errors.

TEST 17—TYLER WHEEL, September 21 and 23, 1871.

149.2 Rev. per m.
20 Circumference of circle

2984.0
300 Lbs.

33000)895200 Foot lbs.)27.13 II. P. of wheel.
66000

235200
231000

42000
33000

90000

Q. per sec.=3.33 (1—0. 1n II) $II^{\frac{3}{2}}$

1.0615 Height of water on weir.
—.0145 Correction for weir level.

1.0470
.2 Number of end contractions $\times$ 0.1.

.20940
6.00000 Length of weir.

5.79060=0.7627236
3,33=0.5224442
1.047=0.0199467 } $II^{\frac{3}{2}}$
0.0099733
60=1.7781513

1239.48=3.0932391=Q. per min.
15.695=1.1957613=Fall.
62.336=1.7947389=Weight of cubic foot.

33000 (a c)=5.4814861=Horse Power.

36.75=1.5652254=II. P. of water.
27 13=1.4334498=II. P. of wheel.

.7383=1.8682244=Ratio, or, percentage.

The formula for correcting the depth for the velocity of the water approaching the weir is

$$II' = \left[\quad (II+h)^{\frac{3}{2}} - h^{\frac{3}{2}} \right]^{\frac{2}{3}}$$

in which the factor

$$h = \frac{v^2}{2\,g};$$

v being the velocity found by dividing the Q per second by the section of the stream approaching the weir. As the flume approaching the weir was 14 feet wide, and the bottom of it was 3-5 feet below the crest of the weir, it follows that the area of a section of the stream, when there was 1.047 feet of water flowing over, is 14 (3.5+1.047)=63.658 square feet.

$$Q \text{ per sec.} = 20.658 = 1.3150883$$
$$\text{Section} = 63.658 = 1.8038530$$

$$\overline{1}.5112353 = v$$
$$\overline{2}$$

$$\overline{1}.0224706 = v^2$$

$$\overline{2}.1916296 = 2 g (a c)$$

$$.0016 = \overline{3}.2141002 = h$$

$$\overline{2}.6070501$$

$$.0001 = \overline{5}.8211503 = h^{\frac{3}{2}}$$
Then H+h=1.047+.0016=1.0486.
$$1.0486 = 0.0206099$$
$$0\ 0103049$$

$$1.0738 = 0.0309148 = (H+h)^{\frac{3}{2}}$$
Then $(H+h)^{\frac{3}{2}} - h^{\frac{3}{2}} = 1.0738 - .0001 = 1.0737$
$$1.0737 = 0.0308830$$
$$0.0102943$$

$$1.0486 = 0.0205887$$
$$1.0486 = H' = \text{corrected depth on the weir.}$$

Substituting H' for H in the weir formula first given above, we find the corrected Q to be 1242.25 cubic feet per minute.

$$1.0486$$
$$.2$$

$$.20972$$
$$6.00000$$

$$5.79028 = 0.7626996$$
$$3.33 = 0.5224442$$
$$1.0486 = 0.0206099$$
$$0.0103049$$
$$60 = 1.7781513$$

$$1242.25 = 3.0942099$$
$$15.695 = 1.1957613$$
$$62.336 = 1.7947389$$

$$33000 \ (a \ c) = \overline{5}.4814861$$

$$1.5661962$$
$$27.13 = 1.4334498$$

Ratio of useful effect .7366 = $\overline{1}$.8672536

To work out the foregoing without the use of logarithms, applying all of the corrections as was then done, would cover many pages of this work. A hundred different weights and speeds were likely to be tried in testing any wheel, each change requiring the same tedious process, so that days, perhaps weeks, were required to ascertain the value of a wheel. It was customary with some engineers to work out a few tests, then to "plot" the remainder on "diagram paper;" but this was found to be unreliable in working out my weir tables, and of course, was equally so in working out tests. With reliable apparatus for testing a wheel, but few corrections are necessary, and only three persons are required in making tests. One having the whole in charge, and who takes weight, revolutions of wheel, and the head and weir gauges, assisted by a "timer," and one who controls the speed of the wheel. A testing flume is filled and emptied so often that it will leak more or less, and this leakage is into measuring pit, so that after a wheel is set ready to test, its gate is closed and sprinkled with sawdust to prevent leakage, that would affect results of trial; then the flume is filled with water, and the leakage of the flume taken at the weir. Suppose the length of weir to be six feet, and depth of leakage to be .183 of a foot; opposite to this in weir table and column for 6 ft. weir will be found 93.28 cubic feet per minute, and this quantity is to be taken from every test made of that particular wheel, supposing the water not to be drawn from the flume during the test; if it is, then the leakage must be taken as before. To illustrate, a test as now taken is here given. The point of attachment of brake to scale beam is ten feet, and each revolution must be multiplied by ten to get correct speed. Look in weir table below for cubic feet discharged. Test of an 18-inch Wetmore wheel, September 30, 1876:

	Head.	Weight.	Rev. per min.	Weir.
No. 7.	18.80	162.5	305	.650

Quantity as per table 624.62—93.28=521.34 cubic feet per minute.

$$\frac{521.34 \times 18.80 \times 62.33}{33000} = 17.91 \text{ H. P. of water.}$$

$$\frac{305 \times 10 \times 162.5}{33000} = 15.02$$

$$\frac{15.02}{17.91} = .8114 \text{ Ratio of useful effect.}$$

Formula for Tabling Wheels.

Q=quantity discharged per second at any head, h.
V=velocity due head h.
Q'=quantity with any head'
V'=velocity due head'
R=relative velocity.
D=diameter of wheel.

The Q having been determined for any given head, to find it for any other head $Q' = \dfrac{Q \times V'}{V}$

The horse power having been determined for any given head, to find it for any other head $\dfrac{H. P. \times V' H'}{V \times H}$

The revolutions having been found for any given head, to find them for any other head $\dfrac{V \times R}{D \times 3.1416} \times 60$ =number of revolutions per minute.

R=relative velocity, determined by experiment.
Having the outlet of one wheel of a certain pattern measured and its power determined, the power of another of similar pattern is approximately obtained by comparing the outlet with the one experimented upon.

Steam and Pressure Gauges.

Is it a matter of importance that such instruments should indicate correctly, and if so, do those using them take pains to verify their accuracy? Recently while testing the turbines used at the water works of St. Johnsbury, Vt., it came in my way, also, to test the accuracy of the pressure gauges used there; these were made by the Utica Steam Gauge Co , Utica, N. Y. The test was made by getting the exact area of the waste valve, using a knife-edged pivoted beam resting on a knife-edged top of valve piston then with scaled weights the pressure in pipe was accurately ascertained, and to be 11 per cent. less than that shown by the pressure gauge.

Elkhart Mills, Power, and the Water Used to Produce It.

MESSRS. MILLER & MAXON.—*Gentlemen:*—Nearly a year since, acting for the manufacturers hereinafter to be mentioned, you employed me to ascertain the power used by the said manufacturers, and the quantity of water necessary to produce the power used and the power deeded.

My only instructions were to do it by the most perfect methods known to me and do it right. A preliminary trial was made in June last, and all interested in such matters were invited to witness all tests, particularly the members of the Hydraulic Company and their attorney, and to all desirous of knowing the matter was fully explained.

Except in cases of indefinitely worded deeds, there is no feature in the use of water, or power in mills, that may not be elucidated and made so plain as to leave no shadow of excuse for litigation except that of a desire to get that which belongs to another.

The deeds in each case to be named give a definite amount of power with right to use sufficient water to produce it, under the conditions specified, a positive condition of which is that measurement of the water shall be after it issues from the wheel.

Two power scales of different capacities were purchased of their manufacturers, Emerson Power Scale Co., Florence, Mass.; these are made upon the same principle as the ordinary Fairbanks scale, but rotary. The largest carries its load nine, the smallest six, feet at each revolution of shaft to which it is affixed.

To operate : the key is removed from driving pulley, thus leaving pulley loose upon its shaft ; the scale is then placed on shaft close to hub of pulley, and rigidly keyed to the shaft. There are spurs projecting from the rim of scale to which the levers of scale connect to the arms of the pulley, so that all of the strain from belt rests upon the scale, and that strain or weight is shown upon scale in pounds as on the ordinary scale beam.

Muzzy's Starch Mill, capacity 1,000 bushels of corn or 24,000 pounds starch per day, 2 Eclipse turbines, one 48, the other 54, inches in diameter.

48 inch or its work weighed January 5, rev. 118x9=1062x875=
929,250 ÷ 33,000...28.15 h. p.
54 inch or its work weighed January 6, rev. 90x9=810x1150=
931,500 ÷ 33,000 ....28.22 h. p.
Total power used, all machinery in full operation..........56.37 h. p.

Globe Tissue Paper Mill, capacity one ton per day, 3 turbines, American 66, Victor 25 and 30 inches.

66 inch American or its work weighed Jan. 15, rev. 99x9=
891x1325=1,180,575 ÷ 33,000.......................................35.77 h. p.
30 inch Victor, washer wheel, Jan. 17, rev. 90x6=540x825=
445,500 ÷ 33,000...13.50 h. p.
25 inch Victor, 84 inch paper machine, paper running 97 ft.
per minute, rev. 44.5x9=400.5x1491=588,735 ÷ 33,000..........17.84 h. p.
Total power for 4 Beating engine, Washer, Jordan,
Pumps, Paper Machine, Rag Cutter and Duster..........67.11 h. p.

Elkhart Knitting Mills.

2 set 48 inch Cards, 3 Jacks, in all 720 Spindles, 2 Parker Twisters, 96 spindles each, 4 Spoolers, Dusters, Dryer and Fan, Stocking Dryer and Fan, Kulp Winders, Hydro Extractor, 60 Knitting Machines.

Power to drive all weighed Jan. 9, rev. 250x6=1500x425=
637,500 ÷ 33,000...19.31 h. p.

Kulp & Umel Planing Mill.

Two Rip Saws, Lathe, Matcher, Resaw, Daniels Planer, 26 inch Fay Planer, Molder, Sand Paper Machine, and Sticker.

Usual machinery running, rev. 200x6 = 1200x630 = 756,000 ÷
33,000...22.90 h. p.
With every machine in mill running, Jan. 12, rev. 175x6=
1050x825=866,250 ÷ 33,000 ....26.25 h. p.

C. G. Conn's Musical Instrument Works.

Every machine in works running, rev., Jan. 21, 130x9=
1170x320=374,400 ÷ 33,000..11.35 h. p.

Sage Brothers' Flouring Mill, capacity 280 barrels per day.

Deeded right to use sufficient water to drive five runs of four foot buhrs to grind 15 bushels of red merchantable wheat per hour, one run to grind 40 bushels of corn per hour, also smut mills and all necessary machinery to prepare flour and meal for market ; as one wheel of same capacity is allowed for four runs of buhrs, the quantity deeded is sufficient practically to drive seven and a half runs each, grinding 15 bushels of hard wheat per hour. Messrs. Kulp & Umel with similar deed to two and a half. A 4 foot buhr driven by spur gears was disconnected from turbine and connected to a horizontal shaft by a pair of bevel gears, the driver having 56, the driven 42, teeth; a belt running horizontally from another line of shafting drove the stone. The power scale was placed on shaft close to gears driving buhr.

Mr. J. W. Lamb, of Constantine, Michigan, an experienced miller, was employed to do the grinding, commencing Saturday, 19th. After making some experiments he had pulleys changed, stones redressed and seemed to take the utmost care to make the tests absolutely accurate, and I believe did so ; four days were expended in making the several trials.

An excellent weir 20 feet in length was used for measuring the discharged water. There was a leakage of 185 feet per minute to be deducted from the quantity flowing over the weir indicated by the depth during each test except the last.

A 48-inch Leffel wheel was used, and nearly at its full capacity during the heaviest tests.

The largest scale was used, making the trials tabled below so that the revolution of shaft must be multiplied by 9 to get feet the load is carried ; that sum must be multiplied by the weight, to find the foot pounds ; dividing those by 33,000 will show the work done in h. p.

Multiply cubic feet by the head, and that sum by 62.34, weight of a cubic foot of water, to find power of water used.

Dividing the work power by the power of water will show useful effect of the turbine.

While making the experiment it required the miller's constant attention to grind fifteen bushels of wheat per hour ; indeed it was evident that it would be impracticable to make a business of grinding that quantity, so it was found necessary to do it upon two stones, requiring 21 h. p. of water per each run, grinding seven and a

Grinding bushels of wheat per hour.	Head Rev. in of Stone Feet	Dep. on Weir.	Cubic Feet.	Rev. of Shaft.	W'g't in Lbs.	Work in h. p.	Power of Water Effect. in h. p.	Useful Effect.
1. Indiana red, 15 bushels	8.85	205	.709	2184	154	590	24.78	36.59	.678
2. Minnesota red spring, 15 bushels	8.70	197	.725	2284	148	680	27.44	37.21	.733
3. Indiana white, 15 bushels	8.87	215	.672	2901	161	440	19.32	33.52	.576
4. Minnesota red spring, 3½ bushels per hour	6.17	201	.544	1014	151	210	8.66	16.42	.527
5. " " 3½ "	9.92	197	.450	1409	148	270	10.87	19.00	.572
6. " " 3½ "	9.50	197	.476	1121	148	345	13.93	21.00	.663
7. " " 7½ "	9.50	181	.740	2340	136	800	29.67	41.99	.706
x. Corn fine meal, 22 "	9.60	200	.740	2340	150	675	27.61	42.40	.651
9. Feed, 38 "	9.95	202			151				
10. Driving stone without work	7.84		1.629	8173				121.00	
Whole machinery in mill, 150 bbls. per day									

*From 21 to 22 bushels were ground, but not fine enough to bolt. While making these tests the head was changed to ascertain the co-efficient of useful effect under such conditions.

half bushels per hour, or 42 horse power for grinding fifteen bushels, and that ten seven and a half runs and two and a half equal runs for machinery, to prepare the product for market, would equal 262½ h. p. of water for the quantities deeded. In grinding corn twenty bushels per hour was all that could be done well with forty-two h. p. of water ; to grind the forty bushels would require at least eighty-four and the full hundred to grind and prepare the meal for market, making for the Sage Brothers' mill 362½ h. p. Indeed I believe it will be impossible under the existing conditions to do that amount of work with the quantity named.

Messrs. Kulp & Umel have the right to two fifteen bushel runs, and machinery equal to five run of buhrs grinding seven and a half bushels of wheat per hour ; the same rate entitles them to one hundred and five h. p. of water.

Allowing the same rate for the other mills, that is, three h. p. of water for each two h. p. of work, Muzzy's starch mills are entitled to one hundred and thirty-five h. p., the Globe Tissue Paper Co., ninety, C. G. Conn and the Knitting mill each forty-five. These are common rates, and the grinding tests show the allowance to be none too much, in fact not enough unless the head can be kept somewhere near the height at which the wheels are set for. A wheel set under nine feet head will of course give more power under ten, but it by no means follows that it will do it with less water.

There were two hundred and forty-seven h. p. of water flowing through a break in the flush boards on the dam January 3, current month, but the mills on the other side of the river were not at work, yet the water in race drew down during the day.

Sage Brothers, Kulp & Umel, Tissue Paper Co., Knitting Mill Co., C. G. Conn and Muzzy Starch Co. still have an unused right to 360 h. p. more of water than they take. If they call for that it is somewhat difficult to conceive where it is coming from.

My record of measurement of discharge from turbines used in the Combination board, Excelsior starch and Elkhart paper mills, proved them capable of using five hundred h. p. of water, which, added to the quantity deeded to the other six mills this side, make for the two-thirds this side the river 1282, plus 641 for the other side, equaling nineteen hundred and twenty h. p. for the whole.

Six inches water flowing over dam falling ten feet evolves about 398 h. p. ; 9 inches, 730 ; 12 inches, 1120 ; 15 inches, 1569 ; 18 inches, 2048.

It should be borne in mind that though the rainfall may be equal now to what it was fifty years ago, yet the cultivation and drainage of the land causes a much more rapid evaporation and clearance of the supply than formerly.

The following results obtained from measurement of water used at different mills will prove my allowance for water to produce the deeded power to be moderate.

The rate of mills is based upon some generally understood matter pertaining thereto.

Cotton mills upon their number of spindles ; woolen mills upon number of sets ; paper mills upon number of tons made per day ; flouring mills upon number of barrels of flour per day. As the rate of mill denotes its value, it is not likely to be underrated, and there is often reason to doubt whether the entire amount of work is done that its rate would indicate. Certainly the rate is rarely exceeded.

To ascertain how much power is required to grind a bushel of wheat, it is simply necessary to measure the water used when the mill is doing its ordinary work, and divide the power of that by the bushels ground per hour.

The least power per bushel used at any mill that I have ever tested was at Lanesboro, Minnesota, White & Beynon: 3.18 h. p. per bushel ; test made in 1874. New mill in perfect order. Head about 24 feet.

The following results made four years ago at Mishawaka will show what a difference there is in such matters, and it is necessary that it should be considered, to understand what is necessary in the case in hand.

ST. JOSEPH MILLING COMPANY, MISHAWAKA, July 6, 1884.

Ordinary discharge of water 8½ feet head 6174 cubic feet per minute, the power of which is 93.35 h. p. Capacity of mill rated 100 barrels per day of 24 hours.

100 barrels at 4½ bushels=450 bushels ÷ 24 hours = 18.75 bushels per hour ; 93.35 h. p. ÷ 18.75 bushels=4.97 h. p. of water per bushel.

RIPPLE MILL, MISHAWAKA, IND., July 8, 1884.

A. & J. H. EBERHART & Co., PROPRIETORS.—Ordinary discharge of water 9540 cubic feet per minute, the power of which is 114.08 h. p. Capacity of mill rated 130 barrels in 24 hours.

130 barrels by 4½ bushels=585 ÷ 24 hours=24.4 bushels per hour ; 114.08 h. p. ÷ 24.4 bushels=4.68 h. p. of water per bushel.

MISHAWAKA MILL, MISHAWAKA, IND., July 11, 1884.

W. & J. MILLER, PROPRIETORS.—Ordinary discharge of water 1634 cubic feet per minute, the power of which is 185.72 h. p. Capacity of mill rated 175 barrels per day of 24 hours.

175 barrels x 4½ bushels=787.5 ÷ 24 hours=32.8 ; 185.72 h. p. ÷ 32.8 bushels=5.66 h. p. of water per bushel.

Highgate, Vt., July 4, 5 and 6, 1885. I measured the water discharged from an excellent tub wheel grinding wheat, the result was to be used in a case in litigation and special care was taken.

To grind the ordinary wheat used there it required 5.2 h. p. per bushel. For the hard red wheat 5.9 h. p. per bushel.

Twenty years ago a revolution was taking place in regard to the best methods of utilizing the power of falling water ; the turbine was taking the place of the earlier overshot and breast wheels, its compactness for its capacity astonished those interested, and the claims for it were so extravagant that manufacturers were bewildered and hardly knew what to do. The deeds of that and earlier times also were often very indefinite.

There were such doubts and conjectures about turbines, milling hydraulics, and dynamics that a series of experiments were instituted for the purpose of making such matters clear. Instruments of the simplest and most accurate effectiveness possible were substituted for the crude devices then in use.

It was a common idea then that a turbine to be really efficient should be built for the head under which it was to work ; that an aperture would not discharge proportionally the same under different heads or different sizes ; that more work could be done with the same wheel in the night than in the day-time, etc., etc.

A testing flume was constructed and for several years turbines were tested under 18, 12 and 6 foot heads. In round numbers the wheel that would give 100 h. p. under 18 feet, would give but 50 under 12 and 20 under 6 feet.

A short experience proved many common ideas to be fallacious, the same apertures discharged proportionally for any head and the turbine that was good under one head was proportionally efficient under all others, and gave the same results night or day.

At that time 73 to 75 per cent. seemed to be a sort of normal efficiency ; almost any aspirant for fame as turbine builder could reach that point.

The deeds of the Elkhart Hydraulic Company are in a measure based upon the merits of the American turbine, and as various kinds are in use under those deeds it is essential to show such to be equally effective.

The following results obtained by tests of wheels built before the system of testing was established will show the efficiency of the ordinary American turbine for a range of sizes :

AMERICANS TESTED THE DATES NAMED:

Test of 48-inch, January 29, 1874.

No. of Test.	Head.	Weight.	Rev. p'r Min.	Horse Power.	Cubic Feet.	Per Cent.
Whole Gate	17.65	1320	107.8	86.24	3418.11	.7598
Part Gate	17.66	1100	110.3	73.53	3010.79	.7316
" "	17.76	960	104	60.51	2594.01	.6948
" "	18.16	500	106	32.12	1690.47	.5548

September 29, 1873, 42-inch, right hand.

No. of Test.	Head.	Weight.	Rev. p'r Min.	Horse Power.	Cubic Feet.	Per Cent.
Whole Gate............	17.93	1200	112.5	61.36	2569.85	.7095
Part Gate.	17.98	990	118.5	53.32	2218.55	.7094
" "	18.30	650	120	35.45	1452.72	.7065
" "	18.45	440	119.5	23.90	1213.58	.5666

October 1, 1873, 42-inch, left hand.

No. of Test.	Head.	Weight.	Rev. p'r Min.	Horse Power.	Cubic Feet.	Per Cent.
Whole Gate............	17.90	1100	118	59.00	2536.02	.6882
Part Gate	18.00	980	120	53.45	2275.17	.6946
" "	18.13	820	121	45.10	1918.04	.6884
" "	18.43	420	116.5	22.24	1160.60	.5479

November 11, 1873, 25-inch wheel.

No. of Test.	Head.	Weight.	Rev. p'r Min.	Horse Power.	Cubic Feet.	Per Cent.
Whole Gate............	18.23	300	212	28.91	1158.24	.7244
Part Gate............	18.30	260	207	24.46	983.53	.7185
" "	18.39	220	205	20.16	880.49	.6565
" "	18.60	110	208	10.40	555.69	.5323

November 12, 1873, 20-inch wheel.

No. of Test.	Head.	Weight.	Rev. p'r Min.	Horse Power.	Cubic Feet.	Per Cent.
Whole Gate..	18.85	130	253.5	14.97	606.54	.6938
Part Gate............	18.55	110	243	12.15	528.55	.6536
" "	18.63	90	244	9.98	448.93	.6313
" "	18.77	50	225.5	5.13	285.15	.5072

August 5, 1874, 60-inch wheel.

No. of Test.	Head.	Weight.	Rev. p'r Min.	Horse Power.	Cubic Feet.	Per Cent.
Whole Gate, 1..........	16.63	3000	88.1	147.27	6358.90	.7315
" 3..........	15.94	2700	80.3	131.40	6220.86	.7028
" 5..........	14.88	2500	80	121.21	5839.12	.7394
" 7..........	14.82	2550	76.5	118.22	5849.43	.6863
" 9..........	14.91	2300	79.5	110.81	5891.40	.6690
" 11..........	14.73	2600	70.5	111.09	5961.55	.6709
" 13..........	14.75	2450	74	109.88	5719.34	.6908
Part Gate, 15............	15.02	2450	74.2	110.17	5719.34	.6800
" 17............	15.12	2150	76	99.03	4049.47	.7018
" 19............	15.08	1850	79.5	82.16	4573.00	.6832
" 21............	16.41	1400	80.5	73.18	3693.02	.6404
" 23............	17.88	950	68.5	39.43	2296.70	.5093
" 25............	15.47	3900	000	000	5700.95	.0000

June 7, 1873, 48-inch wheel.

No. of Test.	Head.	Weight.	Rev. p'r Min.	Horse Power.	Cubic Feet.	Per Cent.
Whole Gate, 1..........	11.91	700	103.5	43.90	2702.80	.7224
" 2..........	11.88	750	99.5	45.22	2725.28	.7398
" 3..........	11.86	800	95.5	46.30	2763.94	.7482
" 4..........	11.92	850	96.5	49.41	2845.02	.7484
" 5..........	11.89	870	90.5	47.11	2835.26	.7383
" 6..........	11.90	900	88	48.00	2841.77	.7525
" 7..........	11.87	920	86.8	48.40	2857.54	.7555
" 8..........	11.88	940	84.5	48.13	2867.85	.7489
" 9..........	11.92	960	83	48.29	2874.38	.7491
" 10..........	11.92	860	88.5	46.43	2812.50	.7535
" 11..........	11.92	880	90.5	48.26	2841.77	.7546

Average per cent. under most favorable conditions, .7232.

Leffel 30-Inch, Tested in 1872.

No. of Test.	Head.	Weight	Rev. p'r Min.	Horse Power.	Cubic Feet.	Per Cent.
Whole Gate, 1..........	15.60	300	201	27.41	1429.12	.650
" 3..........	15.54	320	194	28.22	1455.57	.662
" 5..........	15.48	340	187	28.90	1463.74	.675
" 7..........	15.425	360	181.5	29.70	1469.87	.693
" 9..........	15.41	380	175	30.23	1469.87	.706
" 11..........	15.395	400	175	31.82	1471.92	.743
" 13..........	15.38	420	162.5	31.02	1471.92	.725
" 15..........	15.38	440	151.5	30.30	1471.92	.708
" 17..........	15.37	475	135.5	29.26	1469.87	.686
" 19..........	15.32	405	157.5	28.99	1461.76	.685
" 21..........	15.335	415	151	28.48	1465.78	.655
" 23..........	15.33	415	154	29.05	1465.78	.683
" 25..........	15.33	495	162	20.82	1461.69	.704
" 27..........	15.31	415	154	29.05	1463.74	.687
¾ Gate, 29..........	15.65	300	161	21.95	1106.73	.664
½ " 31..........	16.037	180	165	13.13	637.42	.591

Victor Turbine, Made by Stilwell & Bierce, Dayton, Ohio, Tested the Dates Named.

Test of a 25-Inch wheel, July 25, 1877.

No. of Test.	Head.	Weight	Rev. p'r Min.	Horse Power.	Cubic Feet.	Per Cent.
Whole Gate..............	18.07	625	200	56.81	2214.55	.7533
Part Gate..............	18.04	600	198	54.00	2208.44	.7192
" "	18.13	500	208	47.27	1964.67	.7042

Test of a 26-inch wheel, July 26, 1877.

Whole Gate..............	18.33	500	246	37.27	1387.27	.7777
Part Gate..............	18.41	425	269	34.64	1284.30	.7774
" "	18.43	390	246	29.07	1145.59	.7305
" "	7.97	75	246	5.59	757.93	.4911

Test of a 15-inch wheel, March 26, 1878.

Whole Gate..............	18.34	300	323	29.36	974	.8705
Part Gate..............	18.10	300	321.5	29.22	970	.8808
" "	18.39	160	326.5	15.83	755	.6035
" "	18.74	100	320	9.09	492	.5220

Eclipse Double Turbine, Manufactured by the same Co.

Test of a 30-inch Eclipse wheel.

Head.	ev.p'r Minute.	H. P.	Cubic feet.	Per Cent.
18.79	184.5	33.85	1253	.7628
18.93	170	31.66	1214	.7280
19.10	173.5	24.44	1026	.6497
19.10	165	18.00	862	.5786
19.18	166.6	12.11	699	.4779

It will be seen by the tabled tests of wheels that the American is not exceptionally economical, nor is it possible for any wheel to be economical where there is a variation in the head of one-third, though of course a good part gate wheel is better than one only efficient at whole gate. There is an idea that turbines discharge 60 per cent. of the theoretical quantity due their openings. The idea originated from obsolete wheels of the Fourneyron type. Of the modern wheels I have had care of tabling hundreds, yet have never known of one reaching 55 per cent. of its opening ; 52 perhaps is a fair average, 49 about all the American can do.

An aperture that will measure, will discharge a trifle short of 60 per cent. but such aperture can never be used in a forebay to determine the quantity of water used in a mill ; it is absolutely impracticable for that purpose.

A weir in forebay is also impracticable unless a manager stands beside it at all times to give the proper depth for quantity, and then only at a serious loss of head, and if such weir is placed below discharge of wheel, it also causes such loss of head that wheels subject to such changes can never be economical.

To divide water in proportion to ownership at dam or conduit with weir belongs to the ideas of the past. The water may go through one opening two feet per second, the other six, depending upon the size of wheels below. It is true that the water cannot be drawn below crest of weir by either party, but the one with the most capacious wheel will take water in proportion, and the expense of weir may be saved by fixing upon a mark below which the water shall not be drawn.

A gate and float arrangement may be put in flume or forebay by which proportion or quantity of water may be delivered without perceptible loss of head, the whole working automatically ; and while the quantity due is adhered to the gate will stand open, but if more is attempted to be taken the gate closes in proportion, and a proportional loss of head results, though the full quantity of water is still supplied. The arrangement is simple and more accurate than a weir, and with it the head is invariably kept at a standard height ; if the supply is sufficient, it is given in full, if not, in proportion.

With such an arrangement and good wheels water may be economized to the highest practicable extent.

More precaution will be used in the selection of wheels when the fact becomes understood that turbine building is not a science, it is simply "cut and try." There are some who can do better than others, but the best cannot go to work and be perfectly sure to reach the results aimed at, and however well one may do himself he cannot teach another how to do the same. Owing to uncertain causes, such as warping of patterns, shrinking or expansion of castings, turbines made from the same patterns often differ exceedingly in useful effect. Large wheels in particular are the most likely to fail because the expense has prevented experimenting upon them. A case that almost every manufacturer of twenty years' experience will recall may interest. It is of the Manville, R. I., mill so profusely illustrated in the Leffel circular fifteen years since. The artist drew somewhat upon his imagination. The mill is shown with four 84-inch wheels, while it never had but three, those being helped out by an engine of 430 indicated h. p. The tabled power of the three 84-inch wheels and the 430 h. p. engine rate something like 1700 h. p. The manager ran under those conditions many years, then applied to me about procuring another 84-inch Leffel or some other of like capacity In the conversation that ensued, the question of power was raised and I told him that the whole mill did not need 800 h. p. The idea was poohed at, but a test soon proved that fact, and the only thought since has been to exchange and get better turbines of less size but greater efficiency.

Intelligent co-operation between those who let and those who use power will prevent litigation and increase by far the effectiveness of the power used. But to do this it must be borne in mind that a turbine runs at a relative velocity with the water that propels it, and can only do its best work at one point for a given head, and declines rapidly either way at any deviation from that head ; unless the wheel is exceptionally good at part gate.

Yours truly,

JAMES EMERSON.

ELKHART, IND., January 30, 1889.

Division or Measurement of Water Power.

The time can not be distant when those interested will look back and smile at the crude methods continued in use up to this time to determine the quantity of water used by the different parties taking power from the same fall—methods well enough a half century since, when the most of such power was running to waste, but simply ridiculous now, when the demand is far beyond the supply.

The float method in use at Lowell can hardly be considered anything more than a preliminary to guessing at the quantity used. It, however, does not interfere with the operations of the mills, but any agent may favor his discharge while such measurement is being made, and there were rumors that such cases occurred at times. Mr. Francis has seemed ready to adopt a better plan, whenever such is found, though his many cares have prevented him from experimenting personally for the purpose of developing one.

There are or were various methods in use at Lawrence—wiers here, shanties there; weirs to measure leaks, a weir to test the *tester*—examinations of apparent gate opening, examinations in every conceivable place except, perhaps, the right one. Yet, what would the whole amount to in case those interested should combine for the purpose of deceiving those making the measurements? It is not likely that such a combination exists, but a method that can be affected in that way is a very imperfect one, and the use of such indicates the lack of the "fertility in expedients" necessary to meet emergencies so common in the engineering business. The continued dependence upon old foreign methods is discreditable alike to those having charge of the immense water powers of this country and the ingenuity of our people.

Several years ago, and before any arrangements were made for measuring the power at Holyoke, I advised the agent of the Water Power Co. to arrange to measure the discharge from the mills, then being constructed, in the tail-race of each; also to have all wheels that were to be used in Holyoke tested before being set in the wheel-pits for which they were designed. Reflection soon caused me to abandon ideas so crude. Measurements in the tail-race reduce the head and change the discharge and conditions generally, notifies the party interested of what is being done, and gives a chance to reduce the work and favor the discharge of water.

To attempt to determine the discharge of a wheel in a mill by comparison with a previous discharge in a testing flume, when the wheel was new and in perfect condition, would be unjust to both parties interested. Because a wheel can discharge 5000 cubic feet per minute, it by no means follows that quantity is used in the mill. A larger wheel is invariably put in than actually required, to have a surplus power for emergencies. The buckets and chutes of a wheel soon become rough, get broken, become clogged, or it would require but little ingenuity to so change the gate arrangement as to deceive completely as to the state of gate opening. Any pretense of giving the discharge of one wheel by comparison with that found by test of another of the same make, could only be done by ignoring the knowledge gained from a dozen years of constant experience in turbine testing, namely: That builders are constantly changing their plans; still further, that two wheels designed to be exactly alike, made from the same pattern, often vary wildly in their discharge. In short, the adoption of such a plan for measurement would have been the acknowledgment of such ignorance and incompetency in such matters, that I advised a series of experiments for the purpose of finding an accurate but simple and inexpensive plan for measurement of the water used by manufacturers, free from interference with the work of the mill, or that could be affected by parties interested. The purpose was suggested in the last edition of this work, in the description of the Holyoke

Testing Flume. It is a pleasure to state that that purpose has been accomplished by the finding of a simple automatic method by which the water flowing over any fall may be accurately measured or divided, so that each owner can have the exact quantity belonging to him and no more, unless by consent of the others. The operation is continuous. An illustration of the plan may be seen upon next page.

D represents the ordinary head gate to race, raised sufficiently to supply the mill and keep the water to its proper height. K represents a wicket gate placed in the lower end of race and near penstock, in which the turbine stands. T, a cylindrical tank, with a square recess on one side near the bottom. In this recess there are two openings: one to let the water in, and another to let it out down through the pipe, C, shown by dotted lines. These openings are opened or closed by the swinging cover or valve, e, which works upon the center pivot, t. The valve is connected to the float, F, by a rod connected at s. In the tank, T, there is placed the float, N, which has a rigid central shaft projecting upwards, connecting at the upper end to the wicket gate, K, by the bell crank, A, and rod, B.

OPERATION.

The head gate, D, is raised sufficiently to keep the canal, race or flume filled to a fixed water level, when the quantity agreed upon is being used. The float buoy, F, is half submerged at that time, and both the openings in the tank, T, are closed or opened alternately in a slight degree with the oscillations of the surface water acting upon the float, F. The wicket gate is kept at a fixed opening so long as the draught is constant. Suppose, however, the mill owner attempts to take more than agreed upon, and opens his wheel gates accordingly? The velocity of water in the race instantly increases, the surface level drops, and with it the float, F, which opens the inlet to the tank, T, and, as that fills, the floating buoy, N, rises, and the wicket, K, closes until the velocity is checked and the surface level is restored to its proper position. If it becomes too high, the float, F, opens the outlet and the water in tank, T, is discharged down through the pipe, C, shown by dotted lines. This opens the wicket more, so that the quantity due the mill is always ready, if the general supply is sufficient; if not, then all the head gates upon the fall are to be opened in proportion, so that each mill will invariably get its share. If one attempts to take more, he will simply lose power through loss of head, in proportion to the quantity he unjustly tries to appropriate.

To measure or deliver a given quantity, it is only necessary to adjust the wicket gate in unison with the proper surface level until the discharge is the exact quantity agreed upon, which may be determined by a weir below, or in any manner that may be selected; then, when the discharge is right, secure the wicket gate and floats in a manner beyond chance for change, unless by consent. The method is not theoretical, for I have had it in use many months and have watched its operation daily. It is sensitive far beyond my anticipations when first planned. It may be easily applied to the turbines or other devices used to operate head or overflow gates. With its aid the surface level in a canal or race may be kept constant, so that the most perfect economy is practicable, for it prevents the drawing down of head and the use of an unnecessary quantity of water to make up therefor. It will not *strike* for higher pay, go to sleep, or become careless. I believe it to be perfectly practicable for measuring or dividing water used for power under any condition likely to occur, and far more accurately and cheaply than any other plan known.

By using a hanging balanced gate, like Fig 2, and which may be operated substantially as the wicket described, a perfect aperture discharge may be obtained. Such a gate may be used temporarily at almost any mill, as now arranged, and at any, as they may be arranged; so that wheels may be tested in the mills where they are used, without detention, instead of necessitating a testing flume made purposely for such tests.

A special testing flume in the future can only denote incompetency, for every mill may and should be a perfect testing apparatus by which the slightest defect in efficiency or power should instantly be made apparent. Competition will soon compel greater economy in manufactures, and particularly in the power required; and certainly a vast saving is possible in that, for there are thousands and tens of thousands of tons of coal annually consumed in the New England States alone, to make up for the water power wasted through ignorance or thriftless management.

EMERSON'S WATER GOVERNOR,

OR PLAN FOR AUTOMATICALLY MEASURING OR DIVIDING WATER.

Turbine Against Breast Wheel.

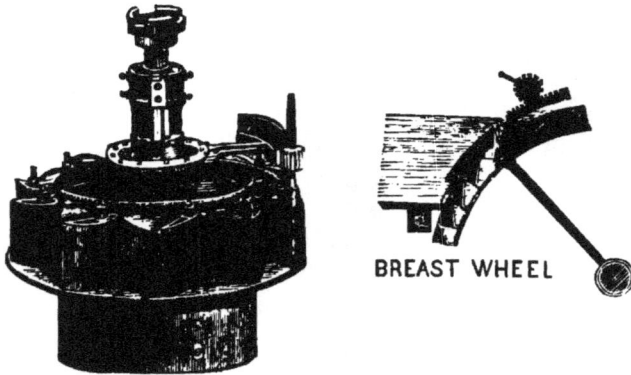

BREAST WHEEL

Messrs. Smith, Northam & Robinson, of Hartford, Ct., have a grist mill four miles from Hartford, that had a breast wheel 16 feet in diameter, 13 feet length of buckets, divided into three sections of 4 feet 4 inches each; the buckets were 18 inches in depth; three gates, in sections to correspond with wheel; the upper gate opening, 5½ inches; the next lower, 3½; the bottom one, 3½ inches; head, 12 feet. The breast wheel was supposed to be so superior to a turbine, that it had been kept in, though it was troubled much by ice during each winter. The firm consulted me upon the subject, and, after months of hesitation, concluded to change, and to follow my directions upon the following terms: The turbines to be selected by me, and tested before acceptance; the plans for change to be furnished by Wm. J. Sumner; my remuneration to be a barrel of bran or flour, according to my success. A weir was constructed in the stream below the mill; the breast wheel and turbine to be tested in the mill, by grinding—the discharge to be measured from each below the mill, under exactly the same conditions. The turbines, 20 and 25-inch New American, were tested by me at Holyoke before acceptance. Results are given below.

25-Inch New American Wheel, Tested Oct. 15, 1880.

Head	Weight	Rev per Minute	Horse Power	Cubic Feet	Percentage
16.29	400	219.5	39.90	1583	.8193
16.31	325	225	33.23	1337	.8256
16.31	325	219	32.35	1249	.8410
16.38	250	218.5	24.82	1032	.7776
16.48	175	224.5	17.86	784	.7318

20-Inch New American Wheel, Tested Oct. 14, 1880.

Head	Weight	Rev per Minute	Horse Power	Cubic Feet	Percentage
15.20	290	260.5	22.89	1001	.7902
15.38	255	258	19.91	840	.8160
15.41	230	258	17.98	754	.8192
15.45	210	252	16.03	671	.8188
15.63	155	253	11.88	522	.7706

Before taking the breast wheel out, it was tested by grinding corn and measuring water below. The stones were sharp and in good condition. The head was 12 feet; gates opened in full. Ground old corn coarse, but very sharp, clean, even grit meal. The change was made; then the turbines were tested in the same way, but I think the corn stones were not in so good condition as when the breast wheel was tried, but may be mistaken. New corn was ground with the 25-inch turbine. The coarsest part of the meal was as near like that ground by the breast wheel as was possible to make it; but it was uneven, much of it being quite fine. This was attributed to its being made from new corn. The miller made every effort to make the trial fair. The results are given below. The rye stone was driven by the 20-inch wheel, the gate being opened about two-thirds—all that could be used. The flour produced was the nicest I have ever seen made from rye.

TEST OF BREAST-WHEEL, AT FULL GATE.

Head, 12 feet; length of weir, 10 feet; depth on weir, 8 13-16 inches; quantity of water, 1239 cubic feet per minute; 28.08 horse power. Ground 2050 pounds per hour, or 1.3 bushels per each horse power of water used.

TEST OF 25-INCH NEW AMERICAN WHEEL,

It having replaced the above-mentioned breast wheel. Head and length of weir the same.

Full Gate.—Depth on weir, 8 15-16 inches; 1266 cubic feet per minute; 28.7 horse power. Ground 3528 pounds per hour, or 2.2 bushels per each horse power of water used.

Gate Opened Two-thirds.—Depth on weir, 7 15-16 inches; 1059 cubic feet per minute; 24 horse power. Ground 2900 pounds per hour, or 2.15 bushels per each horse power of water used.

Gate Half Opened.—Depth on weir, 7 inches; 879 cubic feet per minute; 19.9 horse power. Ground 2400 pounds per hour, or 2.1 bushels per each horse power of water used.

TEST OF 20-INCH NEW AMERICAN WHEEL IN SAME MILL.

Depth on weir, 6½ inches; 789 cubic feet; 17.9 horse power. Ground four bushels of rye in seventeen minutes, or 14.1 bushels per hour. Eighty per cent. of power of water used; 14.3 horse power, or substantially a bushel per horse power.

After a few weeks' time, the proprietors sent me a barrel of "Pillsbury's best."

Burning or Wearing Down of Step

May, and does happen with any make of turbines. Two turbines of the same make, seemingly exactly alike, and placed in a pit side by side, the step of one may wear down monthly, the other not at all. The cause was attributed to pressure from downward discharge; but if eighty per cent. is used to rotate wheel, the other twenty would be no more weight with downward than any other discharge. The Swain was noted for wearing down step. I knew of one 24-inch wheel that had nineteen steps in thirteen months. Others of the make had to be suspended by collars on shaft. A 36-inch wheel of the kind was sent to me to be tested. A collar that should have been on to keep wheel in place was left off. When the gate was opened, the pressure raised the wheel and brake, and it was impossible to test it until one hundred and fifty pounds were added to the brake to keep the wheel *down* upon the step. The Boyden wheels used at Lowell are suspended by neck on shaft, as are the Kilburn & Lincoln wheel of Fall River. The Risdon has a counterpoise above the wheel, drum-shaped (see his new wheel). Many plans have been tried. A common one is to channel top of step; another is to lead water from the penstock through a piece of ¾-inch steam pipe, to bottom of step—a hole up through first being made; some chamber the lower part of step, then make numerous small holes up through, like the top of a pepper-box, taking water from the flume through pipe. Great weight upon the step in the way of shafting, gearing, &c., should be avoided when possible.

Railroad Suggestions.

It may be said that such suggestions are out of place in a work of this kind, but my experience has been gained from experiments made in many parts of the country—often in very distant parts—and the railroads have much to do with my ability to obtain such experience, consequently are part of the instruments I work with.

The rather common practice of roasting car-loads of passengers, when collisions or other accidents occur upon our railroads, has caused an agitation of the subject of car-heating. Safety as well as comfort is desired. The ancient and semi-barbarous plan of placing a stove at each end fails to give either, while such stoves take space for eight seats, disfigure and injure the cars, half roast a few near them, leaving the larger proportion to sit with cold feet and generally uncomfortable throughout the passage. Why not have a boiler for heating placed in the baggage car, to furnish steam for heating the passenger cars with safety?

Another want, is light trains between commercial centres and neighboring cities—trains that may readily be stopped and started, something as horse-cars are; that is, within reasonable distances. Such trains should be made up of light engines and cars, and have commutation fares. With such in operation, there would be no need of the heavy or through trains stopping so often. The manager of the ordinary railroad should feel ashamed to have a horse-railroad run for miles alongside, as from Boston to Lynn, and pay expenses.

Sunday trains on all roads are also much needed. Those who object to such, are impracticable persons, who do it through ignorance, and without consideration of the changes that have taken place since the Jewish Sabbath was instituted. At that time, labor was continued for from fourteen to sixteen hours per day; indeed, it is within the memory of those of middle age, when the hours of labor were nearly the same in the New England States. As a day for rest, Sunday has no such claim as formerly. The God of Moses had reason for requiring such a day; but the God of to-day has not. Besides, ages of experience has proved that He has been cheated constantly, for the most bigoted believer has never hesitated to lie in bed three hours later Sunday than other mornings; then, at evening, say: "Well, boys, we have got a hard day's work to do to-morrow, so we must go to bed early." All nature ignores the day: the billows rage as fiercely, the thunder is as loud, the tempest is as destructive, the blossoms as beautiful, vegetation and animal life as progressive upon that as upon any other day in the week. In Moses' time, families and tribes were separated but little. How different now! Business necessities often separate the nearest relations. The father of a family is often hundreds of miles away. It has happened three times in my own experience that telegrams, announcing the dying condition of members of my family, have been received late Saturday evening, and it is not likely my experience in that way is exceptional. Those who desire to observe the day as sacred, should be allowed to do so without hindrance; but it is very different when such believers try to compel all others to do the same. Our prisons are filled with theoretical believers in the idea, if we may judge from the declarations of those about to be hanged for murder. If such is the effect of believing one day better than another, would it not be better to teach that all days are good, and mix religion with business?

Water Supply for Cities.

Now that there is a general complaint of waste of water, and apprehension that the supply will soon be much less than the demand, would it not be well to have a high service for extinguishing fires, and a low one for domestic use, or have the discharge for the latter retarded? The unthinking user leaves a faucet open just as long, in most cases, where the pressure is a hundred pounds per inch as where it is only five or ten.

Apparatus for Regulating the Flow and Delivery of Water Through Canals, Flumes, and Water-Ways.

Specification Forming part of Letters Patent No. 275,371, dated April 10, 1883.

The object of my invention is, first, to maintain water in a canal or water-way at a uniform height during its passage to the outlet or flume ; second, from this established uniform height of water in the canal or water-way to make a proportional division of the water at the outlets, giving to each consumer of water at his respective outlet the amount of water to which each is entitled, or a proportional amount of the whole to which each is entitled ; third, to measure the amount given to each ; and, fourth, to prevent any one of the consumers from using any more water than he is entitled to, the whole apparatus operating automatically, and being based on the *fixed law that any given velocity of water is acquired through a corresponding loss of head.*

This apparatus is applicable to be used at falls where the water is owned by several parties and is to be proportionally divided between them. It is also applicable for use where the water is owned by one company or owner, and is sold or leased, and a stated quantity is to be measured out to each purchaser or to each party leasing. It is also applicable for use for governing the flow of water from reservoirs, where water is stored for irrigation or for manufacturing purposes, and also for regulating the height of water in rivers or ponds to prevent backflowage in cases where movable dams or flushboards are employed. I accomplish these objects by the apparatus substantially as hereinafter described, and illustrated in the accompanying drawings, in which—

Figure I. is a plan view representing a canal, and showing my invention as applied to the operation of wicket-gates, or those pivoted in a vertical position at one end of the canal, for the head-gates, and also at the other end, or at the flumes, where the water would be drawn from the canal and used for manufacturing or other purposes. Fig. II. is a vertical section of the same at line A of Fig. I. Fig. III. is a plan view representing a canal provided with vertical sliding head-gates at one end, and the gates at the other end or in the flumes, where the water would be drawn from the canal for use, being pivoted to or hung upon a bar placed in a horizontal position. Fig. IV. is an enlarged plan view of a flume and draft-tube, with a swinging gate hung in said flume and operated according to my invention in dividing and measuring the water drawn from the canal through said flume. Fig. V. is a part vertical section of the same at line B of Fig. IV., showing the swinging gate and the lifting-float which operates it and the draft-tube.and also a part vertical section at line D of Fig IV., showing the construction of the governing-float which operates the valve controlling the flow of water into and out of the tank containing the lifting-float. Fig. VI. is a vertical section of the valve and its case, which controls the flow of water into and out of the tank containing the lifting-float, at line E of Fig. V. Figs. VII., VIII., and IX. are sectional views representing details of the valve and its case as applied to and used at the flumes or outlets of the canal.

In the drawings, let 1 represent the side walls of a canal or water-way, at one end of which is made the ordinary bulk-head, as *b*, provided with gates, as 2, to admit the water into the canal or water-way when opened for that purpose.

The ordinary head-gates may be used ; but in this application I have shown pivoted gates, as being more easily operated, this class of gates being shown, as at 4, pivoted at 5 in the bulk-head at one end of the canal, 1, and in the flumes at the opposite end of the canal, in Figs. I. and II. In the use of this pivoted gate to control the flow of water, the gate being set in an upright position to turn upon its post 5 as a pivot, an arm, as 6, is secured to its upper end, to which is attached a horizontal rod, 7, connected with one arm of a bell-crank lever, 8, pivoted at 9, the other arm being connected with a vertical rod, as 10, extending through the top of a tank containing a float, 13, to which the lower end of the rod 10 is secured. A smaller tank, as 15, is made upon or is so connected with the canal or its side wall that the water may flow freely into said tank, either by making the side next the canal-wall open, as at 14, or by connecting said tank with the canal by a pipe, with its end opening into the canal, so that the water of the canal may flow through said pipe into the lower portion of the tank to fill the latter up to the same level as the water in the canal. This tank, as 15, I make preferably of rectangular form, and it contains a float, 16, which I make of a form in horizontal section to fit approximately the interior of the tank, but so that the float may move up and down freely, but not revolve therein. This float may be made of any suitable buoyant material ; but I prefer to make it of some thin sheet metal, and hollow, and perfectly water-tight. A socket, as 24, extends vertically through this float, through which extends a rod, 17, whose upper end has a screw-thread made thereon, adapted to receive a nut, as 18, turned on to the upper end of the rod, with a shoulder, h, above and below the float, and this rod 17 extends down through the bottom of this tank 15, with its lower end attached to an arm, 20, secured to the hub of a valve, 73, inclosed within and fitting a cylindrical valve-case, 19, the hub extending out through the case at its axis. This valve-plug fits the interior of the case, so as to move freely therein, and is approximately of semi-cylindrical form, of sufficient extent in its circumference to cover the inlet and outlet ports in the case, and a pipe, 21, opens at one end into the canal and at the other end into the valve-case, 19, at the periphery, at the upper side, so that the water may pass from the canal through this pipe into the valve-case. The opening of this pipe in the valve-case forms its inlet-port, and the opening of a pipe, 23, into the valve-case, on its lower side and nearly opposite the pipe 21, forms the outlet-port of the valve-case, this pipe or opening 23 being merely to permit the water to flow out of the valve-case and to conduct it away to some waste-conduit, if desired. Another pipe, 22, opening into the valve-case at the side, extends to and opens within the tank 11, preferably in its lower portion.

It will be seen by referring to Fig. VII. that when the arm 20 (shown in dotted lines in that figure) is in a horizontal position the inlet-port or opening of the lower end of the pipe 21 in the valve-case is closed, being covered by the upper end of the valve 73, and the opening of the upper end of the pipe 23 in the valve-case or outlet-port is covered by the lower end of the valve 73.

By referring to Fig. VIII. it will be seen that when the arm 20 (shown in dotted lines) is inclined above a horizontal position the valve is moved so as to open the upper or inlet-port and close the lower or outlet port, and when this arm is inclined below its horizontal position the upper or inlet port is closed and the lower or outlet port is opened, as shown in Fig. IX. Of course with the valve in this position, shown in Fig. VII., water can neither flow into the valve-case through the pipe 21 nor out of it through the pipe 23 ; but with the valve in the position shown in Fig. VIII. water may flow into the valve-case through the pipe 21, and thence through the pipe 22 into the tank 11, to raise the float therein, and with the valve in the position shown in Fig. IX. water may flow out of the tank 11 into the valve-case, and thence out through the pipe 23. It will be seen that by this construction of valve the latter may be moved with the least possible friction in its case, and a very slight change in the height of the water in the canal to change the vertical position of the float 16 will be sufficient to operate the valve to open or close the ports in its case.

Referring to Fig. II., suppose it is desired to maintain the water in the canal at the height indicated by the dotted line L. The permanent or sliding head-gates, as 2, are raised to give the desired opening for the water to flow in, and the nut 18 is turned on to the upper end of the rod 17 until the float 16 in the tank 15 is sustained at the height shown in Fig. II. by the water which flows into said tank from the canal. While in this position the valve is held in the position shown in dotted lines in Fig. II., and the water flows

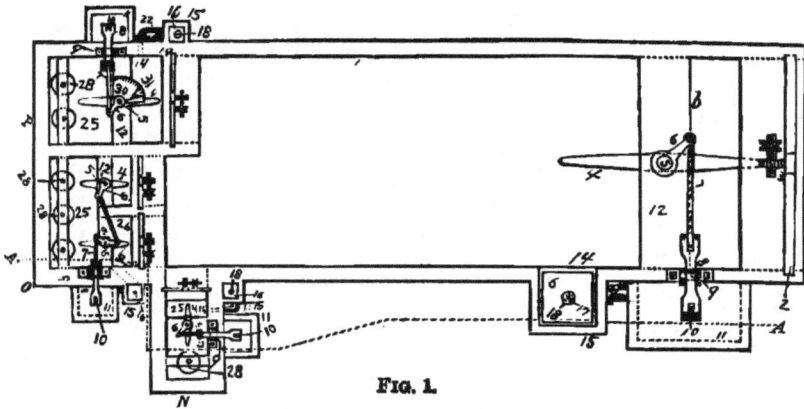

FIG. 1.

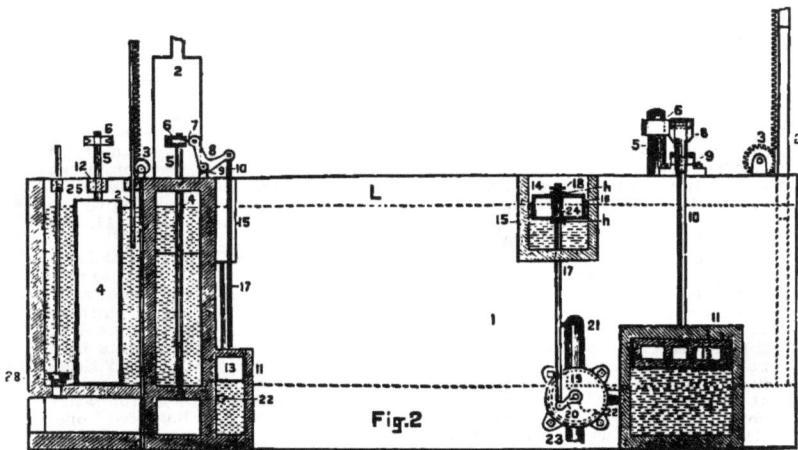

Fig.2

from the canal through the pipe 21, case 19, and pipe 22 into the tank 11, raising the float 13 into the upper part of the tank and holding the pivoted gate 4 wide open, or in a position lengthwise the canal, as shown in Fig. I., so that the water may flow into the canal past the gate 4, on each side the latter ; but as the float 16 is so adjusted, if the water should rise in the canal, the float 16 would be raised, and the arm 20 of the valve would be inclined above a horizontal position and the valve moved into a position to open the outlet-port into the pipe 23 and close the inlet-port from the pipe 21, and the water would flow out from the tank 11 through the pipes 22 and 23, and the valve and the float 13 would fall and close the gate 4, or partially close it, until the water should fall nearly to the desired level at the line L, and when the inlet port or pipe 21 began to open as the float 16 was lowered by the fall of the water the tank would be slowly filled again and the float 13 would rise, and the gate 4 would be gradually opened to keep up the supply of water in the canal. This float 16 may be so nicely adjusted by turning the nut 18 either up or down that the slightest rise of water in the canal, and consequently in the tank 15, will operate the float 16, and the valve and the gate 4 will be shut sufficiently to keep out the excess of water over that required for use in the canal. The gate 4 is always wide open as long as the water remains at the lowest desired level, and when the water rises above this level the gate 4 is partially shut.

The flumes, as 25, at the points along the canal where the water is drawn therefrom, may be supplied with the same kind of gate, 4, each of which is operated by float 16, valve and its case 19, and lifting-float 13 in the same manner as the head-gate is operated, as above described, except that the arm 6 is attached to the post or pivot 5 of the gate in an opposite position from that in which it is attached to said post or pivot at the supply end of the canal. These flume-gates also operate to partially close and prevent any excess of water from passing into the flume over that amount previously determined upon. For example, suppose a manufacturing establishment to be located at any point along the canal, say at N, and to draw the water from the canal through the flume containing the single gate 4 at that point. This flume is provided with a tank, 15, containing a valve operating float, 16, like that hereinbefore described for the head-gate, into which tank the water may flow from the flume through a pipe whose orifice 14 opens into the flume, above the gate, in a direction opposite the flow of the current, and a valve and its case 19, like that above described for the operation of the head-gate, is connected with the float 16 by a rod, 17, with a tank connected by a pipe with said valve-case, and containing a lifting-float, as 13, which is connected with an arm, as 6, on the gate 4 in the flume by rods 10 and 7 and bell-crank lever 8, all as above described for the head-gate at the bulk-head.

It will be seen that in using the valve and its case 19 at the head-gate at the bulk-head b the arm 20 is so attached to the hub of the valve that as the float 16 is raised by the water in the tank 15 and in the canal the valve is moved so that the water may flow out of the tank 11, and by the falling of the lifting-float 13 the head-gate 4 will begin to close ; but at the flumes the arm 20 is attached to the hub of the valve 73 in a reversed position, or as shown in dotted lines in Figs. V., VII., VIII., and IX., so that as the water falls in the flume the falling of the float 16 in the tank 15 would move the valve 73 into a position to permit the water to flow from the tank 11, and the lifting-float 13, in falling, would close the gate 4 in the flume 25.

In the above explanation I have referred to the details of the tanks and valve, as shown in Fig I., at the head of the canal or bulk-head, because precisely the same arrangement is used at the flumes as at the bulk-head, with the exception that the arm 6 is attached to the pivot or post 4, and the arm 20 is secured to the valve 73 in a reversed position when applied and used at the flumes. For illustration, two other manufacturing establishments may be drawing water from the canal—one at O and another at P—and these flumes may be located any distance apart and along the side of the canal, or at its termination. For convenience I have represented them at the latter point, and side by side. Suppose that the party at N owns or has leased one-sixth of all the water which flows through the canal, the party at O three-sixths, and the party at P two-sixths, each flume-opening being of the proper area to permit that quantity of water to flow through at a given velocity—say of two feet per second. These flume-openings being the ordinary head-gates, they may be changed to give different areas of opening at different seasons of the year to meet the usual changes in the supply of water at such times, if found advisable. With the water at the height indicated by the line L the nut 18 on the rod 17 is turned so that the float 16 in the tank 15 holds the valve at the flume N in such a position that the float 13 in the tank 11 at that flume will hold its gate 4 in a position wide open, as shown in Fig. 1. Inasmuch as the amount of water which can be drawn from a flume depends upon the velocity at the outlet of the flume at a given head, this additional use would tend to draw the water down or reduce its height in the flume, and the water in the tank 15 being always at the same level with that in the flume, the float 16 would fall and move the valve into a position to permit the water to flow out of the tank 11 through the pipe 22, valve 19, and outlet 23, and the lifting-float 13 would fall and partially shut the gate 4 in the flume, which would of course reduce the quantity of water passing into the flume at a greater velocity until his proper proportional quantity of one-sixth was reached, when the head and float would rise to their normal condition, allowing him still his proportional quantity, though at a loss of head in proportion to the quantity which he attempts to overdraw. In like manner the other owners or lessees at other points are governed or controlled in their use of water.

In Fig. 1. there are two gates in the flume at O, one of which is provided with a double arm, 6, one of whose ends is connected with the arm of the other gate by a rod, 26, and the other end is connected with the bell-crank lever 8, connected with the lifting-float 13 in the tank 11, so that the movement of said float will operate both gates at the same time.

The flumes may be provided with the ordinary lifting or vertically-sliding gates 2, which may be closed at any time for the purpose of making repairs in the flume, or for any other purpose.

The tanks 15, connected with all the flumes, should all be securely locked and be kept under the charge of one man, so that no other person could have access to them ; or the tanks 15 might be all located in one building or office and each be connected with its flume by a pipe, and all locked and in charge of one person.

If desired, a dial, 30, having a graduated scale, may be placed in any convenient fixed position near the pivot or post 5 of each gate in the flume, with an index secured to the post, as shown at P in Fig. I., so that a glance at the index and dial at any time would show how far open each gate was as to the area of its aperture, so that a slight computation might give approximately the quantity of water passing through.

It will be seen that this apparatus furnishes a very reliable system of maintaining the water in a water-way or canal at a standard height to give a uniform head, and with that head, to divide the water flowing through, giving to each owner or lessee the quantity to which he is entitled, and preventing any attempt on the part of either owner or lessee from using a greater quantity than that to which he is entitled.

In Figs. III., IV., and V. is shown a modification of the same invention as applied to gates arranged to move on a horizontal pivot for the purpose of measuring the amount of water passing through the gate-aperture, Figs. IV. and V., showing an enlarged detailed view, in which 33 represents a horizontal bar fixed in the sides of the flume, to which are hung, so as to swing freely thereon, the arms 34, whose free ends are secured to the gate 36. The outside of this gate should be made convex in its cross-section upon a curve whose radius is the distance from the outside of the gate to the horizontal bar 33, and the gate-aperture 71, made in front of the gate and through the front wall 32 of the flume, should have its ends curved vertically, as at 68, so that the ends of the gate 36 should approximately fit the aperture when the gate is shut.

The tank 11 for the lifting-float 13, when applied to a swinging gate of this construction and used in the position shown in the drawings, is made beneath the floor of the flume, and the lifting-rod 10 in this case extends up through this floor, and may be connected with a cross-bar extending from one arm, 34, to the other, of the gate 36, as shown in Figs. IV. and V.

The chamber 75 for the wheels 28 may be covered by a horizontal partition, 70, if desired, with a small horizontal aperture, as 69, through the front wall 32 of the flume, which would form a draft-tube in which the wheels were located, the water in the flume flowing through this aperture 69 and covering the horizontal partition 70, to pack the apertures to the wheel-

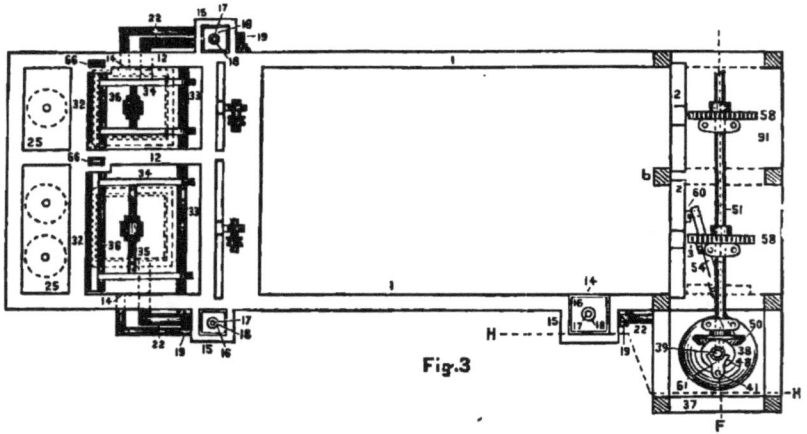

Fig.3

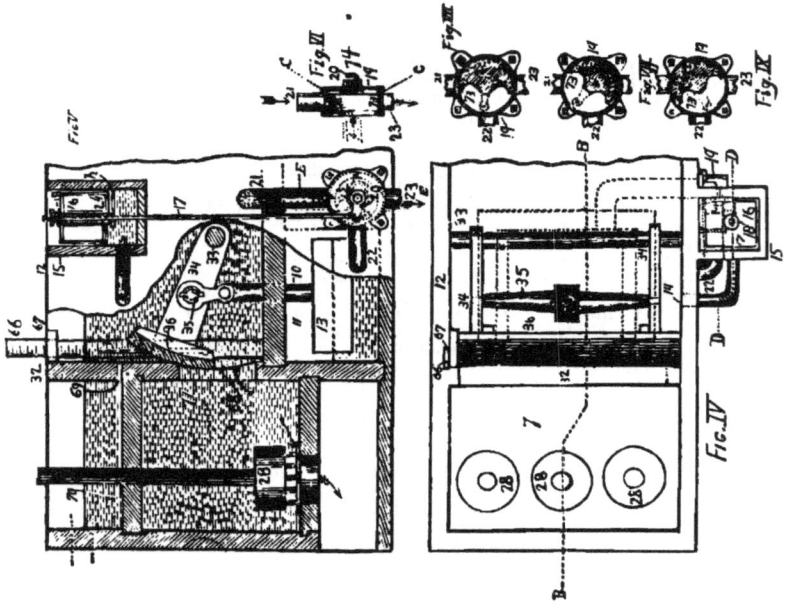

chamber, these wheels representing those used by the establishment located at that point and drawing water from the canal.

A scale, 66, may extend up vertically in any convenient place, with its lower end pivoted to the end of the gate 36, and the graduations on the scale may indicate the vertical opening and fractions thereof of the gate-aperture. Suppose, for example, that the gate-aperture should be ten feet horizontally and two feet vertically, and a glance at the scale should indicate that the lower edge of the gate 36 was just one foot above the lower edge of the gate-aperture. It would require but a few minutes' computation, knowing the area of open aperture and velocity, to ascertain just how much water was flowing through the aperture beneath the gate, so that the quantity of water being used by the party drawing from that flume may be easily and accurately measured at any time by a glance at the scale to see how much it projects above the top of the wall of the flume, or any other horizontal line across the scale as an indicator.

The operation of the float 16 within its tank 15, connected with the flume shown in Figs. III., IV., and V., and also the valve-case 19, connected with said float and with the tank 11 of the lifting-float 13, is precisely like that hereinbefore described as used in Figs. I. and II., except that its action is reversed —that is to say, the tank is so connected with the flume in Figs. IV. and V. that the water may flow freely through the orifice 14 and the pipe leading therefrom into the tank 15, so that the float 16, being properly adjusted by the nut 18 on the rod 17 above the float, will be held at a certain height in the tank 15 by the water therein, the valve in the case 19 being held in a position to retain the proper quantity of water in the tank 11 to sustain the float 13 and gate 36 at such a height as to allow the quantity of water to flow through the gate-aperture 71 at the fixed velocity to which the party is entitled, at the given head which is maintained in the canal by the head-gates, as hereinbefore described. If the party wishes to use more water than that which would flow through the gate-opening 71 at a given velocity—say two feet per second—the water would begin to fall in that part of the flume in which the gate 36 is pivoted, and also would fall in the tank 15, owing to the increased velocity of the water passing through the gate-aperture 71, and the float 16 would fall and change the valve, so that the inlet from the pipe 21 would be opened and the outlet at 23 be closed, permitting the water to flow into the tank 11, raising the lifting-float 13 and opening the gate 36 to give a larger aperture and permit more water to pass through the gate aperture 71 until the velocity was reduced to the stated two feet per second, and this increased opening of the gate-aperture would be accurately indicated by the scale, and the amount of water could then easily be computed. If the water in the reservoir should be exceedingly low, so that the water in the canal should remain at a much lower level than at the line L, the nuts on the rods 17 above the floats 16 are readjusted according to the height of water in the canal, and each party will then be able to draw his proportional quantity of the water, and no more, instead of his full quantity, as when the water is abundant.

It will be seen that when the float 16 is once adjusted for any certain height at which it is desired to maintain the water, by turning the nut 18 on the rod 17 either up or down, the float will operate automatically to move the valve into such a position as to regulate the amount of water retained in the lifting-tank 11 to operate the gate, and keep the proportions of the supply of water in the canal equal to the demand or amount used therefrom.

It will be seen that by merely reversing the position of the arm 20 on the hub of the valve 73 the falling of the water and the float 16 will operate the valve to permit the water to flow into the lifting tank to close a gate, or to open it, according to the position in which the said arm is secured.

It is evident that in cases where a single individual, firm, or corporation owns all the water which runs in the canal or water-way, or owns the entire water privilege, and is only using from one flume, or when it is not desired to divide the water among the different flumes through which it is drawn from the canal to be used, but only to maintain the water in the canal or water-way at a uniform height, it may be done by using the apparatus as connected with the head-gates at the bulk-head alone. In any case, whether used at the head-gates or those in the flume, or both, the tanks 15, containing the operating floats 16, together with their respective valves, and the pipes or water-connections, should all be located under cover to avoid being frozen up in winter, and the tanks 15, with their floats 16, might be located conveniently in some office, and under the control of one man ; and instead of taking the water from the canal into the tank 11 through the valve-case 19 and inlet 21, it may be taken from the reservoir or river by connecting the pipe therewith, if it should be more convenient.

JAMES EMERSON.

Witnesses :
T. A. CURTIS,
N. E. DWINNELL.

QUESTIONS OFTEN ASKED ME IN COURT ANSWERED.

Have I ever been to college or technical school ?

No ; but the teachers and graduates of such institutions often come to me for information.

Have I studied hydraulic works by different authors ?

I have looked through such occasionally.

Have I ever run levels between mills as a surveyor does ?

No ; the cause for effect can better be ascertained by doing it by the water if one knows how to do it.

Why do I answer so positively while others professing to teach the science hesitate ?

Because my answers are based upon knowledge obtained by personal experiments.

How do I know that weir and aperture measurements are correct ?

By catching the discharge from weirs and apertures in tanks, then cubing the contents.

How did I prepare my weir tables, did I work them all personally ?

No ; I never learned the formula for working up such tables, but employed cheap help to work up a set of tables from the Francis formulæ, then cubed the discharge in tanks varying in capacity from two feet up to twenty-five thousand feet.

How do I know that tests of wheels by such tables are correct ?

By testing the same wheels at several different testing flumes remote from each other.

How do I know that float and current meter measurements are worthless ?

By testing the same streams or discharges by weir.

SUGGESTION FOR CAPITALISTS.

As the hours of labor are reduced so that invested capital in mills stands idle two-thirds of the time why not employ two or three sets of hands and keep the work in operation the most of the time, thus making a plant of a million turn out the same quantity now done by one of double that cost ?

Preliminary Proceedings for Legal Division of Water Power.

State of Iowa, } ss.
Linn County. }

To James Emerson of Willimansett, Massachusetts, Samuel Sherwood of Independence, Iowa, and S. N. Williams of Mt. Vernon, Iowa.

GREETING—Whereas, on November 1st, A. D. 1889, in an action now pending in the district court of said Linn county, wherein N. E. Brown is plaintiff, and Susan Brown, W. S. Cooper, Sarah E. Leach, E. E. Leach, Herman D. St. John, and Charles Clay are defendants, it was found by the said court that the said plaintiff, N. E. Brown, is the owner in fee simple of the undivided two sixty-fourths ($\frac{2}{64}$) of the following described property situated in Linn county and the state of Iowa, to wit:—

The water power created, situated on, across and adjacent to the Cedar river at Cedar Rapids, Linn county, Iowa, consisting of a mill dam constructed across the Cedar river at said Cedar Rapids with an abutment or bulk head upon and against either bank of said river, including race ways on each side of said river from said dam, the water power and flowage created by said dam and race ways, and the right to have, build, and maintain said dam, race ways, and power; said dam, abutments, and bulk heads being more particularly described as follows, to wit:—

Said dam being at and between Fractional Block Two (2) in Cedar Rapids, Iowa, and Ely & Angle's addition to West Cedar Rapids, in Linn county, Iowa, one of said bulk heads and the east end of said dam being upon lots "J," "K," "L," and "M" in Fractional Block Two (2) in Cedar Rapids, Iowa, and the other of said bulk heads and the west end of said dam being on lots twenty-three (23) and twenty-four (24) of Ely & Angle's addition to West Cedar Rapids, in Linn county, Iowa, and the street and land adjacent thereto; that the defendant, Susan Brown, is the absolute owner of the undivided fifty-five sixty-fourths ($\frac{55}{64}$), and the one-third ($\frac{1}{3}$) of the two sixty-fourths ($\frac{2}{64}$) in all of the one hundred sixty-seven one hundred ninety seconds ($\frac{167}{192}$) of said property ; that the said W. S. Cooper is the owner of the undivided one-sixteenth ($\frac{1}{16}$) of the said property; that the defendants, Herman D. St. John and Charles Clay, are together the owners of the undivided one sixty-fourth ($\frac{1}{64}$) of said property, and that the defendants, Sarah E. Leach and E. E. Leach, are together the owners of the one forty-eighth ($\frac{1}{48}$) of said property and entitled to the use of the said one forty-eighth ($\frac{1}{48}$) on the west side of said Cedar river.

And it was then and there ordered, adjudged, and decreed by said district court that the said shares and title of the said parties respectively in and to said property be confirmed, and that partition thereof between said parties be made. And that said water power and property hereinbefore described be partitioned and so measured and meted out to the several owners thereof according to their several rights and interests as hereinbefore set forth so that each of said owners shall receive and use of said water power, as developed, his or their own proper share and no more, at any and all stages of the water and in whatever condition said water power and improvements may be, viz.:—

To said W. S. Cooper the four sixty-fourths ($\frac{4}{64}$) of said power; to said Herman D. St. John and Charles Clay together the one sixty-fourth ($\frac{1}{64}$) of said power; to said Susan Brown the one hundred sixty-seven one hundred ninety-seconds ($\frac{167}{192}$) of said power and property; to said N. E. Brown the two sixty-fourths ($\frac{2}{64}$) of said power and property, and to said Sarah E. Leach and E. E. Leach together the one forty-eighth ($\frac{1}{48}$) of the whole of said power and property, the latter to be used on the west side of said river, that each of them may enjoy and use the same severally, and, each to have his or their full use thereof, uninterrupted by interference, invasion, or diminution from the other, and no more.

And whereas, on the 19th day of November, A. D. 1889, and the 13th day of February, A. D. 1890, in said action it was ordered, adjudged, and decreed by said court that to effect said partition, such partition of said property between said parties to said action be made by James Emerson of Willimansett, Massachusetts, Samuel Sherwood of Independence, Iowa, and S. N. Williams of Mt. Vernon, Iowa, referees and commissioners for that purpose; and that to enable such commissioners and referees to make such partition, they were authorized as against any and all persons to enter upon said premises and take control of said water power, dam, and race ways for the reasonable time required to do said work, opening and closing the same at pleasure and as in their judgment may be necessary, stopping any and all water wheels and mills operated by said power and for such time or times as may be necessary and reasonable, and that in making said partition the said referees ascertain the quantity or volume of water now used at and by said power and dam, and the exact power and quantity that each party shall be entitled to draw off or use under the varying stages of the water in the aforesaid river, and said referees are further authorized by said court to make such recommendation in their report as they deem advisable for the future maintenance and use of the interests of the several parties in said action in said water power.

Now, therefore, you are hereby empowered and commanded to make partition of the water power and property above described between the plaintiff, N. E. Brown, and the defendants, Susan Brown, W. S. Cooper, Sarah E. Leach, E. E. Leach, Herman D. St. John, and Charles Clay, by assigning to N. E. Brown, the two sixty-fourths ($\frac{2}{64}$) thereof, to said W. S. Cooper, the four sixty-fourths ($\frac{4}{64}$) thereof, to the defendants, Herman D. St. John and Charles Clay together, the one hundred sixty-seven one hundred ninety seconds ($\frac{167}{192}$) thereof, and to the defendants, Sarah E. Leach and E. E. Leach together, the one forty-eighth ($\frac{1}{48}$) thereof, the said one forty-eighth ($\frac{1}{48}$) to be used on the west side of said river, all in severalty according to law, that each of said parties may enjoy the use and portion thereof belonging to him, her, or them, in severalty, and have his and their full use thereof, uninterrupted by interference, invasion, or diminution from the other, and no more; such partition to be made as hereinbefore provided and directed; and you are further directed to make report in writing of such partition, and your doings under this commission and said decree, and of all expenses and costs pertaining to the same, as soon as can be done with reasonable diligence, to our said district court. You are further authorized to make such recommendations in your said report as you deem advisable for the future maintenance and use of the interests of the several parties to said action in said water power.

WITNESS my hand and the seal of the said court hereto affixed this 9th day of June, A. D. 1890.

O. S. LAMB,
Clerk of the district court of Linn County, Iowa.

N. E. BROWN, Plaintiff,
       vs.
SUSAN BROWN, W. S. COOPER,
SARAH E. LEACH, E. E. LEACH,
HERMAN D. ST. JOHN, and CHARLES
CLAY, Defendants.
         Partition
In District Court of Linn
County, Iowa.

State of Iowa, } ss.
Linn County.

We, James Emerson, Samuel Sherwood, and S. N. Williams, do severally swear, that we will well and faithfully perform the duties of referees in the above entitled cause, and make a just and equitable partition therein, according to the best of our knowledge and ability.

JAMES EMERSON, }
SAMUEL SHERWOOD, } *Referees.*
S. N. WILLIAMS, }

Subscribed and sworn to before me by the said James Emerson, Samuel Sherwood, and S. N. Williams, on this 28th day of June, A. D. 1890.

U. C. BLAKE,
Notary Public in and for Linn County, Iowa.

REPORT OF REFEREES.

STATE OF IOWA, } ss.
LINN COUNTY.

Report of referees in answer to decree of Linn County District Court ordering the partition of the water power at Cedar Rapids of said County ; N. E. Brown, plaintiff ; Susan Brown, W. S. Cooper, Sarah E. and E. E. Leach, Herman D. St. John and Charles Clay, defendants.

We, the referees, met at Cedar Rapids July 29, 1890, and qualified as required. Mr. Emerson took charge of the numerous preparations necessary for dividing the water ; Professor Williams having charge of various tests for ascertaining the cost of steam power at Cedar Rapids and estimated valuation of water power at Waterloo and Cedar Falls and other matters, while Mr. Sherwood, from his general knowledge of water power, and especially for his early acquaintance with the Cedar Rapids water power, was held in reserve as adviser and assistant.

The first act necessary was to put the dam in order that the whole flow of water in the river should pass over its crest for measurement.

The top of dam was raised some ten inches at the lowest point and divided into twenty-nine twenty feet sections and two of ten feet each ; division planks were established between each section and a small post rigidly secured to the dam eighteen inches up stream, from crest of weirs at middle of each. These posts were leveled at the top to correspond to the exact level of weirs : then in case the weight of overflow should cause sectional depressions one end of crest plank would be likely to be as much above the top of post as the other would be below.

The bottom edges of cresting planks were well imbedded in Portland cement, making a perfectly tight joint the whole length of dam. The planking of dam is doubled, the upper ends of top planks are scoured off by ice and overflow so that water flowing over runs down back between the two layers, presenting the appearance of extensive leakage under the dam. The cresting planks are placed up stream from the worn off upper planks, and while the surface of water was below the crest of weirs there was no show of leakage through the planking from end to end of dam ; at the bottom there were three leaks, but so small that there were no whirlpools or other indications of their source above. Gravel would make the dam as tight as a dam built upon seamy rock can be made, but gravel can only be procured at a cost of two dollars per cubic yard, and at that price with difficulty. Sawdust and other debris were used until the leakage was reduced to the lowest stage possible.

The openings to the races were stopped ; on the west side by a temporary dam, on the east side by planks at the openings in wall at its head, and the leakage from each race was measured by weir or aperture.

Wednesday, August 13, water flowing over the crest of dam or weirs seemed to have reached its height and a hasty measurement was made, the result showing a flow of over thirty-nine thousand cubic feet per minute in the river. The next morning Messrs. Sherwood and Williams joined with me in making the most careful measurement possible. An improvised hook gauge and gauge tank for quieting the surface of the water were used, so that the greatest exactness was obtained, the measurements on dam and in the two races aggregating 39,699.43 cubic feet per minute, to which I add one-fourth additional, making the maximum flow of 49,624.28 cubic feet per minute as the largest quantity likely to flow in the river at any season of the year, except during freshets ; or that can be made useful through the head-gate openings on east side of the river.

I make this addition not because I believe there is such quantity that can be utilized under existing conditions, but because the ownership of all the parties aside from Mrs. Brown is so small that it is better to do so than to leave any excuse for further litigation.

St. John & Clay's mill has wheels that under eight feet head will discharge 11,121 cubic feet of water per minute, = 167.9 h.p. of water, of which about 112 effective h.p. may be realized. Their $\frac{1}{64}$ of the whole power is 775.4 cubic feet of water per minute, which falling 8 feet = 11.7 h.p., or 9 h.p. net.

N. E. Brown's wheels with eight feet head will discharge 14,000 cubic feet per minute, or 210.2 h.p. of water, but they are so out of repair that no accurate estimate of net effect can be made. His $\frac{1}{32}$ of the whole power is 1,550.8 cubic feet per minute, which falling eight feet = 23.4 h.p. of water, or 16 to 18 effective h.p.

Cooper's wheels under seven feet head can discharge about 9,800 cubic feet per minute, = 129 h.p. of water, or from 86 to 100 h.p. net. His $\frac{1}{16}$ of 49,624.3 cubic feet = 3,101.5 cubic feet per minute, which falling seven feet = 41 h.p. of which 28 to 35 may be made effective.

Leach's wheels under six feet head will discharge about 7,000 cubic feet per minute, or 79.3 h.p. of water, from which 50 to 60 h.p. net should be realized. His $\frac{1}{48}$ of 49,624.3 cubic feet = 1,033.8 cubic feet per minute, which falling six feet = 11.6 h.p., of which 8 to 10 may be made effective.

Visits have been made to Waterloo and Cedar Falls for the purpose of examining the dams at those places, and to get an estimate of the value of water power there.

At Waterloo the dam is more leaky than the one here. At Cedar Falls the dams were not filled to the crest, yet showed free leakage.

Two owners at the Falls estimated the value of the water there at ten dollars per square inch, and more if free from litigation or diminution through the year.

At Waterloo the water power there is valued at twenty-five thousand dollars, and twice that could it be changed to Cedar Rapids, on account of better facilities here.

It was stated there by several millers, that the power there had been good for the season, though it had diminished somewhat lately—some said ten per cent. ; others thought perhaps a little more, but all said that twenty-five per cent. additional would make a large supply.

One thousand inches of water under eight feet head = 9,450 cubic feet per minute, or less than either of the mills are fitted for using from the east race at Cedar Rapids.

The decree requires a proportional division of the water here. Such division will shut down every mill concerned, except Cooper's, and his much of the time, for the maximum and minimum flow will be divided. The flow to-day, August 21, is but about 37,000 cubic feet, and has been less since the largest measurement was found, and at many times during the year is much less, for the water is often drawn down by the wheels in use two or three feet below the crest of dam.

Preparations will at once be commenced for division according to the decree. The race is ample in capacity to carry several times the amount of water due the mills taking water therefrom, but that of necessity will have to be closed while the bulkheads are being put in and kept so until the work is completed. Wing dam and head-gates will have to be erected on the west side, that the division of water may be made at the head of that race that the loss from leakage of the race may fall upon the proper person. The decree will be carried out with all possible expedition.

But it will take time to complete arrangements for doing it, and soon the water will be so cold that workmen will be unwilling to work in it, besides the closing of the mills without notice has discommoded farmers very much, so that the water is let into the east race this 25th day of August with the distinct understanding that both races will be closed again the 1st day of May next, and kept so until preparations for the proportional partition of the water are completed, then each owner will receive the exact quantity due and no more, until settlement is made for the excess drawn from August 25, current month, to May 1, 1891. At least such will be the course recommended by the referees, for the value of such excess is shown by the tests of

Having had charge of the steam tests, an abstract of the more important is given herewith. I have carefully examined Mr. Emerson's statements of work done, with results, and find them correct. A complete report of details of statements, also testimony taken in connection with the water powers at Cedar Falls and Waterloo, has been prepared and can be furnished if desired.   S. N. WILLIAMS.

Tests have been made with a Westinghouse compound, a Buckeye, and a common slide valve engine, for the purpose of ascertaining the cost of steam power here. These were made by keying a No. 4 power scale to the main line of driving shaft, taking the key from the driving pulley, allowing its arms to rest upon projecting parts of the scale, thus weighing the power in transmission, the scale at each revolution carrying the load nine feet. The number of revolutions per minute, multiplied by nine, that product multiplied by the weight, giving the foot pounds.

TEST OF ST. JOHN & CLAY'S MILL, AUGUST 21. WOODBURY DOUBLE SLIDE VALVE ENGINE, CYLINDER 13½ x 18.

Speed of shaft and scale 189 revolutions per minute.

Power to run shafting and machinery,	20.85 h.p.
Maximum power developed during test,	46.18 h.p.
Pounds of nut coal per horse power per hour.	11.3

11.3 pounds of coal multiplied by 24 = 12,570 pounds, at $2.25 per ton = in round numbers $14.09 for the 46.18 h.p. developed. Two engineers without fireman at $2.50 per day each = $5.00 ; added to the $14.00 = $19.00 per day for running such a mill with steam power. An engineer who fires and runs an engine twelve hours per day, and whose ability is such that his services are worth less than the price named, is a standing menace to the neighborhood.

The tests below were made at the electric light works and were made under more favorable conditions than generally prevail in manufacturing establishments.

WESTINGHOUSE COMPOUND ENGINE, 10 x 18 x 10, HEINE BOILER WITH STOKER. RATED 65 H.P. AT 100 POUNDS STEAM. TEST NO. 1, AUG. 9.

Slack coal ; pump run by separate boiler.

Pounds of coal per horse power by power scale,	11.17
Average net weight as shown by power scale,	385.
Average ampere load,	19.70
Average steam pressure,	101.6
Average speed,	314.8
Average horse power,	33.10

WESTINGHOUSE ENGINE, HEINE BOILER, RONEY STOKER. TEST NO. 3, AUG. 11.

Average steam pressure,	114.6
Average speed of shaft and scale,	312.68
Average net weight,	551.35
Average horse power,	46.16
Pounds coal per horse power, slack at $1.25 per ton,	8.24
Maximum horse power steam at 112,	56.71
Maximum ampere load,	33.5

The maximum load on Westinghouse Engine shows 10.5 16 candle-power lamps to the horse power. This is not by actual count, but is estimated from ampere load after deducting liberal amount for loss in wire and converters.

BUCKEYE ENGINE, HEINE BOILER WITH STOKER, RATED 75 H.P. AT 80 POUNDS STEAM. TEST NO. 5. AUGUST 17.

Cylinder 12 by 24, slack coal, pump run by same boiler.

Pounds coal per horse power,	9.64
Average net weight,	647.30
Average ampere load,	26.40
Average steam pressure,	108.4
Average speed,	434.
Average horse power,	74.57

BUCKEYE ENGINE, BABCOCK, WILCOX & ERIE BOILERS. TEST NO. 4. AUGUST 15.

Average steam pressure,	97.6
Average speed of shaft and scale,	433.9
Average weight, pounds,	667.
Average horse power,	78.9
Coal, pounds per horse power, air-slacked lump,	7.88
Maximum h.p. steam at 96,	92.2
Maximum ampere load,	35.
Maximum number of arc lamps,	42.

I was here over forty years since, about mid winter; the water was lower then than it was the 14th of August, the day the measurement was taken. At that time Mr. Greene had leased his lumber mill to William Harmon from Maine, and in a few days after starting it parties running the mill now owned by Messrs. St. John & Clay said Mr. Harmon was drawing the water so much that it interfered very much with their mill. Mr. Harmon said the same, or that their mill drew the water from his mill. I do not know that there was any other water used at that time. Having been present August 14, the day the measurements were made here at Cedar Rapids, I can unhesitatingly state that they were correct, as they were made in the same way as the measurements were made at Lowell and Holyoke, Mass., in their testing flumes. I have proven these correct beyond a doubt by actual experiment in my own testing flume at Independence, Iowa, which I have had over ten years. I have been present many times at the water-wheel test at Holyoke, conducted by Mr. James Emerson. I have also been familiar with the Cedar River since 1847; have been employed at Cedar Falls; also at this place for forty years since at millwright work. Have seen the river at different stages, and at the time the measurement was made here, to the best of my knowledge, state that Mr Emerson has made a fair and honest report of the quantity of water, also of the power at the time of the measurement, which I am satisfied are as near correct as can be made.

S. SHERWOOD, SR.

Mrs. Susan Brown, defendant and principal owner, is hereby directed to see that all practicable preparations are made for carrying out the decree as soon after the closing of the mills, May 1, 1891, as is possible.

CEDAR RAPIDS, IOWA, August 28, 1890.

JAMES EMERSON,
S. N. WILLIAMS,
S. SHERWOOD, JR.

For more than twenty years parties owning less than one-eighth of this power have used the whole without paying rent, at the same time keeping up a continuous complaint that the principal owner

HOLYOKE DAM.

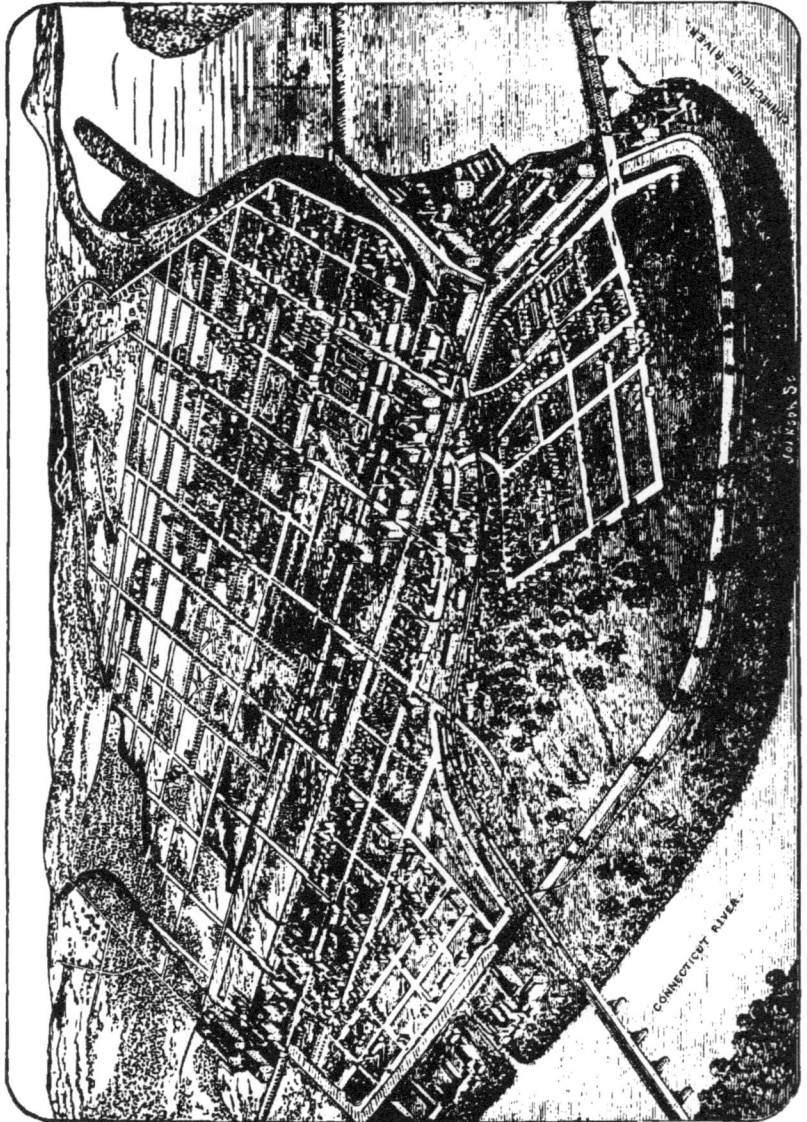

Holyoke and Its Water Power.

Some eighty miles from the mouth of the beautiful Connecticut, almost in the shadow of Mounts Tom and Holyoke, there is a fall of nearly sixty feet in a short distance that once formed what was called the "Great Rapids," near which, from time immemorial, the aborigines of the country gathered in great numbers in quest of fish and game; and until within a few years "Indian fireplaces" have dotted the banks that are now covered with mills; indeed, Indian skeletons, implements and arrow heads are often found in the vicinity at this time. Near the foot of the falls the river makes a sharp turn to the right, and in this curve is situated the city of Holyoke. In 1831 this place was a part of West Springfield, known as "Ireland Depot," with but few inhabitants, and those of but little account. In that year the Hadley Falls Co. was formed, and a small cotton mill of 4000 spindles, (known until recently as Hampden, Jr.,) was erected, receiving its power through a canal, and wing dam running obliquely up the river, which at this point is wide, with rock bottom. A power so immense and convenient to the business centers of the country was not likely to escape the notice of capitalists. The volume of water flowing in the river in ordinary seasons, was found to be about 6000 cubic feet per second, or for the fall about 30,000 h. p.; but allowing one-third diminution for the driest seasons the available minimum was rated at 20.000 h. p. In 1845, it was decided to construct a dam across the river, and one with a base of 60 and a height of 30 feet was completed Nov. 19, 1847, but before filling to the top it rolled over and went down stream; this was a severe loss, but the experience was valuable. The dam now standing was completed Oct. 22, 1849; it had a base of 80 and a height of 30 feet, the upstream incline having a face of 90 feet besides gravel filling at base. The dam is constructed of timber 12 inches square, crossed and bolted, the openings filled with stone. As the bed of the river is rock it was not supposed that the overflow would wear to any perceptible extent, but in 1868 it was found that cavities from

8 to 25 feet in depth had been cut close up to the dam, and in the years 1868, '69 and '70 the Holyoke Water Power Co. made expenditures amounting to $400,000—in the construction of an apron of heavy timber work filled with stone—to fill the space caused by the action of the water. This apron is united to the dam in the strongest manner possible, is 50 feet in width and 52 in height, its base resting 22 feet below that of the dam. Starting from the crest, which is plated with iron, the apron slopes down stream nearly to the water below. The whole structure is 130 feet wide, 30 feet high from bed of river and 1019 feet in length between abutments. There are three levels of canals, with a total fall of 56 feet. The main artery of the system, starting with a width of 140, and a water depth of 22 feet, extends eastward past the great waste weir about a thousand feet and then sweeps southward in a right line for a distance of more than a mile. The second level canal extends northerly for a mile and more, parallel with the first, and 400 feet easterly from it, and thence easterly and southerly for a mile and a quarter more, at a distance of about 400 feet from the river, this marginal portion of the second level affording mill-sites along its whole length, from which the water used passes directly into the river. The third level canal, 100 feet wide and 10 feet deep, is also a marginal canal, with mill-sites along its entire length, and extends 3,550 to the other terminus of the same canal, thus making with the latter, a line of marginal canals, around and near the whole water front of the city.

" Like other commodities which are bought and sold, water-power here has its own unit of measurement, called a mill-power, which is thus defined in the deeds of the Holyoke Water Power Company:—

" 'Each mill-power at the respective falls is declared to be the right, during sixteen hours in a day, to draw from the nearest canal or water course of the grantors, and through the land to be granted, 38 cubic feet of water per second at the upper fall, when the head there is 20 feet, or a quantity inversely proportionate to the height at the other falls.' "

Thirty-eight cubic feet per second under 20 feet head is 86.20 horse-power, 67 per cent. of which is 57.75 horse-power that may be realized. The annual rental per mill-power is 260 ounces of silver of the standard fineness of the coinage of 1859, which is in practice paid in current funds, and amounts to about $300 a year, for 16 hours per day, or $450 for 24 hours per day. The regular supply is now exhausted and only surplus is now leased.

The claims in the foregoing were copied from printed statements at a time when the local idea prevailed that the Holyoke water power was nearly inexhaustible. Interested parties have criticised them, and blamed me for their publication. My desire is to make this work useful to the real engineer of the future, and to all interested in such matters. The Connecticut river, like a small brook, rises and falls rapidly; its extremes of supply are great; the maximum of 6000 cubic feet per second is moderate, as it is often more than six times that. During the eight years in which I have had occasion to notice its flow, the sheet over the dam for a large portion of each year has varied from one to ten feet in depth—often five or six. In two or three of the eight years, the overflow has continued through the entire year; in each of the others, for a brief period in summer, the dam has been dry. The minimum I should estimate at from 2000 to 2500 cubic feet per second; at any rate, the whole of the river passed into the main canal through the twelve head gates, each 8 x 15 feet. The past summer was exceptional — phenomenal, in fact. The water in the river was lower than ever before known; the supply was insufficient for the power required; consequently, the head was continually falling while the mills were running. The dam, when filled, sets the water back several miles; the banks are irregular, so there was no way to measure the supply except to keep the head gates shut a sufficient length of time to make it certain the natural supply was flowing over the dam. This was not done, and no measurement worth the name was made. I was up and down the river many times during the lowest stage of water. In many places the river was easily fordable. There was one place, in particular, some three miles above the dam, that attracted my attention most. The deepest part could not have exceeded three feet, while much of the width was less than a foot in depth. It was deeper above, so that the velocity over the bar was moderate. A cross section of two by two hundred feet, with a velocity of three, or twelve hundred cubic feet per second, I think, would cover the flow; but suppose it to have been fifteen hundred, that quantity, falling fifty-six feet, would evolve 9520 h. p., or, accumulated and used in ten of the twenty-four hours, would produce 22,848 h. p. There were a few days in which the supply was insufficient to run the whole of the mills, even that length of time, I think.

That the water power at Holyoke may at all times equal 30,000 h. p. is merely a question of reservoirs to retain some of the abundant surplus ten months of each year, to be used in the other two.

The haste for large immediate dividends has harnessed the noble river to a business insatiate in its demand for more; the paper manufacturer that has all the power he wants, is a phenomenon. Except the hebdomadal stop, more for repairs than prayers, the draught from the pond is unceasing. The water power sufficient to give employment to a thousand hands in the manufacture of paper would be abundant for the employment of six thousand in the manufacture of cotton goods. The effect of this upon the future of Holyoke is conceivable. The idea of an inexhaustible supply of water at Holyoke caused many of the wheelpits to be made of less depth than that necessary for utilizing the whole head during the dry season, but the greatest loss occurs through the use of poor or ill-adapted turbines—turbines much too large for their capacity in ordinary times, that there may be no detention during backwater. But Holyoke is not exceptional in this, for at least one-third of all the water power of the country that is used is so wasted; and of the three great powers of this state—Holyoke, Lowell and Lawrence—it is safe to estimate the waste at a greater quantity than would be necessary at either of the falls to produce a greater power than that realized from the whole fall of the Merrimac river at either Lowell or Lawrence.

The rates of the principal Water Power companies of the country are here given for the convenience of those interested in such matters. It will be seen that a "Mill-Power" is a very indefinite matter, and it may be well here to give its origin, which is as follows: Early in the present century, there was a mill at Waltham, Mass., containing 3,584 spindles; the company owning that mill colonized Lowell, and the supposed power required at the Waltham mill, was that fixed upon as a "Mill Power" at Lowell, which is here given:

LOWELL, MASS.

Each mill-power or privilege at the respective falls is declared to be the right to draw from the nearest canal or water course of the said proprietors so much water as, during 15 hours in every day of 24 hours, shall give a power equal to 25 cubic feet per second at the great fall, when the fall there is 30 feet, or enough to give the same power at any of the other falls. The whole owned by the Companies, none to let or sell.

OFFICE OF ESSEX COMPANY, LAWRENCE, MASS., June 16, 1877.

JAMES EMERSON: DEAR SIR:—Your letter of this date is at hand. A "Mill Power" at Lawrence is defined to be the "right to draw so much water as shall give a power equal to 30 cubic feet of water per second, when the head is 25 feet," for not more than 16 hours in each day of 24 hours. For this the charge is an annual rent of $1200. and this is at the same rate for small as well as large water takers. This is 85 gross h. p. for $1200=$14.12 per h. p. of water. It might be a fair general statement to say a horse power by steam would cost 50 or $60 a year more than a horse power by water; but this would be modified by circumstances. Yours truly,
HIRAM F. MILLS, Engineer.

U. S. BUNTING COMPANY, LOWELL, MASS., Aug. 1, 1877.

MR. JAMES EMERSON: DEAR SIR:—Replying to your favor of June 16th last to D. W. C. Farrington, I have to say with regard to the subject upon which you made inquiries of him, that it is the custom of the Wamesit Power Company of Lowell to let floor room to their tenants at a stipulated sum, depending upon location, &c., &c; and then the power is hired at $75 per year, per horse power extra. When any question is raised on either side as to the power actually used, we apply a Dynamometer of your make, and measure it as near as possible.
WALTER H. McDANIELS, Supt.

OFFICE OF AMERICAN PRINT WORKS, FALL RIVER, MASS., June 18, 1877.

JAMES EMERSON: DEAR SIR:—Your favor of the 16th inst. is at hand. In reply would say, that the water-power in Fall River is not let, but the stock in the Watuppa Reservoir Co., which controls the water-power, is held by the several corporations using the power in proportion to the height of their respective falls, and no charge is made for use; but the expense of maintaining the Reservoir Company is borne by assessments upon the several Corporations, from time to time, pro rata, according to height of fall of each. There is but a single outlet from the Reservoir. The total fall from Reservoir to tide water is 128 feet; and the mills are located one below another, so that they each get precisely the same quantity of water, as each mill takes just what the one above it delivers. The quantity is about 122 cubic feet per second. I am just now unable to give an answer that would be satisfactory to myself as to the comparative cost of water and steam-power.
Yours truly,
THOS. J. BORDEN, Treas.,
Watuppa Reservoir Co.

OFFICE MINNEAPOLIS MILL CO.,
MINNEAPOLIS, MINN., July 5, 1877.

JAMES EMERSON: DEAR SIR:—With reference to renting power, I would say that rentals are made at so much for Mill Power, which is designated as 30 cubic feet of water per second, with head of 22 feet. Present price for Mill Power $1000 per year, but from this back to earlier dates rates decrease considerably.

Yours truly,

H. H. DOUGLASS, Eng. and Agt. M. M. Co.

BELLOWS FALLS, VT., June 28, 1877.

JAMES EMERSON: DEAR SIR:—Yours of the 22d is just received; a Mill Power, in our lease, is the right to draw equal to 30 cubic feet per second, under 25 feet head. Price for a Mill Power is 387 ounces Troy Weight of silver, of the present standard business of the silver coin of the United States, as an equivalent in gold, which is 450 dollars.

Yours truly,

ROBERTSON, MOORE & CO.

MANCHESTER, N. H., June 29, 1877.

JAMES EMERSON: DEAR SIR:—The rule for a Mill Power here is as follows: Divide 725 by the number of feet fall minus 1, and the quotient will be the number of cubic feet per second for a Mill Power on that fall. For instance: The fall at the upper level is 20 feet; then 725 divided by 19=38.1, which is number of cubic feet per second for that fall. The Mill Powers are let to manufacturing concerns at an annual rent of $800 each. This includes the land necessary to use the power on, together with some room for tenement blocks, but no buildings or machinery.

Yours truly,

JOSEPH B. SAWYER, C. E.

THE OSWEGO CANAL COMPANY, OSWEGO, N. Y.

The Lessees at their joint option may be allowed for each run of classified water, either; 1st.—One thousand cubic feet per minute; or 2d.—So much as will be drawn through a central discharge water-wheel of the kind now used on the Canal, with a spout, the cross section of which shall be 183½ square inches at the smallest point, provided the outlet does not exceed in diameter one-half the diameter of the wheel, nor in clear opening a surface, 3½ times the section of the spout; or 3d.—So much as will be drawn through a Reynolds wheel with a spout, the cross section of which measures 166⅘ square inches at the smallest point, provided the total outlet does not exceed the section of the spout more than 50 per cent., and the superficial center of the outlet is not over 2-3 the radius of the wheel from the center thereof. The second and third alternatives are estimated to give the same quantity of water, and equal to about 1175 cubic feet per minute, when the clear head on the wheel is 16 feet.

DAYTON, OHIO, July 12, 1877.

JAMES EMERSON: DEAR SIR:—Water-power is supplied to the mills on the three levels or falls through *metallic gauges*; calculated and adapted to pass under a certain head so many cubic feet per minute. For example, in the Dayton Hydraulic Company we give a head of 15 inches above the center of the gauge, and with that head give 233½ cubic feet per minute for *one power*. The Company below us, under I believe the same head, over a 10½ feet overshot, give 300 cubic feet per minute for one power. The price per power (or "run of stone," as it was originally called,) is, I believe, uniformly here $200 per year. You wish, also the relative cost between steam and water-power. Water-power at $200 per year for one run of 5½ horse-power, would cost eleven sixty-eight one hundredths dollars per day for 100 horse power. Steam, with the latest improved steam-engine, as tested by experts, will give one horse power with 3 lbs. coal per hour; coal at $3 per ton would be ten eighty one hundredths dollars for 100 horse power per day of 24 hours. An engine of this kind, with boilers, would cost about $5000. Water-wheel, with the same power, "under an ordinary fall 12 or 13 feet," with penstock and flume would cost about $2000. The foundation for steam-engine would cost about the same as it would to dig a wheel pit. With an ordinary slide valve engine, such as we use, costs about $27 per day "24 hours" for 100 horse power. If I can give you further information, will cheerfully do so.

Very respectfully,

JOHN G. LOWE, Sec'y & Sup't D. H. Co.

OUSATON'C WATER COMPANY,
BIRMINGHAM, CONN., July 16, 1877.

JAMES EMERSON: DEAR SIR:—Yours of the 14th is at hand. Our terms for the rent of water, per year, are $250 per square foot, 12 hours per day,—one square foot being a discharge of five cubic feet per second. We use the weir measurement adopting J. B. Francis' formula for the computations. What we designate as a square foot of water under our head is equivalent to 12.5 horsepower, in short $20 per horse-power per year is about the cost of water here. With reference to lot and buildings, the Company offer inducements in proportion to the desirability of the business to be located.

Respectfully yours,

D. S. BRINSMADE, Secretary.

WINDSOR LOCKS, CONN.

Usual head 24 to 28 feet. Water rented so much per inch, yearly, price varying with date of lease; extra water now charged at the rate of from two to two and a half dollars per inch; quantity determined by apertures through iron plate; apertures, parallelograms with parellel interior edges, center of aperture to be 2½ feet below the surface.

UNIONVILLE, CONN.

Water rented as follows: The one hiring to be entitled to such quantity as can be drawn through an opening one foot square, the center of the opening to be under two feet head; I think the power is now owned by the several companies, and that there is none to rent.

COHOES, N. Y., July 14, 1877.

JAMES EMERSON: DEAR SIR:—Your favor of 10th inst. to hand. I understand the charges of the Cohoes Company to be $200.00 per Mill Power per year, or $20.00 per horse power which includes use of water and rent of land. The leases define the term "Mill Power," as "a Water-Power equivalent to the power given by the discharge and use of six cubic feet of water in each second, when the fall is 20 feet."

Yours truly,

WM. T. HORROBIN.

OFFICE OF THE DUNDEE WATER-POWER & LAND CO.,
87 Leonard Street, NEW YORK, July 12, 1877.

JAMES EMERSON: DEAR SIR:—Your letter of 10th inst., received. The Company leases its Mill Sites with one or more "Mill Powers." charging $700 *per year* for *each* Mill Power. This price includes the rental of Mill Site. By one Mill Power is conveyed the right to draw from the nearest race-way or canal 8½ cubic feet of water per second, fall of 22 feet.

Respectfully, &c.,

M. WALKER, Secretary.

TURNER'S FALLS, MASS.

Rent per year for each h. p. of water used $7.50 or about $10 for each h. p. that may be utilized by the use of good water wheels.

2

DISPUTE ABOUT THE QUANTITY OF WATER USED.

In a case at Jordan, N. Y., there was a dispute about the discharge of a wheel. The lease granted the right to use what water could be drawn through an opening 10x17 inches into a scroll wheel. The discharge of the wheel was measured over a weir; the builder objected on the excuse that he knew nothing about such matters; to get over that difficulty a mark was made to indicate the depth from the discharge of the wheel; then its gate was closed and an opening 10x17 inches cut in the bottom of the penstock near the wheel; the water was then let in and the discharge through the opening compared, and was found to be considerable less than that of wheel. Of course there was no chance to dispute that point.

DISPUTE ABOUT WORK DONE.

In a recent case at North Sunderland, Mass., that had been in dispute four years, and quite a sum had been expended in litigation, my services were required in court, where the *expert* testimony was so scientific that it was beyond my comprehension. A proposal was made and adopted, that the court adjourn to meet at the mill, where the case could be settled so that all could understand. The case was as follows: A turbine had been put into the mill, under the agreement that with 15 inches of water, under 62 feet head, it should grind 35 bushels of corn per hour. Arriving at the mill, a weir was constructed below the wheel, the gate was then raised until 15 inches or 394.6 cubic feet of water per minute flowed over the weir; with that quantity the wheel ground 61 and a fraction bushels per hour. The suit ended there, and the owner seemed pleased that he owned a more valuable power than he had thought. Other tests were made, from which it appeared that 2.07 bushels were ground per each horse-power utilized. The buhr was five feet in diameter, and kept down to 145 revolutions per minute.

DISPUTE AS TO WHICH USED THE MOST WATER.

To settle a case at Auburn, N. Y., where a fine power is owned by L. W. Nye and the Auburn Mf'g Co., weirs were put in above their mills, the wheel gates opened in full, then a thousand cubic feet per minute was allowed to flow into each tail race through flume and wheel; permanent marks were made on iron scales, firmly secured to the wall of each tail race, then marks were added for 1500, 2000, 2500, 3000, 3500, 4000, 4500, 5000, and finally 5236, as the maximum the Manufacturing Company's wheel could discharge. The discharge of Mr. Nye's wheel had stopped at 3906 cubic feet per minute. The scales in the tail races remain and denote at any time the quantity of water used by each party. The weirs above the mills were removed as soon as the scales were marked. In well constructed tail races the quantity used may be very accurately denoted, though, of course, the plan will not answer where the water from different mills is discharged into the same pit, or where there is backwater.

Backwater Suits.

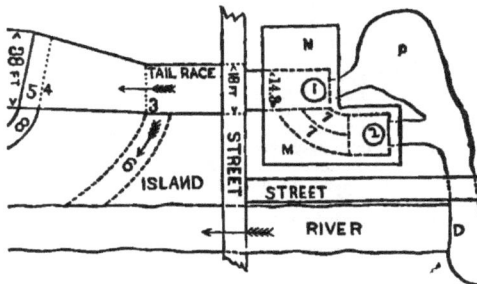

There are mill-owners in all parts of the country, who believe themselves injured by backwater from dams below; to such, the case of L. L. Brown & Co. vs. H. N. Dean & Son of South Adams, Mass., will be of interest. Where Brown's paper mill, M. M., now stands, was started 60 years since a saw and grist mill;

as may be seen in sketch; this is near the head of one of the branches of the river which forms the island. Dean's tannery was afterwards located on the race 6; the dotted line 3 represents the dam therefor; the crest of this dam was about level with the bottom of the saw mill wheel pit; flush boards were used to raise the dam still higher, as the bed of the stream above was then so high as to prevent flowage back into saw-mill pit as claimed by Mr. Dean. There seems to have been a dispute about the right to use the flush boards, though it was conceded that they were almost constantly in use, though at times removed when power was not required at the tannery. Afterwards Mr. Dean purchased an old fulling mill privilege, and moved his tannery further down stream; erecting a new dam which is marked 5, the water being conveyed to tannery through the race 8. The dotted line 4 represents the fulling mill dam but little of which remains, though there is sufficient to show that it was at least 5½ inches higher than the new dam, the crest of which is six or seven inches higher than the floor of wheel pit in what is now Brown's paper mill; the stream over the new dam is 38½ feet in width; at the old tannery dam it is considerably narrower. The crest of the old dam is removed, still the foundation is but a little lower than the crest of the new dam. Two 48-inch Swain turbines, 1 and 2, have been placed in Brown's mill to drive the machinery; these take the water from the pond p, through the sluices shown; the discharge from the upper one passes down through arched races 7, 7, and is discharged below the lower turbine into the main race, which is here but a little, if any, over 14 feet in width; this race has rough stone side walls. These wheels unitedly discharge from 125 to 138 cubic feet of water per second; and the depth in race is 25 inches where the width is 14 feet; 23½ inches where the width is 18 feet, and still less as the width increases; as it flows over the new dam it is but nine inches in depth; the velocity is much greater below the old tannery dam than above. Mr. Brown claims that the new dam backs the water on to his wheels; to prove this, witnesses testify that until the new dam was constructed there never was any water in his wheel pit when his gates were closed, but now there always is. It was proved by Brown's witness that in race 6 Dean had 5½ to 6 ft. head, and he now has but 5, while he discharges into the river much lower down. Mr. Brown denies that the race has ever been lowered, but the bottom is now composed of small pebbles and gravel, while for miles, above or below the mill, the bed of the river is literally paved with stones rounded by attrition, varying in size from two inches to as many feet. With a discharge from good wheels of 138 cubic feet per second, the depth over a 14 foot weir would be 25 inches, so that it is plain that Dean's dam is not the cause of the depth in the tail race of Brown's mill. That there was no water in the saw mill pit while Dean's mill was at race 6 is readily accounted for from the fact that that race drained the saw mill pit while it was open, but that race was filled up when the new dam was constructed. The water in the pit since, when wheel gates are shut, is simply *standing*, not backwater. Though denying that the race had been lowered it was not denied by Mr. Brown that the boulders had been cleared out of the race, and of course it would have been useless to remove these boulders unless they had obstructed the discharge from the mill above. From the character and tone of Brown's witnesses it was evident that they were sincere in their statements; but nature furnishes better evidence that the tail race had been lowered, also, that if Dean had a right of 5½ to 6 feet head at race 6 he could not possibly encroach upon the privilege above, with a 5 foot dam at race 8. I was not called into the case until the day before the trial commenced, and had no knowledge of the place before, so that I was unable to account for the water standing in Brown's pit after Dean's new dam was constructed, until it happened to be mentioned that the race 6 was filled up as soon as the tannery was moved to the new dam; then the cause became plain, but it was too late to explain, and the fact is only mentioned that lawyers engaged in such cases may understand that STANDING water in a wheel pit is beneficial instead of injurious. As the wheels in Brown's mill discharge double the water used by Mr. Dean, it would have been much less expensive to have furnished Mr. Dean with a larger wheel so that he could have obtained more power even with less head.

Vexatious Waste of Water.

One of the most vexatious grievances suffered by manufacturers arises through the following circumstances. Suppose a dozen mills to be located within a short distance upon the same fall, one above the other; eleven of them have wheels with which the natural flow of the stream is amply sufficient to keep their machinery in constant motion; but the upper mill of the dozen has wheels of the poorest kind, so that they require double the water necessary to do the work of mill, and the owner, through mulish perverseness continues their use, each day exhausting his pond by noon, then as half of the water has flowed over the dams below, all of the mills have to stand idle the rest of the day. Of the equity in such a case there can be but one opinion; no engineering skill can aid, and only the strong arm of the law can remedy the matter. Such cases are very common.

"Efficiency, Useful Effect, or Percentage."

Are terms used to denote the economy of a wheel in its use of water, or the number of gallons it will pump back into the pond for each one hundred gallons drawn therefrom to drive the wheel There are wheels that for each hundred gallons used will return but twenty-five, others will return fifty, while medium wheels return seventy-five, a better class eighty to eighty-five; the very highest, under favorable circumstances will return something over ninety per cent., and of course, other merits being equal, are by far the most desirable.

What is the Real Working Head?

The term "Head" as used in connection with water-power means the difference in height from the surface of water in wheel pit to the surface in the penstock above, when the wheel is running.

What is a Square Inch of Water?

A square inch of water means a stream exactly an inch square, its length depending upon the head from which it issues; for a head of four feet, it means a stream an inch square, 16.04 feet in length, per second; for a head of a hundred feet, a stream an inch square, 80.35 feet in length, per second. To turn this into cubic feet, multiply by 12, then divide by 1728.

Pressure of Water on Dams and Boilers.

The pressure depends upon the length of dam and depth of water. It makes no difference whether the pond extends back a rod or a mile. So of steam boilers—the large boiler requires thicker iron, simply because there are more square inches of surface.

What Power is Required to Drive a Run of Stones.

A more difficult question to answer, because the quantity ground in a given time has much to do with it; experienced millers west do not use more than fifteen horse-power per run. including receiving grain, flouring and delivery in barrels. White & Beynon. Lanesboro, Mi n. have six run of stones; have 89 horse-power of water, about 72 horse-power actual; keep five run at work, the sixth being stopped for sharpening. White, Nash & Co. of the same place have the same power, five run of stones. four kept constantly running; use their wheel at part gate. It will be seen by examination of the Dayton, Ohio, water renting rate that 5¼ h. p. has been considered sufficient for a run of stones, while with the 1000 cubic feet allowed at Oswego, N. Y., used on a Reynolds wheel would not realize more than 15 h. p., so that 15 h. p. for each run of stone and necessary machinery is a liberal allowance.

Loss of Head through the Use of Small Conduit.

A belief prevails among turbine builders that where the water approaches a wheel with perceptible velocity that there is a corresponding loss of head so that the wheel can not transmit the power due the head. Such is not my belief, for there seems to be no good reason for ignoring the momentum gained by such velocity, that is within reasonable limits. The woolen mill of Beebe, Webber & Co., of Holyoke, is located below the second level. Head varies from eleven to twelve feet. Originally the use of only five sets of machinery was contemplated. The water is brought to the mill through a round wooden trunk 75 feet in length, with an inside diameter of 57 inches. The wheel pit is circular, 14 feet in diameter, and 2½ feet in depth. A five foot Tyler Scroll wheel had been used fourteen years, but was unable to transmit sufficient power to drive the eight sets of machinery now in the mill, and it was found to be necessary to obtain more power, but the small size of trunk and shallowness of pit caused wheel builders to hesitate, through fear that the loss from head would more than equal any gain that could be obtained through increase in size of wheel. I recommended the use of a 60-inch turbine, and the builders, Messrs. Fales, Jenks & Sons were induced to guarantee eighty-five horse-power under the existing conditions. The wheel was set and my brake applied. Before the gate was opened the difference between the level of the two canals was found to be 11 feet, 8 inches. For that head I calculated that 4000 pounds should balance the force of the discharge with the gate open and the wheel held stationary by the brake, and on opening the gate that weight which had been put on the beam was found to exactly balance, though the head on the wheel was less than ten feet. Under ordinary conditions the wheel used will carry at its best speed exactly half what it will balance when held stationary by the brake; but the velocity of the water seemed to change its character somewhat, for it gave its highest result, 88.66 h. p., carrying 1900 pounds at 77 revolutions per minute; its tabled speed was several revolutions less; at that speed the head as shown by a glass tube inserted in penstock directly over the wheel was found to be 10½ feet.

Turbine Builders' Theories.

It is an old theory in turbine building that turbines should carry about half what they can lift when held stationary; with gate opened in full, the Houston wheel almost invariably does so, and there are a few others that approach that rate, while there are many that do not. Many of the Risdon wheels run with three-fourths of what they can lift. Some wheels will run with, say, nine hundred pounds, and only lift one thousand. A few days since a wheel was brought to be tested; it was set and tried first while held by brake; gate opened in full, it balanced 470 pounds, discharging 928 cubic feet per minute. It was started with 300 pounds making 178 revolutions per minute, and discharging 1241 cubic feet of water; weight was gradually added, the speed decreasing with each addition, while the discharge increased. Discharging 1289 cubic feet, it made 124 revolutions per minute and carried 475 pounds. It was stopped by brake, then of course could not start until partially unloaded. It will be obvious to all that the more surplus lifting power a turbine has the steadier it will run under sudden changes produced by adding or throwing off machinery; the wheel was a central discharge. Builders starting with such are behind the age.

A Proposition of Seeming Equity that has no Merit.

A common proposition, and to those unacquainted with the subject a seemingly fair one, is that two turbines shall be connected together and their merits determined by ascertaining which shall drive the other. Such a test would be perfectly worthless. The pitch of the buckets of one might be such that it would under the head tried carry 100 pounds, and make 200 revolutions per minute, while those of the other might be such that it would carry 200 pounds and make 100 revolutions per minute, both using the same quantity of water during the trial. Of course the slow wheel would drive the fast one, but other things being equal the fast wheel would be the best.

Backwater.

Turbines of any make are not perceptibly effected by backwater except through loss of head. I think a slight difference was found by a commission appointed by the French government to experiment with the Fourneyron wheels. I have in two or three cases where long draft-tubes were used, thought the loss greater than should occur from the loss of head, but have had no chance to determine the matter by actual test.

Submerging Turbines.

Many builders insist that it is essential that a turbine should discharge under water, but it is doubtful for the same head whether it makes any difference if the wheel is properly made, though it prevents trouble from ice and generally extra head is gained by submerging lower part of wheel.

Draft Tubes.

If a draft-tube for any considerable proportion of the head is used, its lower end should be submerged to such depth as to render its immersion constant, otherwise when first starting up only the head above the wheel will be available until the discharge has exhausted the air from the tube, then when it does take hold, unless the gate of the wheel works very quick the speed is wild for a short time. Where there is backwater some length of time, a short draft-tube renders it convenient to get at the wheel in case it is necessary to do so, but in most cases I should prefer to have the lower part of a turbine stand in the tail water.

Percentage of Discharge.

The discharge of a turbine in proportion to its openings depends upon its construction. With those of a central discharge it is the least; with such wheels of fair efficiency it is likely to range between 40 and 50 per cent., with outward discharge. 60 per cent. and upwards, while with those discharging the water downwards it averages about 55 per cent. The chutes of a curb are made much larger at their outer than their inner ends, consequently, can pass much more water than the wheel will discharge, though the openings of the wheel may be somewhat the largest, so that the openings of the wheel govern the discharge. In the past, engineers have expended more time inventing impossibilities and hair splitting theories than in determining by simple tests points in dispute easy of solution. It is hardly possible that a case can ever arise in milling matters that a really intelligent engineer cannot readily solve the difficulty, and make it so simple and plain as to give no excuse for litigation, and what is more to the point, in many cases both parties can be benefited at a tithe of the expense caused by a suit at law. If there is a difference of opinion about power used, the matter may readily be determined, as may be the case if the dispute is about the quantity of water used; and the power of steam is as readily determined as that of falling water. A few plans tried by myself are here given:

"DISPUTE ABOUT EFFICIENCY OF TURBINE."

Thomas Harris, of Providence, R. I. expended something like $9,800 experimenting with four Leffel wheels in a mill at Putnam, Ct., head of 28 feet. A 40-inch wheel was tried first, then a deeper wheel, same size, then a 48-inch wheel, then a second 48-inch of extra depth; the speed of looms could not be got above 126 picks per minute. I was called in to test the power and select a suitable wheel. By stopping eleven spinning frames the rest of the machinery was brought up to speed. The wheel was then tested and found to give 186 h. p. Allowing 17 h p. for the eleven spinning frames, and 20 additional for cold mornings and backwater. I selected a wheel of 220 h. p. Since that wheel was placed in the mill, the production has been increased 1000 yards per day. 40-inch sheeting, while the discharge of water has been one-fifth less than required for the Leffel wheel. The expense of changing, my charge included, was $1,500.

Highest Possible Results Guaranteed.

For years past turbine builders of a certain class have unhesitatingly promised what they well knew at the time their wheels could not do. The practice has been so general that even in court it has been offered and rather accepted as an excuse, that though the wheel only accomplished one-half what was promised, the guarantee was no more extravagant than the average turbine builder would give, simply because there was no means within the reach of ordinary builders for determining such matters. The case is very different now and purchasers are less inclined to submit or juries to excuse, and builders will do well to take heed accordingly. It has been my lot within two years to be employed as expert in four different cases in which the same builder has been interested.

"Chipping Buckets."

Has been mentioned frequently in these reports; the plan has been tried with many kinds, not always successfully; it does not have much effect on the [Risdon wheel, the reversed curves of the buckets of that wheel seeming to answer the same purpose. Chipping away the edge of buckets reduces diameter of wheel above the bottom of chutes, so that its speed is usually increased thereby. (See Tyler's tests.) While increasing whole gate results it usually injures the wheel at part gate. It would seem that where the edges of the buckets extend close to end of chutes that they act like a fan or rotary pump and draw the water into the wheel. Chipping the buckets away often reduces the discharge. Increasing gate opening does not increase discharge beyond a certain limit, though it may have good effect by changing direction of water through the chutes.

Tight Gates, or Good Part Gates.

Probably a hundred objections have been made to wheels with leaky gates where one has been made to those only reasonably efficient while working with gate opened in full, which can never be the case if a governor is necessary. The most leaky, fly-trap gate in use can not waste more than four or five per cent., while the Boyden, Houston, Collins. Hunt, Geyeline and many other wheels of the same nature waste from 25 to 50 per cent. daily, if run from one-third to three-fourths gate as wheels are often used.

Variation of Turbines.

One of the most difficult matters in relation to turbines, is to make purchasers realize the fact that wheels made from the same patterns vary exceedingly in useful effect; yet it has been well understood for twenty years past that a turbine doing well in a mill affords no guarantee that another of the same make will give equal satisfaction in another mill; hence the uncertainty that has prevailed for years past. My report of tests will show this to be the case with wheels of all makes. But a few special cases are given here: The Tyler wheel first; a 30-inch flume wheel tested April 20, 1876.

Remarks.	No. of Test.	Head.	W'ht.	Rev.	H.P.	Weir.	Cubic feet.	Per Cent.
Leakage, 70.77 Cubic feet,	April 20, 1876. Whole Gate.	18.43	375	168.5	28.72	765	1245.64	.6618

The buckets were cut back to first white line shown on diagram of wheel, (see next page), then it was tested again.

| Leakage, 59.32 Cubic feet. | April 21 | 18.65 | 375 | 202 | 34.43 | 753 | 1226.55 | .7970 |

The buckets were then chipped back to the second line; the gate, an inside register, had six openings 2¼ x 12 inches; these openings were increased to three inches in width, to twelve inches in height; then the wheel was tested a third time, April 22.

	Head	Weight	Rev.	Horse Power	Weir	Discharge	Per Cent.
Whole Gate.	18.50	640	000	000	.786	1302.57	.0000
Leakage, 67.83 cubic feet.	18.50	375	219	37.32	.742	1190.25	.8966
Length of Weir 10 feet. Temp.	18.50	385	215.3	37.67	.738	1180.19	.9127
of water, 45 Fah. Weight of	18.48	395	209	37.52	.745	1197.81	.8966
water per cubic foot, 62.378.	18.49	400	205	37.27	.745	1197.81	.8904
Circumference of Circle 15 ft.,	18.50	390	211	37.40	.743	1192.77	.8966
application of two pounds at	18.50	380	215	37.13	.740	1185.22	.8958
the periphery rotated wheel.	18.50	370	220	37.00	.738	1180.19	.8966
Part Gate	18.60	325	.215	31.76	.680	1037.38	.8709
" "	18.62	300	209	28.50	.650	965.60	.8386
" "	18.67	275	212 5	26.56	.628	914.01	.8234
" "	18.76	220	213.5	21.35	.562	764 49	.7880
" "	18.85	160	215	15.63	.486	602.53	.7280
" "	58.93	105	213.5	10.18	.417	465.71	.6109
" "	19.01	60	197	5.37	.345	334.25	.4471

The tests of the 22d were too regular to allow of doubt as to their accuracy; they were not made in haste; the wheel was stopped after the third test, result worked out and the matter considered.

The wheel was returned to shop and refinished; the edges of the buckets being smoothed up, holes were drilled in the heavy side of wheel and plugged with wood to balance it, then it was sent to Centennial, afterwards returned to me for re-test. The moment its gate was opened after it was set for test, it was evident it had been changed; it was so sensitive that it was almost impossible to control it with the brake. It could not be made to work easy, though tried in various ways. The data and results below are the best obtained:

Head.	Weight.	Revolutions.	Horse Power.	Weir.	Discharge.	Per Cent.
18.38	375	221	37.67	.794	1318.42	.8242

The leakage into pit from flume was 72.73 cubic feet per minute; adding ten pounds to the weight to make up for the difference required to rotate the wheel, would have increased the power to 38.67, and percentage to 84.62. After the trial the step was found to be canted over; the wheel was taken to machine shop and changed three times after the first trial, making four trials in all, varying but slightly from the first test. The last time it was taken to the shop the lower rim was reduced by a chip 1-32 of an inch all around it, causing an increased discharge. The data and results of best tests of the trial, before and after reducing rim, are given below to show the accuracy of weir measurements compared with theoretical discharge due the increased area of opening. Results of test before the rim was turned off, then after it was reduced:

	Head.	Weight.	Rev.	Horse Power.	Discharge.	Per Cent.
Before	18.40	375	218	37.16	1328.45	.8061
After	18.39	385	214.4	37.52	1353.80	.8010

Actual increase as per weir measurement, ... 25.36 cubic feet
Theoretical discharge due the increased opening, . . . 25.31 cubic feet

Test of a 43-inch Risdon wheel, April 28, 1874. Same brake used as for testing the Tyler. Correction for leakage into pit 77.74 cubic feet. Weight of water per cubic foot, 62.38. Length of weir, 10 feet. Temperature of water, 40 Fah.

	Head.	Weight.	Rev.	Horse Power.	Weir.	Cubic feet.	Per Cent.
Whole Gate,	17.91	1200	151	82.36	1.256	2664.03	.9132
" "	17 93	1200	148	80.72	1.260	2676.91	.8877
" "	17.92	1200	148.3	80.89	1.261	2680.14	.8910
" "	17.90	1250	144.5	82.10	1.264	2689.82	.9021
" "	17.98	1150	146.5	76.58	1.195	2469.92	.9121
" "	18 00	1200	137.5	75.00	1.203	2495.13	.8834
" "	18.17	1000	147	66.82	1.127	2258.84	.8613
" "	18.29	850	150	57.95	1.045	2012.02	.8331
" "	18.30	700	138.6	44.10	.932	1686.47	.7559
" "	18.43	650	148	43 72	.932	1686.47	.7439

The report of the foregoing test caused Otto Troost of Winona, Minn., to order one like it. The order was to get one as good, let the cost be what it would. Mr. Risdon built one from the same patterns and sent it to me to be tested. Eight pounds rotated the wheel. The results are given below:

Whole Gate, July 8,	17.83	1200	142.5	77.73	1.290	2795.31	.8264
	17.82	1220	138.5	76.80	1.291	2798.25	.8159
	17.82	1240	136	76.65	1.290	2795.31	.8153
Leakage 56.57 cu.ft.	17.80	1180	143.5	76.95	1.286	2782.27	.8157
Wgt.of water 62.285	17.79	1160	145.5	76.72	1.284	2775.77	.8231
	17.79	1140	147.5	76.43	1.282	2769.27	.8220
Part Gate,	17.84	1100	142.5	71.25	1.230	2601 93	.8136
" "	17.84	1125	139.7	71.50	1.232	2608.31	.8140
" "	17.92	940	148	63.23	1.160	2381.74	.7886
" "	17.92	960	146.5	63.92	1.162	2387.94	.7915
" "	17.92	980	143.5	63.95	1.164	2394.16	.7997
" "	17.91	1000	142	64.54	1.166	2400.38	.7954

Taken to machine shop, then re-tested July 9th; required 11 pounds to rotate wheel.

18.00	141.5	77.18	1.289	2782.45	.8168

Again taken to shop, then re-tested July 13.

17.97	146	76.32	1.286	2772.69	.8119

Taken to machine shop a third time, re-tested July 16.

17.97	152.5	76.24	1.284	2766.19	.8131

I will here explain about slight changes mentioned in report of Risdon's tests. First, a 25-inch Risdon wheel was tested. June 16, 1874; it gave 75 per cent. useful effect. Mr. Risdon had it taken to shop and the rim of wheel reduced the lightest chip possible; the wheel was re-tested the next day and gave .8704 per cent. A second 25-inch was tested July 20.

Head.	Weight.	Rev. per M.	Horse Power	Cubic feet.	Per Cent.
18.41	320	232	22.49	845.93.	.7655

The wheel was taken to the shop and the bridge tree lowered one and a half inches; then re-tested July 21.

Head.	Weight.	Rev. per M.	Horse Power	Cubic feet.	Per Cent.
18.61	320	256	24.82	823.06	.8593

A 54-inch Risdon was tested, July 12, 1876.

Head.	Weight.	Rev. per M.	Horse Power	Cubic feet.	Per Cent.
17.06	1900	107	123.21	5047.72	.7586

When the wheel was put together the chutes projected too far inward, and the inner ends were cut off leaving them square across, and about half an inch in thickness; after the test the wheel was taken out and the back side of the inner end of chutes were chipped away, leaving the ends a "quarter round;" this added ten square inches to the openings; the wheel was re-tested Aug. 1.

Head.	Weight.	Rev. per M.	Horse Power	Cubic feet.	Per Cent.
8.66	1400	78.5	49.95	3742.54	.8177

Similar variations will be found in testing any make of wheels. When the system of testing commenced some ten years since, there was hardly a wheel tried that was in a condition to run until various alterations had been made; the step was out of place, or the followers were made of seasoned wood and would swell and bind the wheel as soon as wet. Few balanced their wheels, and it really needed a machine shop to put wheels in order before they could be tested; days, sometimes weeks were required to test a wheel. Builders do better now, still many wheels are yet sent to me that are in no condition to be tried in a testing flume or mill. The test of an Eclipse wheel is given on next page to show the effect of tight followers and swollen step; these were loosened before second trial:

THE POWER REQUIRED TO GRIND WHEAT, CORN, Etc.

When the testing system commenced it was supposed that there was great waste of power, as the opinion prevailed that a bushel of wheat per hour could be ground for each horse power expended.

Much pains was taken to obtain data to determine that point, and several imperfect trials were reported in the second and third editions of this work.

The first, a trial at North Sunderland, Mass., a half insane miller doing what he called the grinding, which simply consisted of cracking the corn, leaving it about the same as what is called *hominy.* The same was the case in the turbine against breast wheel test, near Hartford, Conn. "What power is required to drive a run of stones," is another case often referred to. The value of the TURBINE AGAINST BREAST WHEEL tests consists in showing the comparative merit of the breast wheel and turbine, and the difference in power required for cracking corn or grinding it, and to show that where everything is in perfect condition for a spurt, a bushel of rye or wheat perhaps may be ground per hour per horse power, as a horse may trot a mile in two and one-fourth minutes, but the same horse would be killed in a week if compelled to do ten miles per hour ten hours per day. So I will refer those desirous of information about grinding to the report of the Elkhart, Indiana, tests. These were made by the most exhaustive method possible.

No pains or expense was spared in preparation or testing, then the results were carefully worked up while every point was fresh in mind, and it was my urgently expressed wish that the opposing party, upon the completion of the making up of the report of results, should immediately be furnished freely with copies thereof; other counsel prevailed and the reports were held for two years, then brought into court. In that time I had patented many inventions, acted as expert in numerous cases of hydrodynamics, and the Elkhart case had passed from my mind as though it had never been there. The printer in making up the report seemed to have pied the form containing the table of results of grinding, after I had read the proof, so there were numerous errors, but that was a matter of little consequence, for the summing up for the report was done from the results of tests before the table was made, and I here state that those results are as perfect as I believe it is possible to make such.

While in charge of a testing flume, it was my invariable rule to refuse to make any test that could not be openly witnessed and publicly reported; the Elkhart tests were the first and will be the last made by me that cannot be reported as soon as made.

It was a practice for wheel builders to have wheels tested, hold them awhile, repaint, then send them as new wheels to be retested. Messrs. Stout, Mills & Temple of Dayton, Ohio, Messrs. Fales & Jenks of Pawtucket, R. I., and John Tyler of Claremont, N. H., did so, quite likely others. I presume that any party who did so will give their experience upon application.

A retest of the same wheel, unless some change has been made therein, should not vary over one-fourth of one per cent.

Water Wheel Royalties.

Upon what system of reasoning does the turbine patentee claim royalty upon plans of no certain value? To render a patent valid the inventor must have plans so well defined that he can describe them so that those skilled in the art may readily build from such plans or description thereof; if he can not do so, what right has he in the invention? There has been too much sympathy for the "poor inventor" and not enough generally for those who find means to carry out such inventions; an arrangement generally that the inventor shall find experience or plans, the capitalist money for the inventor to live upon and experiment with. For months, perhaps years, the inventor slashes away with little consideration for anything but his own fancies, and if a true inventor, enjoying much in witnessing the development of his ideas; while the capitalist too often finds that he has changed positions, that in fact he has the experience, while his money has turned to moonshine, or something as unsubstantial. Capitalists do not invest in 70 per cent. turbines, and there are no good reasons for expecting royalties for such.

Numerous Sizes of Turbines.

In looking over the piles of circulars issued by the hosts of turbine builders one is surprised at the numerous sizes tabled by each; and when it is understood that these tables represent both right and left hand wheels, the question arises as to how any man could ever expect to do a profitable business where so many expensive patterns are required, unless such wheels can be sold at an immense profit. A list so numerous acknowledges the fact, that such wheels can only be used economically when exactly adapted to a fixed quantity of water; in short, that they are extravagant in its use unless working with gates completely opened; this has been the case since the first introduction of the turbine, and in some cases may now be done more through habit than necessity; but if necessary, then it is plain that such wheels can not economize the power of our variable streams; either there must be a waste of one-half of the power during eight or nine months of the year, or a total stoppage through the dry season. Then, again, what earthly use is there for "right and left hand" wheels of the same size? By turning the teeth of the crown gear up or down, the shafting is rotated in the direction desired. With thirty-two sizes of turbines to work up to a desirable percentage, farewell hope! Manufacturers and turbine builders must consider and work together, if wheels of high useful effect are invariably to be expected. Numerous sizes add much to cost and the purchaser has to pay for it. If left hand wheels were impossible they would soon be found unnecessary, for preparations can easily be made to meet the case. Seven or eight sizes only, would allow the builder to work them up right, and the purchaser would soon be able to procure a turbine that would utilize the whole power of his stream, either summer or winter.

Hard Running Wheels.

For several years after the testing system began there was hardly a builder who took any particular pains to have his wheels run easy. "Oh it will go, only put the water to it," would be the reply when the subject was mentioned. I can recall several that would have gained very different results had their wheels been in proper condition, but the matter was not so well understood then as now. Even now it requires constant attention to avoid errors in that way, for it is very common for wheels to turn perfectly free at the start, then after running a few minutes become bound through swelling of step or followers, so as to lose a number of revolutions per minute, carrying the same weight as at first starting, Risdon's highest result 91.32 and Tyler's 91.27 were supposed to be erroneous, because neither could be repeated, but from the cause named above they could not be rejected.

Test of Wheel to Determine Loss of Power in Transmission Through Gears.

In making the experiments to determine the loss of power in transmission through gears, mitre gears twenty-seven inches in diameter, five inch face, fifty-seven teeth, were used on wheel and "jack-shaft," the last being six feet in length, and three inches in diameter; a spur gear twenty-four inches in diameter, four and one half inch face, forty-four teeth, was secured upon the "jack-shaft," which worked into another gear of the same size upon a second horizontal shaft, same size and length as the first; the second representing the main line of shafting through a mill, both horizontal shafts worked in common babbitted bearings. The dynamometer was placed upon the end of shaft representing the main line, and the wheel tested through the two pairs of gears; then upon the wheel shaft.

	Tests.	Head.	Revolutions.	Horse Power.	Percentage.
Dynamometer on horizontal shaft,	1st test,	16.03 feet,	160 per minute	26.55	75.90
Dynamometer on Wheel shaft,	2d "	16.08 "	168 "	26.73	77.40

Important Tests to those Gearing Wheels where the Head Varies.

The best speed for each head is first given : 20-inch wheel.

Head.	Weight.	Revolutions per minute.	Horse Power.	Cubic feet.	Percentage.
18.44	500	249	39.92	1400.31	.7753
7.85	200	164	9.94	869.84	.7724
18.35	640	161	31.22	1418.94	.6663
7.99	75	246	5.50	757.93	.4911
48-inch wheel.					
*17.55	1100	121	80.66	3586.83	.6733
9.79	600	90.5	32.90	2540.80	.7018
17.47	1525	90	83.18	3618.81	.6982
10.00	200	120.5	14.60	2199.34	.2522

*121 revolutions per minute was found to be the best speed for whole and part gate.

Turbine Buckets.

Ten years since turbine builders added much to the cost of their wheels by making the buckets of sheet iron, steel, brass, or bronze; shaped in iron moulds. The best turbines yet produced have been made entirely of cast iron. Wrought iron is decidedly the poorest material that can be used for that purpose.

A Word to Aspirants for Fame as Turbine Builders.

The incentive to turbine building is probably its supposed profit. A wood-sawyer, so little of a mechanic as to be unable to file his own saw, unhesitatingly rushes into the business, yet it is one requiring the highest possible skill; experience soon causes the adventurer to regret his haste. *A strictly honorable turbine business under existing circumstances, can not be made to pay ;* that is, to sell every wheel by test on its real merits would leave half the number made on the builders' hands, for purchasers require the highest results at the lowest prices, and there are scores of builders ready to guarantee such so far as talk is concerned.

Professional Experts.

If those acting as above could see themselves as those thoroughly acquainted with the subject in hand see them, it would have a tendency to lower their pretensions. Could the arts be put back to what from our standpoint they seem to have been 3,000 years ago, one of our best mechanics might prepare himself to act as general expert without seeming presumptious, but to pretend to be able to do so now, when the mere word mechanic covers a thousand occupations, each having numerous variations, renders the pretense ridiculous. Yet we have such, and those who have great influence in court, particularly in patent suits. The turbine has been studied for more than a half century by the best mechanics, and the matter is not sufficiently understood to fairly allow of its being considered a science; yet the professional expert will look the matter up in a day, then go into court and testify to points of which it is simply impossible that he can know anything about. No one man can be an expert in all kinds of business, life is too short. The most intelligent and skillful telegraph operator must be the best expert in a telegraphing case, so of the shoemaker, the blacksmith, the miller, merchant, turbine builder, or engineer. In either of these callings an apprenticeship of years is required to render a person proficient; then is it reasonable to suppose that the professional expert can master any of them in a few hours' study? We would not go to a shoemaker to inquire about a turbine, or the turbine builder to learn about telegraphing. If the matter is simple and plain, an expert is unnecessary; if difficult to be understood, then certainly one skilled in the matter is the best qualified to make it plain. In cases where litigation is contemplated an expert well versed in the matter should be employed first, then if he understands his business, in three cases out of four, he will cause the matter to be settled, often advantageously to both parties interested; if he can not cause it to be settled, he can prepare it for the lawyer, so that it may be legally determined expeditiously and at the least expense. To employ the lawyer first is like trying to learn a child to read without learning it the letters; that, however, would be no more absurd than to suppose that any one man can be proficient in all kinds of business.

Faith in expert testimony is undoubtedly decreasing, simply because those called as experts are generally mere theorists, or perhaps edit some so-called scientific paper that is published on speculation—the editor, like the paper upon which it is printed, being picked up where it can be had the cheapest. A graduate from our technical schools might readily study up horse-shoeing, and testify in such a learned manner as to astonish the court with his profoundity, yet his shallowness would at once become apparent could the cross-examination be conducted by an ordinary blacksmith, as I have often wished I could do with hydraulic experts. Yet, in almost any case in litigation relative to milling matters, the testimony of such men as A. M. Swain, George A. Houston, F. H. Risdon, Wm. M. Mills, and others that could be named, would be very valuable; but such men would require time to consider the matter before testifying. "Why, I thought you experts were so full of knowledge upon such matters that you were always ready to gush over," said an applicant for my services. Such may be the case with others; it is not with myself. I want, invariably, to hear both sides of a case, and time to compare the circumstances with facts gained from my own experience, before acting for any one.

Slip of Belt.

The speed of machinery is computed from size of pulleys or gears in connection with the driving shaft; in such computations the slip of belt is seldom or never taken into consideration, yet that slip is an important item. In testing the power of a steam-engine, the counter of my dynamometer showed such a difference from the engineer's estimate, that the matter was thoroughly investigated. The driving pulley on the engine was 12 feet in diameter, that on the main line of shafting 6 feet; running light or simply driving shafting, the fly wheel making 75 revolutions the main line made 150, but with weight applied to scale beam of brake, the belt began to slip, the slip increasing with each weight added; at the maximum power of engine, the main line made 144 revolutions while the fly wheel made 75. Belt and pulleys were in perfect condition.

Gearing Turbines by Tables.

The practice of gearing turbines from tables prepared by guess, has been productive of much loss of power. In testing wheels it is a rare thing to find two of the same size and make, that do their best at the same speed; the best speed of the Leffel wheels is invariably wi le from their tabled rate. At Bridgton, Maine, Pondicherry mill, a 54-inch Leffel wheel has been in use for ten years, working under twelve feet head and running at ninety revolutions per minute; the mill has six sets of woolen machinery, but from lack of power only five sets have been used. By test a short time since it was found that by running the wheel at seventy-eight revolutions instead of ninety, it would give twelve h. p. more than it ever had done; so that for ten years it had been running at four-fifths of its capacity, and at a time when its greatest capacity was much needed.

Testing Curbs.

The fact is well established now that the chutes of a turbine have as much to do with giving high results as does the wheel itself; also, that each part of the complete turbine has relation to all of the other parts, so that a change of one piece may have a serious effect upon the whole. Builders have prepared several turbines with interchangeable parts in order to test understandingly; but it would seem better to make a testing curb with changeable chutes, so constructed that their number or direction might readily be changed, and their capacity of discharge increased or diminished. With such a curb it should be possible to determine the merits of any wheel that could be tested therein.

V Shaped Belts, Cable Transmission, etc.

Some time since there was a mania for driving machinery with belts of the above named shape, but experience soon cured the desire. Transmission by wire cable is another matter that should be well considered before adoption; it will answer the purpose in places where shafting can not be used, but it is a very poor substitute still at the best. Light shafting is still another subject for consideration; if used, the pulleys should be placed close to the hangers, for if placed any distance therefrom, the shafting will spring, and require a much tighter belt, which soon gets the shaft out of line. There is a proper limit either way.

The Metric System.

And why the metric instead of that so generally in use wherever the English langauge is spoken? Does the practical mechanic or engineer desire such change, or do the comparatively few who use that system surpass us in mechanism or general intelligence? Taking the foot as the unit, divide it into tenths, hundredths, etc., and the most perfect measurements possible may readily be made and expressed thereby. Then why change for new terms, when our langauge is now so unwieldly and overburdened with useless words and synonyms, that it would be a blessing if one-half of its words could be obliterated, and the other half simplified in spelling. Simplicity should be the aim, that all may comprehend; change has not always been improvement. It would be well if the engineers and professors, who are so much better known th ouch their pretensions than achievements, could be made to understand that muddiness does not always denote depth. A change to the metric system would cause immense confusion in our standards, boundaries and records, without bringing a shadow of benefit in return. Our language now is almost the *universal* language dreamed of, and it seems idiotic to change for that of a people occupying less of the earth's surface than is covered by some of our states.

THE DAY OF THE CHURCH.

Torquemada persuading the doubting to give their souls to God and their property to the Church.

Calvin persuading Servetus to believe in the Holy Trinity.

Procession of Corpus Christi.

Workingmen of that time.

THE DAY OF THE MAN.

HOME. PEACE

THE MECHANIC TO THE FRONT.

THE REAL CREATIVE MECHANIC APPROACHES NEAREST TO THE ATTRIBUTES OF HIS CREATOR

THE PRIEST, LAWYER, DOCTOR, AND PROFESSOR TO BE VALUED ACCORDING TO THE USEFULNESS OF EACH.

HIS POSSIBILITIES HAVE HARDLY BEEN CONCEIVED OF. LET HIM REACH OUT TO THE UTTERMOST UNKNOWN.

AND MAN AND WOMAN TO FURNISH BRAINS, MACHINERY TO DO THE ROUGH AND HARD LABOR.

THE LIBERTY TO REASON HAS OPENED NEW WORLDS OF THOUGHT.

STEAM, THE TELEGRAPH, TELEPHONE, ELECTRICITY ALL ARE IN THEIR INFANCY, AND SURE TO DEVELOP NEW FIELDS OF PROGRESS.

OUR CREATOR IS ADMIRED FOR HIS WONDERFUL WORKS, NOT PROFESSIONS

THE FAR-REACHING TELESCOPE, THE NEAR-SEARCHING MICROSCOPE, EACH OPEN UNKNOWN WORLDS AS THEY BECOME MORE PERFECTED.

NON-SECTARIAN BUT UNIVERSAL EDUCATION.

Doctor Dodimus Celebrated Case of

Duckworth's Open Mouth Lockjaw.

Sir Alec. McMuttonhead — Oh, positively it is not a cancer. He died the next week of cancer, all the same.

Dr. Hamiltongue—Oh! undoubtedly the ball is here. Haven't we all said so from the first, and hasn't the electro-detector located it here sure? Dr. Doodle — Of course it is here!

Patsey McGrath—Loik at the pair of them doinkeys saking for the ball there, when it is here be my hoind, in the shoulder.

PROGRESS IN MEDICINE.

"It is a fact that the number of healthy men and women is growing less every year, and the sick more numerous. In the face of these facts, it might be noted that this country is full of doctors and full of drug stores; that these doctors and drug stores increase every year, and in heavy ratio the sick and dying increase also.

"It would seem like ignorance and arrogance combined for any physician or school of physicians to claim a monopoly in the practice of medicine, when all physicians of all the schools of medicine combined are powerless in curing but a fair percentage of acute, and still less of chronic, diseases.

"And instead of doctors opposing new discoveries, condemning new systems of practice, they should welcome them, for no one knows better than the doctor himself how powerless he frequently is to cure, or even aid, in the sick room."

Two thousand years ago Cato wrote of physicians precisely as we do to-day. He said: " If they attempt to treat of the practice in any other language than the Greek, they are sure to lose credit, there being all the less confidence felt by our people in that which so nearly concerns their welfare if it becomes intelligible to them. In fact, this is the only one of the arts in which the moment he declares himself an adept he is at once believed. Besides, there is no law to punish the ignorance of a physician. It is at our peril they experimentalize, the only person that can kill another with impunity."

Pliny speaks of Rome trying, then condemning, the employment of physicians and going without six hundred years.

At about eight years of age I had a tumble and fall of fifty feet from the top of a building; the shock was severe, but energy, elasticity, and self-will set me on my feet in half an hour. The doctor said such strength was unnatural, so he bled me nearly to death.

At about twelve years of age, I with two others was thrown from a carriage on to a pile of ragged rocks. All of us were terribly cut, my thigh broken, skull fractured, etc.; the rocks were fearfully drenched with blood. Three doctors came; their first act was to take a quart of blood from each of us.

At the age of nineteen I was stricken down with yellow fever at St. Jago, Cuba. My last recollection there is of a doctor prodding away at my arm with his lancet. Our ship left with the assurance of burying me as soon as free from port. The doctor was left behind; the fresh sea breeze cured me. That was in 1841, and the last time a physician has been employed for myself.

In 1859 leaving my family in East Boston one morning to go to my business in the city, a little idolized daughter kissed me with her Good bye, papa. On my return that evening I found her senseless, with a bag of ice bound upon her head. A kind neighbor, we being strangers there, had called in his family physician, a homeopathist. At eleven o'clock that night, the child being in convulsions, another neighbor called the nearest doctor, who happened to be of the old style; on arrival and learning who had previously been employed, he simply became brutal and left. He did not neglect to send his bill. The child died at midnight of scarlatina.

A year later my wife sickened, and for weeks was attended by a physician of my early acquaintance; suddenly he stopped calling, and was absent a week; my wife at once came to the conclusion that he had abandoned her as past hope. The seventh day of absence our physician returned, decidedly under the influence of liquor, and the question was plainly asked why he had absented himself.

" Well, hum! well, the fact is, I believe your wife has consumption, and I can do her no good !" Then, after a moment's hesitation, he continued: " Damn it, Emerson, we doctors don't know half so much as folks think we do; we guess as to what the matter is, then go to our books for the remedy; if the first does not help, we try another."

Experience has established the fact, that discoveries and improvements almost invariably come from those outside of any regular line of business, and it would seem to be a crime, besides being unconstitutional and unjust, to make a law that would give any school of medicine a monopoly of the business. If, however, the law is to be called in to give such monopoly it should fix the price, say, at twenty-five cents per visit.

The *Freethinker* of London, England, contains the following: "*Christian Life* gives the following figures: In the common gaols of Ontario (Canada) 11,810 persons were locked up last year. No less than 2,448 were unable to read or write." The religious denominations were represented as above. Will the clergy of this country use their influence to procure similar information from our prisons that the influence of the various religions upon the morals of the masses may be ascertained? Come, brother Talmage, will you try?

THE AMBITIOUS FISHERMAN.

THE BIBLE IN SCHOOLS AND A GOD IN THE CONSTITUTION.

The Bible is a book useful for the student, but is of the past; its worship has been the cause of oceans of bloodshed. Aside from errors of translation, words often are changed in their meaning by change of locality, so that there is no certainty that we have the writer's true meaning in the Bible stories; but we can readily see from its contradictory statements that it is merely a history of that people, and through its tribal conceit all others were ignored. Cain feared that some one meeting would slay him, which could not have been the case had there been none to meet, and the very form of statement proves that the mark would be understood. Still further to find a wife at Nod, there must have been people there. Its nine hundred year lives must have meant dynasties. Its fish story most likely belonged to the class of myths common at that time in connection with the stories of the gods and goddesses. A book that cannot be opened at random and read in society is not suitable to be put into the schools to be read by children.

It is but a few years since George Francis Train was imprisoned for publishing obscene literature. Unfortunately for the complainant he was so ignorant of the Bible contents that he was unaware that the dirty literature consisted of extracts from that *sacred work*. The arrest became a boomerang. If brought to trial, the character of the contents of the Bible would be ventilated; so that Train was brought into court, and pronounced insane, consequently irresponsible, so discharged; but as that would leave him irresponsible if he saw fit to shoot the complainant, he was the next day again brought into court and pronounced sane.

Think of the fool Freeman stabbing his five year old daughter to the heart in this State and age through his insane fanaticism for emulating old Abraham.

Ideas are changing rapidly. Success in keeping the Bible in the schools or getting a God in the Constitution is likely to result something like the success of the ambitious fisherman at the head of this article.

SILK.

Silk consists of the pale yellow, buff colored, or white fiber, which the silkworm spins around about itself when entering the chrysalis state. Silkworms are divided into two classes, the mulberry-feeding worm, from the cocoons of which is reeled the ordinary raw silk, and the wild silkworms which feed upon certain kinds of oak, ailanthus, castor-oil plant, etc. The product of the latter specimens (amongst which the *Tussah-worm* is found, producing the *Tussah-silk*) was little heard of in this country and Europe until recently, and but for the outbreak of the silkworm disease in Europe would probably have remained in India and China, although it had been utilized in both these countries for many centuries. The date when the use of silk for textile purposes was first discovered is not exactly known. Some of the Chinese historians claim that it was about 2700 years B. C., whereas others only go as far back as about 1703 B. C., or the reign of *Hoang-ti*, the third of the Chinese emperors. He, the legend tells us, was desirous that his legitimate wife *Si-ling-chi* should contribute to the happiness of his people, so he charged her to examine the silkworms and test the practicability of using the thread. In accordance with this wish, she collected insects and feeding them in a specially prepared place commenced her studies and examinations, discovering not only the means of raising them, but also the manner of reeling the silk and its use for textile purposes. It is claimed that even to the present day the empresses of China on a certain day go through the ceremony of feeding the silkworms, and rendering homage to *Si-ling-chi* as *Goddess of Silk Worms*.

The principal countries for carrying on the silkworm culture are Southern Europe, China, Japan, and India. In our country silk culture is only in its infancy, yet it is rapidly assuming proportions of importance.

When full grown the worm ceases to feed, climbs up from the feeding tray to the bush, or whatever may have been prepared for it, and commences to form itself in a loose envelopment of silken fibers, gradually enwrapping itself in a much closer covering forming an oval ball or *cocoon* about the size of a pigeon's egg, generally requiring from four to five days in its construction.

RAW SILK OR REELED SILK

constitutes the raw material for the American silk manufacturer. When imported the same generally comes in picul bales of one hundred and thirty-three and a third pounds. Such as come from China are made up in bundles weighing from eight to twenty-five pounds each and are protected at the corners by floss or waste. The Italian silk comes in bales made up in skeins. Before it reaches the loom this raw silk must pass several manipulations and processes. First the same is taken to the sorting-room, and the various sizes of thread, or, in other words, the different degrees of fineness, are assorted each by itself. The next process is the transferring of the silk from the skeins (which are of irregular length) to the bobbins. A parcel of skeins enclosed in a light cotton bag is soaked in water having a temperature of 110° F. for a few hours so as to soften the gum. After taking these bags out of the water they are submitted for from 5 to 10 minutes to the action of a hydro-extractor to liberate the superfluous water, and the silk with its gum thus sufficiently softened is ready for winding. The next manipulation the silk thread undergoes is cleaning.

In this process the thread is simply transferred from one bobbin to another and passes during the transfer through the cleaner, which consists of two sufficiently close parallel plates to catch any irregularity upon the silk. Chinese silk always requires cleaning, whereas Italian silk does not usually.

WILD SILKS.

The most important of them is Tussah, and is principally found in India. This silk has until lately been greatly neglected, but at present commences to attract great notice. The cocoons are larger than those of the Bombyx mori, have the shape of an egg, and are of a silver-drab color. The outside silk of the cocoon is slightly reddish, and consists of separate fibers of different lengths, while the remainder of the cocoon is generally unbroken to its center. In India the report compiled by that government gives particulars of no less than thirty-six varieties of wild silkworms feeding upon different forest trees and shrubs

"SPUN SILK."

It is to be understood that the raw silk of commerce is spun by the worm as the spider spins its web, but in reeling this there is waste; then there are cocoons from which the worm has eaten its way out, of course spoiling the cocoon for reeling; then much of the product of the wild worm cannot be reeled. All such silk has to be carded and spun substantially the same as cotton, and as the fiber is short it has to be twisted hard to make it strong, so that hose or other goods made of spun silk have not the soft feeling of the raw silk, though the silk itself may be of quite as good quality.

MOIRE ANTIQUE AND WATERED SILKS.

For these the silks must be broad and of substantial make. They are first wet and then folded with particular care to insure the threads of the fabric lying all in the same direction; they are then submitted to great pressure. By this pressure the air is slowly expelled, and in escaping draws the moisture into curious waved lines, which leave the permanent markings called watering. Moire antique silk is streaked in veins like the veins in the antique marbles. Figured silks are woven in Jacquard looms. Very heavy silks are often made so by dye-stuffs. Honest manufacturers will say that two dollars per yard at retail should purchase the best dress silks that can be made.

MODERN SILK SPINNING FRAME AND TWISTER MADE BY THE
W. G. & A. R. MORRISON CO., WILLIMANTIC, CONN.

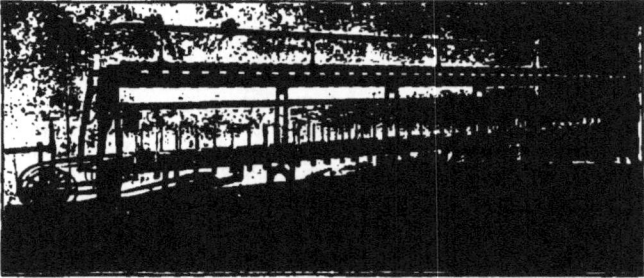

EMERSON'S

◄Duplex Piano Stool►

Stool closed.　　　Opened showing second Stool.　　　Second Stool removed.

The present period may be denominated as the Musical Age. Almost every family of ordinary culture has its Piano or Cabinet Organ, in many cases both.

To develop the capabilities of these Instruments in orchestral effect, much of the popular music of the day is arranged in Duets, requiring four hands for its proper execution, of course necessitating the use of two stools or seats for the players. Probably there are few persons of ordinary observation and experience, who have not seen a chair filled with bound volumes of Music, " Webster's Unabridged," or other material to supply a seat for the second player.

For the most of the time but one Stool is required in a family, so that a second Stool is an encumbrance, except for the short time it is needed.

To obviate this objection, many attempts have been made during the past fifteen years to produce a Stool suitable for either one or two players. Numerous patents have been granted for such devices, but these generally have been conspicuous as to their double nature, and very inconvenient either as single or double stools.

The plan herewith illustrated is believed to be the long sought convenience for the purpose named,—insurpassable in beauty as a single Stool, or in convenience for teacher and pupil while giving and receiving instruction in music, or the execution of four hand pieces by two players.

FLAX, ITS CULTURE AND MANUFACTURE.

Early in the present century almost every farmer in the Eastern States raised flax, its product then being a necessity for all. The apothecary depended upon its seed for soothing the ailing; the painter for its unequaled oil for his paints; the farmer for the fiber of its pachydermatous stalks for his clothing; his wife for her bedding, laces, embroideries, etc.; the ship owner and sailor for sails and cordage. To prepare the flax it was pulled and cured, then in bundles submerged in water until the woody outside rotted or became so brittle as to readily separate from the fiber when dry and beaten in the "flax breaker." The farmer then with a "swingle," a sort of two edged heavy wooden sword, in his right hand, seized a handful of the broken stalks in his left, held the stalks over the top of the swingling plank, striking them close to the side of the plank with the edge of swingle. The swingling plank was thin at the top, made of hard wood, standing about three feet above the floor, to which it was firmly secured. The repeated blows of the swingle caused the woody shell to fly off in minute pieces or "shives." Every few blows the fiber would be drawn through the teeth of a "hatchel" or comb as a woman clears her hair. The tow trousers of the Continental times were produced from the coarse refuse combed from the flax while it was being hatcheled. This hatchel was formed by placing a gross of smooth sharp pointed steel spikes firmly in a square base secured to a bench, the spikes projecting vertically upwards six inches above the base. The fiber thus prepared was taken by the wife and wound upon the distaff of the linen spinning wheel, at which she sat and produced the thread for the shoemaker, tailor, sailmaker, and other artisans too numerous to mention, also all necessary for household use. The little flax spinning wheel was a very different affair from that of the spinning wheel for wool, as may be seen in illustrations on opposite page. At the former the woman sat and operated the wheel with her foot, using the fingers of both hands to draw the thread, the spindle being of the flier pattern; while with the wool spinning wheel she stood at its side, turning the wheel with her right hand and drawing the thread from the roll of wool with her left, the twisting being done on the end of the plain spindle.

The spinning for the fabulous laces, linens, edgings, lawns, etc., of the earlier times was done substantially in the manner described or in a still more primitive way.

Spinning street yarn is not a figment of the imagination; in South America, near the equator, the writer has often seen the native women walking the street, talking, and spinning on the way, the cotton being carried under the arm, the thread being drawn by the dropping of the bobbin on which it was wound as spun, a twirling motion being given to the bobbin as it was dropped; then it was skillfully caught at the arm's length, without seeming effort.

Flax is raised in the Northwestern States and Canada, but mostly for its seed, though its fiber is in some demand for manufacture into thread, and is beginning to be used in the manufacture of paper.

Early in the century, all farmers' wives were supposed to be capable of attending to all of the duties indicated by the implements shown opposite,

spinning both flax and wool, carding the latter from the fleece for the spinning, after which going through a series of preparatory processes such as spooling, reeling, sizing, warping, drawing in, etc., etc., from the spinning wheel to the loom, where from the coarsest to the most delicate fabric for family use was produced, often very intricate patterns of bed coverings, carpets, and other ornamental designs.

The man that can realize the multifarious duties accomplished by the wife of a century since and then consider her sex inferior in constructive or mental ability to that of his own must be conceited indeed.

A half century since, our farm houses and mills contained as fair women and girls as could be desired, dressed perhaps in homespun, but their nimble fingers, in their leisure moments, were ever busy making edging, embroidery, or fancy trimmings for their underclothing or household use ; in their place we have *ladies*, outwardly dressed fine but with ten cent undervests. The washings of the former weekly displayed volumes of refinement, of the latter sweat-stained and often ragged undervests that indicate continued use without change. The tobacconist ornaments his goods with beautiful forms clad in diaphanous and delicately trimmed under garments, but does not seem to take to the lady and ten cent undervest.

The woman with her heelless shoes, white stockings, and zephyr step, had feet that were things of beauty, while the lady of to-day, with her high heels and distorted feet that require large bay windows on her boots to accommodate her abnormal toe joints which intimate the evolution of thumbs and a return to the quadrumana family, is certainly less attractive.

If the continuation of the robber tariff, which has so benefited the rich, has not reduced wages to the standard common in all highly protected countries, it is solely because the irrepressible inventor has by improved machinery reduced the cost of manufacturing. It certainly has been the cause of a much lower grade of working men and women than formerly, but a revolution is taking place in the status of woman from which a progression may spring.

JUTE AND ITS MANUFACTURE.

Jute is raised in India, having, while growing, something the appearance of oats, though much larger, as it reaches a height of fifteen feet or more, but like rushes it grows in water, two crops each year ; its fiber, the reverse of that of flax, is on the outside of the stalk.

The ground is prepared and seeded, then flowed; with plenty of water, the growth is very rapid. At maturity the stalks are cut, then, like flax, are immersed in water to soften the fiber ; the process is then similar to that of flax ; the ends are cut even to prepare the fiber for baling, the ends cut off being known as "jute butts."

There are various places of its manufacture in this country, one, quite extensive, at Ludlow, Mass., from whence my information has been obtained.

The machinery used is similar to that of cotton manufacture but coarser and much heavier. The product of the Ludlow mills is the covering matting used by furniture dealers for their furniture in transit.

VIEW OF MINNEAPOLIS, AND FLOURING MILLS IN THE FOREGROUND.

THE ROLLER PROCESS OF FLOUR MAKING.

Revised for 1891 by The Edward P. Allis Company of Milwaukee, Wisconsin, U. S., Flour Mill Builders and Furnishers.

To prepare wheat for milling, it is good practice to run it through a dustless receiving separator to free the grain of coarse trash like straw-joints, corn, oats, etc. ; then through a dustless milling separator to remove finer trash, like cheat, screenings, oats, sand, and seeds, and through two separate scouring and polishing machines to remove dust, and scour off smut, the fuzz on the ends of the berry and as much of the outer woody bran coatings as may be easily detachable. The most complete flour mills attach dust collectors to the exhaust air trunk of the above grain cleaners for the sake of cleanliness. To remove metallic particles the wheat should be passed through an automatic magnetic separator. During dry winter weather the bran of the wheat often becomes brittle and consequently easily pulverized when passed through the rolls. To obviate this, a wheat heater is employed, using steam at about 98 degrees Fahrenheit, as a heating agent. This attracts the latent moisture from the interior of the berry to the surface, thus toughening it. Some varieties of wheat require steaming in place of heating, and wheat raised by irrigation generally requires wetting down in bulk for 24 to 48 hours before being used.

Briefly speaking, the roller system has for its object. 1st, the gradual reduction of wheat into middlings, 2d, the purification of the middlings, and 3d, the gradual reduction of the middlings into flour.

This method of flour making is divided into two systems popularly known as the "long" and the "short" system. These terms apply principally to the number of reductions used to convert the wheat into middlings. The long system is used in the larger mills, especially those doing a merchant or shipping business, and produces a maximum amount of middlings, and for this reason is the more profitable system. Not less than five reductions on wheat are employed in the long system. Each reduction is technically known as a "break." Each break is made on a pair of corrugated or fluted rolls. The corrugations of the first break rolls are rather coarse, but they are finer on each succeeding break. The number of corrugations on each pair of break rolls varies with the kind and condition of the wheat and the number of reductions employed, so that exact information on this point cannot be given here. One roll of each pair of break rolls has a speed $2\frac{1}{2}$ or 3 times greater than its mate. After each reduction of the wheat on the break rolls it is bolted or "scalped" on coarse mesh wire or silk cloth to separate the middlings and flour from the broken wheat so that the latter may be sent to the succeeding break and be further reduced. These scalpers are of a revolving hexagon or round reel form, or on the reciprocating sieve design. The miller in charge so graduates the breaks from first to last that the bran issues after the last break (and the subsequent scalping operation) free of flour, as long as the grain is in good milling condition. Should it be damp or tough, or the weather be murky, a bran duster is necessary to remove all remaining traces of flour from the bran. The middlings and flour derived by the foregoing process is collected from the various scalpers and sent to a grading reel, clothed with silk cloth of varying fineness. This reel separates the flour from the middlings. The flour is bolted on round reel flour dressers or centrifugal bolts to prepare it for the market and forms a commercial grade known as "bakers' flour." It constitutes about 50 to 70% of the entire floury product when made from winter wheat, or about 20 to 40% when made from spring wheat. The middlings are divided into three or four grades or sizes and sent to middlings purifiers, which, by means of reciprocating sieves and a graduated air suction, remove all free bran particles and fiber. Three grades of middlings are formed by this operation, viz.: middlings free from impurities, middlings containing a small amount of fine adhering bran particles, and coarse middlings attached to germ or bran. A fine fiber or cellular tissue permeates each particle of middlings, which is of a white color and undistinguishable from flour until it is wet or baked into bread, when it imparts a dark color. To remove this fiber success-fully it is necessary to gradually reduce the middlings in size, by successive passages through smooth rolls, separating the flour derived before the mid-dlings pass to the following reduction. This sizing operation also liberates adhering bran and germ impurities. The various grades of middlings are at first reduced separately on individual pairs of rolls, according to their size and quality, but, at an advanced stage in the process, when a similarity in size and quality is reached they may be mixed and worked to a finish. The flour from the foregoing operations is bolted on flour dressers and is known as patent flour, and commands the highest price on account of its pureness. Spring wheat produces from 50 to 75% of patent flour, and winter wheat 15 to 35%. In finishing up, a small percentage of flour results, varying from 3 to 10% of the entire flour product, which is too dark in color to be incor-porated with the other grades. This forms the low grade. In large mills, any or all of the above three grades of flour may be subdivided according to quality and sold as separate brands.

In the short system, it is, as its name implies, a curtailing of the above process. For instance, where five breaks on wheat and seven or more crushes on middlings are used in the long system, only three breaks on wheat and five crushes on middlings would be used in the short system. It is claimed by excellent authority that owing to the more abrupt method of reducing and crushing as practiced in the short system, a smaller percentage of middlings results and consequently a reduced percentage of high grade and high-priced flour. Short system mills are usually of small capacity, ranging in size from twenty-five to seventy-five barrels per day, and usually mix all the flour to form one straight grade, and are more adapted to winter wheat than to spring wheat.

A WORKING FLOURING MILL.

According to the Plans of The Edward P. Allis Co., Milwaukee, Wis.

The engraving opposite shows a perspective view of a working flour mill having a capacity of 50 barrels of flour in 24 hours. This engraving and the description thereof is given in connection with the adjoining article on "The Roller Process of Flour Making." A small mill is selected for these modest sized pages in preference to one of large capacity, to enable us to show the details on as large a scale as possible.

The operation of the mill commences by putting the wheat, as it comes from wagons or cars, into the hopper scale seen in the right hand corner of the first floor in the engraving, which weighs 40 to 60 bushels per draft.

From the scale the wheat descends to the bin shown in the basement directly underneath, which will hold sufficient grain to operate the mill one day. The adjoining elevator serves to elevate the wheat, as needed, to the milling separator shown on the second floor. Here the wheat is relieved of all foreign particles and shrunken grains unfit for milling. The grain is now re-elevated to the upper one of the two adjoining smutters and scourers, and after the wheat has been acted upon by these two machines it is stored temporarily in a bin on the second floor, not shown in the engraving, where it is ready for passage through the rolls and bolts in its conversion into flour, as described in detail on other pages. The shrunken wheat, taken out by the separator is spouted to a screenings grinder placed against the far side wall and is converted into feed for horses or cattle. The dust from the three wheat cleaners is blown into Cyclone dust collectors, those conical affairs shown near the ceiling of the second floor, which separate the air from the dust, discharging the dust at the bottom and the air at the top. In the background of the second floor are shown the various flour dressers, centrifugal finishers, and middlings purifiers. On the first floor are shown the four double roller machines with automatic feeders, each machine containing two pairs of rolls, each pair working entirely independent of the other. Three pairs of these rolls are corrugated for the purpose of gradually reducing the floury part of the wheat to middlings, while the remaining five pairs are smooth to gradually reduce the middlings, after purification, to flour. Near the side wall in the foreground is shown a flour packer with its flour storage bin on the floor above. In the rear is the power room, containing the engine, boiler, pumps, and heater. A bushel of 60 lb. wheat produces 38 to 44 lbs. of flour, 6 to 10 lbs. of bran, 6 to 8 lbs. of ship stuff, 1 to 3 lbs. of screenings, and ¼ to ½ lb. invisible loss during milling. These quantities vary with the kind and condition of the wheat, the condition of the weather, the size, kind, and condition of the mill, and the skill of the miller in charge. From 6 to 14 horse power per barrel per hour is required as motive power, depending on the size of mill and the proximity of power to the machinery.

Following is a list of the flouring mills of Minneapolis, with names of owners and capacity of each per day :—

Pillsbury (A) Mill, 7200 bbls., Pillsbury Washburn Flour Mill Co.; Pillsbury (B) Mill, 2500 bbls., Pillsbury Washburn Flour Mill Co.; Anchor Mill, 1600 bbls., Pillsbury Washburn Flour Mill Co. ; Palisade Mill, 2000 bbls., Pillsbury Washburn Flour Mill Co. ; Lincoln Mill, 1000 bbls., Pillsbury Washburn Flour Mill Co.; Washburn (A) Mill, 4200 bbls., Washburn-Crosby Co. ; Washburn (B) Mill, 1300 bbls., Washburn-Crosby Co. ; Washburn (C) Mill, 3000 bbls., Washburn-Crosby Co. ; Crown Roller Mill, 2100 bbls., Northwestern Consolidated Mill Co. ; Columbia Mill, 1600 bbls., Northwestern Consolidated Mill Co. ; Northwestern Mill, 1600 bbls., Northwestern Consolidated Mill Co. ; Galaxy Mill, 1500 bbls., Northwestern Consolidated Mill Co.; Zenith Mill, 1000 bbls., Northwestern Consolidated Mill Co. ; Excelsior Mill, 1100 bbls., Minneapolis Flour Mfg. Co. ; St. Anthony Mill, 650 bbls., Minneapolis Flour Mfg. Co. ; Standard Mill, 1700 bbls., Minneapolis Flour Mfg. Co. ; Humboldt Mill, 1150 bbls., Hinkle, Greenleaf & Co. ; Dakota Mill, 350 bbls., H. F. Brown & Co. ; Holly Mill, 500 bbls., Holly Mill Co. ; Minneapolis Mill, 1200 bbls., Crocker, Fisk & Co. ; Cataract Mill, 800 bbls., D. R. Barber & Son ; Phœnix Mill, 275 bbls., Stamwitz & Schober. Total 38,325 bbls.

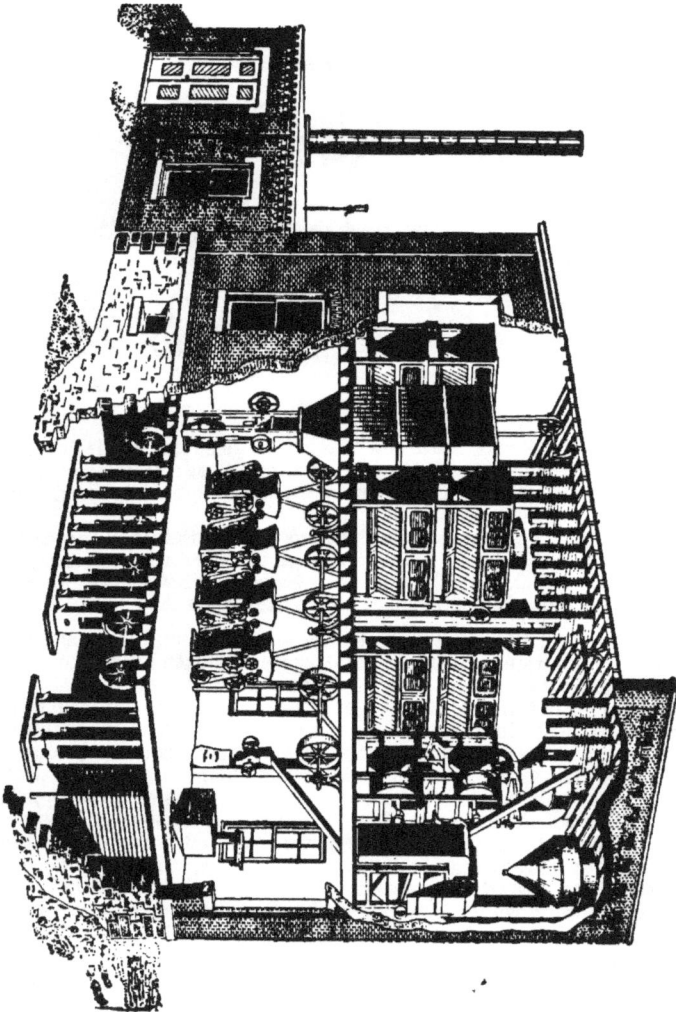

THE EDWARD P. ALLIS COMPANY, MILWAUKEE, WIS

MARRIAGE, DIVORCE, NUDITY.

YOUTHS' PREPARATORY EDUCATION FOR POLYGAMOUS ACTS.

ARABIAN NIGHTS,

THE BIBLE,

AND CLASSICS.

Camaralzaman was proclaimed king, and married on the same day with the greatest magnificence; being thoroughly satisfied with the beauty, wit, and affection of the princess Haiatalnefous.

The two queens continued to live together in friendship and union, and were each well contented with the equality which king Camaralzaman observed in his conduct towards them in sharing his bed with them alternately.

From time immemorial, theoretically, love has been represented as heavenly, in practice almost invariably gross. Death in any form for a woman before dishonor. Lucretia has been the model, but it should be borne in mind that the other side of that story has never been told. From the earliest history down to Anthony Comstock the clergy have been the most strenuous promoters of such ideas and, unless sadly belied, the most common violators of them, not because naturally worse than others but because of having leisure and opportunity. The wise man of the Bible requiring a thousand women, the Lord taking his share of captive virgins, Lot and his buxom daughters, Camaralzaman and his two wives, and the classics describing the loves of the gods and goddesses are not reading likely to inculcate monogamy in the youthful mind, yet society as described by Rabelais when the clergy had entire control was far worse; humanity is better off with less of that control.

The marriage laws are unequal and unjust, often causing the innocent to suffer for the fault of others. The "for better or worse" is a device of evil because of it the beautiful bride soon becomes the dowdy wife; the passionate lover, the indifferent husband.

Marriage by equitable contract should produce equality and continued effort to please. Give both the same right to propose such partnership. Motherhood is a natural right, its desire inherent from infancy, proved by the craving for dolls. This right is often denied to the best through lack of self-assertion. Free women from her bondage of conventionalism and long petticoats, encourage her to think and talk of something besides dress, give her equal rights with man. Protect by making all children legitimate and have their rights secured, but allow of separation on breaking of contract by either party. Parties properly mated will need no law, those compelled to remain together often would be better apart.

Nudity is a matter of custom; the innocent mother readily exposes her bosom nursing her babe, unconscious of evil. The fashionable lady displays both ends at the beach, and all women wearing long skirts frequently make a liberal show of underwear walking in the mud, windy weather, getting into car seats, etc.

Why, says Mrs. Grundy, you would put us down equal with the cattle! My dear madam, the cattle have neither lawyers nor priests, yet their sexual relations are a world above those of the human animal. Your implication is an insult to them. Have less conceit and more self-examination.

TWO WOMEN.

I know two women ; and one is chaste
And cold as the snows on a winter waste;
Stainless ever in act and thought
(As a man born dumb in speech errs not).
But she has malice toward her kind—
A cruel tongue and a jealous mind.
Void of pity, and full of greed,
She judges the world by her narrow creed.
A brewer of quarrels, a breeder of hate,
Yet she holds the key to " Society's " gate.

The other woman, with a heart of flame,
Went mad for a love that marred her name.
And out of the grave of her murdered faith
She rose like a soul that has passed thro' death.
Her aim is noble, her pity so broad
It covers the world like the mercy of God.
A healer of discord, a soother of woes,
Peace follows her footsteps wherever she goes.
The worthier life of the two, no doubt ;
And yet " Society " locks her out.

The other woman for me.   —*Ella Wheeler Wilcox.*

WOMAN SUFFRAGE.

With the manifest destiny so plainly marked upon the face of the age that woman suffrage is bound to come, it seems strange to see the ordinary republican seven by nine rural member so readily join the Irish statesman in defeating the measure. A biped with ordinary manhood should freely grant such equality of right, and certainly the American woman is likely to vote as intelligently as the newly manufactured citizen from any foreign country.

Politics are not likely to be reduced in quality by the addition of a more reputable class of voters. Massachusetts is not doing itself credit in the matter.

Cotton Manufacture.

The manufacture of cotton goods, now so enormous in quantity and so varied in multiplicity of uses, is of comparatively recent date. The fibre was first introduced in England about 1640; a century latter, or in 1741, but 1,160.000 pounds were used there—a quantity that would but partially have loaded a single ship of that period, or a freight train of to-day. The invention of the spinning frame, by Arkwright, in 1768–71; the spinning jenny, by Haregrave, about the same time, and the combination of the two by Crompton, thus forming the mule, gave the first great impetus to the business, which was enormously increased by Eli Whitney's invention of the cotton gin, and of the card setting machine by Amos Whittemore, both Massachusetts men. The really successful power-loom seems to have been invented by the Rev. Edmund Cartwright of England. Fear that such machinery would render their employment unnecessary, caused the working class to gather in mobs, and destroy it, so that, as late as 1813 it was supposed that there were only about 2400 power-looms in use in all England. The war of 1812 with that country made it more necessary to manufacture cotton goods in this country, and it was done in several of the states, but with what would now be considered a ludicrous division of labor, as the spinning was done with water or horse-power, then distributed among the farmers' families, and there woven in hand-looms. Power-looms were tried in various places, but without being able to compete with hand-looms. Probably the first mill ever constructed for taking in raw cotton, and turning it out as finished cloth, was completed in Waltham, Mass., in 1813. This had 1700 spindles, and all the other machinery necessary for the purpose named. The enterprise proved successful, and another and larger mill, having 3584 spindles, was soon added to the first. [See Lowell water-power rate.] From such beginnings have grown the immense cotton manufactures of the country. Six million one hundred thousand bales of cotton were raised in this country the past year. Spinning is the heavy work of the business, and upon the spindle is based the estimates of cost, value, capacity and power required for the mill. Circumstances in each case, of course, affect such estimates. Suppose a new mill to be constructed where a dam and canal are ready to take the water for power, the cost for race, wheel-pit, mill and machinery would be estimated, under favorable circumstances, at about $14.50 per spindle. Sixteen dollars per spindle at this time should fit up such a mill with machinery of the most perfect kind. If canal, dam and boarding houses were to be added, the cost would probably be $20 per spindle. At Fall River, where steam is used for power, the estimate at this time is $17 per spindle, but Fall River does not furnish boarding houses. A mill 45 x 100 feet, four stories and attic, would require one floor for spindles. A spinning frame, having 128 spindles, requires 56 feet floor space, 16 x 3½ feet, or about 2.3 spindles per foot; but a passage way is required each side of the frame, so that 5000 spindles would be a good outfit for such a mill or floor. Six thousand might be used, but would hardly be advisable, unless room was scarce for the power at command. The power should equal two h. p. for each one hundred spinning spindles in a mill. There are light running spindles and machinery that could be driven with something less than that rate—others that would require more; but the rate is a fair average estimate. Warp and thread mills require more power in proportion to the number of spinning spindles than mills that make cloth. Two h. p. per each hundred spinning spindles is a fair estimate for the power required in silk mills. Thirty-five to fifty spindles in cotton mills are required per loom, the number depending upon the No. of yarn used or fineness of cloth produced.

The census for 1880 will give the number of spindles in the Southern states as 714,078; looms, 15,222, or about forty-seven spindles per loom, which would indicate that their product is quite fine cloth, or, what is more probable, that many of the spindles are employed in making yarn that is not woven there.

COTTON MANUFACTURE.

All of the illustrations, except for Picker, were kindly furnished by the Lowell Machine Shop, Lowell, Mass.

CHARLES L. HILDRETH, SUPT.

From the bale the cotton goes to the Pickers, Opener, Breaker, and Finisher. All similar in appearance.

PICKER.

From Finishing Picker in "laps" the cotton goes to Breaker Card.

CARD.

Thence in slivers from sixty to a hundred cards to the Doubler.

DOUBLER.

Which turns it into laps ; these are taken to the Finishing Card ; this is similar to the Breaker Card in appearance. From Finishing Card cotton in slivers goes to the Railway Head, which delivers into cans, from six to ten cards, to Railway Head.

RAILWAY HEAD.

From Railway Head to Drawing Frame, which leaves it in cans.

DRAWING FRAME.

From Drawing Frame it goes to the Slubber, which turns it into bobbins.

SLUBBER.

Thence to the Intermediate or Fly Frame.

FLY FRAME.

From Intermediate or Fly Frame on bobbins to the Fine Frame, similar to Fly Frame or Slubber. From Fine Frame to the Ring Spinning Frame and Mule. The Ring Frame makes the warp, the Mule makes filling, the filling going direct from the Mule to the Loom. In general appearance the Mule resembles the illustration of mule in woolen manufactures on another page.

RING SPINNING FRAME.

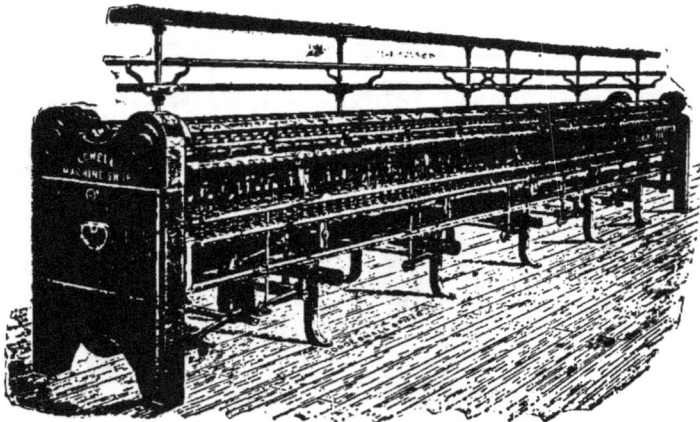

From Spinning Frame the warp goes to the Spooler.

SPOOLER.

From the Spooler to the Warper.

Warper.

From the Warper to the Slasher
Dresser.

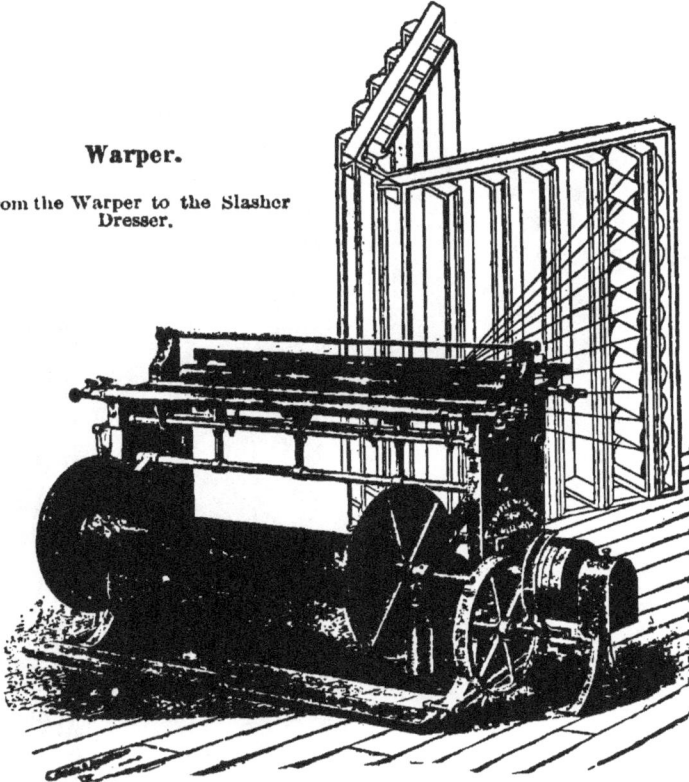

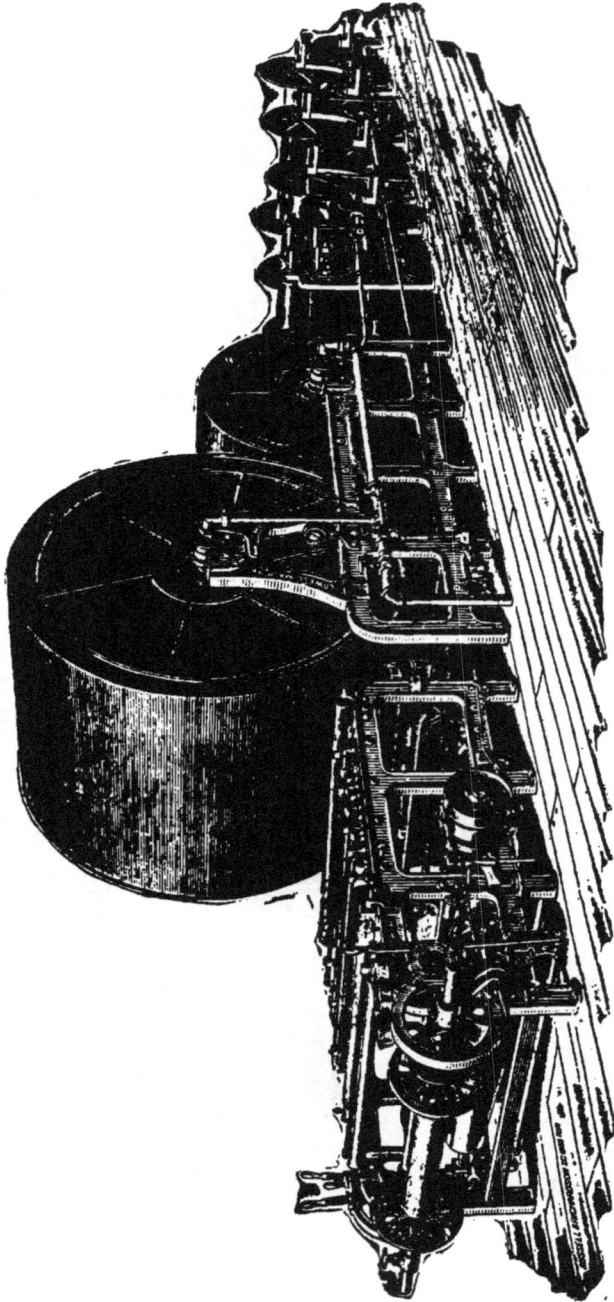

SLASHER DRESSER.

From Slasher to Drawing-in Girls (Hand Work).

From Drawing-in to Loom.

LOOM.

For spool cotton the process is similar up to the spooler, then there are doublers, twisters, thread spoolers, etc., etc. Bleaching, calico printing, and a thousand other varieties of work such as tape, fringe, counterpanes, each requires special machinery, looms, etc.

Ribbon, Webbing and Tape Loom.

Manufactured by L. J. Knowles & Brother, Worcester, Mass.

TESTS MADE AT THE MILL OF EDWARD O. DAMON DURING THE MONTH OF SEPTEMBER, 1880.

The machinery consisted of 73 Tape Looms, carrying 2008 shuttles; 2 Quillers of 36 spindles each, and 1 of 18 spindles; 2 Warp Dressers; 1 2-ply Warper; 1 Yarn Spooler; 1 Tape Reel; 1 Yarn Warper, single; 2 Tape Spoolers; 2 Tape Presses; 30 Counter Shafts, average length, six feet.

The power for the above machinery was taken from below, through the floor, on to a short main shaft about ten feet in length, the main pulley being forty-one inches in diameter and about eighteen inches face; belt, fourteen inches wide. One-ply. From this shaft the power was transmitted through a pair of bevel gears (about 30-inch diameter, 4-inch face) to the short counter-shafts. The power scale was applied to the main driving pulley, and examined and tested at short intervals, to see that it was working smoothly and correctly. Gen. Theo. G. Ellis, of Hartford, made a very thorough examination and test during the month, and reported, as the result of these tests, that the amount of power transmitted to the machinery, at 145 revolutions of the main shaft, was 10.61 h. p. At 136 revolutions of this main shaft, he reports the approximate power as 9 h. p. The tests of the same machinery, leaving the dressers off, was found to be about 2 h. p. less.

H. A. FOSTER, *Supt.*

TESTS MADE AT MILL OF NASHAWANNUCK MF'G CO., EASTHAMPTON, MASS.
[Elastic Goods, Suspenders and Ribbons.]

The machinery consisted of 149 Looms (Knowles' and various kinds), 15 Spoolers, 14 Warpers, 11 Quillers.

Power required to drive all the above machinery to speed, 23.2 h. p.

EDWARD PAINTER, *Supt.*

TESTS MADE AT MILL OF GLENDALE ELASTIC FABRIC COMPANY, EASTHAMPTON, MASS.

The machinery consisted of 100 Looms, 10 Spoolers, 12 Quillers, 775 Braiders. Power required to drive all to speed, 25.4 h. p.

E. C. KOENG, *Supt.*

Test of Turbine Wheel and Power Required to drive Machinery in Mill at Natick, R. I.

To ascertain power required to drive machinery, the gate was opened until designated machines ran at regulator speed, then the power of wheel was found, with same head and gate opening.   The turbine replaced breast wheels, and the discharge from the turbine as shown by the old water mark in tail-race 35 feet in width was 8 inches less in depth than from breast wheels.

Test of Machinery, March 14, 1874.

The first test, shafting alone.   The gate 3½ turns open, with 21 feet, 3½ inches fall.   Wheel making 77 revolutions per minute, horse power, 43.

The second test, all the shafting and 457 Mason Looms, (print goods, 64 sq., 150 picks per minute.)   The gate 5½ turns open, with 21 feet, 1½ inches fall. Revolutions of wheel 77, horse power, 88.

The third test, all the above and 77 ring spinning frames, of 9,856 Rabbeth spindles, 6750 revolutions per minute, also 8 warpers, 8 spoolers of 64 spindles each, and 17 mules with 10,364 spindles.   The gate 9¼ turns open, with 20 feet, 9 inches fall.   Revolutions of wheel 77, horse power, 192.

The fourth test, all the machinery in the mill, or in addition to the above, 1 Kitson opener, 2500 revolutions, 6 30-inch Whitin's lappers, 3 beaters, each 2200 revolutions, 70 30-inch breaker cards with 125 revolutions of cylinders, with 5 Mason Railway heads, 2 doublers, 70 30-i .ch finisher cards with 125 revolutions of cylinders. with 5 Lanphear railway heads, 10 drawing frames with 59 deliveries, 6 slubber speeders with 420 spindles, 554 revolutions of flyers, 12 fine speeders, 1248 spindles, 770 revolutions of flyers.   The gate 10½ open, 20 feet, 5¼ inches fall.   Revolutions of wheel 76, horse power, 263.   Gate opened in full to get power of wheel, 20½ feet full, 291½ horse-power.

Nelson Mill, Winchendon, Mass.

Denims, Sheetings, and Colored Goods—

	H. P.
4 pickers, 64 cards, 7300 spindles, 2 drawing frames and 180 looms,	158.80
All the above. except pickers,	130 10
All except pickers and cards,	89.46
Only looms running,	57.85
Shafting,	39.33

Monohansett Mill, Putnam, Conn.

Two hundred horse-power drives two hundred and ninety two 40-inch wide looms to 140 picks per minute, 5632 frame spindles, 6768 mule spindles with all the other necessary machinery.

Eagle Mill, Connecticut.

This is to certify that I weighed up the power for John L. Ross, of the following machinery and shafting at his mill, in Eagleville, Conn., with a Dynamometer on main shaft, and the power developed was found as follows, to wit:

Test No.   1—Run the shafting, 1 dresser, 1 spooler, and 12 frames, indicating 27.64 h.p.

Test No. 2—shafting, 1 dresser, 1 spooler, 15 spinning frames,								30.81 "
Test No. 3—	"	1	"	1	"	18	"	34.86 "
Test No. 4—	"	1	"	1	"	18	"	34.86 "
Test No. 5—	"	1	"	1	"	15	"	31.12 "
Test No. 6—	"	1	"	1	"	12	"	27.27 "
Test No. 7—	"	1	"	1	"	12	"	27.27 "
Test No. 8—	"	1	"	1	"	15	"	31.12 "
Test No. 9—	"	1	"	1	"	18	"	34.86 "
Test No. 10—All of shafting connected to run the above machinery,								10.89 "

Experiments at Massachusetts Cotton Mills.

LOWELL, MASS., MARCH, 1872.

Trial of power required to drive 15 stretchers. (3d speeders) 52 spindles each =780 spindles. Speed main shaft of machine, 396 revolutions. Speed of flyer. 1121 revolutions. Frames driven by a train of 8 counter-shafts—two frames by each, except the last, which drives one. These shafts are driven, the first from the main line, and the others in succession from each other. 1st. Machines and shafting required 8056 lbs. per sec.=14.65 horse-power=537 lbs. or .976 horse-power each=10.3 lbs. per spindle=53.24 spindles per horse-power. 2d. Shafting and loose pulleys, 2000 lbs.=3.64 horse-power. 3d. Shafting alone, belts off, 732 lbs.=1.33 horse-power.

Trial of power to drive 6 throstle spinning frames, (warp), 5 having 128 spindles each, and one 112 spindles,=752 spindles, driven by a train of 6 counter-shafts, the first belted from the main line, and the others in succession from each other. This being an odd row of frames, only one frame is belted from each shaft. Spinning No. 20 yarn, cylinder running 750 revolutions, and flyers 4312 revolutions per minute. 1st. Shafting and loose pulleys, 1150 lbs.=2.09 horse-power. 2d. Shafting alone, machine belts off, 767 lbs.=1.39 horse-power. 3d. Frames and shafting, 6900 lbs.=12.54 horse-power.

Trial of power required for 112 looms, weaving 36-inch sheetings, No. 20 yarn, 60 threads to the inch each, warp and filling. Speed, 130 picks per minute. These looms are placed in the back part of the middle portion of No. 1 mill—one-half in the basement and half in the room above—being belted from 5 lines of shafting in the lower room. These shafts are driven in succession, one from the other, the first from the main line. Size of shafting, 2 3-16 inches, except the first piece in each line, on which the counter-pulleys are placed; these are of several different sizes, but about 2¼ inch on an average. The driving pulleys are 12 inch diameter, and the loom pulleys 14 inch. 1st. 112 looms with shafting lubricated with tallow. Average of several trials: 8870 lbs.=16.13 horse-power =79.20 lbs. per loom=7.24 looms per horse-power. 2d. The same, after oiling the journals of the shafting: 8492 lbs.=15.44 horse-power=75.82 lbs. per loom= 7.24 looms per horse-power. 3d. Trial of shafting and loose pulleys, lubricated with tallow. Average of several trials: 2876 lbs.=5.23 horse-power. 4th. Same after freshly oiling: 2245 lbs.=4.08 horse-power. 5th. Shafting alone, belts off: 913 lbs.=2.40 horse-power.

Trial of power required to drive 8 Lowell Machine Shop Mules, 624 spindles each, with Emerson's Dynamometer. Five mules were running on No. 22 yarn, spindles making 5500 revolutions per minute, and three mules on No. 37 yarn, spindles making 6230 revolutions per minute. 1st. The 8 mules including shafting. 12,250 lbs.=22.25 horse-power,=2.45 lbs. per spindle,=224 spindles per horse-power. 2d. Shafting alone, 17.10 lbs.=3.11 horse-power,=14 per cent. of the whole power. 3d. 8 mules without shafting, 19.16 horse-power=211 lbs. per spindle=260 spindles per horse-power.

Test of Machinery at the Alpaca Mill, Holyoke, Mass.

Looms made by George Hatterly & Sons, Keighley, Yorkshire, England. These looms were supposed to require but one-tenth of a horse-power each to drive them; 250 of them in use there. Two sets of four each were tried, each set taking exactly the same power.

	H. P.
Four looms (plain,) 40-inch reed space, 180 picks per minute,	1.13
Spinning frame, 144 flyer spindles, 2500 revolutions per minute,	2.60
Lister Comb, 18 inch nip, combing long wool,	.68
Preparer for comb, second of five, fair average of the set,	.69
Dandy roving frame, 24 spindles, 1300 revolutions per minute,	.78
Six spindle way box,	.68
Six spindle finisher,	.56

Many patents have been taken out for the purpose of protecting devices supposed to produce very light running spindles, but there are spinning frames in this vicinity (with unpatented devices,) 128 spindles each that run lighter than any frames that I have seen elsewhere; these are driven with 5-8 of an inch belt, and can and have been driven with belts of but 1-4 of an inch in width.

Test of Turbine and Power Required to Drive Machinery.

Clyde Bleachery, River Point, R. I.

To ascertain power required to drive machinery, the gate was opened until certain machines ran at speed, afterwards the power of the wheel was tested with the same gate opening, head and speed.

	H. P.
1st Test. Gate open 2 1-2 turns,	15.47

Driving shafting of mill and small pump.

2d Test   Gate opened 6 turns, — 53.18

One 5 bole water mangle, 1 Scotch starching mangle, 2 boles, 1 spindle calendar, 5 boles, 1 3-bole calendar, 1 5-bole calendar and 1 cloth winder.

3d Test. Gate opened 8 turns, — 62.73

All the machinery in the bleaching room, viz: 3 washing machines, 10 feet log, 2 washing machines, 6 feet log, 2 souring machines, 4 feet log, 1 chemic machine, 4 feet log, 1 liming machine, 4 feet log, and 3 squeezers.

4th Test.   Gate opened in full, — 75.34

All the above, with machinery in drying room additional.   The latter is 1 drying machine, 11 cylinders. 30x120 inches, 1 squeezer, 1 opening mangle, 2 shearers. 4 sets knives each, and 1 Canroy winder.

The 15.47 h. p. required to drive shafting must be deducted from the second, third and fourth tests to get the power required to drive the machinery named.

Memoranda of power required for operating certain bleaching, finishing and dyeing machines, at S. H. Greene & Sons' Bleach and Print Works, Riverpoint, R. I., tested with Emerson's Lever Dynamometer, April 1874.

	H. P.
Washing Machine with 2 boles—21 inches diameter. 10 feet long with squeezers attached; consisting of 2 boles—21 inches diameter, 12 inches long,	13.60
Limer, brown sour, chemic and white sour machines—2 boles each—21 inches diameter, 4 feet long, each required	3.01
Water mangle—5 boles,	11.39
Friction mangle—2 boles,	16.38
Calendar—5 boles,	7.53
Calendar—3 boles,	5.91
Calendar—4 boles, (one bole being a 4-inch spindle,)	8.07
Shearing machines—4 sets knives,	9.98
Burrows' patent dye beck—40 ps.,	3.86
Washing Machine, Madder Dye House, with 2 boles—10 feet long, 20 inches diameter, with squeezers—2 boles, 12 inches long, attached with cloth loose in water pit,	7.97
Hot water machines—2 boles, in dye house,	1.79
Canroy Winder—for printing machines,	6.32
Power to drive shafting and spring water pumps of bleachery, drying room and mangle, and finishing rooms for white work,	18.18

All the above were trials while the machines were at work. cloth threaded in.

A number of trials were made.   The above give the average in practical work.

HENRY L. GREENE.

Tests of Various Kinds of Machinery.

During the past ten years I have tested the power required to drive a great variety of machinery, but have kept no record of such until recently, because such tests to others are of but little value unless the conditions are exactly the same, which is unlikely to be the case.

The following were taken in the mills named and represent the power required to drive the machines while doing their regular work; by the tests it will be seen that the greater the number of spindles in a frame, the greater the number is likely to be per horse power. It will also be apparent that much depends upon the make of the frame.

DWIGHT MF'G CO., CHICOPEE, MASS. J. W. Cumnock, Agent, Nov. 1878.
MANUFACTURE SHEETINGS SHIRTINGS AND P. K'S.

Test of Lan phear frame, 128 Rabbeth spindles.

To drive empty spindles, required	1.06 horse power.
To drive spindle and bobbin, without connection,	1.14
Mean, from empty to full bobbins, required	1.30
Revolutions of drum per minute,	810
Computed revolutions of spindle per minute,	7800
Revolutions of front roll,	72
No. of yarn,	40
Length of travers on bobbin in inches,	5
Foot lbs. per spindle when at work,	336.5
Number of spindles per horse power,	98

Another Rabbeth frame, supposed to be exactly like the above required more power. Spindles per horse power, 92

Lowell frame, 202 light long spindles No. 4 mill.

Mean, from empty to full bobbins,	2.52 horse power.
Revolutions of drum per minute,	1025
Computed revolutions of spindles,	7800
Revolutions of front roll,	97
Length of travers on bobbin in inches	5¼
Number of yarn,	22
Foot lbs. per spindle,	412
Spindles per horse power,	78

To drive the cylinder and spindles, rolls stopped, required 1.96 horse power.

Lowell frame having 208 short spindles in No. 4 mill.

Mean, from empty to full bobbins,	3 31 horse power.
Revolutions of drum per minute,	1025
Computed revolutions of spindles,	7800
Revolutions of front roll,	97
Length of travers on bobbin in inches,	5¼
No. of yarn,	22
Foot lbs. per spindle,	525
Spindles per horse power,	63

These spindles were reduced in weight, then required 2.84 h. p. or 73.8 spindles per h. p. To drive the cylinder and spindles, the rolls being stopped, required 2.2 horse power.

Lowell frame, (old) 208 long spindles.

Mean, from empty to full bobbins,	3.54 horse power.
Revolutions of drum per minute,	1025
Computed revolutions of spindles,	7800
Revolutions of front roll,	95
Length of travers on bobbins in inches	5¼
No. of yarn,	22
Foot lbs. per spindle,	563
Spindles per horse power,	59

Whitin frame, 128 long spindles, in No. 1 mill.

Mean, from empty to full bobbins,	1.45 horse power.
Revolutions of drum per minute,	720
Computed revolutions of spindles,	5040
Revolutions of front roll,	82

Length of travers on bobbin in inches,   5½
No. of yarn,   14
Foot lbs. per spindle,   375
Spindles per horse power,   88

Another frame, same row, supposed to be exactly like the above, carried 98 spindles per horse power.

Biddeford frame, 144 long spindles, No. 5 mill.
Mean, from empty to full bobbins,   1.57 horse power.
Revolutions of drum per minute,   789
Computed revolutions of spindles,   5523
Revolutions of front roll,   78
Length of travers on bobbin in inches,   5¼
No. of yarn,   22
Foot lbs. per spindle   359
Spindles per horse power,   73

Biddeford frame, 144 Pearl spindles, No. 5 mill.
Mean, from empty to full bobbins,   1.91 horse power.
Revolutions of drum per minute,   797
Computed revolutions of spindles,   7000
Revolution of front roll,   92
Length of travers on bobbin in inches,   5¼
No. of yarn,   22
Foot lbs. per spindle,   439
Spindles per horse power,   75

Whitin one rail frame, 128 Buttrick & Flanders' spindles, in No. 1 mill.
Mean, from empty to full bobbins,   1.16 horse power.
Revolutions of drum per minute,   720
Computed revolutions of spindles,   6720
Revolutions of front roll,   100
Length of travers on bobbin in inches,   6¼
No. of yarn,   14
Foot lbs. per spindle,   307.5
Spindles per horse power,   107

Another frame in the same row supposed to be exactly like the above, required more to drive it only carrying 94 spindles per horse power.

Biddeford one rail frame, 144 Buttrick spindles in No. 5 mill, using Pearl bobbins.
Mean, from empty to full bobbins,   1.77 horse power.
Revolutions of drum per minute,   825
Computed revolutions of spindles,   7300
Revolutions of front roll,   98
Length of travers on bobbin in inches,   5¼
No. of yarn,   22
Foot lbs. per spindle,   405
Spindles per horse power,   81

Biddeford two rail frame of 144 Buttrick spindles, using Pearl bobbins.
Mean, from empty to full bobbins,   1.68 horse power.
Revolutions of drum per minute,   825
Computed revolutions of spindles,   7300
Revolutions of front roll,   98
Length of travers on bobbin in inches,   5¼
No. of yarn,   22
Foot lbs. per spindle,   385
Spindles per horse power,   85.5

Lowell Doubler, doffing 64 cards. **Required**, 4.28 horse power.

Howard & Bullock Slasher, 16 inch fans, making 1200 revolutions per minute. Yarn moving 35 yards per minute. **Required**, 5 04 horse power.

Lowell Machine Shop Looms. 12 on 36-inch goods, 64 picks per inch. 11 on 40-inch goods, 76 picks per inch, 145 picks per minute. **Required**, 4.08 h. p. or 5.6 looms per horse power.

Lowell Coarse Speeders, 40 spindles, .36 hank roving. Eight and half inch space, 12-inch travers. 1¼ roll, making 196 revolutions per minute. Flyers, 625 revolutions per minute. Required, 1.41 h. p., 117 foot lbs. per spindle or 283 spindles per h. p.

Intermediate, 56 spindles. .90 hank roving. 6¼-inch space. 9¼-inch travers. Front roll, 1¼ inches in diameter, making 200 revolutions per minute. Flyers 940 revolutions per minute. Required, 1.43 h. p., 340 foot lbs. per spindle or 89.2 spindles per horse power.

Fine, 72 spindles, 5-inch space, 8¼-inch travers, 2.83 hank roving. Diameter of front roll 1¼-inch, making 140 revolutions per minute. Flyers 1215 revolutions per minute. Required, 1.68 h. p., 783 foot lbs. per spindle or 42 spindles per horse power.

Two Drawing Frames, 3 to 1, 4 deliveries each. Roll 1¾ inch diameter, making 308 revolutions per minute. Required, 1.09 horse power.

Two Pawtucket Spoolers, 80 spindles each, or 160 per pair. Revolutions of cylinder 165 and of spindles 786 per minute. No. of yarn 22, warp. Required, .74 h. p. Spindles per horse power, 217.

Five Howard & Bullock Warpers (English.) Cylinder making 45 revolutions per minute. Width of section, 54 inches. Average No. of threads to each warper, 350. Required, .83 h. p., or .16 h. p. per warper.

CHICOPEE MF'G CO., CHICOPEE FALLS, MASS. George H. Jones, Agent, Nov. 1878.

MANUFACTURE COTTON FLANNELS, QUILTS AND SHEETINGS.

Test of frame having 256 Sawyer spindles, in a mill of that company.

To drive the empty spindles, required	1.26 horse power.
To drive bobbins before connection with yarn, required	1.46
Mean, from empty to full bobbins,	2.09
Revolutions of drum per minute,	863
Computed revolutions of spindle,	7612
Revolutions of front roll,	88
No. of yarn,	25
Length of travers on bobbin in inches,	5
Foot lbs. per spindle when at work,	269.6
Spindles per horse power,	122

WARP MILL, HOLYOKE, MASS. J. L. Burlingame, Agent, Dec. 1878.

MANUFACTURE WARPS.

Lowell Frame, 160 Sawyer spindles (old frame.)

Mean power required	1.87 horse power.
Revolutions of drum.	860
Calculated revolutions of spindles,	7166
Revolutions of front roll,	103
No. of yarn,	18
Travers on bobbin in inches,	5¾
Foot lbs. per spindle,	387
Spindles per horse power,	85.2

Another in same mill; New Lowell Frame, 160 Sawyer spindles.

Mean power required	1.74 horse power.
Revolutions of drum,	935
Calculated revolutions of spindles,	7480
Revolutions of front roll,	84
No. of yarn,	28
Travers on bobbin in inches,	5¾
Foot lbs. per spindle,	358
Spindles per horse power,	92.1

HADLEY CO., HOLYOKE, MASS. William Grover, Agent, Dec. 1878.

MANUFACTURE YARN, THREAD AND TWINE.

Whitin Frame, 144 Buttrick spindles, ring 1½ inches.

Mean power required	1.58 horse power.
Revolutions of drum,	913
Revolutions of front roll,	93
Calculated revolutions of spindle,	7606
Travers on bobbin in inches,	5¼
No. of yarn,	22
Foot lbs. per spindle,	361
Spindle per horse power,	91.2

Whitin Frame, 144 common long spindles, ring 1¼ inches.

Mean power required	1.62 horse power.
Revolutions of drum,	940
Revolutions of front roll,	87
Calculated revolutions of spindles.	6043
No. of yarn,	22
Travers on bobbin in inches,	5¼
Foot lbs. per spindle,	368
Spindles per horse power,	90

Two Whitin Frames, 160 long light spindles each, or 320 spindles per pair.

Mean power required	2.60 horse power.
Revolutions of drum,	941
Revolutions of front roll,	68
Calculated revolutions of spindles,	6761
No. of yarn,	40
Travers on bobbin in inches,	4½
Foot lbs. per spindle,	268
Spindles per horse power,	123

Whitin Frame, 144 Sawyer spindles, ring 1½ inches.

Mean power required	1.30 horse power.
Revolutions of drum,	926
Revolutions of front roll,	78
Calculated revolutions of spindles,	7408
No. of yarn,	30
Travers on bobbin in inches,	5¼
Foot lbs. per spindle,	298
Spindles per horse power,	111

Whitin 9-inch Slubber of 72 spindles, hank roving one-third.

Mean power required	.59 horse power.
Revolutions of roll,	95
Revolutions of spindles,	582
Foot lbs. per spindle,	268
Spindles per horse power,	123

Whitin Intermediate Frame, 120 spindles, hank roving 4¼.

Mean power required	.47 horse power.
One and one-eighth inch rolls Revolutions,	96.5
Revolutions of spindles,	850
Foot lbs. per spindle,	130
Spindles per horse power,	254

Whitin's Jack Roving Frame, 144 spindles, hank roving 15.

Mean power required	.48 horse power.
Revolutions of roll, 7½. Diameter of same in inches,	1½
Revolutions of spindles,	1086
Foot lbs. per spindle,	110
Spindles per horse power,	300

Whitin's Drawing Frame, 16 ends, 4 cans.

Power required	.58 horse power.
Revolutions of roll,	280

Fales & Jenks' Frame, 272 Rabbeth spindles, ring 1½ inches.

Mean power required	2.36 horse power.
Seven inch drum. Revolutions,	725
Revolutions of front roll,	100
Calculated revolutions of spindles,	6767
Travers on bobbin in inches,	5½
No. of yarn,	20
Foot lbs. per spindle,	286
Spindles per horse power,	115

Another Fales & Jenks' Frame, 272 Rabbeth spindles, ring 1½ inches.

Mean power required	2.15 horse power.
Revolutions of drum	725
Revolutions of front roll,	73
Calculated revolutions of spindles,	6767

No. of yarn,   30
Travers on bobbin in inches,   5¼
Foot lbs. per spindle,   261
Spindles per horse power,   126

Fales & Jenks' 1876, Twister, 248 Rabbeth spindles, two cylinders.
Mean power required   4.80 horse power.
No. of yarn,   40—3 ply.
No. of Traveler, 14. 2-inch ring.
Diameter of drum, 8 inches. Revolutions of same,   750
Three inch roll. Revolutions,   27¾
Diameter of whirl, 1 5-16 inch. Revolutions of spindles,   4562
Foot lbs. per spindle   639
Spindles per horse power,   51.6
    Two cylinders in the same frame can hardly be desirable.

Fales & Jenks' 1872, Single Cylinder Twister, 144 Rabbeth spindles.
Mean power required   1.74 horse power.
No. of yarn,   40—2 ply.
No. of Traveler,   16
Seven inch drum. Revolutions,   823
One and one-half inch roll. Revolutions of spindle,   43
One and one-sixteenth whirl. Revolutions of spindles,   5435
Foot lbs. per spindle,   400
Spindles per horse power,   82.5

Higgins' Sons & Co. Slubber, 60 spindles.
Revolutions front roll,   118½
Revolutions of spindles,   676
Required   1.02 horse power.
Spindles per horse power,   58.6

Higgins' Sons & Co. 7-inch intermediate frame, 128 spindles, hank roving 3¼.
Mean power required   1.58 horse power.
Diameter of roll 1½ inch. Revolutions   128
Revolutions of spindles,   1118
Foot lbs. per spindle,   408
Spindles per horse power,   81

Higgins' Sons & Co. (English) 5½ inch Jack Frame, 144 spindles, hank roving 11.
Mean power required   1.45 horse power.
Revolutions of roll, 83. Diameter of same in inches,   1¼
Revolutions of spindles,   1400
Foot lbs. per spindle,   333
Spindles per horse power,   99

English Twister, 286 Rabbeth spindles, 1¼ inch ring.
Mean power required   3.97 horse power.
No. of yarn,   36—2 ply.
No. of Traveler,   15
Eight inch drum. Revolutions,   600
Three inch roll. Revolutions,   23.5
One and one-fourth inch whirl. Revolutions,   3840
Foot lbs. per spindle,   458
Spindles per horse power,   72

Crighton & Son (English) Doubler, 16 ends, Lap 187 pwt. to the yard.
Driving pulley, making 600 revolutions per minute. Required, .55 horse power.

Boyd's (Glasgow) Spooler or winding machine, 50 spindles or drums. One side winding from three bobbins; the other side winding from three cops.
Driving pulley and drums, making 228 revolutions per minute.
Mean power required   .15 horse power.
Foot lbs. per spindle,   99
Spindles per horse power,   333

Pair of Dobson & Barlow (English) Mules, 832 spindles each.
Ten stretches in 4 minutes, 25 seconds.
Diameter of front roll, 1 inch. Revolutions of same per minute, 72
No. of yarn, 70. Calculated revolutions of spindles,   5663
Maximum force required,   7.83 horse power.
Spindles per horse power,   212.5

Pair of Mason Mules, 832 spindles each.
Ten stretches in 3 minutes, 55 seconds.
Revolutions of front roll,   78
No. of yarn, 70. Calculated revolutions of spindles,   6000
Maximum force required,   4.40 horse power.
Spindles per horse power   375

French Comber made by Hethrington & Sons, Manchester, England.
Making 62 strokes per minute. Required, .24 horse power.

Platt Bros' Jack Frame, 144 spindles, hank roving.
Mean power required   .73 horse power.
Roll 1⅛ inch diameter. Revolutions of same   61
Revolutions of spindles,   1181
Foot lbs. per spindle,   167
Spindles per horse power,   198

Kitson Picker (changed) using Whithead & Atherton's.
Whipper beater.
Diameter of roll, 9 inches. Revolutions of same per minute,   8¾
Revolutions of 24-inch whipper,   1130
Revolutions of 16-inch beaters,   1545
Revolutions of fans,   2000 and 1500
Yards of lap per minute,   6.86
Maximum force required,   10.24 horse power.

Whitehead & Atherton's Picker.
Diameter of Rolls 9 inches. Revolutions of same per minute,   8½
Revolutions of 24-inch whipper,   1070
Revolutions of 16-inch beater,   1380
Revolutions of fans,   1900 and 1340
Yards of lap per minute,   6.67
Maximum force required,   9.35 horse power.

Kitson's 2d Picker or Finisher.
Diameter of rolls 9 inch. Revolutions of same per minute,   7
Revolutions of 1st beater, 16-inch,   1475
Revolutions of 2d beater, 16-inch,   1410
Revolutions of fans. 1430. Yards of Lap,   5.5
Maximum force required,   7.8 horse power.

Whitehead & Atherton's 2d Picker or Finisher.
Diameter of rolls, 9 inch. Revolutions of same per minute,   7½
Revolutions of 1st beater, 16-inch,   1410
Revolutions of 2d beater, 16-inch,   1410
Revolutions of fans,   1374
Yards of lap per minute,   5.9
Maximum force required,   6.64 horse power.

The Kitson picker had a six inch belt, the Whitehead & Atherton a four inch;
by timing the two and weighing laps, a difference of more than ten per cent. was
found in favor of the Kitson, but this was done away with by soaping the pulleys
and belt of the Whitehead & Atherton machine. As arranged, for doing the
same amount of work, each required the same power.

THE EVOLUTION OF ONE OF EMERSON'S PATENTS.

We print herewith an article from the *Boston Advertiser* of Nov., 1889, describing a business that originated with James Emerson, of Willimansett, and which was under his control until 1860. He commenced in 1852. The windlass was so radically different from all previous devices for the purpose, that it was laughed at by seafaring men, particularly naval officers, etc. Four years of persistent effort and a gift to an impecunious ship-owner gained the privilege of putting one on a ship. The war through the improvised battle ships from the merchant service introduced it into the navy. Perhaps some of the readers of this will recollect about a year since reading of the "Gov. Ames," a five masted schooner, being dismantled on the "Georges" and that her salvation depended on her windlass. The patterns for that windlass were made by Mr. Emerson or from his plans. A 2¼ inch chain weighs 15 tons or 40 pounds to a link, the two chains and anchors 37 tons; the windlass has to sustain not only that weight but the entire strain the two chains will hold, and such chains often part and let vessels go ashore. Yet after nearly 40 years of continued labor upon devices for ships, mills, hydraulics, dynamics and steam heating devices, it is a pleasant thought that of the numerous lives and millions of property often dependent upon his judgment, no life nor serious loss of property has ever occurred.

The circular of Emerson, Walker & Thompson, of 11 Leadenhall street, London, Eng., of 1885, claims to have fitted up 6000 vessels with the windlass.

A little more than 12 years ago travelers across "Red Bridge," in the eastern suburbs of Providence, noticed a small wooden building erected not far from the bank of the Seekonk river. A modest sign over the door told that this was the new plant of the American Ship Windlass Co. The building soon became too small. In six months a second fully as large as the first went up by its side. The next year there was another enlargement and the next year still another. Thus, year by year, the plant has grown, until, at the present time the value of the land and buildings of the American Ship Windlass Co. is fully nine times that of the original plant. Extensions are still in progress, for the business is still increasing rapidly, and to-day the sound of the hammer is heard as a new building is in process of erection upon the site of the old. Its windlasses and capstans were well known, while they were manufactured under the old regime. As long ago as 1856 the Massachusetts Charitable Mechanics' Association* awarded a gold medal to the Emerson patent windlass. The present American windlass is based upon the Emerson patents.

Since 1856 the windlass has received many medals and other awards from fairs and expositions and has always taken the highest award or prize offered for windlasses whenever exhibited. More than 20 years after the Emerson windlass received the gold medal of the Massachusetts Charitable Mechanics' Association, the same society again recognized its merits in a similar manner. A gold medal was also awarded it by the World's Industrial and Cotton Centennial Exposition held at New Orleans in 1884-5. The North, Central, and South American Exposition of 1885-6 granted to it the first degree of merit. The only award given for windlasses and capstans at the U. S. Centennial Exposition was granted to the American Ship Windlass Co.

The best proofs of the complete success of this windlass is found in the fact that the finest steam and sailing vessels afloat are fitted with these machines. The U. S. government has repeatedly recognized their merit. The new steel U. S. cruisers, the Chicago, Boston, and Atlanta are furnished with them, as are also the dispatch boat Dolphin, the Thetis, Bear, Baltimore, Vesuvius, Yorktown, and Petrel; the coast survey vessels Hassler and Blake; the lighthouse boats Haze, Dahlia, and Myrtle, and so great a number of the U. S. revenue cutters that to enumerate them would be to write almost a complete list of these vessels. Steamers of the Mallory, Pacific Mail, Ward's, Ocean, Clyde, Morgan, Old Colony, New Brazil, Cromwell, Norwich, Winsor, and many other lines, transatlantic and coastwise, are furnished with the "American" windlasses, which have always given the fullest satisfaction. At present at least 95 per cent. of the windlasses made and sold in the American market come from the works of the American Ship Windlass Co.

---

*The same Association also awarded a gold medal to the Emerson Power Scale, an instrument that now has no competitor.

Ship's Windlass.

It has often happened, when low results have compelled me to report unfavorably of turbine plans, that the designers have intimated that if I had experienced the vicissitudes of an inventor's life, more leniency would be shown. The Patent Office Reports will show that quite a number and variety of patents have been granted to me, and the records of the office will show a still larger number of applications for others, some of which were rejected, others granted, then abandoned. Two causes have prevented me from realizing much pecuniary benefit from patents. First, because my inventions have been a generation before the age. Secondly, because my plans have been very expensive to develop. I have never cared to immortalize myself by the invention of a mouse trap, pie fork or clothes pin. One patent I have ever felt ashamed of; it was taken out under the following circumstances. A lady friend as a joke asked me to get up a device to keep her husband's mustache out of his coffee. A plan was readily found, consisting of a peculiarly shaped comb with guarders and nippers. Two young ladies asked to have a patent taken out and assigned to them. It was applied for. The model proved so attractive that it was purloined, and the commissioner had to send for another; in the meantime, the man for whom the plan was devised took his comb to the Fifth Avenue Hotel, N. Y., and exhibited it; in less than a month several hundred dollars' worth of orders were received from fancy goods dealers. By that time the joke had become stale and the matter was dropped in disgust, though I believed then, and continue in the same belief now, that more money could have been made from that than from any other patent granted me. It is not my purpose to go into a general history of my inventions but there are several, now very popular and lucrative devices, patents of others, that were offered to leading men thirty years since by myself; the plans were pronounced chimerical. The self coupling for cars, steam brakes and heating cars by steam—the plans, almost identical with those now so common, were urged by me upon the managers of the several railroads as early as 1850, but in vain. My experience in introducing the ship's Windlass, herewith illustrated, will be sufficient to show my turbine friends that I have known something of an inventor's troubles. Readers who are not acquainted with such matters, may, by looking in "Webster's Unabridged," see an illustration of a ship's windlass; such as was in use on all merchant vessels of any size forty years since. Such windlasses were made of a single oak log, varying in length from six to twenty-five feet, according to the size of the vessel; three or four turns of the cable would be wound around the windlass, the inner or loose end of cable next to the bitt; in heaving in, the chain, like a nut or screw, would work towards the middle or pawl bitt, so that after a few turns the cable would have to be made fast forward of the windlass, then the three or four turns of the chain slipped back towards the bitt. The cables were stowed below by the mainmast in order to have a long stretch of chain back of the windlass to help hold it from slipping when icy or muddy. Now, by considering that the largest chain cables are made of round iron, 2¼ inches in diameter, the links being eight inches wide and twelve in length—fifty pounds to each foot in length of chain, each cable five hundred and forty feet in length with an anchor of three tons in weight at the end—and it will readily be understood that a crew had a hard job to handle such a cable, more particularly in deep water; besides, it was often impossible to get an anchor ready to let go before a ship would be ashore, for it was always necessary to haul up sufficient length of cable from the chain locker to reach bottom before the anchor could be let go; for to drop a heavy anchor and chain in ten fathoms, or sixty feet of water and allow it to bring up on the windlass, would endanger the safety of cable, windlass or bows of the ship; consequently, sufficient length to reach bottom had to be *ranged* forward of the windlass as a preliminary step, the turns of the cable around the windlass adding much to the labor. A careless word drew my thoughts to this matter, and in 1850 some of the plans in the illustrations were presented to seafaring friends, and by them very coolly received: "What! Trust lives and such immense amounts of property to cast iron gears? might as well have a glass windlass. How are you agoing to handle the swivels and shackles, placed at every fifteen fathoms of cable?" said another. A capitalist offered to assist me, if a certain old sea captain approved of my plans; they were submitted to him; he was one of the old school, a regular old salt. He examined the plans, a model in fact, worked it, hove in and let go anchor for an hour; then got up, came to me and exclaimed: "Well, I have seen a good many d——d fools, but you seem to be the biggest one of the lot. What! do you

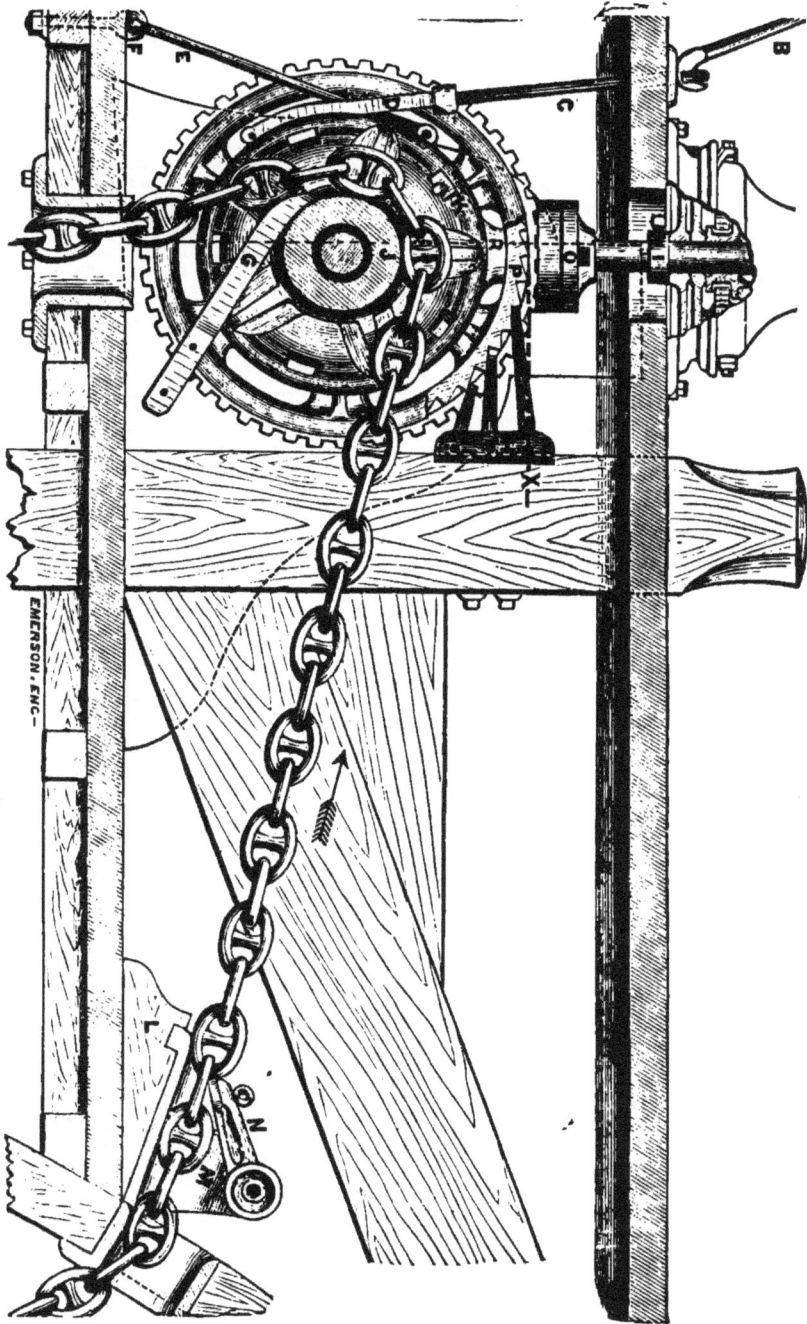

EMERSON · ENG ·

want to commit murder by the wholesale, with your d——d cast iron jimcrack? Why, let a ship anchor in a gale, and ship and crew would go to h–ll together." My capitalist declined to go into the business. Finally, one was found willing to help; then the objection was raised that the links of cables varied in length so much that it would be impossible to handle them in the way proposed. My plans were modified, and a device designed and patented for obviat ng the difficulty; then an owner was found willing to furnish his ship with chains of a better make—the links being sufficiently equal in length to work on the *grubs* or chain wheels illustrated. This, of course, rendered all of the trouble and expense of the special plan patented useless. A windlass, costing some eight hundred dollars was constructed, but before being finished the ship owner had been frightened so that he did not dare risk its use; it was offered as a gift to Donald McKay, Paul Curtis and other leading ship builders of Boston, New York and other places. One day while listlessly wandering around, hoping against hope, I met a Captain R. B. Forbes, a man who through various causes had been flattered until he had got a high idea of the value of his own opinion. Timidly approaching him I asked if he would be so kind when passing by as to take a look at my windlass, and give an opinion of its merits. "What!" he said, "that big coffee mill? I have seen it and can give my opinion now; which is, that it is worth nearly a cent per pound for old iron, less cost of breaking it up and carting it to the foundry." After months of waiting a place was found for it on a large ship being built at Kennebunk Port, Me. (Wm. Lord, Jr.) When the ship was ready to sail the captain insisted that I should go with the ship to St. John, N. B, and work the windlass. The tide there is strong and the ship had to be moored in fifteen fathoms of water; we arrived an hour before daylight; the morning dark and foggy; the port captain came on board and took charge. Some one said: "Captain, we have a patent windlass and expect to moore quick." "The windlass ain't worth a daum," was the reply—he supposing it to be an English capstan that some one had put upon a few ships. A steam tug took us to our berth, and the order was given to let go starboard anchor. In less than three minutes we were riding with forty-five fathoms of chain out; the tug towed hard to port; the starboard chain was eased away to eighty fathoms; the port anchor was let go, and in twenty minutes we were safely moored and the tug called alongside for the captain of the port, who, before leaving, held his lantern to my face, grunting ont, "d—ned yankee, saved me half a day's time." The captain of the ship congratulated me as being sure of having made my fortune by the invention. But prejudice is not so easily overcome. To ask a builder or owner of a ship to use one of my windlasses, was certain to bring some sneer as to whether I proposed to send an engineer or machinist with it. Pilots and insurance agents were strongly opposed to it; after much urging one was placed upon the insurance agent's steamer as a gift, the old windlass to be replaced if mine was not liked. Impecunious ship builders, who found it hard to get the old windlass upon credit, favored mine, and were ready to pay for it in large promises, and it was really through such that it gained a place. In time, the better builders would listen, but were still shy; an engineer was necessary, was the cry; besides, if they lost their cables it was generally impossible to replace them with others of the same length of link. This continued until the convenience of the windlass had become so apparent that commanders of ships began to importune for them. In the mean time, one had been placed upon the Pomona, ship of a thousand tons, belonging to the "Dramatic Line," from New York to Liverpool. As those ships brought large numbers of immigrants, I had watched her proceedings closely because of what had been said about trusting lives and property to the strength of cast iron gears. Suddenly a rumor came that the Pomona had been lost, and that four hundred passengers had gone down in her; little was known, only that she had been wrecked on the coast of Ireland. It was two or three weeks before particulars were received, and in that time it seems as though I never slept. Four hundred lives were more of a responsibility than I felt capable of carrying in peace; but the time named brought relief. The ship struck before there was thought of danger. It may as well be stated here, that while I had control of the manufacture of the windlass, no loss ever occurred through its use; on the contrary, ships were often saved through the immense strain that could be brought to bear on the cables when heaving them off shore. In only a single instance was a tooth from a gear broken, and that was when two boat crews from a man-of-war was added to the ship's crew for the purpose of heaving up the anchor, while it was a-foul of the man-of-war's anchor; both were hove up together. The merits of the windlass had become so well established previous to 1860, that I had furnished that and other devices to the Russian and Egyptian governments; had had orders from China, Spain, Italy, England, Scotland, and throughout this country wherever ships were built. The following certificates will show the change of opinion.

MASSACHUSETTS CHARITABLE MECHANIC ASSOCIATION, 1856.

Emerson's Patent Windlass, worked by slow or fast power by a Capstan on the forecastle. This machine can perform with four men, the work usually requiring a dozen, and is a valuable element in the safety of life and property, more especially in these days of "ordinary seamen." To this valuable machine the Committee award a *Gold Medal.*

R. B. FORBES,
JOHN S. SLEEPER,
BENJAMIN L. ALLEN,   } *Committee.*
JOHN H. GLIDDON,
ELIAS E. DAVIDSON,

BOSTON, April 10, 1860.

This will certify that after a careful inspection of Emerson's Patent Windlass, together with some acquaintance with its working on the steamer R. B. Forbes, we are satisfied that it is superior to any modern Patent Windlass that we have seen. It has great power and can apparently be used with ease and safety.

CHARLES PEARSON, }
EBENEZER DAVIS,   } *Marine Inspectors.*
RICHARD BAKER,

BOSTON PILOTS.

The undersigned, having known the Emerson Patent Windlass for several years, believe it to be superior to any Windlass in use. Its great power or speed renders it peculiarly applicable for getting under way in heavy weather, or where there is but little room, and the improved lever and screw nippers render it perfect for bringing a ship to.

JACOB K. LUNT,   SAMUEL C. MARTIN,   JOHN T. GARDNER,
H. A. TEWKSBURY,   WM. F. TEWKSBURY,   JONATHAN BRUCE, Jr.
STEPHEN BURROWS,   WM. CRISPIN,   P. H. CHANDLER,
ALFRED NASH.   W. G. BAILEY,   A. F. HAYDEN,   ROBERT KELLY.

BOSTON, March 28, 1860.

MR. EMERSON:—Recently at the Cape of Good Hope, I had many chances to test the power of your patent Windlass. As I had both anchors down it was often necessary to heave up to clear the chains, and I have no hesitation in saying, that for power or speed, or for general convenience, your Windlass is far superior to any other that I have ever seen.   JAMES HALL,
Master of Bark Wm. G. Anderson.

MR. EMERSON:—I readily join Captain Hall in speaking of the satisfactory working of your patent Windlass on board the Wm. G. Anderson, I can say also, that the Windlass put by you on the bark Ethan Allen, has been very severely tested (the bark having parted her largest chain) and has given entire satisfaction. I recommend them to ship owners with great confidence.
BOSTON, March 28, 1860.   EDWARD BOYNTON, Owner.

BUFFALO, December 10th, 1867.

CAPTAIN JAMES AVERELL. DEAR SIR:—I have used the Emerson Windlass purchased of you for the barque Annie Vought of Buffalo (one thousand tons), and have always found it work to my entire satisfaction. For strength, compactness and convenience, it cannot be excelled. Its motions are simple and positive, hence there is no lost motion; it does away entirely with the old tedious way of ranging chain before letting go anchor. With the Emerson Windlass the chains are always ready for letting go, and as a matter of great importance, can never get foul on the windlass, which is frequently the case with the old style windlass. Its power is unlimited. We had occasion to use our best bower anchor (3500 pounds), with 65 fathoms chain (1 3-4 bar link). It required only one man to let go anchor and veer away chain, whereas, with the old style windlass it would require the whole crew. And in heaving up, it required only 35 minutes till the anchor broke ground, then half our crew (five men) were sufficient to work the windlass, leaving the others free to work the yards, make sail, or be wherever required. It is superior to any windlass I have ever seen in use, and would unhesitatingly recommend its general adoption for large class vessels especially.
Respectfully,   JAMES G. ORR, Master Barque Annie Vought.

Commodore Stringham and Gregory were very friendly and aided me in many ways as did several of the naval constructors; with others, John Lenthal, chief of the bureau of construction at the navy department; but the prejudice was too strong to allow of the use of my windlass on naval vessels. "If we should lose a chain in some out of the way port, we could not replace it, perhaps, with anything near what would be required." So I went to work and got up a plan that would take any sized chain, spending much time and money in doing it; then carried it to the naval constructor who had been the strongest in that objection. "By George! that is simple, I didn't think it could be done; but, after all, the other plan is best; a chain is not often lost," was his comment. Such was the frivolous treatment experienced for years. Owing to the war, many merchant vessels having my windlass in use were turned into naval or war vessels. The following certificate will show how the windlass answered its purpose:

U. S. STEAMER SOUTH CAROLINA, OFF GALVESTON, Aug. 20, 1861.

SIR:—In accordance with instructions from Flagg Officer Mervine, which direct me to inform the Department as to the merits of the "Emerson Windlass," now in use on board this vessel, I have the honor to report that it has been used by us constantly for the past three months, and that our opportunities for judging of its utility have been amply sufficient. We find it certain and quick in its operations, not only in heaving in, but also in veering; it is strong and compact, taking up less room than any thing of the kind I ever saw. In fact, it reduces the tedious, old fashioned, and I may say, often dangerous way of handling our heavy anchors and chain cables, to the simple process [in heaving up] of walking around with the capstan, the chain taking care of itself as it comes in; while in veering, a small "plug" is removed, leaving the whole control of the heaviest chain in the hands of one man, who by the aid of a "lever" on a friction band, manages it with perfect ease. Besides, it is always read'. I have been lying with fifteen fathoms of chain out, on several occasions, and have, without giving previous notice to any one, been under weigh and steaming along at the rate of four knots, in five minutes after the order was given to man the capstan. It will be seen, therefore, that the facility thus afforded for getting underweigh is a positive saving of fuel in a blockading steamer, for otherwise, she might deem necessary, for entire efficiency, to keep underweigh almost all the time. Respectfully, I am sir, your ob't serv't.

JAMES ALDEN, Commanding U. S. Steamer South Carolina.
Hon. GIDEON WELLES, Secretary U. S. Navy, Washington, D. C.

In 1858 there was no chain making in this country, our cables all being imported. The following circular will explain itself. The lengths named were readily adopted, and I presume still continue to be the standard lengths.

To CHAIN MANUFACTURERS OF GREAT BRITAIN. *Gentlemen:*—Being engaged in the manufacture of Windlasses which hold the chains by the links instead of by a turn around the windlass, I often find a great difference in the length of links of the different manufacturers' chains. This seriously affects the working of the windlass, and is sometimes very inconvenient in replacing a lost chain. As this kind of windlass and capstan is fast taking the place of the wooden windlass, it would be much better to have some regular length of link for each size chain. I herewith give a graduated scale of lengths for different sizes, which is very near the same as the scale of the Messrs. H. Wood & Co. of Liverpool; it, however, is a little more even than theirs. The shackle is another cause of difficulty. These should be made so that the inside of them, that is from the inside of the bolt to the inside of the other end, should be the same length as the inside of a link, and then the shackle link in the end of the chain which the bolt of the shackle goes through, should be long enough to make up for the butt of the shackle. There should be a long link at one end only, of each piece of chain, which should be for the bolt end of the shackle. There should also be a good swivel next the anchor shackle in all cases.

JAMES EMERSON.

Inches.		Stud Link.	Length.	Inches		Short Link.	Length
1		. . . .	5 7-8	1-2		. . . .	2 3-8
1	1 16	. . . .	6 1-4	9-16		. . . .	2 3-4
1	1-8	. . . .	6 1-2	5-8		. . . .	3
1	3-16	. . . .	6 3-4	11-16		. . . .	3 1-4
1	1-4	. . . .	7 1-8	3-4		. . . .	3 1-2
1	5-16	. . . .	7 3-8	13-16		. . . .	3 7-8
1	3-8	. . . .	7 3-4	7-8		. . . .	4 1-8
1	7-16	. . . .	8 1-8	15-16		. . . .	4 3-8
1	1-2	. . . .	8 1-2	1		. . . .	4 5-8
1	9-16	. . . .	8 7-8	1	1-16	. . . .	5
1	5-8	. . . .	9 1-4	1	1-8	. . . .	5 3-8
1	11-16	. . . .	9 5-8	1	3-8	. . . .	5 3-4
1	3-4	. . . .	10	1	1-4	. . . .	6 1-8
1	13 16	. . . .	10 1-4				
1	7-8	. . . .	10 1-2				
1	15-16	. . . .	10 3-4				
2		. . . .	11 1-8				
2	1-8	. . . .	11 3-4				

Our views correspond with the above.
FEARING, THACHER & CO.,
WHITON, BROWNE & WHEELWRIGHT,
BAXTER & SUMNER,
J. NICKERSON & CO.,
J. BAKER & CO.,
Imp'ters of Chain Cables, Anchors, &c.

BOSTON. August, 1858.

The American Ship Windlass Company.

The following article copied from the Boston Commercial Bulletin of August 24, 1878, will show what has become of my windlass: "The American Ship Windlass Company, of Providence, R. I., seems to be a good illustration of the results which are achieved at the present day by division of labor, and by the devotion of all the skill and capital of an entire establishment, as far as practicable, to a single branch of manufacture. The productions of this company, comprising windlasses for every size and class of vessels, have attained a marked degree of excellence, and at our Centennial Exposition they received the only award given for windlasses and capstans.

The American Ship Windlass Company was established in 1857 and incorporated in 1860, and up to the present time, they have made nearly 3000 windlasses. John Roach & Son, of Chester, Pa., use their windlasses exclusively, putting them into all their vessels. Nearly all of the United States revenue cutters are now provided with them.

The company are now building windlasses for vessels which are being constructed at all of the different points along the coast of Maine, and for the steamers "Miantonomo," and "Puritan," which John Roach & Son have now in process of construction at their yard, and Wm. Cramp & Sons, of Philadelphia, are putting in the American Company's windlasses upon the steamers which they have built for the Russian Government.

The windlasses of the American Company are made to be operated either by hand, messenger chains or steam, and six different kinds of windlasses are manufactured for either of these motors. The windlasses are also made in eleven different sizes, for cables varying from $\frac{1}{2}$ inch to $2\frac{1}{2}$ inches in size; and the company are consequently able to provide windlasses for the smallest yachts as well as for the largest ships. Their works are model ones and are supplied with all of the latest and most improved machinery and other appliances, including many tools specially designed and constructed for the company. They are located on East River Street, near the Red Bridge, and are under the active management of Frank S. Manton, agent, and George Metcalf, treasurer; and one evidence of the executive ability of the managers is the perfect system which pervades the establishment throughout.

Hydraulic Mortars and Cements.

Certain limestones, which contain upward of 10 per cent. silica, possess the property, when burned, of forming a cement or mortar which hardens under water. Such limestone is called hydraulic lime, and the mortar is called hydraulic mortar. This stone, before burning, consists of a mixture of carbonate of lime and silica, or a silicate, chiefly a silicate of alumina. The latter is insoluble in hydrochloic acid, hence remains undissolved when the stone is treated with this acid, but in burning this silicate is fluxed by the alkaline carbonates and becomes soluble in acid, the carbonic acid being expelled. When common lime is slacked it swells enormously and develops a great deal of heat; this is not the case in slacking hydraulic lime, which absorbs water without any considerable increase of temperature.

If ordinary lime be mixed with a suitable quantity of silica or sand, an artificial hydraulic mortar is obtained, to which we apply the name of cement. These cements may be either natural or artificial. The former are found in volcanic regions, having been produced by the terrestrial heat. Pozzuolana, found at Pozzuoli, near Naples, is a natural cement of the following composition: Silica, 44·5; alumina, 15·0; lime, 8·8; magnesia, 4·7; oxide of iron, 12·0 (with oxide of titanium); potash and soda, 5·5; water, 9·3; total, 100·8.

The quantity of lime, is, however, so small that it requires to be mixed with ordinary lime to form hydraulic mortar. It was employed in combination with an equal quantity of lime in building the Eddystone Lighthouse.

Artificial cement, also called "Roman cement," has been manufactured in England on the Thames and in the Isles of Wight and Sheppey, since 1796. It is made by burning the calcareous nodules which overlie the chalk in that country. A sample analyzed by Michaëllis contained: lime, 58·38; magnesia, 5; silica, 28.83; alumina, 6·40; oxide iron, 4.80. When mixed with water it hardens in fifteen or twenty minutes, and possesses great firmness and strength.

Portland cement was patented in England by Joseph Aspdin in 1824. He took the limestone of Leeds, pulverized and burned it, then mixed it with water and an equal weight of clay to a plastic mass. When dry this was broken up and burned again until all the carbonic acid was expelled. It was then pulverized and ready for use. Pasley made it from chalk or limestone with Medway river clay, which contains salt. Pettenkofer suggests that cement is improved by soaking the clay in salt water.

Portland cement is now made, says Wagner, by making bricks of an intimate mixture of limestone and clay, drying them in the air and burning them in a tall shaft furnace from 45 to 100 feet, 12 feet in diameter, with a strong grate 4 feet from the bottom. It is charged with alternate layers of coal and cement stone. The properties of the cement are largely dependent on the temperature employed in burning; a white heat is best, but if the temperature is too high it will no longer unite with water, and may even be melted to a glass. If the temperature does not exceed a red heat it unites readily with water and gets hot like ordinary lime, but possesses very little strength. The color changes with the burning and forms a criterion for judging the quality. In normal condition it forms a gray, sharp powder, with a shade of green, but not glassy.

The manufacture of Portland cement is now carried on in every part of the world where limestone and clay are to be found. In order to obtain a good cement, not only must the proper heat be employed in burning, but the proper proportion of clay, usually 25 per cent., must be used, and the clay must have certain properties, such as a large proportion of silica, must be very finely divided, and must be very intimately mixed with the limestone. Analysis of Portland cement from various sources show the percentage of lime to vary from 55 to 62; silica, 23 to 25; alumina, 5 to 9; oxide of iron, 2 to 6; soda and potash, usually less than 1 per cent.

Horse Power and other Matters.

When Watt began to introduce his steam-engines, he wished to be able to state their power as compared with that of horses, which were then generally employed for driving mills. He accordingly made a series of experiments, which led him to the conclusion that the average power of a horse was sufficient to raise about 33,000 lbs. one foot in vertical height per minute, and this has been adopted in England and this country as the general measure of power.

A waterfall has one-horse power for every 33,000 lbs. of water flowing in the stream per minute, for each foot of fall. To compute the power of stream, therefore, multiply the area of its cross section in feet by the velocity in feet per minute, and we have the number of cubic feet flowing along the stream per minute. Multiply this by 62½, the number of pounds in a cubic foot of water, and this by the vertical fall in feet, and we have the foot-pounds per minute of the fall; dividing by 33,000, gives us the horse-power.

For example: a stream flows through a flume 10 feet wide, and the depth of the water is 4 feet; the area of the cross section will be 40 feet. The velocity is 150 feet per minute—40x150=6000=the cubic feet of water flowing per minute. The fall is 10 feet; 10x375,000=3,750,000=the foot-pounds of the waterfall. Divide 3,750,000 by 33,000, and we have 113.63 h. p., as the power of the fall.

The power of a steam-engine is calculated by multiplying together the area of the piston in inches, the mean pressure in pounds per square inch, the length of the stroke in feet, and the number of strokes per minute, and dividing by 33,000.

Water-wheels yield from 50 to 91 per cent. of the water. The actual power of a steam-engine is less than the indicated power, owing to a loss from friction; the amount of this loss varies with the arrangement of the engine and the perfection of the workmanship.

To compute the number of teeth in a pinion to have any given velocity. Multiply the velocity or number of revolutions of the driver by its number of teeth or its diameter, and divide the product by the desired number of revolutions of the pinion or driven.

To compute the diameter of a pinion, when the diameter of driver and the number of teeth in driver and pinion are given. Multiply the diameter of driver by the number of teeth in the pinion, and divide the product by the number of teeth in the driver, and the quotient will be the diameter of pinion.

To compute the number of revolutions of a pinion or driven, when the number of revolutions of driver and the diameter or the number of teeth of driver and driven are given. Multiply the number of revolutions of driver by its number of teeth or its diameter, and divide the product by the number of teeth or the diameter of the driven.

To ascertain the number of revolutions of a driver, when the revolutions of driven and teeth or diameter of driver and driven are given. Multiply the number of teeth or the diameter of driven by its revolutions, and divide the product by the number of teeth or the diameter of the driver.

WHAT IS POETRY?

The best explanation that occurs to me may be found in Paine's Age of Reason; but what seems poetry to one may seem trash to another. The gloomy Puritan liked that of the "Hark from the tombs" order, while the unperverted nature admires something more human.

Popularity has much to do with the average taste in poetry as it has with dress.

We often see in some standard print an essay, say by Jonathan Dubkins, bursting with admiration for the versatility of Shakespeare's works, or of the intense beauty of Milton's Paradise Lost; but it is rare to find a copy of either that seems much worn, while the popular seal skin cloak or an imitation may be seen upon the form of every girl or woman that can procure it, from which fact it would almost seem that the pretense of admiration for those authors is less than claimed, and that the purpose of such essays is more to display the greatness of the Dubkinses than those written of.

For myself, admiration for poetry only comes as it touches my feelings, and it may be found in prose as well as in verse; much of the book of Job seems poetry to me. I would sooner be the author of Pope's Essay on Man, than of any other English work, because I believe it to be an inspiration from a higher source, as I also do of Uncle Tom's Cabin.

Ninety-nine per cent. of the pretended admiration for Shakespeare, Tennyson, and many other popular heroes, may justly be attributed to pure flunkyism. It is true that many popular sayings may be found in works of Shakespeare, and equally true that the same may be found in works written two thousand years ago. The Comedy of Errors is taken in the lump from Plautus Comedies, the two Dromios being added to bring it down to an Englishman's idea of humor. There may be immense invention in his works, but such have not caught my attention.

"'Tis not so deep as a well, nor so wide as a church door; but 'tis enough, 'twill serve: ask for me to-morrow, and you shall find me a grave man," may be witty, but certainly is not common with those wounded to the death, any more than it is for those in deep sorrow, as were Juliet and her nurse, to make puns, and dirty puns at that.

The author of a dime novel would scorn the conception of such wretched stuff as makes up the Taming of a Shrew.

Lavinia, in Titus Andronicus, is certainly a marvelous creation. A young lady of our time having her tongue cut out and hands cut off would feel sick, to say the least; not so with Shakespeare's maiden, "Good uncle Marcus, see how swift she comes!"

Shylock has been a butt for general execration, but in excuse it should be borne in mind that for centuries his people had, through superstitious prejudice, been treated worse than the dogs of the street, and it is hardly to be wondered at that he turned upon one of his oppressors. The sagacity of Portia was not phenomenal.

LISLE THREAD.

Lisle thread proper is prepared from *pure cotton*—the finest staple that can be had, the best quality of Sea Island being generally used. However, of late years it has been found by observation and experience that the softness and pliability necessary to the easy and safe working of this yarn or thread in hosiery and glove frame, as well as in the machinery making fine imitation laces of it, are best secured by the use of South American (Pernambuco) cotton, the latter being less harsh, softer, more elastic and regular in fiber, as well as being very fine in quality. The peculiarity of this thread, says the *Economist*, is its hard finish and the peculiar twist or manipulations which it undergoes before being ready for use. Each thread or strand passes through a flame, which divests it of all attaching fiber. This thread is also more elastic than the finest linen thread and breaks less. It also gives the finished article a more brilliant appearance, and is less costly than the latter. It derives its name from Lisle, a town in France, where it was first manufactured to a large extent, and, like many of the industrial arts, was originally brought from the East. It is now not only extensively produced in France, Belgium, and in other portions of Continental Europe, but in Great Britain as well, and is sometimes called "Scotch thread," when made in that country, in contradistinction to that made on the Continent. It is not only used largely for gloves, hosiery, and trimmings, but also quite extensively in the manufacture of imitation laces, embroideries, etc. We believe some few years ago a suit was before the United States court of this district, which involved the question of what constituted Lisle thread gloves, and was decided in favor of the importer, who proved that Lisle thread proper was made of the purest and finest cotton, and not of flax, as some maintained who had not investigated the subject of its manufacture. As far as we can learn, none of this thread is made in this country, although we understand attempts have been made to manufacture it, but from the cost and light demand prevailing were abandoned. The imports of it are also light, being confined chiefly to a few of our hosiery manufacturers.

En passant it may not be amiss to state that all the dictionaries fail to give a definition of the word "Lisle," which is not in reality the proper word after all, but a corruption of L'Isle, Ryssel, in the French Netherlands (called the island, from its standing in a kind of lake formerly ; but the waters are now drained off), situated in east longitude 3°, latitude 50° 42″, on the river Deule, twenty-five miles north of Arras, and twelve miles from Tournay. It is a large, populous city, the capital of French Flanders, beautifully built, and was once strongly fortified. It has been noted for its silk manufacture, and fine linen or cambric, which have been made to great perfection there, as well as for its camlets, which are much admired.

PROTECTIVE TARIFF.

Of all the fallacies that ever became embedded in the brain of an intelligent people, none was ever greater than the idea that high protective duties will permanently help the manufacturer and employee. In all highly protected countries, wages are low and manufactures primitive. We at times see children phenomenally precocious in growth or intellect, but such usually die young or shrink below the average ; so of manufactures, if the profits are large, home competition keeps pace, each tries to produce at the least cost without regard to real quality so long as shoddy can be made to appear fair on the surface.

Combinations are formed by which the lowest class of help can be brought in to compete with our native employees. Through this combination workmen are transported from Bremen or Liverpool to Chicago, for ten dollars. Inventors, who have done so much in advancing the country's prosperity, have little chance comparatively under such conditions. Manufacturers will not bother with new devices while their profits are from twenty-five to a hundred per cent. Necessity is the mother of invention. Free competition is productive of efforts to excel in devices that enable the production of the best goods at the lowest cost. If protection is right for the manufacturer, then the employee should be protected by a high duty on imported labor.

FIRE ESCAPES.

Constant travel with its concomitant hotel experience has brought me into proximity with many styles of fire escapes, but the only kind I believe to be reliable are the balconies or towers with fixed iron stairways ; but such stairways should never be placed against windows through which the fire from the inside can flash out upon them. I think where there are adjoining buildings of the same height it will always be well to have stairs from the upper stories of hotels and manufactories lead upwards to walks leading along the roof to the other buildings for the employees to escape upon.

TIDE POWER.

Tide power once quite common in this country when land and space was of little account is now hardly known though often called to mind by the various trade papers.

It has seemed to me that where the rise and fall of the tides is considerable large tanks as weights might be made to develop convenient power for light manufacturing purposes at little expense if properly suspended.

MEDDLING WITH THE MAILS.

I think up to the time of Postmaster-General Holt the mails had been considered sacred, to be used by all unquestioned. Slavery had then become rampant and a demand was made that all matter inimical to that barbarism should be searched for and excluded; a subservient North yielded. Since then Anthony Comstock in the interest of a hierarchy, a twin barbarism, has insisted upon deciding what shall and shall not be excluded. Suppose some intelligent person should insist, as well might be done, that the Bible, Shakespeare's works, and plenty others should be excluded. There could only be another travesty of trial, as in the George Francis Train case, or the Bible and many popular works would have to go. Who is to say where the exclusion is to stop?

The better way is to follow the Creator, serve all alike and allow no meddling whatever with the mails.

Before trying to purify the world by law, first purify the law, so that it may not be necessary for a coterie of old grannies at Washington, four years behind their work, to brood over the decisions of Alfred the Great or Edward the Little in order to ascertain whether John Doe or Richard Roe owns a stray jackass.

Blot out all laws once in twenty years, re-enact the few necessary, then select judges from the most intelligent men or women, never from lawyers. Make the law conform to the right; faugh! law and plows of Edward's time.

CHOOSING ALL OFFICIALS BY THE PEOPLE.

Members of Congress are chosen and paid for doing certain specific duties. Why are they allowed to spend so much time electioneering for themselves and others, and what business have they to meddle with appointments? What right has a president or other official to appoint to an important office one who has been rejected by his own constituents? We have ten officials where one would be better, and nominally the highest are selected because they can be used rather than for their ability—fourth rate men. Who can remember who was governor of this state three years since?

Why not elect all officials, from president down, directly by the people, elect yearly, have but few, and make those responsible? Have it understood that such officials are really servants instead of masters. Eschew lawyers generally, take business men, but for merit. Allow no official consecutive re-election, high or low.

UNDESIRABLE NAMES AND FLUNKYISH TITLES.

Owing to fanaticism, predilection, interested motive, or lack of taste, parents often load children down with names that prove an incubus through life; as such children become legally responsible at a fixed age, why not at that time make it customary for children to select names to suit? Think of being loaded down with Peleg, Ichabod, Nehemiah, etc., etc.

Why has the publication of a newspaper become so low a business that the editor now prefers to be called colonel rather than editor?

Why does any man of brains desire to be known by any prefix or suffix to his name?

Think of Mr. Washington, Abraham Lincoln, Esq. or Ph.D., Prof. Benjamin Franklin, Royal Lightning Catcher to her Majesty, etc. It would seem that the smaller the mind the greater the desire for titles.

ADULTERATIONS AND SHORT MEASURES.

Adulteration of almost every commodity sold is now so general as to hardly cause comment. The same is the case with goods sold by the piece as so many yards, or so many articles in packages.

Why not make a law to confiscate all such goods wherever found?

THE REPUBLICAN PARTY.

As I have never desired office my vote has invariably been cast for what to me has seemed to be for the best result.

I voted for Fremont, and for Lincoln twice, and still believe that the latter was the best man ever elected to the presidency. The abolition of slavery was caused by the spontaneous rising of the masses to blot out an institution of such barbarism. Nominally it was done as the Republican party, but men of all parties united for that purpose, then withdrew as it was accomplished ; it was then the residuum crystallized into the real Republican party, a party for power and plunder. Its carpet-bag governments were the reproach of the civilized world.

The North was quite as much to blame for the rebellion as was the South, and the settlement should have been magnanimous and universal, instead of which the small minded leaders made heroes of Jeff Davis and others. Grant was nominated for president, not because the party believed him a hero, statesman, or republican, but because the leaders expected to ride into power on his popularity. And here it may be stated that a great wrong has been done to the earlier generals who really took the brunt of the fighting, in giving so much credit to Grant and Sherman, but it is a fact worthy of notice that their greatest eulogists now were called copper-heads during the war. From the beginning the party has been honey-combed with corruption up to the present time. Its carpet-bag governments, its stealing of the presidency, its outrageous pension acts for selfish influence, its gerrymandering of congressional districts, its favoring of monopolies, its numerous commissions for the centralization of power, the giving of a cabinet office as a reward for a corruption fund, its unseating of members and admissions of unpeopled states for party purpose regardless of honor, honesty, or the country's welfare, show that the party is under the control of a class unworthy of respect. The selfish, downward tendency of the party is well represented by its known men.

Henry Wilson went first for Henry Wilson, then for the Republican party, leaving as residuary legatee the country. George D. Robinson goes first for George D. Robinson, second for George D. Robinson, and as residuary legatee, George D. Robinson.

May the shadow of the Republican party become less and a better take its place.

THE NORTH POLE.

As some years have passed since any expedition has been sent out to look for that long-sought but yet unreached place, it is likely some ambitious country, institution, or person will soon be urging the matter upon the notice of the public, and I will suggest a plan that to me has long seemed practicable.

It is to construct fifty houses of light, non-conducting but strong material, to be sent in parts in ships, as far north as possible, to be put together on the ice and placed upon runners. The ice there is undoubtedly rough, but boats with few hands, and those in feeble condition, have been moved long distances. Sledges for the transportation of coal, food, clothing, bedding, and oil for lights should be provided in plenty, with plenty of men to handle them. A house accompanied by the sledges should be started, and continue north, then return to repeat the operation. Several gangs should be employed so as to work and rest alternately; stations twenty miles apart would insure relief and safety and give confidence. I believe road engines might be made to do the leveling of rough ice, and drawing the houses and sledges.

Of course this would require a million or more of dollars, but there are plenty of men in the country that could easily furnish all the necessary cash and not mind it, or the country could easily do it. If younger, such a job would suit me to a dot. Who of our millionaires will undertake it instead of endowing some college of which we already have more than are needed?

EVOLUTION.

Evolution is an idea as old as history and was well considered in Chambers' "Vestiges of Creation" long before Darwin rode the hobby. That man evoluted from the monkey is an old idea and one of the earliest that I can remember to have heard expressed, uttered by a hard-shell Baptist minister who cobbled shoes week days and preached Sundays under the inspiration of rum and molasses.

"The survival of the fittest," good in itself, offers no proof of evolution though it may of progression.

The mollusk of the earliest times is the mollusk of to-day. The old idea that man contains the pith of every previous product is perhaps correct, and to me it seems reasonable that the spirit or germ of life may evolute step by step from the lowest to the highest, also that man may so stultify his intellect that at the change called death his spirit will naturally gravitate to the body of a flea in order to find a suitable home. As for physical evolution it will be time to believe in that when a single instance in proof can be offered.

Progression will be more rapid when the brawling multitude that think but little yet invariably condemn everything out of the ordinary rut, think more and object only from conviction, and less credit is given to those who brood upon eggs that never hatch. What good ever came from the brooding of an old monk sitting in a dark cell or cave, or a dervish sitting upon the top of a column? Thought, like steam heat, to be useful requires ventilation. There are plenty who thus sit and brood, look profoundly wise and think that they *think*.

The prefix of professor, or any title added to a name, is more than likely to be the reverse of a guarantee of ability.

DIET.

As it was a rule in ancient times for those who had been sick to publicly state how they had been cured that others might benefit thereby, I will state how for nearly a half century I have lived without being sick.

First, my diet has always been spare, at the same time I have invariably eaten anything that I have desired and at any time without any regard to regular hours, often at midnight or later if restless ; a piece of mince pie or a biscuit well buttered soon brings sleep to me. Very little meat, pork never, raised bread is an abomination to me. Hot biscuit, hot doughnuts, pies of all kinds, puddings, strawberry short cakes, buckwheats, fruit, and a few of the ordinary vegetables constitute my ordinary meals, with hot tea or coffee, no liquor, beer, or tobacco in any form.

Think of firing a boiler three times a day instead of as required.

Notes on Water Flow, &c.

NOTE FIRST.

Water, like all other bodies when in motion, dislikes to change the direction of that motion and this resistance to change increases with the square of its velocity. For instance, to turn a quarter circle in a pipe which is bent on a circle of ten times its own diameter, requires additional force or "head"; when the water moves but one foot per second; this additional head is but the one-thousandth of a foot, but at a velocity of ten feet per second the resistance is one-tenth of a foot head (100 times as much), this is an easier bend than is generally found in mill work; when the circle is 2½ times the diameter of the pipe this resistance is double; and here begins the heavier resistance, for from this to turning a square corner it has increased to 16 times that first noted; and as will be easily seen, in the case of short turns with high velocity, destroys much of its power.

One of the commonest and easiest turns which we see given to water is in the scroll of an ordinary wooden wheel. Supposing this scroll to be 72 inches in diameter with a 12-inch spout leading to it; that is, the diameter of the scroll is 6 times that of the spout and the velocity of water 25 feet per second(=10 feet head). To maintain this velocity requires an additional head of 2½ feet, but as this loss is hidden by the reduced velocity of the water caused by its impact on the buckets, and also rapidly grows less with its reduced velocity as shown in the first part of the note, it is very generally ignored and sometimes denied altogether.

NOTE SECOND.

As a corollary of note 1st we see that as an abrupt change of direction requires power to overcome, the less we have of it in the chutes which admit water to the wheel, the better, as any force expended h re is so much taken from the amount which can reach the wheel; while changing the direction of the water by the form of the wheel itself, is applying this force where it does its work.

NOTE THIRD.

Loss of head from insufficient conduit. Water wheel builders lay great stress on this and generally give rather exaggerated views. The error is on the safe side, and when practicable it is well to follow their suggestions. It sometimes becomes necessary, however, to use trunks for supplying wheels which from original construction or want of room have less size than would be desired. It therefore becomes necessary to know what this loss is. Here comes the mooted question, whether this loss is that due to the head necessary to produce the required velocity or only that necessary to maintain this velocity in the conduit. Without entering into the arguments on the subject, some of which are rather more curious than useful, it is sufficient to say that but little if any loss is found to exist, except that due to the frictional resistance of the conduit, and this is measurable.

The following table, abridged from "Beardman's Manual of Hydrology," covers most of the cases required in ordinary practice.

Table of slope or fall in feet, and cubic feet discharged by pipe running full.

Size of Pipe in inches.	Slope 1 foot in 528.		Slope 1 foot in 264.		Slope 1 foot in 150.		Slope 1 foot in 66.	
	Vel'y in feet per minute.	Cubic feet per minute.	Vel'y in feet per minute.	Cubic feet per minute.	Vel'y in feet per minute.	Cubic feet per minute.	Vel'y in feet per minute.	Cubic fee. per minute.
12	130	102	196	155	243	192	392	310
18	160	282	226	399	300	528	452	798
24	184	580	261	820	345	1.085	522	1.640
30	206	1.013	292	1.432	386	1.895	584	2.864
36	226	1.598	319	2.259	423	2.989	638	4.518
42	244	2.350	345	3.321	457	4.395	690	6.642
48	261	3.281	369	4.637	488	6.137	738	9.274
54	276	4.403	391	6.223	518	8.237	782	12.446
60	291	5.731	412	8.100	546	10.720	824	16.200

From this table the mill owner can find what he can do with different sized conduits; making these square instead of round would be an ample allowance in size for roughness or irregularity of construction.

G. W. PEARSONS, C, E.

SUPERSTITION, IDOLATRY OR WORSHIP, WHICH?

Devil worship.

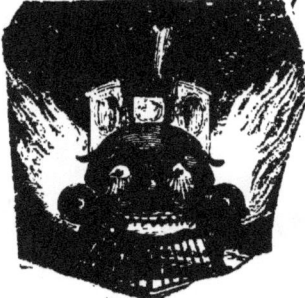

Dragon and Joss worship.

Pagan worship.
Jupiter.

Oh ho! My lord Jupiter, Jupiter!
Oh my hatchet, my hatchet,
my hatchet!

Paganism and Buddhism
in disguise.

Jehovah.

Bible worship.

David.

Worship of the Cross.

Oh Virgin Mother, ask your dear
son to intercede for us and bear
our sins, it saves us so much tro
uble, but caution him to be careful.
You remember that after making
the earth he thought it was flat,
had ends and that the eastern
continent was all there was of it.

Oh, thou great and
fearful God, hold not
thy peace but do as I,
request, Psalms CIX.,
your holy word.
Amen.

Wonderful things in the Bible I see,
This is the dearest, that Jesus loves
me." So glad she finds a
lover! But how about
Jesus? Can it be
pleasant for him to
shelter all of the hypo-
crites, thieves, mur-
derers and sour old
maids that propose to
rest in his bosom?

Card Setting Machine.

Manufactured by Samuel W. Kent, Worcester, Mass.

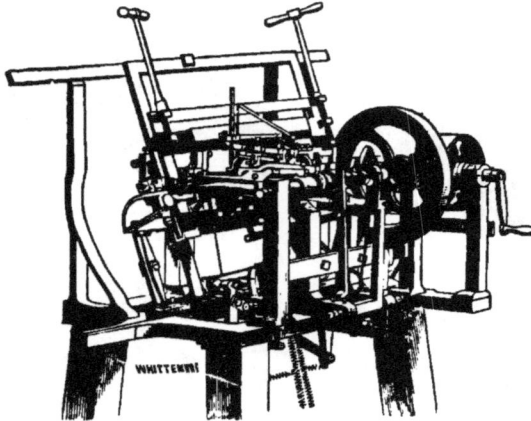

This machine holds the leather used for the base of card clothing, feeds and cuts the wire, bends it for the teeth, pierces the holes, places the teeth therein, then clinches them tightly. Few machines so perfectly demonstrate the possibilities of mechanical movements as does this; but, as an invention, there are many others far superior, for the card machine is a combination of several separate devices. Aside from the device for cutting and bending the wire for the teeth, the other movements are mostly of a feed character, but the adjustments are so numerous that a mind of great organizing, rather than inventive ability, was required to bring them into harmonious operation. Amos Whittemore, of Cambridge, Mass., obtained the patent for the machine, but Eleazar Smith, of Walpole, Mass., claimed to be the inventor. It is difficult, however, to obtain much information about its early history, though it has had an immense influence upon the textile manufacture. Mr. Kent has been engaged in the manufacture of the machine for nearly a half century. He informs me that when he commenced, a separate machine, worked by hand [here illustrated],

was used to make the teeth, from which it would seem that the complete machine was many years in working its way into general use.

Noble's Wool Comb.

Manufactured by John Crossley & Sons, Halifax, England.

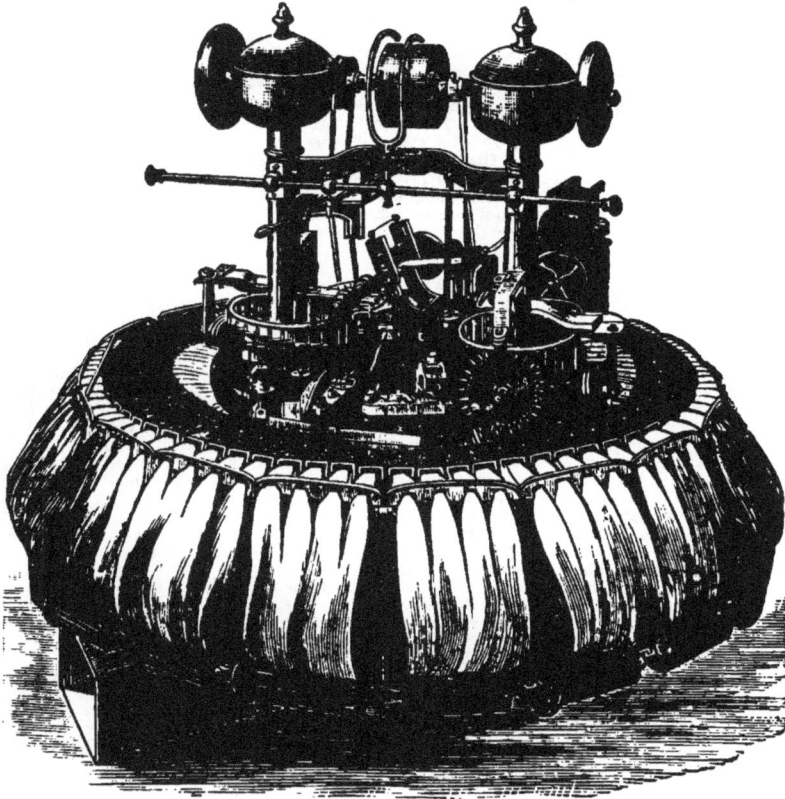

Woolen Manufacture.

The production of wool and manufacture of woolen goods constitute an old, perhaps the oldest, industry carried on among us. The fabled search by Jason for the golden fleece of Colchis typifies the esteem of the ancients for wool. Wherever we turn, in sacred or profane history, we find the lamb the symbol for tenderness, and wool, the cherished product of the sheep, always highly prized by man. Wild sheep are found everywhere, but all domestic breeds are derived from the Asiatic variety, which was developed from the argah, or big horn of Siberia. Originally, all were covered with long hair, and wool beneath; the hair has been bred out, but appears when the animal is neglected. The merino is the most valuable of all, dating back some two thousand years. The word shows in its own structure the choice nature of this sheep. The Spanish noun means judge or inspector of the *transhumance* (pasture-changing) flocks. Merino, the adjective, means wandering or the pasture-changing and best chosen flocks. Thus, the word brings down the process by which the flocks were selected, and also by which they were managed and developed. It is the treasury of fine fibre for all varieties, and was introduced here in 1801-12. The Saxon, the finest variety, is too delicate for common use. Merino furnishes the best clothing or card wool and the fine or soft combing varieties. Leicester developed by Bakewell in the eighteenth century, Cotswold and similar coarse, long, bright English varieties yield the lustrous worsted. The mauchamp, a variation from pure merino in France half a century since, has a lustrous fibre almost equal to the silky Cashmere goat. Carpet wools are long, rough and coarse, generally from South America, East India and the Mediterranean. The great pastoral districts for the merino and its crossing are now Australasia, River La Plata and Cape of Good Hope. These lands produced last year nearly 600,000,-000 pounds in the grease, or 298,000,000 of pure wool. California produces largely for us, rather more than 50,000,000 grease pounds. The manufacture falls into two great divisions. First, woolens, which are carded and generally felted; second, worsteds, named from a village near Norwich, Eng., which are combed, the lustre of the wool preserved, and are finished without fulling.

Christopher Columbus was the son of a wool comber. But it is probable that the combing of wool at that time was but a simple process of carding or getting the wool ready for twisting into yarn upon the rude hand machines of that day. It has been the work of later years to perfect the art of wool combing or the separating of the long worsted fibres or hairs of the wool from the short down, or noils, as the combing waste is now called. The wool of commerce is now divided into three distinct classes—clothing wools, worsted or combing wools and coarse or carpet wools.

In order to fully understand the difference between the ordinary old-fashioned woolen goods and the more modern worsted fabrics, turn for a moment to the yarn from which each is woven. Place a bit of ordinary woolen yarn under a microscope, after untwisting it. You observe that the yarn was made up of numerous minute fibres, running in every direction, interlaced, hooked and curled together in such a snarl that it would not be possible to tell in what direction a majority of the fibres run. Give a little twist to the snarl and it is ordinary yarn again. Put a bit of worsted yarn under the glass, after taking out the twist in the same manner as before. You now observe that the hairs or fibres all run in the same direction; that they are all nearly straight or much more so than those of the ordinary yarn; that each fibre presents, instead of a downy appearance, almost a transparent lustre.

Until within a few years the separating of the worsted fibres from the short wool or noils was all done by hand, and a very tedious and unsatisfactory process it was; but by the more recent invention of very curious and almost life-like machines, an illustration of one of which may be seen at the head of this article, this separation or combing has reached such a stage of perfection as to have greatly increased the demand for and consumption of goods made of wool. The prices of goods of the finest texture and most beautiful lustre have been reduced to within the reach of people of moderate means.

ORDINARY WOOLEN MANUFACTURING

Is carried on in mills with machinery classed as "sets," the cost of which at this time are about $8,000 each. A mill building 50 x 160 feet, with four stories and an attic, gives room for ten "sets," though this does not include room for sorting, washing, drying, dyeing, picking, and boiler for heating. Such a mill driven by water-power would cost somewhere about $150,000, or $15,000 per set, varying somewhat, according to the conditions, cost of land, dam, &c.

Sets are based upon the number of cards used. These cards are of various lengths, but those of 48-inch are used most now.

MACHINES NECESSARY TO MAKE UP A SET OF WOOLEN MACHINERY.

Wool and Waste Duster answer for six sets.
Wool Mixing Picker answers for six sets.
Cards—three per set: first and second Breaker and Finisher.
Mule—four hundred spindles per set.
Spoolers—two per set.
Dresser, Reel and Beamer answer for six sets.
Looms—five broad or ten narrow per set.
Fulling Mill—two per set.
Washer answers for eight sets.
Hydro Extractor answers for six sets.

Gig—two per set.
Shears answers for four sets.
Brush answers for four sets.
Press answers for six sets.

The manufacture is conducted in the following manner: The wool must be sorted with reference to weight, softness, fineness, strength, color and cleanness. Wools of the best kind are separated into sorts, technically called *picklocks*, *prime*, *choice* and *super*. The first named is the most superior, and the others

Wool and Waste Duster.

Manufactured by Davis & Furber, North Andover, Mass.

follow in the order of their gradation to the last, which is the most inferior in quality. Inferior wools are sorted into *downrights*, *seconds*, *abb*, *livery* and *short coarse*. *Seconds* is wool grown on the throat and breast, and *livery*, that grown about the belly of the animal. *Abb* is an inferior kind of *seconds*, and *short coarse* is also derived from the breast. This operation is performed by hand and by skilled sorters. It is then scoured with a weak aqueous solution of alkali, then thoroughly rinsed in pure water and dried. It then goes to the dye-vats and is colored or "dyed in the wool"; then oiled, to prevent matting or felting; then goes to the Picker, which prepares it for carding.

Wool carding by machinery was first accomplished at West Riding, Yorkshire, England, about 1787. John and Arthur Scholfield, from that vicinity, came to this country in 1793 and, a year later, commenced to card wool by machinery in a mill at Byfield, near Newburyport, Mass. At that time, and, in fact, for many years later, each farmer in the New England States kept a sufficient number of sheep to supply his family with clothing, and a spinning-wheel and loom were to be found in each family; indeed, were often a part of the outfit of daughters, when they were married, instead of the piano, now required. There are many now living that can well remember the process of carding wool by hand, and from that the practice of sending the family supply of wool to the carding mills, where it was carded, and left in rolls about two feet in length to be spun upon the old spinning-wheels by the farmers' wives, daughters, or the "hired girl." In the

Wool Mixing Picker.

Manufactured by Davis & Furber, North Andover, Mass.

earlier years of machine carding, the machines were very crude—merely strips of card clothing nailed upon flat surfaces beneath a single large cylinder. Such cards, however, were equal to the spinning-wheels, but the production of the spinning-jenny necessitated continuous rolls. For a time the short rolls were pieced by children as they were spun, the rolls being carried on the left and joined by the right hand—an operation that wore the skin from the fingers, and often caused the blood to flow therefrom. A piecer commonly supplied twenty spindles, so that three were required for each machine of sixty spindles. The "Finisher Card" is the outcome of the persistent efforts made to do away with the really crude, expensive and inefficient system of piecing. Many patents have been granted for different plans, and numerous prejudices have had to be overcome in order to accomplish the purpose.

The covering of card cylinders, known as "card clothing," consists of wire teeth, of suitable form, set in a base of leather or its equivalent, made in many degrees of fineness, so as to meet the wants of carders of every variety of wool. The first breaker card has clothing made with coarse wire; the second breaker has finer; the finisher card the finest. From this card, the wool is delivered in numerous continuous soft cord-like rolls ready for spinning. The wool is weighed out and spread upon a feed-apron to the first breaker card. The licker-in presents it to the main cylinder, where it is worked by various devices; then, by what is called the Apperly feed, it is taken to the second breaker, and from that to the Finisher Card.

From the finisher card, the wool goes to the Mule. The illustration following the Card represents a Self-acting Mule, a machine that would have been looked upon with wonder a century since, and certainly with reason, if compared with the spinning-wheel of that period. A brief extract from the builders' circular will give their claims for the special merits of their mule:

"This mule has a low carriage, an improved acceleration speed motion for spindles for spinning warp or other yarn requiring much twist, and is adapted for spinning all kinds of stock, and grades of yarn. It has a patent adjustable draft scroll, which can be so changed in a few minutes as to adapt it for giving any desired motion to the carriage when running out, whether for long draft or for twisting yarn without drawing at all, whereby much time and labor are saved that would be required to change scrolls."

Messrs. Johnson & Bassett make mule building a specialty. Their mules usually have four hundred spindles each, or enough for a set each.

From the mule, the yarn is taken to the Spooler, which is used for transferring the yarn from the bobbins on to jack or dresser spools for forming wool warps and also for doubling two or more threads together, the twist being put in on

Set of Wool Cards, in Position.

Manufactured by the Cleveland Machine Co., Worcester, Mass.

SELF-ACTING MULE.

Manufactured by Johnson & Bassett, Worcester, Mass.

Dead Spindle Spooler and Bobbin Stand.

Manufactured by Davis & Furber, North Andover, Mass.

the jack. A Dresser, Reel and Beamer are next required, in order to get the yarn on beams for weaving.

Dresser, Reel and Beamer.

Manufactured by Davis & Furber, North Andover, Mass.

From the beamer, it goes to the Loom. For plain goods, the ordinary cam loom is sufficient, but the competition for superiority of styles necessitates looms capable of producing new patterns at will. This want seems to have been met by the production of the Knowles' Chain Loom, in which from two to forty harnesses and seven shuttles may be used, and of course capable of weaving an almost endless variety of styles, from plain to the most elaborate of patterns. The illustration annexed shows one of their looms, with any length of pattern required, easily changed to any style desired in a few moments, and so convenient as to be likely to supersede the ordinary cam looms in future, though for common use the twelve to twenty-five harness looms are sufficient.

Twenty-five Harness, Open Shed, Fancy Loom.

Manufactured by L. J. Knowles & Brother, Worcester, Mass.

From the loom, the fabric goes to the Fulling Mill, where it is fulled. It then

Rotary Fulling Mill.

Manufactured by R. Hunt Machine Co., Orange, Mass.

goes to the Washer, where it is soaped, scoured and rinsed. It then

Cloth Washer.

Manufactured by R. Hunt Machine Co., Orange, Mass.

goes to the Hydro-Extractor, and is dried. This machine is also used in the preliminary operation of drying the wool after it is scoured and dyed.

Hydro-Extractor.

Manufactured by the Cleveland Machine Co., Worcester, Mass.

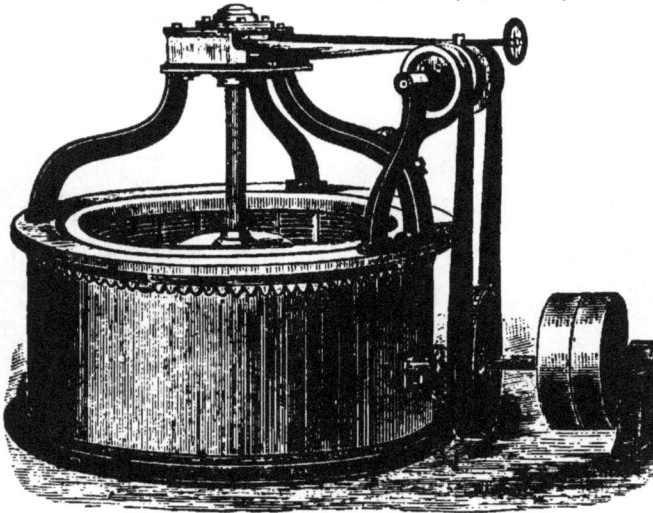

From the hydro-extractor, it goes to the Gig, in which the nap is raised with teasels, the natural hooks of which many attempts have been made to equal by mechanical substitutes; but, up to this time, without success. In this gig the

Quadruple Acting Gig.

Manufactured by Parks & Woolson, Springfield, Vt.

cloth is acted upon at four different points while passing through the machine. If a lustre like broadcloth is desired, the cloth is boiled or steamed to lay the fibre of the nap. From the gig, the cloth goes to the Shearing Machine, which has revolving blades working against one that is stationary, or a "ledger blade."

Shearing Machine.   Brushing Machine.

Manufactured by Parks & Woolson, Springfield, Vt.

DOUBLE-ACTING.

In this machine the nap is made even by shearing it. From the shearing machine, the fabric is sent to the Brushing Machine. From the brushing machine, it is sent to the Press, where it is passed between hot rolls to lay the nap.

Press.

Manufactured by Harwood & Quincy, Boston, Mass.

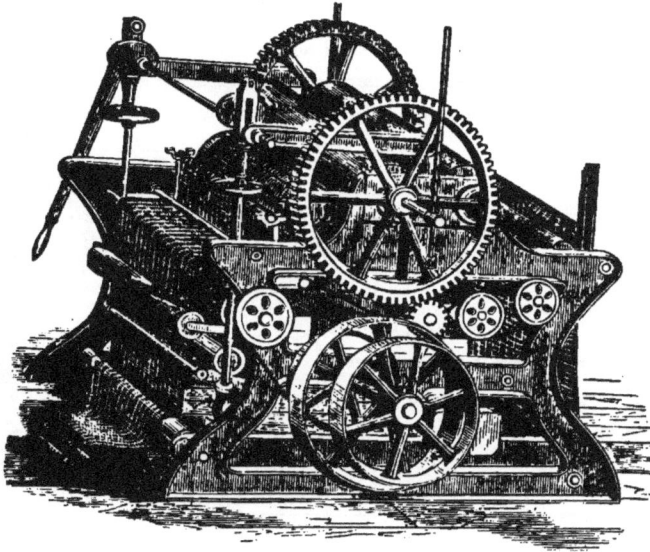

This Rotary Press leaves no press folds, nor does it require press plates or papers.

WILLIMANSETT, MASS., May 21, 1881.

JAMES DUGDALE, LOWELL, MASS.

Dear Sir: I think you have been engaged for a number of years in the manufacture of worsted yarn, and that formerly you combed the wool by hand on instruments or devices substantially the same as the old hatchel used for combing flax, and that you now comb by machinery or machine combs. That the inventive and liberal patent system may be compared with that of the conservative or older method, will you be so kind as to state the difference in cost and efficiency of the two plans, and oblige,

Yours truly, JAMES EMERSON.

LOWELL, MASS., May 25, 1881.

JAMES EMERSON, WILLIMANSETT, MASS.

Dear Sir: In reply to your letter of 21st inst., I have to say—The price paid for combing wool by hand was governed by the quality and length of staple. In 1863–64, the price paid in Lowell was 17 cents per pound for medium quality. A good workman was able to comb only from ten to twelve pounds per day—about twelve slivers weighing one pound. The first cost of the Improved Wool Combing Machinery is very high; consequently, repairs are very expensive. Still, the average cost is about five cents per pound, with the advantage of the sliver weighing from twelve to sixteen pounds. I shall be pleased to hear from you again.

Respectfully yours, JAMES DUGDALE.

Germania Mills, Holyoke, Mass.

Test of machinery with Emerson's Portable Dynamometer.   Tables prepared by A. M. Swain.

WEAVE ROOM.

DESCRIPTION OF MACHINERY.	TIME.	W'GHT.	SPEED.	H. P.
	A. M.			
	6.20	52	190	2.99
Shaft 136 feet long, 150 revolutions. 21	6.30	55	182	3.03
Broad Crompton Looms driven from the	6.45	55	192	3.19
line. An average of 10 looms were prob-	7.	50	190	2.87
ably in operation. Counted them in	7.15	60	190	3.45
rapid succession over and over again.	7.30	53	190	3.05
The least number in operation was 5 at	7 45	54	190	3.10
one time. The most was 15; 9, 10 and 11	8.	60	188	3.41
was the usual count.	8.15	55	188	3.13
Goods, heavy doeskin and cassimeres,	8.45	50	192	2.90
76 inches wide, 56 picks to the inch, 26	9.	60	192	3.49
ounces to the yard, in a portion of the	9.15	60	192	3.49
Looms.	9.30	86	192	5.
April 12, 1873.	10 30	80	190	4.60
	10.35	60	192	3.49
	10.36	65	193	3.80
	11.40	15	192	.87
	12.15	651	188	2.90

TESTS IN PICKING AND DRYING ROOMS.

DESCRIPTION OF MACHINERY.	TIME.	W'GHT.	SPEED.	H. P.
	P. M.			
2 Fans, 8 vanes each.	12.15	80 ll s.	180	4.36
2 Fans, 8 vanes, 2 Fans, 5 vanes,		122½	180	6.68
4 Fans, 1 Sargent's Burr Picker,	1.30	230	181	12.61
4 Fans, 1 Sargent's Burr Picker,	1.45	222	180	12.11
4 Fans, 1 Sargent's Burr Picker,	2.	230	180	12.54
4 Fans, 1 Burr, 1 Kellogg Picker,	2.05	Belt	Slipped	
1 Sargent's Burr, 2 Kellogg Pickers,	2.30	180	179	9.76
1 Sargent's Burr. 2 Kellogg Pickers,	2.45	180	179	9.76
4 Fans, 1 Burr, 2 Kellogg,	3.	290	178	15.64
4 Fans, 1 Burr, 2 Kellogg,	3.15	205	178	15.91
4 Fans, 1 Burr,	3.45	231	178	12.46
1 Burr,	4.	125	180	6.81
1 Burr, 4 Fans,	4.15	235	179	12.74
1 Burr, 4 Fans,	4.25	234	180	12.76
2 Kellogg Pickers,	5.	65	182	3.58
1 Kellogg Picker, large,	5.05	40	181	2.19
1 Kellogg Picker, small,	5.10	30	181	1.64
Counter Shaft and loose Pulleys for above machinery.		25	180	1.36
April 9, 1873.				

Steam Engine.

New, made by Brown of Fitchburg, Mass. 18-inch cylinder, 42-inch stroke, rated 75 horse-power with 60 pounds of steam; tested by Prony brake, steam pressure ranging from 65 to 70 pounds during the trial; the power varied from 60 to 65 h. p. according to pressure.

Putnam Machine Co. Engine.

New 15-inch cylinder, 3-feet stroke, guaranteed to give 60 h. p. with 60 pounds of steam; tested by Prony brake; gave 44 h. p. with 65 to 70 pounds steam. Such has been my general experience, and I doubt whether a steam-engine can be found that realizes more than 3-4 of its claimed rate. Indicator cards may give the pressure in cylinder, but the only way to get the efficiency of an engine is to take it from shaft.

Compound Engine.

Steam working first in a 6-inch cylinder, from that into one of 12 inches, 20 h. p. was claimed; dynamometer on shaft showed 7. Then it was found that the most of the force was used in working the engine.

Power Required to Drive Woolen Machinery.

THE POWER required to drive sets of woolen machinery depends upon the quality of goods and number of sets in a mill; the more sets the less power in proportion is required. I have tested the power used at many mills, but a few cases will show the general average.

VASSELBORO WOOLEN MILLS, VASSELBORO, ME. 22-set mill, light cassimeres; required, 135 horse-power.

WM. WALKER & CO., LOWELL, MASS. 4-set mill, flannels; required or used, 30 horse-power.

JAMES O. INMAN, PASCOAG, R. I. Heavy doeskin, pant goods, 4 sets; used, 40 horse-power.

BEEBE, WEBBER & CO., HOLYOKE, MASS. Pant goods, eight sets; 64 horse-power.

Power Required to Drive Elevators.

These elevators were in Boston stores, the belts when not at work running on loose pulleys. To operate the first kind tried, without load, when running at the common speed, 1.89 horse power. With a load of 1.06 pounds, 3.92 horse power.

The second was a Tuft's elevator, running at the same speed as the first, without load, 2.46 horse power. With a load of 1004 pounds, required 5.29 horse power.

Hydro-Extractor.

Extractors start hard unless started very slowly, but lose their resistance instantly; three-fourths of a horse power would be a liberal average for such as I have tested, though from one to two horse power may be expended for a moment, if started hastily.

CONSTANTINE, THE FIRST CHRISTIAN EMPEROR.

The Lord's Day or Sunday, as its name implies, was the pagans' day for the worship of their god, the sun, and every idea or ceremony of the Christian religion, except its thirst for blood, is paganism disguised by change of name.

Sun Worshipers.

Constantine as Sun Worshiper.

The good Constantine presiding at the council of Nice, A. D. 326.

Constantine as the murderer of his nephew twelve years of age.

Constantine as the murderer of his son Crispus, A. D. 326.

His slaughtered victims, Maximian, Bassianus, Sopater, Licinius and others.

"Oh, Fausta! pray don't make a fuss, the good Constantine, your noble husband, has mildly decreed that you shall be boiled in your bath, so please come on and be boiled."

Baths.

'Oh, Sop, just absolve me and oblige, yours truly, Con."

"Absolve you, oh, of course," says his Christian reverence. "Yours were mere peccadillos, but holy grease is expensive, so you must endow a few churches. Let us roast doubters and furnish our Mary with a new petticoat dyed with the blood of those who dispute our word. Amen."

Sop.
" The gods have no absolution for crimes like thine."
"I will have your head, Sopater, then turn Christian!"
And he did both.

PAPER MANUFACTURE.

Like the ordinary historian, I might draw upon my imagination for my facts, and give time and place where the first idea of paper was conceived, but the reader will be quite as well informed if the truth is given instead; and that is, that I do not know anything about it. It is evident, however, that it must have been centuries upon centuries ago. Writing would necessitate paper or a substitute. Writing, from the nature of the case, must have been understood before the commencement of history, for, without writing, there could have been no record. A mark was placed upon Cain for the purpose of warning those he might meet that he was not to be molested. The statement plainly implies that such mark or writing was generally understood, or it would have been useless; and it furnishes a plausible pretext for the Irish historian's genealogical tree springing from an Irish root, with Adam placed high among the branches, and the statement that the Irish had a written language at the time of Adam, all of which may be true; but if Old Israel was the son of an Irish emigrant, it would be an interesting study for the scientist to trace out the cause of such a radical change in the form of the nose. Evidence bearing upon that point might be difficult to find; but such would hardly be the case about paper, for the word is derived from that of papyrus, and papyrus was paper essentially the same as the paper of to-day, though crude and coarse, perhaps, in comparison with the best now made: the interior part of a reed or flag indigenous to Egypt, and places where papyrus was known.

Its preparation for use was similar to that of paper. The part of the reed to be used was selected; it was then sized or glued, then subjected to heavy pressure. Sheets of any size desired could be made, as is proved by the fact that it was carried or kept in rolls. Vellum, often mentioned in connection with the early manuscript copies of Scripture and the printing of the first books, was white, finely prepared calf-skin. The object of this article, however, is more for the purpose of briefly describing the manufacture of paper now than to treat of its use in the past.

Until within a generation past, paper, or the finer qualities of paper, has been produced from rags, and the pulp has been worked into sheets by hand. Forms or sieves of the size of sheets required were used to take up the pulp; as the water drained out, the sheet formed, and when dried it was pressed. John Ames, of Springfield, Mass., now living, invented the cylinder paper machine, which is still in use in some mills where a cheap grade of paper is made. The Fourdrinier improvement has since been added.

Paper is in such demand now that constant investigation is going on for the purpose of discovering new fibre suitable for the purpose. Many kinds of stock are now used: rags, ground wood pulp, wood pulp chemically prepared, waste of many kinds, old rope, hemp, manila, fishing lines, jute, jute butts, straw, etc. Clay of various kinds is used, but by a neighboring manufacturer the individual does not do so.

Rag Duster.

Manufactured by Holyoke Machine Co., Holyoke, Mass.

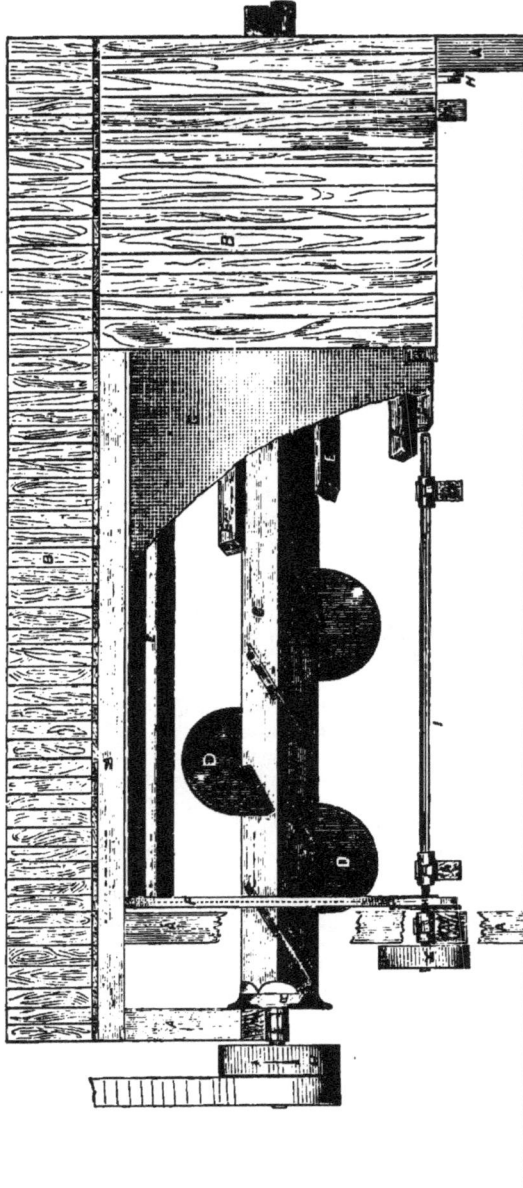

Large quantities of rags are imported. Italy and Egypt furnish many of them, but rags from these countries mean rags, clothing worn until the strength is gone. There are certain qualities imported that are used for fine writing papers, but domestic rags are the best. An American has a very indefinite idea of what rags mean. One morning, in my early boyhood, my slumbers were broken by a chattering and confusion of sound that would have matched Babel itself. Starting from my berth, a collection of rags and patchwork

Bleach Boiler.

Manufactured by D. F. Coghlin, Holyoke, Mass.

was presented to my sight that exceeded anything of the kind my imagination had ever conceived of. It is a blessing to the right kind of a Yankee to travel, for in many places he can see much to make him thankful that he is a Yankee. If he can get rid of some of his conceit, he will also see much that it is desirable to learn. The manufacture is conducted as follows:

The rags for fine paper are first dusted by running them through the Rag Duster. They are then cut into pieces, two or three inches in area of extent; this is done by women, each one of whom has the point half of a scythe firmly fixed vertically in a bench in front, the edge of the scythe being from her. With this instrument she cuts large handfuls of rags in various directions, until the mass is reduced into pieces the size required. She has to cut off buttons, hooks and eyes, seams, hems, and everything objectionable. After being cut, the rags are sorted, and everything rejected that is likely to injure the quality of the paper. They are then again dusted, and then placed in the Bleach Boiler—a

Gould's Improved Beating Engine.

Manufactured by Holyoke Machine Co., Holyoke, Mass.

Rag Engine.

Manufactured by Holyoke Machine Co., Holyoke, Mass.

horizontal boiler, varying from five to eight feet in diameter, and from fifteen to twenty feet in length. Lime is put in with the rags. The boiler is rotated slowly for twelve hours, steam being introduced through the hollow journals; the pressure of steam being kept up to sixty pounds during the whole time. This is done to soften and aid in the disintegration of the rags. From the bleach boiler, the rags are dumped into large boxes on trucks, in which they are taken to the Washers, almost identical in their operation and appearance with the rag engine above. Indeed, in many mills the washing and beating are done in the same engine. All the water that can be used is applied during the process of washing, but the roll is not pressed down so hard as it is while beating. The washing is continued from four to eight hours—usually about six, by which time the rags have become soft, pulpy stuff, which is then let down into the Drainers —tanks with perforated bottoms. Chloride of lime is added here to bleach the mass perfectly white. From the drainers, the stuff is taken up and put into the Beating Engines, where it is kept in constant motion, and continuously passing between the beating or tearing knives and roll. Coloring is here added to give the paper the desired tint. The mass is kept in the beaters until it is reduced to the condition required—usually about six hours. Then it is discharged into the "Stuff Chest" below.

For a long time—perhaps a century, more or less—there has been little change in the general character of the beating engine, except increase in size. Recently, attempts at improvement have been made, and now the Gould Engines are gaining favor from their increased productiveness, saving of labor and even quality of pulp.

Gould Beating Engine.

Manufactured by Holyoke Machine Co., Holyoke, Mass.

A charge for the engine is about 60 barrels, which is prepared for paper machines in about three hours. The centrifugal force keeps the pulp in constant motion, rendering stirring by hand unnecessary.

Experiments made for the Messrs. Stanwood, Tower & Co., at their paper mill at Gardiner, Me., a four ton mill, manila paper, jute stock. Regular speed of Beater 108 revolutions per minute, but during a test trial of 12 hours it varied from 106 to 112, requiring 55.36 horse-power as a maximum; during the trial 2400 pounds of excellent paper was weighed off, while there was a perceptible gain of pulp in stuff-chest at the close. Four 40-inch engines of the ordinary style in the same mill beating the same stock were then tested; with rolls hard down it required 109.64 horse-power to drive them to a speed of 160 revolutions per minute; this included shafting from wheels to the engines.

Ames Cylinder and Fourdrinier Paper Machine Combined.

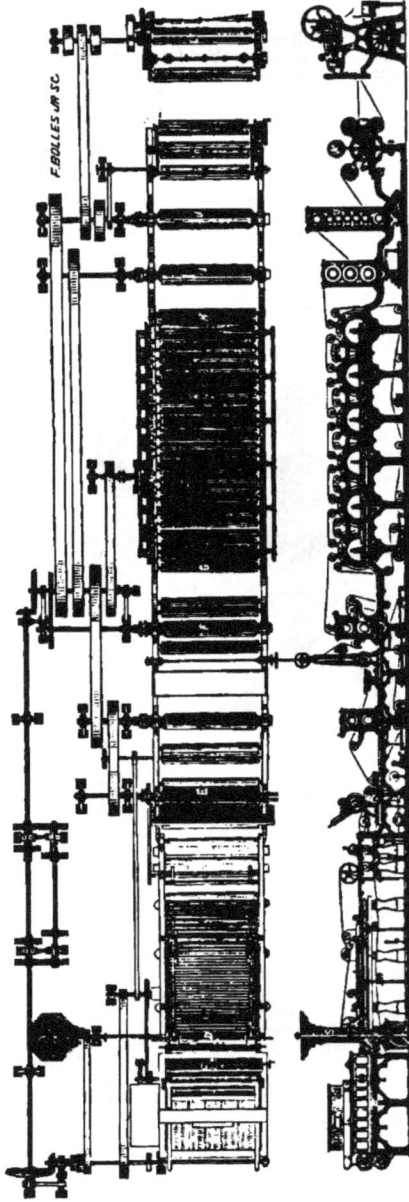

F. BOLLES JR SC.

From the stuff chest, the pulp in a liquid state—the color of the paper being made—is pumped up into the Stuff Box, A. From that, it runs down into the Vat, B, and is strained off by passing down through the fine Screen or Strainer, C; then flows out on to an endless apron, made of fine wire cloth. working round the Roll, E. In connection with the urn-shaped column, there is a device called the Shake, which keeps the wire apron in constant oscillation or tremor. This part of the machine, from the Vat, B, to the Roll, E, is the Fourdrinier improvement of the Ames Cylinder Paper Machine. As the pulp moves along on the wire apron, the water is drained and extracted from the sheet of pulp by exhaust pumps beneath. It then passes on to the Cooch Rolls, E, and from there to an endless apron of felt, and is carried along to and between the Press Rolls, F, which leave the sheet in a condition to bear its own weight until it arrives at the Drying Rolls, G, along and around which it is carried by another endless apron of felt, and is dried, the rolls being kept hot by steam. After passing the drying rolls, the sheet or web passes through a solution of hot liquid glue or animal size, and then between rolls that extract or press out the superfluous size; then, by rotary cutters, it is divided into sheets the size required, and is then placed upon a table, in regular square packs, by a mechanical device called a Lay-boy, invented by J. C. Kneeland, of Northampton, Mass.

Sheet Super Calender.

Manufactured by Holyoke Machine Co., Holyoke, Mass.

These packs are then taken to the **Drying Loft**, separated into sheets, which are hung evenly upon poles to dry, the loft being kept hot by steam. When dried, the sheets are sent to the finishing room, and are passed between rolls under great pressure. The process is called calendering, the Sheet Calender being used.

Ordinary paper for writing or commercial purpose is cut into sheets known as Flat Cap, 14 x 17½; Foolscap, 13 x 16; Letter, 10 x 16; Note, 8 x 10 inches. These sheets are counted into reams of 480 sheets each, folded, then trimmed in the **Trimming Press**. See cut, next page.

Trimming Press, or Paper Cutter.

Hydraulic Press.

Manufactured by the Holyoke Machine Co., Holyoke, Mass.

It is then subjected in packages to a pressure of several hundred tons in Hydraulic Press.

Lever Plater.

Manufactured by Holyoke Machine Co., Holyoke, Mass.

After pressing, the packages are boxed, ready for delivery.

A finer grade of paper, used for wedding or fancy cards, and various purposes, is calendered in the sheet calender; then placed between metal plates, and passed between the rolls of the Lever Plater; then cut into sheets the size required, and boxed for shipment.

Book paper, often quite fine and nice, is of a somewhat inferior grade—often, if not generally, made of mixed stock: rags and wood pulp, sized with resin size in the beating engine, instead of with animal size in the paper machine.

The process in the paper machine at the commencement is the same as before described, but instead of being divided into sheets, it goes in the web through the stack of Chilled Rolls, J, near the right end of the machine, which give what is called "machine finish." It is reeled or rolled, as represented on the Rolls, K. If a finer finish is desired, it is super-calendered. (See cut on next page.) It is then divided into sheets, the size required, by rotating cutters.

Newspaper is made of a cheaper grade of stock: rags, ground wood pulp, straw, waste of various kinds, etc.

Cheap wrapping paper is also made of straw, or something cheaper.

The best manila paper is made of jute, jute butts, old rope, hemp, manila, fishing lines, etc.

Web Super Calender.

Manufactured by Holyoke Machine Co., Holyoke, Mass.

Fine tissue paper is also made of jute, but the process of beating requires twenty-four instead of six hours in order to disintegrate the stock more slowly, leave the fibre longer, and the product more tenacious.

The cost of a paper mill of course depends upon circumstances to a certain extent. The rough estimate of cost for a one-ton fine paper mill would be $75,000 to $100,000; larger capacity, in proportion. [A ton mill means one capable of producing a ton of paper per day.]

Whiting Paper Co., Holyoke, Mass., No. 1 Mill.

4-Ton Mill, Fine Writing Paper.

Following machinery driven by the main wheel, which by test gave 180 h. p.
2 1250 pound washing engines.
2 1200 pound beating engines
2 800 pound beating engines.
2 6 inch Littlefield pumps.
1 Andrews pump.
4 rag dusters; 2 rag boilers.
1 Elevator, 2 boiler pumps, 1 engine lathe, 1 sheet calendar, 5 rolls, 1 small
pump, 1 circular saw for box work.

Finishing room wheel,   42.92 h. p.
Drives 6 5-roll calendars, 2 platers. 5 ruling machines, 3 trimming presses, 1
elevator, 1 grind stone.
These two wheels do the work named, but 20 horse-power additional would be
acceptable on large wheel.

Test by Emerson's Dynamometer.

Experiment upon an 800 pound paper engine for rag stock; furnished with 800
pounds of bleached stock in the evening of March 26, 1875, at the Housatonic
Mill of the Smith Paper Co. at Lee, Mass.   The roll was 46 inches long by 40
inches diameter   Experiment began with a stock nearly finished, which was
finished, discharged and the engine replenished.

Time. P. M.	Rev. of Roll.	Rev. of Dynamom.	Weight.	Horse Power.
7.00	118	284	165	14.20
7.30	124	294	131	13.45
7.35	124	300	135	12 27
7.45	124	288	131	13.17
*9.30	124	274	184	15.27
*9.35	124	274	184	15.27

*Roll down and stock half finished.

Experiments upon a 300 pound paper engine for rag stock : furnished with 300
pounds of bleached stock on the afternoon of March 24, 1875. at the Housatonic
Mill of the Smith Paper Co., at Lee, Mass.   The Roll was 33 inches long by 28
inches in diameter.

Time. P. M.	Rev. of Roll.	Rev. of Dynamom'r	Weight.	Horse Power.
3.50	131	230	28	1.95
4.00	131	230	63	4.39
4.15	143	250	57	4.31
4.20	149	260	73	5.75
4.25	152	270	49	4.00
5.00	166	291	93	8.20
5.05	150	264	93	7.44
5.30	146	257	94	7.31
6.00	143	250	95	7.19
6.45	149	260	104	8.19
7.00	144	252	107	8.17
7.30	149	260	118	9.26
8.00	126	220	119	7.93
8.15	133	233	122	8.61
8.30	146	255	119	9.19
8.45	149	261	105	8.30
9.00	123	215	105	6.84
9.15	137	240	104	7.56
9.30	150	264	101	8.08
9.45	137	240	101	7.34
10.15	132	232	100	7.03
10.30	137	240	26	1.89

Experiment on a 62 inch paper machine making news print from rag stock. This machine is ordinarily run with a speed that will deliver the paper at the rate of 90 feet per minute; but during these experiments it delivered 61 feet per minute the first experiment and 78 feet per minute during the last experiment.

Time. P. M.	Rev. of Dynamom'r	Weight.	Horse Power.	H. Power of Pump.	Table Power.
4.00	200	101	6.12	2.78	8.90
4.20	230	104	7.26	3.56	10.82

The main line of shafting makes 108 revolutions per minute when 90 feet of paper is delivered per minute. From this main line the agitator, the water pump and the shaker at the head of the machine are driven and are not included in the test by the Dynamometer; but are calculated from the speed and width of belts by which they are driven, on the theory that a belt 1 inch wide, running 1000 feet per minute is a horse-power.

	H. P.
The Shaker belt moves 600 feet and is 3 inches wide, equals 1800,	1.80
The Agitator belt moves 329 feet and is 4 inches wide, equals 1316.	1.32
The Pump belt moves 251 feet and is 6 inches wide, equals 1506,	1.50
	4.12

But as this pump is single acting, only acting during one-half of the revolution, I have called it two-thirds of the apparent power equals 1.00 h. p., and deduct ½ a h. p., then leaving 4.12 h. p. for the paper moving 90 feet per minute. Then by simple proportion of 78 to 96 with paper moving 78 feet per minute equals 3.56 horse power; with paper moving 61 feet per minute equals 2.78 horse power.

[*Copy.*]   L. M. WRIGHT, C. E.

Holyoke Paper Co., Holyoke, Mass.

Four 500 pound beating engines took the whole power of a wheel that by test gave 80 horse-power; even with that power care was required in furnishing or they would not run to speed; after running so for some years, the Beaters were altered or put into better condition, so that the wheel now gives a large surplus of power. Mill makes fine writing paper.

Test of a 72-Inch Wheel and Machinery, Fitchburg, Mass.

These experiments were made to determine power required to drive Beating-engines, 36-inch rolls, paper and rag stock. Before testing the wheel, the speed of the main shaft was taken under different conditions to ascertain the power required to drive machinery at the following speeds, the water in the pond being one inch below the lowest part of the crest of the dam.

1ST TRIAL.—3 Engines beating, 1 washing. and all machinery attached. Speed of main shaft, 120 revolutions per minute. 49 h. p.

2D TRIAL.—2 Engines beating, 2 washing, all machinery attached. Speed of main shaft, 146 revolutions per minute, 49 h. p.

3D TRIAL.—2 Engines beating, 2 washing, duster thrown off. Speed of main shaft, 160 revolutions per minute, 48.3 h. p.

During the above trials the head was about 14 feet. The dynamometer was then applied to the end of main shaft, and the power of the wheel, at nearly same speed, obtained.

With the flush-boards off, leaving 13 feet head, under which the wheel was designed to give 60 horse-power, its power would have been 43.16.

No attempt was made to measure the water, it simply took the whole river.

Capacity of Beater 450 pounds.

Paper and Shoemaking Machinery.

Report of a test to determine the power required to run one of the Rag Engines at Bacon's Paper Mill, in North Lawrence, Massachusetts.

LOWELL, December 16, 1870

J. A. Bacon, Esq.:

DEAR SIR:—I have worked up carefully the tests made yesterday with Emerson's Dynamometer, at your mill in North Lawrence. When the engine roll made 145 revolutions per minute, the dial hand of the Dynamometer made 3.8 revolutions per minute. I have estimated the speed of the roll, upon the supposition that it varied during the different tests in the same proportion as the speed of the dial hand. I give the results obtained, in the order in which the tests were made.

Number of Test.	CONDITION OF THE ENGINE.	Revolutions of Roll per minute.	Horse-Power indicated by Dynamometer.
1	No paper in................	137	2.5
2	Paper being put in..........	149	7.26
3	" " " "	141	3.36
4	Washing paper......	153	4.
5	" "	145	4.03
6	" "	147	4.41
7	" "	144	4.57
8	" "	145	5.19
9	" "	153	5.2
10	" "	148	4.71
11	Beating pulp..............	149	5.08
12	" "	147	5.02
13	" "	149	5.08
14	Brushing the paper	149	3.9

While the paper was being put in, the power indicated gradually rose from 2.5 horse-power to*7.26 horse-power. It stood at 7.26 horse-power for about three minutes, after which it gradually fell to 3.36 horse-power. From test 4 to test 8, the roll was gradually set down harder and harder. At test No. 7, the roll was down as hard as is usual in making paper. At test No. 8, the roll was down harder than is common.

Very respectfully yours,

(Signed,)   CHANNING WHITAKER,
                            Mechanical Engineer.

Report of a test to determine the power required to drive Shoemaking Machinery, at the State Prison, in Charlestown, Massachusetts.

LOWELL, July 13th, 1871.

Rodney S. Tay, Esq., Treasurer Tucker Mf'g Co., Boston :

DEAR SIR:—On the 13th inst., I made a test with Emerson's small Dynamometer, of the power required to drive Mr. Blanchard's Shoemaking Machinery at the State Prison, in Charlestown. In Mr. Blanchard's lower room there are, besides the counter-shafting, 12 sewing machines, 2 peggers. 2 skivers. 1 heel trimmer, 1 bottom roller, 1 buffer, 1 roller, 1 splitter. All of the machinery is not in use at any one time. But making such allowance for this fact as seems to be fair, there is required for driving the machinery and counter-shafting in this room, 4.9 horse-power. In Mr. Blanchard's upper room, there are, besides the counter-shafting, 2 brushes and 4 buffers. There is required, for driving the machinery and counter-shafting in this room, 2.3 horse-power. Making a total of 7.2 horse-power used by Mr. Blanchard.

Very respectfully yours,

(Signed,)   CHANNING WHITAKER,
                            Mechanical Engineer.

*I tested an ordinary 450 pound Beater in same mill that took something over 13 horse power.
                                                            J. E.

[From American Engineer.]

A Man of Courage.

We publish elsewhere a communication from Mr. James Emerson giving some further facts in regard to warming of railway cars. Mr. Emerson has given the last five years of his life to this work, and so far, as we believe, he is the only man who has made extended experiments in heating by steam. He is well known as an hydraulic engineer. What he accomplished in that department of engineering is well told by a writer in the December, 1885, number of the *Milling Engineer*. The writer says:—

It will be sixteen years on April 1st next, since James Emerson, an inventor, of Lowell, Mass., issued a small, one-page circular, saying that he had purchased of the Swain Turbine Company their testing flume, built for the purpose of privately testing their own water-wheels, and that he was about to open a series of public competitive tests. It marked the commencement of an era of wonderful progress in turbines.

Mr. Emerson was a man of irreproachable integrity. He could not be bribed. He was too independent to be held as the tool of any one. He was fearless in his criticisms, and many a poor miller who had been defrauded by some unprincipled water-wheel agent, rejoiced to find that at last a man had arisen who knew and was not afraid to publish the truth. When he attacked a certain water-wheel builder, who circulated most elegant pamphlets, and who loudly claimed that his wheel was the best in the country, and that it had an efficiency of 90 per cent, although in reality it was worthless,—when Mr. Emerson drove him out of hydraulics into the patent medicine business, the whole fraternity of water-wheel users rejoiced. When he stated that the wheels of several loud talking, ignorant men had so passed out of use that they were more likely to be found at the junk-shop than anywhere else, and of a certain inventor, who claimed his wheel gave 135 per cent., that he had no doubt of his sincerity, but he had much doubt of his intelligence, there was great popular sympathy with a man who could so fearlessly say what he thought. The influence of his tests was marvelous. Nine-tenths of the water-wheels brought to him that first year only gave three-fourths of the power which their builders claimed and represented that they would give. At the present time all the leading water-wheels honestly give the power they claim, and the reason is because Mr. Emerson taught builders to estimate power correctly. Then nearly all the leading firms claimed and published that their turbines possessed the same economy of water at every stage of gate. None of them claim it now.

The influence of these tests was beneficial to every honest builder. The first wheel tested by the Stilwell & Bierce Manufacturing Co. only gave 68 per cent., although they honestly believed it could be relied upon to give 85. When they discovered the truth they commenced experimenting and improving their wheels until they gained records of over 90 per cent. A similar improvement was made by Stout, Mills & Temple, T. H. Risdon & Co., the Holyoke Machine Co., and many others. The effect of his tests, in the introduction of the best forms of water-wheels, was also remarkable. The attachment of a plate to a cylinder gate to raise and lower with the gate and to form the top of the stationary water course was then used by no builder of prominence. Now every firm building a cylinder gate wheel uses it to obtain good results at the part gate. He was the first to establish the fact that the discharge of water through a wheel of a given diameter could be increased to double the amount then customary, without injuring the efficiency of the wheel, and now there is hardly a prominent builder in the country who is not making use of that discovery.

I have not written this article as an eulogy of James Emerson, but because his name is inseparably linked to the recent progress of water-wheel science. Like every other prominent man, he was not perfect. The time had come when a better water-wheel, and more accurate information about the weakness and excellencies of the various systems in use, was demanded. Mr. Francis' valuable formulæ, upon which the whole system depended, were a locked-up mystery of little benefit to the majority of water-wheel builders. Location, experience, and remarkable fitness to the requirements of that special work made Mr. Emerson the means of creating such an improvement in a certain class of machines as few men have ever accomplished.

Water Wheels.

In treating of water-power, means for its utilization is an important feature to be considered. As a motor, running or falling water was used back in the earliest ages of which we have authentic history; and the various devices employed for transmitting its power were hardly more crude than many that are patented for the same purpose at the present time. Volumes would be required to illustrate and describe the multitudinous plans that have been devised, but a very few pages would suffice for describing the principles of all. Our country is lavishly supplied with this natural motive power; and, as might be expected from a race so energetic, many devices have been produced for utilizing it advantageously. I have before me the copy of a patent granted to Benjamin Tyler, grandfather of John Tyler, of the well known Tyler wheel, which reads as follows:—

By the President, THO. JEFFERSON.
JAMES MADDISON, *Secretary of State.*

City of Washington—To wit:
I DO HEREBY CERTIFY, that the foregoing Letters Patent were delivered to me on the twelfth day of March in the year of our Lord one thousand eight hundred and four to be examined; that I have examined the same and find them conformable to law, and I do hereby return the same to the Secretary of State, within fifteen days from the date aforesaid, to wit:—on this nineteenth day of March in the year aforesaid.
LEVI LINCOLN, Atty-Gen. of the United States.

THE SCHEDULE referred to in these Letters Patent and making a part of the same, containing a description in the words of the said Benjamin Tyler himself, of the Wry Fly, which may be applied by wind or water to various machines, viz.: Grist mills, Hulling mills, Spinning mills, Fulling mills, Paper mills, and to the use of Furnaces, etc.
The Wry Fly is a wheel which, built upon the lower end of a perpendicular shaft in a circular form, resembling that of a tub. It is made fast by the insertion of two or more short cones, which, passing through the shaft, extends to the outer side of the wheel. The outside of the wheel is made of plank, jointed and fitted to each other, doweled at top and bottom, and hooped by three bands of iron, so as to make it water-tight; the top must be about one-fifth part larger than the bottom in order to drive the hoops, but this proportion may be varied, or even reversed, according to the situation of place, proportion of the wheel, and quantity of water. The buckets are made of winding timber, and placed inside of the wheel, made fast by strong wooden pins drove in an oblique direction; they are fitted to the inside of the tub, or wheel, in such a manner as to form an acute angle from the wheel, the inner edge of the bucket inclining towards the water, which is poured upon the top, or upper end of it, about twelve and a half degrees; instead of their standing perpendicular with the shaft of the wheel they are placed in the form of a screw, the lower ends inclining towards the water, and against the course of the stream, after the rate of forty-five degrees; this however may be likewise varied, according to the circumstances of the place, quantity of water, and size of the wheel; over this wheel, and exactly fitted to the top of it, is a cup, or short cylinder, made fast and immovable by timbers connected with other parts of the building. Said Wry Fly may be used with or without said cylinder.

BENJAMIN TYLER.

P. HENDERSON,
SAMUEL HITCHCOCK, } *Witnesses.*

From the description of the Wry Fly it will be seen that, except the chutes, it contained the principal features of the modern turbine, the merits of which are due to many minds; while still greater skill is required to bring it to that state of perfection it is undoubtedly destined to attain. The increasing importance of the manufacturing interest necessitated the improvement of devices for utilizing to the greatest possible extent the water-power of the country. An article suggested by the change of wheels at Lowell, Mass., is here quoted from the *Courier* of that city published in 1871.
"The removal of the last in the city (except two or three on the Concord River) of the old-fashioned and unwieldy breast-wheels suggests to us that a chapter of information on the hydraulic motors now in use here, and the history of their improvement and adoption, may prove of interest

Devices of the Past.

FLUTTER WHEEL.

UNDERSHOT WHEEL

IMPACT WHEEL.

OVERSHOT WHEEL

BREAST WHEEL

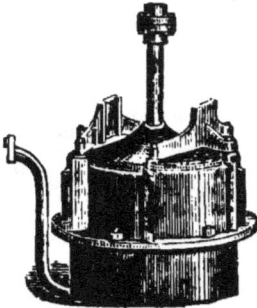

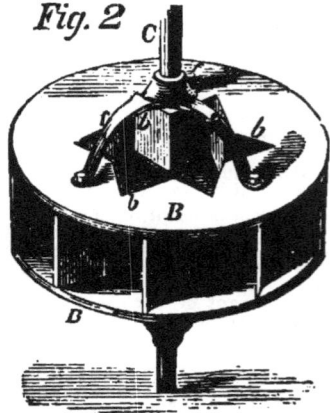

Fig. 2

to all the readers of the *Courier* who take a just pride in whatever aids Lowell to substantiate her claim to the title of "the first manufacturing city in America." And first, let us glance at the old style of wheels, of which those now being removed from the mills of the Lawrence and Prescott companies are fair examples. These, generally known as "breast-wheels," are dependent fo their useful effect simply upon the weight of the water, admitted to the buckets near the top, and retained as long as possible, or until nearly at the bottom of the wheel, where its force is spent and it is discharged. These wheels have in Lowell been constructed of wood, and or great size, varying in diameter from 13 to 30 feet, and usually about 12 feet long. Wheels of this class are still in use to a great extent, and in rare instances reach the enormous size of 70 feet in diameter. From the starting of the first mill (Merrimack) in 1823, up to the year 1845, when the number of spindles was about one-half that present running, the breast-wheels alone were in use, and were considered the most perfect in all respects of the kinds generally known. But although held in such high estimation, they were very extravagant in the use of water; for although the proportion of the useful effect given by the wheel to the power expended sometimes reached as high as 75 per cent., the average performance fell far below this point, being only about 60 per cent. And the importance of overcoming this radical defect becoming more and more obvious, improvements were gradually devised which resulted finally in the invention of a class of wheels known as turbines.

The word turbine is derived from the Latin *turbo*, which means among other things, a top; and also, the whirling or spinning motion of a top. The name, though sometimes given a wider range of meaning, is properly applied to a re-action wheel with vertical axis. The wheel itself is a French invention, dating back to 1830, or thereabouts; and it was introduced into this country several years later by an eminent engineer of Pennsylvania, Mr. Ellwood Morris, who built and put in operation two of these wheels, and published the results of his experiments upon them about the year 1843. The advantages of the turbine were found to be mainly these; a greater economy in the use of water; adaptation to any fall; greater velocity, compactness and durability, and that it was not obstructed by backwater. Since Mr. Morris' experiments there have appeared before the public almost innumerable varieties of turbines, each inventor claiming for his wheel some advantage over all its predecessors; and up to the present time several hundred patents have been granted in this country alone for modifications and alleged improvements of the turbine as first invented. Many of these wheels are quite popular, and are in use in small establishments all over the country; but being roughly and cheaply made, none of them have yet been found to compare with the original Fourneyron turbine as improved by the inventions of Uriah A. Boyden, whose name is familiar to every one who is at all acquainted with the history of our city.

In 1844 Mr. Boyden designed a 75 horse-power turbine for the Appleton Company's Picker-house, introducing, as has been said, several changes of his own devising. This wheel was tested immediately after its completion, and found to give a useful effect of 78 per cent. of the power of the water. Encouraged by this success, Mr. Boyden proceeded in 1846 with the construction, for the same company, of three more turbines of 190 horse-power each, which upon being similarly tested gave the remarkable result of a useful effect of 88 per cent. In experiments since that time results have been obtained as high as 92 per cent.; but it is considered that a fair average for these wheels is about 75 per cent. against 60 for the breast-wheels as above stated. From the date of the Appleton Company's adoption of turbines, they have come rapidly into use; being substituted for the clumsy affairs first used as fast as the latter became unserviceable from wear and decay.

One of the advantages of the turbine, as already stated, lies in the fact of its occupying so much less space, in proportion to the power, than any other wheels. And this will be more fully realized when it is considered that there are in actual use for manufacturing purposes turbines of only 6 *inches diameter;* and though these, it must be owned, are rare, those of 10 and 12 inches are not unfrequently met with; usually operating, however, in localities where the amount of water is limited, while the fall is considerable. Of the 70 powerful turbines in use in the mills of Lowell, the smallest has a diameter of 5*, and the largest of 11 feet, and the capacity of a single wheel reaches, in several cases, 675 horse power.

*At the time the foregoing was written there were many turbines in use at Lowell of less diameter than 5 feet, though perhaps none of the Fourneyron style; since that time the large companies there have taken the Swain turbine in preference. The following article gives the origin of the Fourneyron wheel, and will enable the reader to consider Mr. Boyden's claim as inventor, understandingly.

FOURNEYRON WHEEL.

MacAdam Bros.', Belfast, Ireland.

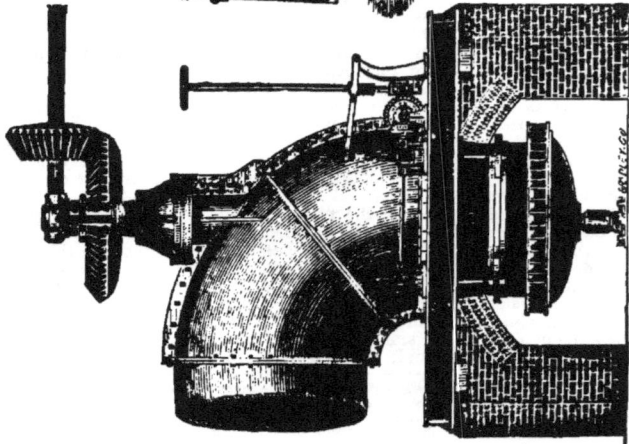

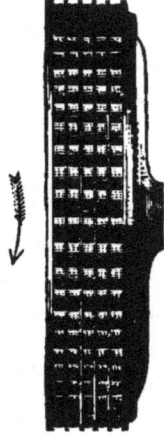

Holyoke Machine Co., Holyoke, Mass.

The whole power given by the fall of the Merrimack at Lowell, of 33 feet, is estimated at about 10,000 horse-power, the entire amount of which is already leased to the corporations. In addition to this, there are in the mills 31 steam-engines, furnishing 5000 horse-power additional; and besides these sources there are the three falls of Concord River, the power of which we have no means of estimating.

---

*Fourneyron Wheel.

[Extract from a Treatise on the power of water, by Joseph Glynn.]

M. Fourneyron, who began his experiments in 1823, erected his first turbine in 1827, at Pont sur l'Ognon, in France. The result far exceeded his expectations, but he had much prejudice to contend with, and it was not until 1834 that he constructed another, in Franche Comté at the iron-works of M. Caron, to blow a furnace. It was of 7 or 8 horse-power, and worked at times with a fall of only 9 inches. Its performance was so satisfactory that the same proprietor had afterwards another of 50 horse-power erected, to replace 2 water-wheels, which together, were equal to 30 horse-power.

The fall of water was 4 feet 3 inches, and the useful effect, varied with the head and the immersion of the turbine, 65 to 80 per cent.

Several others were now erected: 2 for falls of 7 feet; 1 at Inval, near Gisors, for a fall of 6 feet 6 inches, the power being nearly 40-horse, on the river Epté, expending 35 cubic feet of water per second, the useful effect being 71 per cent. of the force employed.

One with a fall of 63 feet gave 75 per cent.; and when it had the full head or column for which it was constructed—namely, 79 feet—its useful effect is said to have reached 87 per cent. of the power expended.

Another, with 126 feet, gave 81 per cent.; and 1 with 144 feet fall, gave 80 per cent.

At the instance of M. Arago, a commission of inquiry was instituted by the Government of France, for examining the turbine of Inval, near Paris, the total fall of water being 6 feet 6 inches, as has been before mentioned. By putting a dam in the river, below the turbine, so as to raise the tail water, and diminish the head to 3 feet 9 inches, the effect was still equal to 70 per cent.; with the head diminished to 2 feet, the effect was 64 per cent.; and when the head was reduced to 10 inches, it gave 58 per cent. of the power expended, notwithstanding the great immersion of the machine.

In the year 1837, M. Fourneyron erected a turbine at St. Blasier (St. Blaise,) in the Black Forest of Baden, for a fall or column of water of 72 feet (22 mètres). The wheel is made of cast-iron, with wrought-iron buckets; it is about 20 inches in diameter, and weighs about 105 pounds; it is said to be equal to 56 horse-power, and to give an useful effect equal to 70 or 75 per cent. of the water power employed. It drives a spinning-mill belonging to M. d'Eichtal. A second turbine, at the same establishment, is worked by a column of water of 108 mètres, or 354 feet high, which is brought into the machine by cast-iron pipes of 18 inches diameter of the local measure, or about 16½ inches English. The diameter of the water-wheel is 14¼, or about 13 inches English, and it is said to expend a cubic foot of water per second; probably the expenditure may be somewhat more than this.

The width of the water-wheel across the pier is .225, or less than a quarter of in inch. It makes from 2200 to 2300 revolutions per minute; and on the end of

---

*The Fourneyron wheel receives the water from the inside, discharging it outwards. The gate, a thin hoop somewhat deeper than the wheel, is placed between the chutes and wheel, and is opened by being raised With such an arrangement, economical part gate results are impossible; and M. Fourneyron and many others have made the wheel with divisions in the buckets as shown in the MacAdam plan. The "quarter turn" of the Holyoke Machine Co. is the invention of Mr. Boyden. MacAdam places the wheel at the small end of a vertical cone-shaped tube. Valentine and others have placed it in scroll and various kinds of curbs. It has been constructed so as to receive the water from below by many parties. It has been made with register gate inside of chutes, between chutes and wheel, and in one case in my experience, with two register gates, one inside of chutes, the other outside of wheel. It has been made with short straight chutes, also long curved ones. It has been suspended by the upper end of its shaft in various ways, instead of resting upon a step. It has been made of iron in the coarsest and cheapest style, and of bronze at an enormous cost. It has proved as variable in useful effect as any of the other kinds of turbines.

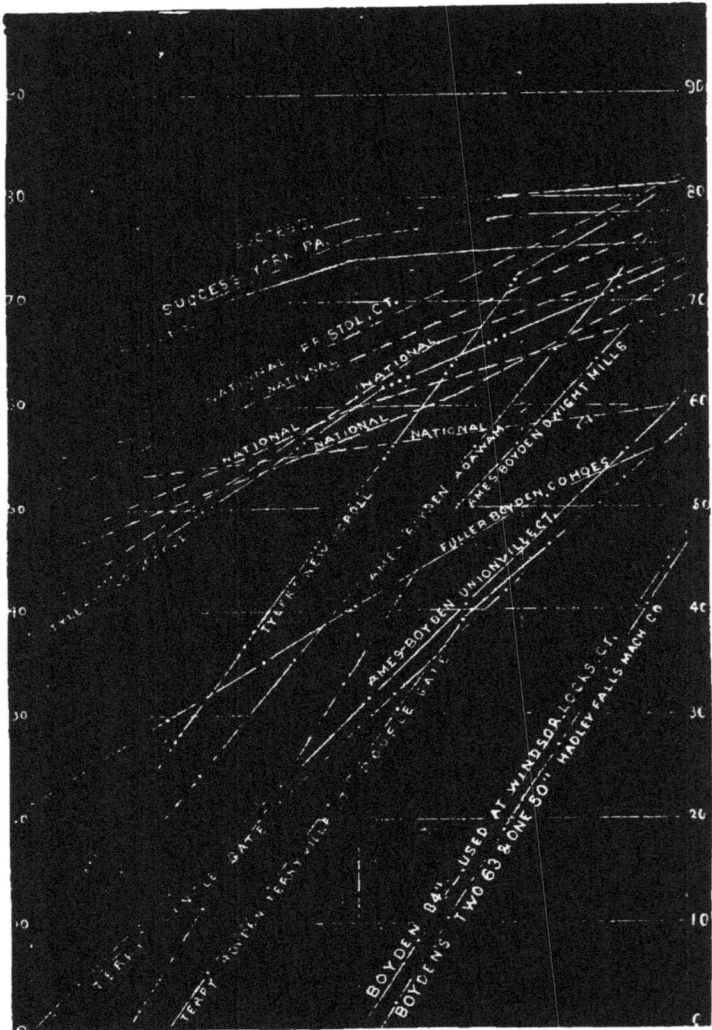

The Boyden turbine is highly recommended by its builder, and is much admired by the corporation superintendent, in kids, whose responsibility is remote; but the practical manufacturer, who has his own bills to pay, lets it severely alone. As may be seen by the diagrams, wheels constructed upon any of the popular plans represented in this work, may be made far superior in every respect, at one half its cost. Its continued use is owing solely to the low state of intelligence in milling engineering. It is idle to expect perfection or any constant efficiency in turbines until purchasers become sufficiently awake to their own interests to be willing to pay a fair price, and then to insist upon knowing exactly what the very wheel that is to be purchased will do before accepting it. The same quality of workmanship will make any other kind of turbine as durable as the Boyden.

the spindle or upright shaft of the turbine is a bevelled pinion, of 19 teeth, working into two wheels, on the right and left, each of which has 300 teeth. These give motion to the machinery of the factory, and drive 8,000 water spindles, roving frames, carding engines, cleansers and other accessories. The useful effect is reported to be from 80 to 85 per cent. of the theoretical water-power. The water as filtered at the reservoir before it enters the conduit pipes; and it is important to notice this, since the apertures of discharge in the wheel are so small as to be easily obstructed or choked.

The water enters the buckets in the direction of the tangent to the last element of the guide-curves, which is a tangent to the first element of the curved buckets. The water ought to press steadily against the curved buckets, entering them without shock or impulse, and quitting them without velocity, in order to obtain the greatest useful effect; otherwise a portion of the water's power must be wasted or expended, without producing useful effect on the wheel.

It is difficult to imagine that a machine so small as this can give motion to the works of a cotton mill on so large a scale. Professor Ruhlmann says, that when he saw it actually doing so, he could not for some time credit the evidence of his senses; and, although he went purposely to examine it, his astonishment prevented him from comprehending, in the first instance, that the fact was really as it appeared.

The Jonval Turbine

[From J. E. Stevenson's Circular.]

By referring to our certificate on another page, it will be seen that it is impossible to construct a turbine greatly to exceed in useful effect a "Jonval," when properly constructed and well finished; and by reference to the table of experiments here inserted, it will be noticed that the efficiency of the Jonval turbine depends not upon the name "Jonval," neither upon the simple fact that one wheel is placed above another—as from 6 Jonvals tested, but ONE gave 90.77 per cent., that being the *one made by us;* whereas one other gave only 50.34 per cent., the *lowest of all tested.* And why this difference? Simply because one builder knew what he was doing, and the other did not. There are many parties, purporting to manufacture the Jonval turbine, who state in their circulars that "at a trial of turbine wheels, at Fairmount Water Works, at Philadelphia, in 1859 and 1860, the Jonval wheel gave the highest percentage of all tested;" and they would have the public believe that with their rough, unfinished castings, guide and bucket curves, of whatever form they may happen to be, they give this wonderful result, when none of them possess more than one feature of the Jonval turbine; and these experiments show that a "Jonval," made by a man of experience, and tested under the most favorable circumstances, gave the *poorest result of all,* simply because he failed in the application of the principles embodied in its construction.

The following is a table of the experiments at the Fairmount Water Works, at Philadelphia, in 1859 and 1860, as taken from the report of the Chief Engineer. The table explains itself.

At a trial of water wheels, at Fairmount Works, by order of the Select and Common Council of the City of Philadelphia, a Jonval turbine, made by "J. E. Stevenson," of Paterson, New Jersey, was tested March 9th, 1860, and produced a co-efficient of useful effect of .8777 per cent. under the following circumstance : 925 pounds were raised 25 feet by 70.25 cubic feet of water under a head and fall of 6 feet. To this must be added the friction of the transmitting machinery, estimated at 3 per cent., making a total useful effect of .9077 of the power employed.

O. H. P. PARKER,
Chairman of the Water Com.

(SEAL OF CITY.)

HENRY P. M. BIRKINBINE,
Chief Engineer.

In attestation of the above signatures of O. H. P. Parker and Henry P. M. Birkinbine, I set my hand and affix the seal of the City of Philadelphia, this, 3rd day of April, 1860.

ALEXANDER HENRY,
Mayor of Philadelphia.

Jonval Wheels, with Variations.

BODINE JONVAL.

BLACKSTONE.   ELMER.

HOUSTON.   CURTIS.   DELPHOS.

 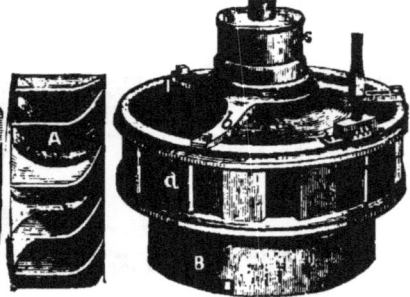

CASE.   VANDEWATER-BURNHAM.

Table of Experiments.

NAME OF WHEEL.	Kind of Wheel.	Per cent. of effect.	3 per ct. added for friction.	Where Built.
Stevenson's 2nd wheel......	Jonval..	.8777	.9077	Paterson, N. J.
Geyelin's 2nd wheel.......	Jonval..	.8210	.8510	Philadelphia, Pa.
Andrews & Kalbach's 3rd wheel..................	Spiral...	.8197	.8497	Bernville, Pa.
Collins' 2nd wheel..........	Jonval..	.7672	.7972	Troy, N. Y.
Andrews & Kalbach's 2nd wheel..................	Spiral...	.7591	.7891	Bernville, Pa.
Smith's Parker's 4th trial....	Spiral...	.7569	.7869	Reading, Pa.
Smith's Parker's 3rd trial ...	Spiral...	.7467	.7767	Reading, Pa.
Stevenson's 1st wheel.......	Jonval..	.7335	7635	Paterson, N. J.
Blake	Scroll...	.7169	.7469	East Pepperell, Mass.
Tyler......................	Scroll...	.7123	.7423	West Lebanon, N. H.
Geyelin's 1st wheel.........	Jonval..	.6799	.7099	Philadelphia, Pa.
Smith's Parker's 2nd wheel..	Spiral...	.6726	.7026	Reading, Pa.
Merchant's Goodwin........	Scroll.	.6412	.6712	Guilford, N. Y.
Mason's Smith..............	Scroll...	.6324	.6624	Buffalo, N. Y.
Andrew's 1st wheel	Spiral...	.6205	.6505	Bernville, Pa.
Rich	Scroll...	.6132	.6432	Salmon River, N. Y.
Littlepage..................	Spiral...	.5415	.5715	Austin, Texas.
Monroe....................	Scroll...	.5359	.5659	Worcester, Mass.
Collins' 1st wheel...........	Jonval..	.4734	.5034	Troy, N. Y.

Turbine builders may object to my classification of the various wheels represented upon the opposite page; but because M. Jonval defined certain lines for a turbine, he no more proved that those lines covered the principle than he would have proved that the only place to walk upon a street is exactly three feet from its centre on a line parallel therewith, had he defined such a line. The wheel itself was common and known as the Tub wheel. Two wheels made upon the Wry-Fly specification, placed one above the other, would have covered the plan of M. Jonval; placing a fixed wheel above the wheel proper would have little originality unless done before any other builder had made an application of chutes to turbines. The experiments of D. P. Blackstone, show plainly that the vertical part of the buckets of the Vandewater-Burnham wheel, represented with the others, is of little practical utility; indeed, the vertical part of such buckets have often been proved to be decidedly injurious. Wheels constructed in that way, however, render it more convenient to apply the water economically at part gate. Many plans for gates have been tried with the Jonval, but none that has not in some way proved objectionable. Many have been made without any gate, simply letting the water on from the head gate of flume. Geyelin of Philadelphia has a telescopic tube below the wheel, the bottom thereof being lowered to the apron beneath, in order to stop the water. Wicket gates have also been tried in a tube below the wheel, but both plans cause an extravagant use of water, unless the wheel runs at whole gate. "Outside register gates" are the most common; these also render it impossible to economize water at part gate. The inside register, like that of Gates Curtis is far better in that way, but like the other register gates it works hard. Downward discharge wheels were objected to because they were supposed to press heavily upon the step; such an idea could only have gained a place through very superficial reasoning, for if 75 or 80 per cent. of the weight of the water forced the wheel ahead, the balance of the weight could only press down upon the step, whether downward, central or outward discharge,

Perpetual Motion.

For ages past the above idea has been the constant dream of a certain class of minds, and is as prevalent to-day as in the past. To save such minds from the trouble of re-inventing for the thousandth time the same old wornout devices, a few of the most common are here sufficiently illustrated to show those engaged in such efforts that their plans are old. It is rather singular that such hydraulic geniuses almost invariably select the poorest kind of wheels to be combined, in order to get 175 per cent. from a double use of the water. The plan of Mr. Jones only contemplated increased capacity for diameter and part gate economy, but his plans have long been abandoned by more intelligent builders.

Little Giant.

George H. Jones, Auburn, N. Y.

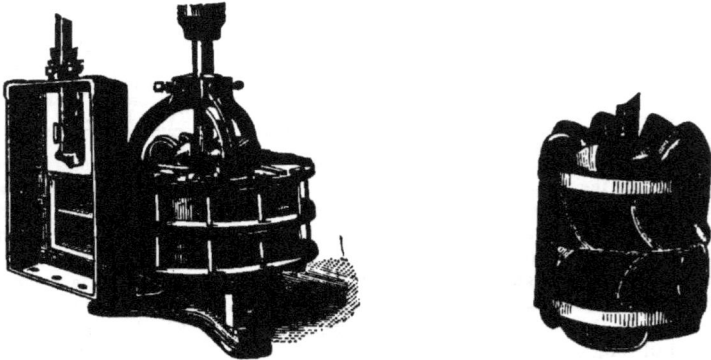

"Our wheel discharges its water inward, downward and outward, and discharges as much inward as any central discharge wheel of same diameter; as much outward as a Fourneyron, and downward as much as any Jonval, &c., &c."

There is not the slightest reason to doubt the ability of such devices to discharge an abundance of water, but years of experience and demonstrations by decisive tests prove beyond chance for dispute that all double arrangements are less effective than simple single turbines. Various kinds have been tested and invariably with the same results; the single wheel has proved the best in every way. The Leffel turbine has been continued in its original form simply because all the claims hinge upon the use of the double wheel, and to give up that would invalidate the whole patent.

Double Wheels.

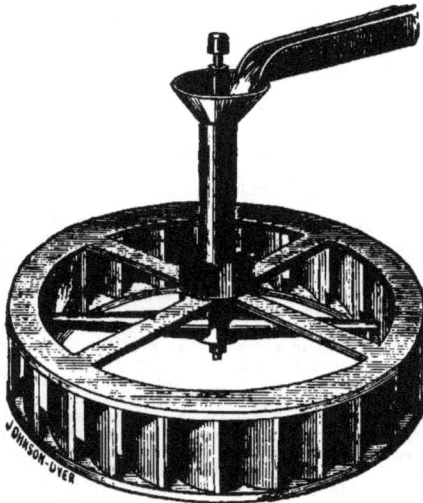

DEVICE BROUGHT FROM MEXICO. Barker Mill discharging into a Fourneyron wheel; connected by gears same as Wynkoop; small gear on rim of the funnel.

Same device used in a tanning mill. See Weisbach.

The Plan Represented below has caused the Expenditure of much Time and Money.

The plan consists simply of placing several Jonval wheels in a tube, one above another, each pair rotating in opposite directions. H. Twitchell of Pulaski, N. Y., furnished a set for trial, three wheels; the upper one stationary, acting as chutes; the two beneath rotating in opposite directions, being connected together by gears, hollow and solid shaft, arranged the same as those connecting Wynkoop wheels. First test was with upper wheel, lower wheel removed.

Test of upper wheel with lower wheel removed.				Test of the two wheels connected by gears.			
Head.	H. P.	Cubic feet.	Per Cent.	Head.	H. P.	Cubic feet.	Per Cent.
18.59	6.96	353.04	.5615	18.49	4.97	325.12	.4486

L. D. WYNKOOP'S
Double Power Water Wheel.

Patented January 30, 1866.

In this improvement we have a device for combining wheels driven by the force of running water, and also by the weight of the fluid, both acting in the same direction, and the latter using the water which has already given power to the former.

Fig. 1 shows the external appearance of the case of the wheel, and Fig. 2 the two motors with their gearing. The stream is received at A, Fig. 1, and, by the spiral form of the case, is forced to receive a rotary motion as in the common Turbine. This water acts directly on the buckets, B, Fig. 2, which radiate from the center. They are connected to a hollow shaft, which carries the large bevel gear, C, gearing into the pinion, D, on the horizontal shaft.

Passing through the inside of this main shaft is the shaft, E, to which the scroll wheel, F, is secured at the bottom, and a bevel gear, smaller than C, at the top. This gear meshes with the pinion, G, on the horizontal shaft. After the water, by its rotary force, has done its work on B, it falls and operates F, giving it twice the speed of B. By this combination we claim that this device has twice the power of an ordinary wheel with the same weight and force of water. It has been tested by practical men with even greater results. There is now one of these wheels in successful operation in the machine shop of Messrs. CLAPP & HAMBLIN'S, of this City, who have the exclusive right for the manufacture of the same in this State, and the public are invited to call and test it for themselves. The proprietors, L. D. Wynkoop and S. P. Stone, are now prepared to negotiate with any responsible parties for the right to manufacture the same in any State of the Union. In offering this to the public, we are aware that the cry of humbug will be made, but we guarantee all we claim for it, and we wish no one to engage in it until he is satisfied that what we claim is true. Any inquiries addressed to the proprietors, at Owosso, Mich., will be promptly attended to.

L. D. WYNKOOP.
S. P. STONE.

We the undersigned have seen the Wynkoop Wheel in operation and believe it to be all the inventor claims for it.

J. B. BARNES, Mayor,	JAMES W. STEDMAN,
A. BARTLETT, City Marshal,	P. M. ROWELL,
E. D. GREGORY,	J. H. CHAMPION,
C. W. CLAPP & CO.,	EDWARD SMITH, Machinist,
D. R. STONE,	A. J. PATTERSON,
N. McBAIN,	C. A. BALDWIN,
GREEN & LEE, Editors "Press,"	E. SALSBURY,
DANIEL LYON,	H. S. GALUSHE,
WILLIAM FLETCHER,	C. OSBURN.

CERTIFICATE:

☞ From the experiments performed with the Wynkoop Wheel in the Foundry of Messrs. Clapp & Hamblin of this city, I find it to utilize more than 175 per cent. of the absolute weight of water used; probably nearer 200. This I regard as no violation of the principle laid down in our natural philosophy, viz: that no wheel can be invented which will utilize 100 per cent., as the wheel in question is not a single one, but such a combination of wheels, as can not fail to give a vast increase of power.

I. C. COCHRAN, Principal of Owosso Union School.
HENRY GOULD, Millwright, Owosso City.

WYNKOOP'S

DOUBLE POWER WATER WHEEL.

EFFICIENCY CLAIMED IN TABLES.   **EFFICIENCY OBTAINED BY TEST.**

Head in feet.	Horse Power.	Cubic ft. Disch'd.	Percentage.	Head in feet.	Horse Power.	Cubic ft. Disch'd	Percentage.
15	36.18	944	135	15.12	36.51	2318.23	.5513

The debut of this wheel furnished ample proof that "Perpetual Motion" theories take as readily with those ranking with the learned, as with those having little knowledge of books. Several College Professors, (one at the head of a State Board of Education,) endorsed the claims of Mr. Wynkoop, furnished means to develop the merits of his device, and were present at its test and quietus.

The Economy Water Wheel.

COMPLETE AND READY FOR SETTING UP.

We offer a challenge of $1000 to the country to produce a Turbine Water Wheel of same diameter and under same fall that will furnish one-half as much power as our wheel.

Fulton, Myers & Co.,

SOLE MANUFACTURERS,

Indianapolis, Ind.

The Great Compound I.-X.-L.-- Turbine Water Wheels.

JOSEPH HOUGH, Sole Patentee,

BUCKINGHAM P. O., BUCKS CO., PA.

My father was a practical miller and a thorough millwright; he contended that there was no water wheel then in existence that utilized but a little over half the power of the water that passed over, under, or through any water wheel in use; and how to utilize this lost power was a difficult problem to an inventive mind to solve; nevertheless, perseverance and a determination to conquer all obstacles in the way, I finally invented and completely overcame this great difficulty, the utilizing of this otherwise lost power of water. I will give a complete description of the construction and action of the water on my Double Right and Left Reacting Turbine Water Wheels The one-half diameter of their size are blocked or filled up in the centre, to cause the water by a tapering centre, a suitable height above the chutes, to spread all around from the inlets above, striking every bucket at the same time, at the farthest part from the wheel's centre, at the point of the wheel where the water exerts the greatest force and power. The chutes and also the buckets of both water wheels are straight blades, set at an angle of 45 degrees, that the current of water from the chutes above may strike the buckets of the upper wheel squarely at right angles, and as these buckets recede from the force of the current, the water escapes off these blades with still greater force by adding the second wheel of the same dimensions immediately under the first, with blades set also at 45 degrees in an opposite direction to again receive this otherwise lost force of the current of water, turning said wheel to the left hand. This is the important feature in my great improvement on all single turbines, by utilizing this otherwise lost force, utilizing the water twice over.

Smith's Upper Spring Valley Mills.

This is to certify, that the undersigned, millers and millwrights of Buckingham, were present at the testing trial of J. Hough's Great Compound I.-X.-L. Double Right and Left Reacting Turbine Water Wheels, and we are free to say, they far exceeded our most sanguine expectations by doing one-third more work with the same amount of water as it takes for the old ordinary single turbine water wheel. This we were eye witnesses to.

WM. B. SMITH,
LEWELLEN FRIES,
SAMUEL DEHAVEN,
JOHN C. VANDERGRIFT,
JAMES M. VANDERGRIFT,
ELI DOAN.

Efficiency of Turbines.

In reporting the efficiency of the many water wheels brought to be tested during the past ten years, it has been a very difficult matter to suit all that have been interested, yet no builder has ever expressed a doubt that any other builder has ever received a less favorable report than he deserved; but in their own particular case something a little more favorable should have been said, or something unfavorable left unsaid. Thousands have asked my advice in turbine matters, and many hundreds, if not thousands of turbines have been selected upon the advice given; yet not a single complaint has ever been made that the wheel recommended proved unworthy of the recommendation, nor in the ten years, has a wheel that I have reported poor, proved by practical use to be good. Time will determine whether my opinions, statements and reports relative to turbine matters have been well founded, and to that decision I am willing to trust.

In making up the following reports, my purpose has been more to aid builders in selecting the best plans than to sell wheels constructed upon any of those now existing; to do so, requires a knowledge of the lowest as well as of the highest results obtained by test of each, and such are given.

The extreme variations in the results obtained from every kind tested, should convince purchasers that there is no certain way of procuring a good turbine otherwise than by testing, before acceptance, as they would do if purchasing a horse.

The wheel of B. J. Barber might properly have been placed in the group with the Wynkoop and others of that class—not that Mr. Barber believed in perpetual motion or 175 per cent. wheels; but he believed that an unexpended force remained in the water discharged by any single turbine, and that that force could be utilized by adding a second wheel below the first. His plan, however, carried out as shown, simply produces the ordinary downward discharge wheel. Mr. Barber erred in using the central discharge at all, for much better results are possible with the plain downward discharge than can be obtained from the central, or his combination.

The other wheels with double discharge, reported in the following pages, such as the Swain, Leffel, Eclipse, Angell, Walsh and others, were so constructed, under the expectation of obtaining increased capacity for a given diameter, but a comparison with the capacity of recent plans will show that such expectations were not well founded.

A. M. Swain, North Chelmsford, Mass.

SWAIN TURBINE.

One of the earlier high class wheels, made with many buckets and small openings, placed in "quarter turn" or "flume curb." Mr. Swain had much to do about starting the testing system. Quite a number of these wheels, ranging in size from 18 to 42 inches in diameter have been tested.

Test of a 21-inch.	Head.	Weight.	Rev. per minute.	H. P.	Cubic feet.	Per Cent.
Whole Gate,..............	17.91	300	281	25.55	936.55	.8072
Part Gate,..............	18.25	275	282.5	23 54	864.94	.7902
"	18.34	230	280 5	19.55	742.22	.7611
"	18.44	165	241.5	12.08	562.20	.6175

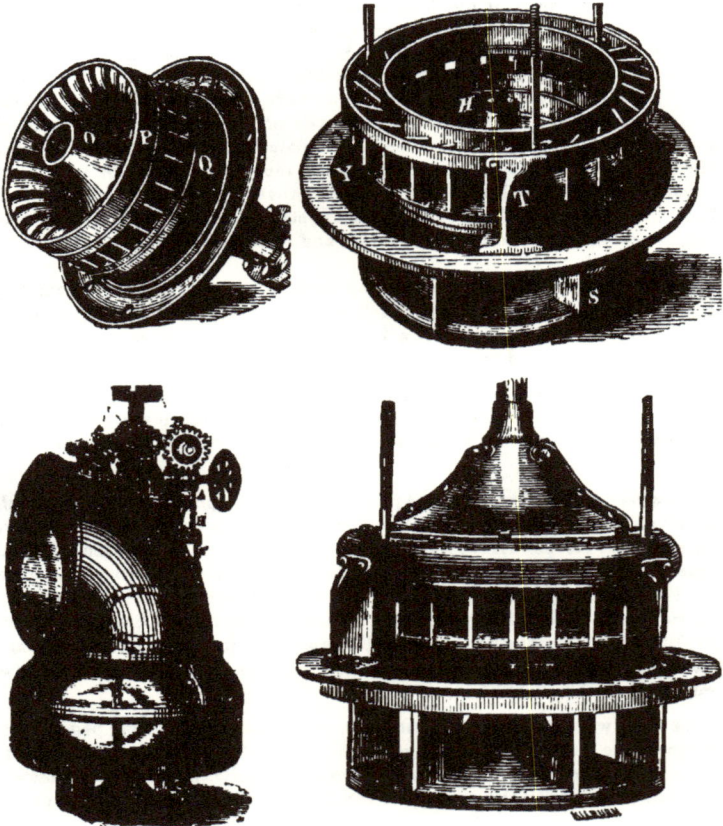

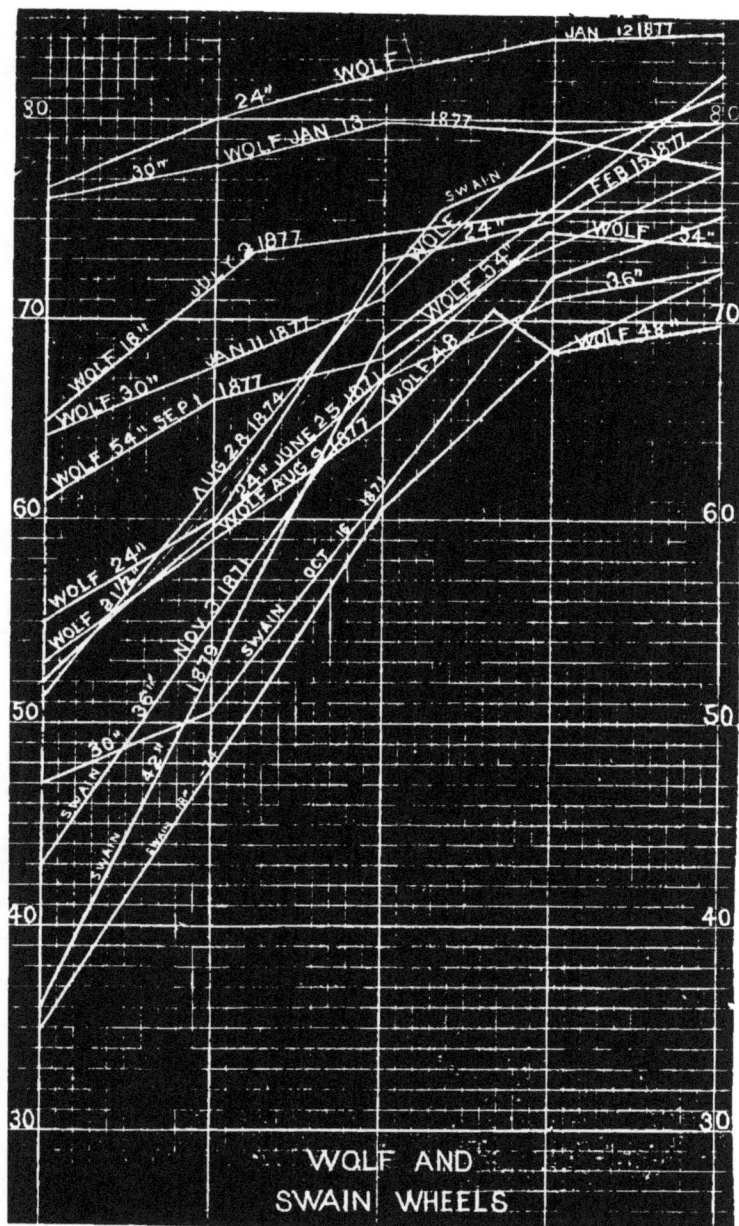

WOLF AND
SWAIN WHEELS

C. B. Walsh's Double Turbine, Waupaca, Wis

Test of a 35-inch wheel.					Test of a 13-inch wheel.			
Head.	R.v.	H. P.	Cub. ft.	Pir Ct.	Head.	Rev.	H. P.	Cub ft.	Per Ct.
17.97	166	61.19	2298	.784	18.18	368.5	9.49	508	.544
18.03	154.5	56.18	1980	.830	18.28	410	9.93	443	.649
18.22	151.5	44.76	1564	.833	18.37	377	8.00	355	.650
18.33	156.5	35.56	1291	.800	18.41	394	8.09	346	.672
18.49	151	24.02	935	.735	18.51	420	6.36	275	.661
18 57	154.5	17.55	755	.663	18.52	409	6.20	271	.656
18.68	152	10.36	523	.562	18.56	410	5.59	257	.618

Wetmore Wheel, Upham Machine Co., Claremont, N. H.

Test of a 36-inch, Sept. 17, 1873.

	Head.	Weight.	Rev. per minute.	H. P.	Cubic feet.	Per Cent.
Whole Gate,..............	17.94	810	146	53.76	1913.68	.8291
Part Gate,..............	17.98	650	143,5	42.40	1794 63	.6959
" "	18.09	450	151	30.88	1441.34	.6270
" "	18.18	320	141	20.51	1201.11	.4975
" "	18.28	150	143.5	9.78	946.76	.2970

Test of an 18-inch, Sept. 29, 1876.

Whole Gate,..............	18.69	150 .	303	13.77	501.32	.7781
Part Gate,..............	18.71	120	307	11.16	423.66	.7454
"	18.79	75	305	6.93	304.45	.6414
"	18.82	60	292.5	5.31	262.16	.5699

Same wheel, buckets having been chipped.

Whole Gate,..............	18.71	150	320,6	14.57	507.41	.8125
Part Gate,..............	18.87	80	304	7.46	323.50	.6632
"	18.90	55	313.5	5.22	263 18	.5577

Test of a 24-inch, Oct. 2, 1876.

Whole Gate, ...	18.34	350	228	24.18	960.21	.7270
Part Gate, ...	18.58	185	230	12.89	582.25	.6300
"	18.66	135	229.5	9.35	483.83	.5611

Second test of same, buckets having been chipped.

Whole Gate,..............	18.30	350	249.5	27.21	963.51	.8170
Part Gate, ...	18.49	225	234.5	15.98	671.85	.6810
"	18.67	100	237.5	7.20	424.49	.4810

Test of a 48-inch, Oct. 5, 1876.

Whole Gate,..............	10.15	1550	83.5	51.23	3713.65	.7195
Part Gate,..............	11.06	1100	80.5	40.25	3201.64	.6018
"	10.63	700	83	26.41	2354.21	.5588
"	11.02	400	82	14.90	1827.39	.3834

Perry Turbine.

Perry & Taylor, Bridgton, Maine.

Downward discharge, with inside register gate. Messrs. Perry & Taylor have provided themselves with apparatus for testing their wheels before delivery, and guarantee the results furnished at each sale.

Sept. 1, 1877.	Head.	Rev'n Per Min.	Horse Power.	Cubic feet.	Per Cent.
Whole Gate,..............	13.12	271	6.15	300.65	.8288
" "	13.12	247.3	6.18	302.70	.8223
" "	13.12	274	6.22	300.65	.8382
" "	13.13	303.7	6.21	299.29	.8382
Part Gate,......	13.22	261	4.74	246.85	.7703
" "	13.31	248.5	3.38	191.31	.7039
" "	13.33	280.5	3.18	184.13	.6871
SECOND WHEEL.					
Whole Gate,..............	11.80	162.5	19.94	1086.89	.8246
" "	11.78	158	20.11	1093.57	.8280
" "	11.77	154	20.30	1090.23	.8390
" "	11.76	151	20.59	1100.25	.8440
" "	11.74	145.7	20.53	1110.30	.8354
" "	11.73	141.5	20.58	1115.34	.8343
Part Gate,......	12.13	150	10.52	709.16	.6486
" "	11.98	164	11.18	762.55	.6492
" "	12.06	150.5	11.97	780.61	.6589
" "	12.23	151.5	8.60	628.50	.5887
" "	12.36	148	6.39	525.33	.5221
" "	12.42	139	4.73	464.96	.4341

C. G. Mullikin, Lansing, Iowa.

Test of wheel, 28 inches in diameter, the buckets extending to edge of the crown plate, and filling bore of curb. (The first test in all of the tables is at whole gate; the others at part gate.) The following results were obtained.

Head.	Rev.	H. P.	Cub. ft.	Per'tge
18.54	178.5	30.32	1080.39	.7950
18.55	188	29.64	1038.32	.8141
18.59	189.3	28.65	960.35	.8492
18.65	182.5	24.72	818.23	.8569
18.73	183.5	20.69	689.91	.8470
18.83	181.8	14.71	513.39	.8049
18.96	186	9.90	358.68	.7702

The next trial was with a 44-inch wheel, with buckets of different curve or pitch from those of the 28-inch.

Head.	Rev.	H. P.	Cub. ft.	Per'tge
18.16	125	48.90	1994	.7157
18.27	127	40.41	1735	.6754
18.37	121.5	33.13	1504	.6353
18.48	120	23.45	1187	.5660
18.67	125	11.36	764	.4125

The buckets were then cut away to line marked 2, and again tested.

Head.	Rev.	H. P.	Cub. ft.	Per'tge
18.06	133.5	63.71	2435	.7670
18.07	137.5	60.93	2326	.7676
18.17	131.5	53.79	2048	.7652
18.29	135.5	44.33	1759	.7295
18.44	136.3	30.47	1370	.6490
18.58	137.5	20.00	1021	.5594

Buckets cut to line 3, third test.

Head.	Rev.	H. P.	Cub. ft.	Per'tge
18.02	135.5	63.71	2389	.7837
18.22	136.5	62.04	2276	.7919
18.32	138.5	56.65	2075	.7891
18.46	136	43.27	1671	.7443
18.62	132	30.60	1319	.6594
18.48	133	19.64	969	.5809

30-inch wheel, buckets on line 1.

Head.	Rev.	H. P.	Cub. ft.	Per'tge
18.65	194.5	32.71	1188	.7824
18.65	199.2	28.97	1081	.7621
18.70	196.5	25.42	965	.7740
18.77	190.5	19.48	811	.6772
18.84	193	15.35	799	.5309

Buckets chipped to line 2, re-tested.

Head.	Rev.	H. P.	Cub. ft.	Per'tge
18.74	205	37.27	1218	.8646
18.58	200	26.36	982	.7649
18.70	199	15.82	699	.6405
18.77	202	10.55	549	.4941

Buckets of 28-inch chipped to line 2, and again tested.

Head	Rev.	H. P.	Cub. ft.	Per'tge
18.48	171	30.70	1189.57	.7393
18.50	179	29.29	1184.49	.7077
18.53	178	28.31	1074.58	.7489
18.60	175	24.59	934.59	.7489
18.68	172.5	17.66	729.08	.6865
18.80	179	10.57	498.86	.5967

Collins' Wheel.

Manufactured by J. P. Collins & Co., Norwich, Conn.

[FROM MY FOURTH ANNUAL REPORT.]

Is local in reputation and only made to order. A 24-inch, brass bucket, nicely finished wheel in a curb similar to the above was sent to me to be tested. The sender stated that $900, had been paid for it and that Mr. Collins had sold it as the very best he could make. Mr. Collins was notified of the matter with the time fixed for the trial. Two days in advance he put in an appearance, very plea-antly remarking: "*I acknowledge the right of every purchaser to ascertain by actual trial the value of any wheel purchased;*" he further stated, that he had brought his overalls in order to put the wheel in order if it was not in good condition. Indeed, he was very genial, and one may judge of my surprise the next morning when just as the wheel was deposited at my flume, Mr. Collins, accompanied by sheriff, three or four appraisers, and other appurtenances of the law, stepped in with a writ of replevin and demanded the wheel. As I knew the purchaser had offered to sell it for a third of its price, I thought it an excellent sale, and, of course, made no unnecessary objections, but recalled to Mr. Collins his acknowledgement of the right of purchaser, to ascertain the value of wheel pu·chased; and was met with the statement that that particular wheel was not his present wheel at all; that he did not make such now; that he had no objection to his regular wheel being tested, &c., &c. Now, at the Philadelphia test, 1860, Mr. Collins produced the wheel that gave the lowest result of all, or about 47 per cent., as reported there, and as the wheel sent here was made somewhere about two years since, it becomes a rather interesting point to ascertain when he commenced to make good wheels.

American Turbine.

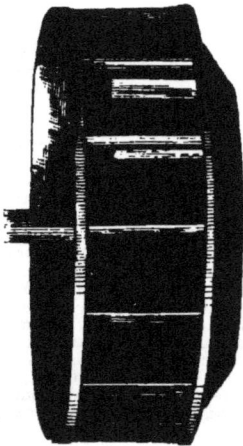

MANUFACTURED BY

Stout, Mills & Temple, Dayton, Ohio.

Test of a 36-inch Dayton wheel, Nov. 20, 1872.

	Head.	Weight.	Rev. per Min.	Horse Power.	Cubic feet.	Per Cent.
Whole Gate, ...	19.00	750	144	49.09	1780.14	.768
Part Gate, ...	19.11	615	140.3	39.22	1350.02	.804
" "	19.21	500	139.7	31.44	1123.75	.770
" "	19.30	350	146.5	23.31	888.29	.719

November 13, 1873, 36-inch wheel.

Whole Gate, ...	18.26	630	146.5	42.01	1625.77	.7503
Part Gate, ...	18.36	600	137.5	37.50	1440.20	.7501
" "	18.46	450	148	30.27	1174.92	.7400
" "	18.66	290	137.5	18.12	790.83	.6496

June 11, 1873, 48-inch wheel.

Whole Gate, ...	18.10	1530	107.7	99.86	3514.90	.8314
Part Gate, ...	18.18	1320	109	87.20	3068.46	.8280
" "	18.41	1130	109.5	74.99	2647.48	.8149
" "	18.60	880	108.2	57.71	2200.89	.7467
" "	18 86	640	108.7	42.16	1772.46	.6964

Test of 48-inch, January 29, 1874.

Whole Gate, ...	17.65	1320	107.8	86 24	3418.11	.7598
Part Gate, ...	17.66	1100	110.3	73.53	3010.79	.7316
" "	17.76	960	104	60.51	2504.01	.6948
" "	18.16	500	106	32.12	1690.47	.5548

September 29, 1873, 42-inch right hand.

Whole Gate, ...	17.93	1200	112.5	61.36	2569.85	.7095
Part Gate, ...	17.98	990	118.5	53.32	2218.55	.7094
" "	18.30	650	120	35.45	1452.72	.7065
" "	18.45	440	119.5	23.90	1213.58	.5666

October 1, 1873, 42-inch, left hand.

Whole Gate, ...	17.90	1100	118	59.00	2536.02	.6882
Part Gate,	18.00	980	120	53.45	2275.17	.6946
" "	18.13	820	121	45.10	1918.04	.6884
" "	18.43	420	116.5	22.24	1160.60	.5479

November 11, 1873, 25-inch wheel.

Whole Gate, ...	18.23	300	212	28.91	1158.24	.7244
Part Gate, ...	18.30	260	207	24.46	983.53	.7185
" "	18.39	220	205	20.16	880.49	.6565
" "	18.60	110	208	10.40	555.69	.5323

November 12, 1873, 20-inch wheel.

Whole Gate, ...	18.85	130	253.5	14.97	606.54	.6938
Part Gate, ...	18.55	110	243	12.15	528.55	.6536
" "	18.63	90	244	9.98	448.93	.6313
" "	18.77	50	225.5	5.13	285.15	.5072

Rotary Engine, or Water Wheel.

John Lucas, Hastings, Minn.

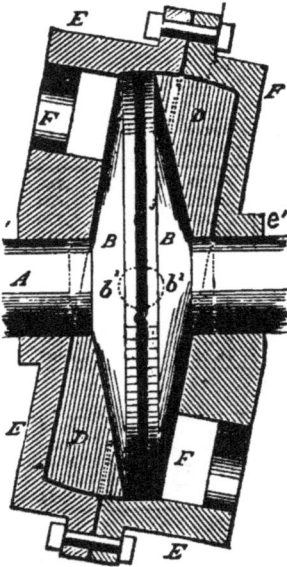

A slight examination of this device renders it obvious that it will utilize the full power of the water used, less loss from friction and leakage. In case E the Piston wheel B, works in bearings in the case, on shaft A. Wheel and shaft are slotted through the center; in this slot hangs on its pivot C, the Piston D, which oscillates in line with shaft A, as the Piston wheel is rotated by the passing water or steam. Used as a water wheel it does not need packing, consequently seems likely to prove durable, particularly as it does best when running very slow. In testing one, 12 inches in diameter at my flume under 18 feet head, it was found that the percentage increased rapidly as the speed decreased; the screw for tightening the brake was so coarse that the speed could not be got below 126 revolutions per minute without stopping it; at that speed it gave 87 per cent. For driving sewing machines, church organs, printing presses and other light machinery where a high head is available it seems to be the best device yet produced, as, unlike the turbine, it requires but a small supply or discharge pipe, it is noiseless, runs slow and utilizes the full power of the water used whether working at the maximum or minimum of its capacity. It may be placed upon the shaft of the sewing machine and driven by a supply through a small, flexible pipe connected to the sink faucet, or other convenient place. One the size of the illustration herewith would be abundant in capacity under the ordinary city pressure.

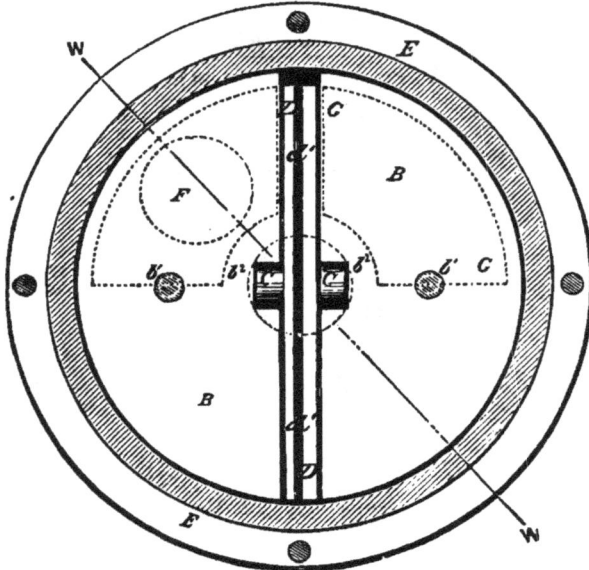

Success Turbine, S. M. Smith, York, Pa.

Downward discharge. Only one wheel tested, that tested several times; first as it came to the flume, then it was taken to machine shop, put in proper condition and again tested.

Head.	Weight.	Rev. per minute.	H. P.	Cubic feet.	Per Cent.
18.22	330	197	29.55	1137	.7564
18.32	300	198	27.00	1022	.7647
18.43	240	204.5	22.61	877	.7419
18.56	175	193.7	15.41	662	.6653

Second Test of Same.

Head.	Weight.	Rev. per minute.	H. P.	Cubic feet.	Per Cent.
18.30	340	198 5	30.67	1095	.8119
18.35	315	203.7	29.16	1025	.8198
18.55	220	202.5	20 25	742	.7800
18.64	160	202.5	14.72	593	.7064

Bollinger Turbine, O. J. Bollinger, York, Pa.

Central Discharge.	Head.	Weight.	Rev. per minute.	H. P.	Cubic feet.	Per Cent.
Whole Gate, ...	18 21	370	193	32.45	1343.25	.7034
Part Gate, ...	18.24	340	187.5	28.97	1196.20	.7042
" "	18.40	250	190	21.59	948.11	.6577
" "	18.58	160	193.5	14.07	678.43	.5784

Delphos Turbine, Delphos, Ohio.

Test of a 24-inch Wheel.

Head.	Weight.	Rev. per minute	H. P.	Cubic feet.	Per Cent.
18.38	280	248·	21.04	786.52	.7726
18.43	205	246 5	15.31	696.78	.6329
18.60	115	248˙5	8.65	540.57	.4567
18.76	55	204	3.40	373.06	.2582

Another 24-inch Wheel.

	Head.	Weight.	Rev. per minute.	H. P.	Cubic feet.	Per Cent.
Whole Gate, ...	18.38	275	246.6	20.55	806.70	.7350
Part Gate, ...	18.42	235	238	16.94	750.58	.6498
" "	18.47	170	246.5	12.69	654.01	.5570
" "	18.56	95	247	7.11	505.31	.4025

Test of a 36-inch Wheel.

	Head.	Weight.	Rev. per minute.	H. P.	Cubic feet.	Per Cent.
Whole Gate, ...	18.18	700	175	55.68	2069.54	.7856
Part Gate, ...	18.25	510	173.5	40.22	1785.56	.6552
" "	18.38	320	176	25.60	1483.77	.4983
" "	18.54	120	173	9.24	1041.36	.2540

National Water Wheel Co., Bristol, Conn.

J. T. CASE WHEEL.

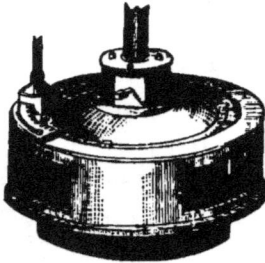

Made with sixteen chutes, in groups of four each, as shown in cut. Thin outside register gate, so arranged that four. eight, twelve, or the whole sixteen may be closed as desired, in order to utilize one-fourth, one-half, three-fourths or the whole discharge advantageously. The wheel has central and downward discharge, and is claimed to be like the Swain, but in reality has little resemblance to that wheel. The Company have had about a dozen different wheels tested at my flume; the results may be found below and on next page. The "part gates" revolutions are those that gave the highest results.

Test of 30-inch, Aug. 19, 1872.

	Head.	Weight.	Rev.per minute.	Horse Power.	Cubic Feet.	Per Cent.
Whole Gate,...............	18.35	500	133	30.23	1477.64	.610
12 chutes opened,........	18.61	265	153	18.43	900.34	.583
8 chutes opened,.........	18.78	150	154.5	10.53	588.99	.505
4 chutes opened.........	18.19	65	141	4.16	336.80	.377

Test of 20-inch, August 22, 1872.

	Head.	Weight.	Rev.per minute.	Horse Power.	Cubic Feet.	Per Cent.
Whole Gate,...............	18.63	187.5	237.2	20.78	773.98	.763
4 chutes opened,........	19.03	30	249	3.39	215.66	.438
8 chutes opened,........	18.88	95	247	10.36	492.99	.590
12 chutes opened,........	18.80	160	237.5	17.67	700.59	.712

Another 20-inch, in same curb, but with different shaped buckets.

	Head.	Weight.	Rev.per minute.	Horse Power.	Cubic Feet.	Per Cent.
Whole Gate,...............	18.63	180	249	20.37	765.59	.758
4 chutes opened, ...	19 07	27 5	242.5	3.03	218.16	.386
8 chutes opened, ...	18.86	80	248.5	9.04	424.76	.599
12 chutes opened,........	18.70	130	252	14.89	610.48	.692

Another 20-inch, different from the others, same curb.

	Head.	Weight.	Rev.per minute.	Horse Power.	Cubic Feet.	Per Cent.
Whole Gate,...............	18.61	155	342.5	17.08	700.80	.695
12 chutes opened,........	18.81	110	254.5	12.42	545.32	.642
8 chutes opened,........	18.95	75	247	8.34	386.29	.604
4 chutes opened,........	19.12	25	242.5	2.75	199.21	.383

Another wheel in same curb, but with compound register gate.

	Head.	Weight.	Rev.per minute.	Horse Power.	Cubic Feet.	Per Cent.
Whole Gate,...............	18.91	140	247.5	15.75	616.96	.716
12 chutes opened,........	19.05	70	249	7.92	373.37	.591
8 chutes opened,........	19.13	40	248	4.31	257.32	.464
4 chutes opened,........	19.23	20	197.5	1.79	150.85	.327

Test of a 24-inch wheel in shape like the Houston, Sept. 27, 1872.

	Head.	Weight.	Rev.per minute.	Horse Power.	Cubic Feet.	Per Cent.
Whole Gate,...............	18.84	180	175	14.31	843.14	.477
Part Gate,...............	18.91	135	186.5	11.44	699.47	.456
Part Gate,...............	18.99	95	178.5	7.71	544.45	.395
Head Reduced, Whole Gate,	12.43	100	655.1	7.52	647.55	.495

Another of the same kind and shape.

	Head.	Weight.	Rev.per minute.	Horse Power.	Cubic Feet.	Per Cent.
Whole Gate,...............	18.79	160	195	14.18	832.16	.480
Head Reduced, Whole Gate,	12.25	110	139.5	6.97	665.39	.453

Test of a 40-inch, January 21st, 1873.

	Head.	Weight.	Rev.per minute.	Horse Power.	Cubic Feet.	Per Cent.
Whole Gate,...............	17.76	650	163.5	48.30	1974.81	.729
12 chutes opened,........	18.40	450	158	32.31	1510.78	.616
8 chutes opened,........	18.85	245	158	17.59	1031.50	.479
4 chutes opened,........	19.26	65	156	4.61	618.24	.205

J. T. CASE WHEEL.

No. of Test.	Head.	Weight.	Rev. per Minute.	Horse Power.	Weir.	Cubic Feet.	Per Cent.
Whole Gate, 1...	17.85	600	174	47.42	1.046	2038.74	.689
" 2...	17.82	620	178	50.16	1.050	2050.56	.726
" 3...	17.78	650	168.5	49.78	1 054	2061.36	.719
" 4...	17.75	675	165	50.62	1.057	2070.32	.729
" 5...	17.73	700	162.5	51.70	1.064	2092.16	.737
" 6...	17.70	725	160	52.72	1.070	2110.10	.747
" 7...	17.69	750	157	53.52	1.074	2122.10	.754
" 8...	17.67	775	155	54.60	1.080	2140.06	.764
" 9...	17.63	800	153	55.63	1.084	2152.08	.776
" 10...	17.62	825	150	56.25	1.088	2163.51	.781
" 11...	17.62	850	146	56.41	1.088	2163.51	.783
" 12...	17.61	875	144	57.27	1.090	2170.14	.793
" 13...	17.60	900	142	55.09	1.096	2188.26	.774
" 14...	17.58	925	132	55.50	1.100	2200.36	.759
" 15...	17.58	950	123	53 11	1.104	2212.48	.722
Head Reduced.							
Whole Gate,17...	12.15	650	111	32.79	.972	1825.86	.782
" 18...	12.14	625	117	33.24	.970	1819.74	.796
" 19...	12.15	600	118.5	32.32	.968	1814.02	.776
" 20...	12.17	575	122.5	32.01	.965	1805.47	.771
" 21...	12.15	615	117	32.71	.969	1816.97	.784
" 22...	12.14	635	115	33.19	.972	1825.86	.792
Head Reduced.							
Whole Gate.24...	6.74	350	84	13.36	.802	1361.90	.770
" 25...	6.71	360	80.8	13.22	.803	1364.51	.765
" 26...	6.70	370	77.5	13.00	.805	1369.84	.749
" 27...	6.73	340	86	13.29	.801	1359.89	.769
" 28...	6.75	330	87	13.05	.800	1356.78	.754
Part Gate, 12 p'ts op'n, 30...	17.84	800	132.6	48.22	.944	1743.95	.820
8 " " 31...	18 04	400	134	24.36	.730	1173.96	.609
8 " " 32...	18 41	350	142	22.48	.712	1129.28	.560
8 " " 33...	18.32	500	120	27.27	.749	1221.74	.646
4 " " 34...	18 94	150	139	9.48	.512	670.49	.395

Test of 50 inch, Aug. 1873,	Head.	Weight.	Rev. per minute.	Horse Power.	Cubic Feet.	Per Cent.
Whole Gate,...............	16.39	1500	114	103.63	4627.66	.7242
"	16.37	1600	111	107 63	4721.55	.7381
"	16.37	1700	105.6	108.80	4801.70	.7356
"	16.35	1800	101.5	110.72	4836.17	.7422
"	16.34	1900	97	111.70	4840.00	.7486
"	16.35	2000	92	111.50	4874.55	.7415

At Unionville, Conn., Platner & Porter Mf'g Co., I tested the power of one of the latest style National wheels; it was far below its tabl.d rate for power, and so extraordinarily extravagant in the use of water that it was immediately removed to make room for one of a better kind.

Angell Wheel, Providence, R. I.

Double discharge, gates similar to Leffel, though each alternate piece forming side of chutes is stationary, as represented in the diagram. Buckets bolted to hub of wheel, and often shear off.

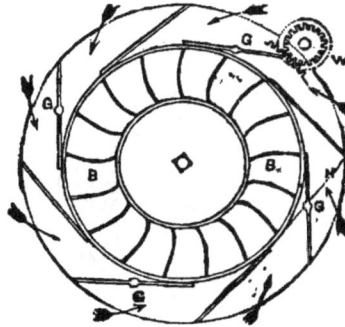

TEST OF A 36-INCH WHEEL.

No. of Test.	Head	Weight.	Rev. per Min.	Horse Power	Cubic feet.	Per Cent.
May, 1873, Whole Gate,....	17.96	880	147.3	58.92	2328.86	.7451
⅞ Gate,.....................	18.12	790	143	51.35	2052.24	.7305
¾ "	18.29	650	145.5	42.99	1760.51	.7063
½ "	18.47	475	140	30.22	1362.77	.6366
½ "	18.48	370	142	23.85	1179.99	.5786

A 30-INCH WHEEL SENT TO BE TESTED.

No. of Test.	Head.	Weight.	Rev. per Min.	Horse Power.	Cubic feet.	Per Cent.
Nov. 3, Whole Gate,.......	18.27	520	184	43.50	1475.36	.8539
⅞ Gate,.....................	18.31	490	184	40.98	1398.86	.8465
¾ "	18.46	380	182.5	31.69	1094.11	.8114
½ "	18.54	270	187	22.95	902.17	.7260
7-16 "	18.65	190	184 5	14.25	677.38	.5968

On report of results, a wheel of the same size and made in same lot was sold to Otto Troost, of Winona, Minn., sent to me to be tested for verification. Results are here given.

No. of Test.	Head.	Weight.	Rev. per Min.	Horse Power.	Cubic feet.	Per Cent.
Dec. 15, Whole Gate, ...	18.40	520	176.6	41.74	1583.40	.7579
"	18.43	500	180.6	41.04	1575.21	.7479
"	18.42	480	186.5	40.66	1564.21	.7465
"	18.47	550	167	41.11	1616.24	.7285

Of course the wheel was rejected; then the wheel tested, Nov. 3, was repurchased by the Angell Company, and returned to fill the order. It was placed in the flume, and found to run so hard, that I at once refused to test it until put in order; it was taken out and reset some five or six times, but could never be got in condition to run, so that it would not require thirty or forty pounds to start it, where ten should have done so, and the wheel was returned to the shop for inspection.

January 14, 1874, a 48 inch wheel was sold conditionally to take the place of a Leffel, where at least 80 per cent. average useful effect was required; it was sent to be tested, and gave the following results:

No. of Test.	Head.	Weight.	Rev. per Min.	Horse Power.	Cubic feet.	Per Cent.
Whole Gate,...............	17.62	1450	108	94.90	3802.70	.7667
Part Gate,...............	17.88	1180	108.2	77.38	3158.79	.7249
"	18.02	1000	105.4	63.88	2765.80	.6780
"	18.21	750	108	49.09	2271.58	.6279
"	18.42	500	102	30.90	1698.00	.5228
"	18.46	450	108.2	29.50	1709 18	.4947

TEST OF A 40-INCH WHEEL, AT FITCHBURG, MASS., JULY 2, 1872.

Head in Feet.	Weir.	Rev. per Minute.	Weight.	Horse Power.	Cubic ft. Discharg'd	Percentage.
21.80	10.85	240	260	28.36	1757.47	.3999
21.63	11.18	215	310	30.29	1837.70	.4034
20.99	14.00	156	510	36.16	2555.08	.3502
Tabled rate, same head,	280			about 103	about 2800	.9000

Flenniken Brothers, Rockford, Ill.

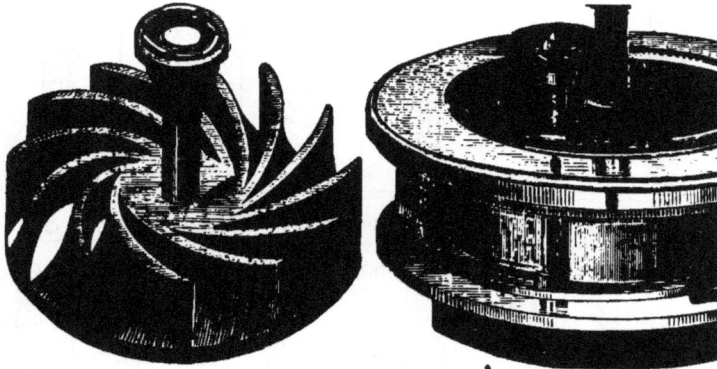

Test of a 20-inch wheel.

	Head.	Weight.	Rev. per minute.	Horse Power.	Cubic feet.	Per Cent.
Whole Gate, ...	18.37	110	342	17 10	605.44	.8083
Part Gate, ...	18.49	90	315	12.88	524.43	.7025
"	18.53	60	321	8 75	448.92	.5566
"	18.63	35	351	5.58	362.70	.4368
"	18.72	20	280	2.54	278.14	.2244

Gardiner Cox, Ellsworth, N. Y.

Furnished a wheel that he called double; it consisted of a hub, with a Jonval wheel around its lower end, the buckets above being continued by sheet iron spirals to the top of the hub, forming a twelve threaded screw, the pitch being twelve degrees from line of rotation.

T. H. Risdon & Co., Mount Holly, N. J.

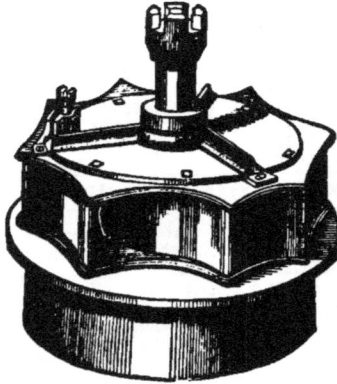

Mr. Risdon seems to have a passion for the turbine business, and has continued to experiment for many years to an almost unlimited extent; his first experiment at my flume was in 1871, with a 30-inch Vandewater wheel in a curb here represented. It gave a useful effect of .7714 per cent. He then tried a 36-inch of the same kind, which gave .7871 per cent. His next effort was with a wheel of his own designing; (see next page, Fig. 2;) but in a curb similar to that used by the National Water Wheel Co., of Bristol, Conn. The test is given in full in the second table below. As may be seen, the part gates were not proportionally good, while the rim of the curb was so nearly divided by the ports that it was too fragile for durability, consequently it was abandoned and a new one, represented by cut B was constructed; in that, the following results were obtained by the test of a 43-inch wheel.

Head.	Weight.	Rev. per min.	H. Power.	Cubic feet.	Percentage.
17.91	1200	151	82.36	2664.03	.9132
17.93	1200	148	80.72	2676.91	.8897
17.92	1200	148.3	80.89	2680.14	.8910
17.90	1250	144.5	82.10	2689.82	.9021
17.98	1150	146.5	76.58	2469.92	.9121
18.00	1200	137.5	75.00	2495.13	.8834
18.17	1000	147	66.82	2258.84	.8613
18.29	850	150	57.95	2012.62	.8331
18.30	700	138.6	44.10	·1686.07	.7559
18.43	650	148	43.72	1686.07	.7459

No. of Test.	Head.	Weight.	Rev.	Horse Power.	Weir.	Cubic Feet.	Per Cent.
Whole Gate, 1...	18.25	700	163.5	52.02	1.193	1749.38	.863
" 2...	18.25	750	151.5	51.65	1.201	1766.64	.850
" 3...	18.28	705	162.5	52.07	1.196	1755.81	.859
" 4...	18.28	710	162	52.28	1.196	1755.81	.863
" 5...	18.30	715	160	52.00	1.199	1762.29	.854
" 6...	18.32	720	159.6	52.20	1.199	1762.29	.856
" 7...	18.34	725	158.6	52.26	1.200	1764.47	.854
" 8...	18.33	730	157.6	52.29	1.201	1766.64	.855
" 9...	18.32	735	156.5	52.29	1.202	1768.81	.854
" 10...	18.32	740	155.5	52.30	1.201	1766.64	.856
" 11...	18.34	745	155	52.48	1.202	1768.81	.857
" 12...	18.32	750	153	52.16	1.203	1770.98	.832
·1 13...	18.34	760	149.5	51.65	1.203	1770.98	.842
" 14...	18.34	690	167	52.37	1.199	1762.29	.858
" 15...	18.34	680	169	52.24	1.198	1760.13	.857
" 16...	18.35	670	171.5	52.22	1.197	1757.97	.858
1 Pt. closed, 18...	18.51	560	154	39.20	1.047	1442.53	.779
" 19...	18.51	545	158	39.14	1.045	1438.45	.778
2 Pts. closed 21...	18.85	190	162	13.99	.723	831.05	.473
Whole Gate, 22...	18.37	745	154.6	52.35	1.200	1764.47	.855

T. H. Risdon & Co., Mount Holly, N. J.

RISDON'S WHEEL.

Of the many Risdon wheels tested by me, quite a number of them have ranged along in the seventies in percentage, but through some slight change after a first trial every wheel tested, (except two or three of the 20-inch size) has been made to return a useful effect of over eighty per cent. before delivery to purchaser, quite a number from eighty-five to ninety, and a few even higher than ninety. Figure 3 represents the curb Mr. Risdon now considers the best, but he also furnishes wheels in the register gate curb, represented at the head of the opposite page.

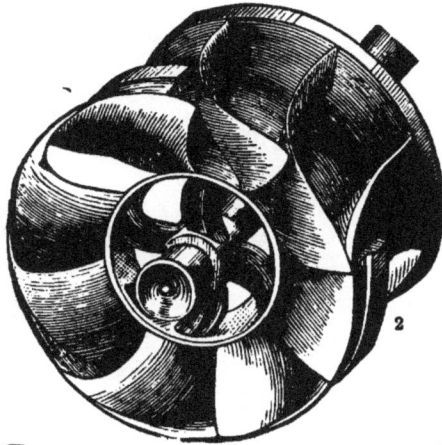

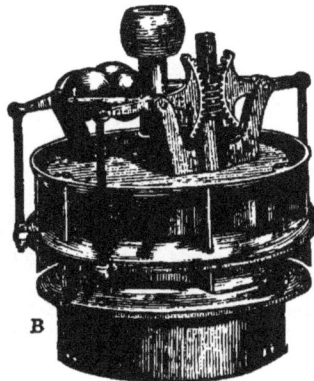

Test of a 20-inch.	Head.	Weight.	Rev. per minute.	H. P.	Cubic feet.	Per Cent.
Whole Gate,..............	18.58	175	295.5	15.67	640.26	.7131
Part Gate,...............	18.62	165	300	15.00	593.93	.7175
" "	18.67	165	272.5	13.62	540.42	.7143
" "	18.68	140	308	13.06	532.32	.6936

Tyler's New Scroll Wheel.

Scroll wheels are passing away, still there are many places yet where they may be advantageously used; of the many plans devised for this class of wheels, John Tyler of Claremont, N. H., has undoubtedly produced the very best, and decidedly so.

Test of a 30-inch, Sept. 19, 1873.

Whole Gate.	Head.	Weight.	Rev.per minute.	Horse Power.	Cubic feet.	Per Cent.
Whole Gate, ...	18.30	385	189	33 07	1208.65	.7917
Part Gate, ...	18.32	330	196.5	29.78	1121.64	.7674
"	18.41	200	187	17.00	900.47	.5433
" ....,...........	18.50	60	192	5.24	718.35	.2088
"	18.54	17.5	183	1.46	680.58	.0612

Wheel reset and made to run easier, then re-tested, Sept. 23, 1873.

	Head.	Weight.	Rev.per minute.	Horse Power.	Cubic feet.	Per Cent.
Whole Gate, ...	18 48	390	191	33.86	1188.49	.8164
Part Gate, ...	18.55	330	195.5	29.33	1097.36	.7630
"	18.62	250	191	21.71	954.50	.6469
"	18.65	200	197.5	17.95	892.28	.5713
"	18.72	100	185	8.41	730.75	.3256

This wheel was kept as a sample wheel; its gate raises 13¼ inches; the opening to scroll being 13¼x12, while the openings in wheel equalled 218 square inches; raising the gate three inches caused a discharge of one-half that could pass through the wheel at whole gate. February 1st, 1877, the buckets of this wheel were slightly chipped on the edge, then it was sent to me to be tested, but without any intimation that it was the same previously tried. The weather was bad at the time and it stood exposed for two weeks; was then set and tested with the ordinary care.

	Head.	Weight.	Rev.per minute.	Horse Power.	Cubic feet.	Per Cent.
Gate opened 13¼ inches, ...	18.47	360	202.2	33.08	1179.66	.8050
" " 9 inches, ...	18.52	300	204	27.81	1074.91	.7407
" " 8 inches, ...	18.59	225	201.5	20.61	927.18	.6421
" " 7 inches.	18.65	150	193.3	13.17	795.14	.4709

Tyler's Flume Wheel and Curb.

John Tyler, Claremont, N. H.

This curb has an inside register gate, one side of each chute being cast in chute rim; the other part is bolted to the register hoop or gate. The wheel is the same whether used in scroll or flume curb.

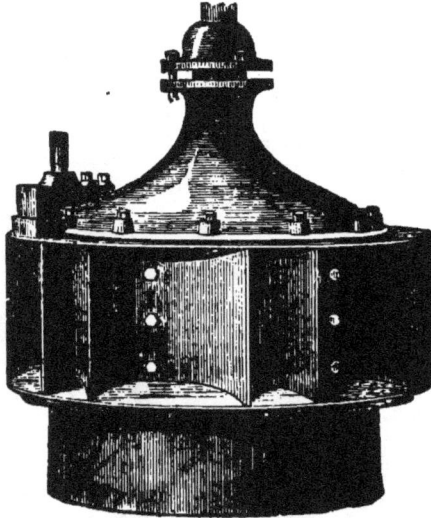

Test of 30-inch Flume wheel.

April 20, 1876.	Head.	Weight.	Rev. per minute.	H. P.	Cubic feet.	Per Cent.
Whole Gate,................	18.43	375	168.6	28.72	1245.64	.6618
Part Gate,................	18.44	360	169.2	27.68	1207.50	.6677
" "	18.65	185	167.5	14.08	722.18	.5530
Buckets were chipped back. Re-tested April 21.						
Whole Gate, ...	18.65	375	202	34.43	1226.55	.7970
Part Gate,................ ...	18.62	325	196.5	29.01	1094.45	.7531
" "	18.84	190	202	17.45	755.40	.6487
Buckets chipped back more and gate opening enlarged. Again tested April 22						
Whole Gate, ...	18.50	385	215.3	37.67	1180	.9127
Part Gate,................	18.60	325	215	31.76	1037.38	.8709
" "	18.67	275	212.5	26.56	914.01	.8234
" "	18.76	220	213.5	21.35	764 49	.7880
" "	18.85	160	215	15.63	602.53	.7280
" "	18.93	105	213.5	10.18	465.71	.6109
" "	19.01	60	197	5.37	334.25	.4471

Chipping the buckets threw the wheel out of balance so that it was returned to builders, where it was balanced by drilling holes on heavy side and filling them with wood; the wheel was smoothed up generally, then sent to Centennial test, then sent to me for re-test; on trying it again it was found that some change had been made that rendered it almost impossible to control it with brake. Re-tested Feb. 13, 1877.

	Head.	Weight.	Rev. per minute.	Horse Power.	Cubic feet.	Per Cent.
Whole Gate, ...	18. 3	375	221	37.67	1318.42	.8242

The wheel was taken to machine shop, altered and re-tested four times, but without material change in results.

Oct. 13, 1877, tested 42-inch Tyler Flume wheel.

	Head.	Weight.	Rev. per minute.	Horse Power.	Cubic feet.	Per Cent.
Whole Gate, ...	18.10	1000	146	66.36	2619.73	.7409

Taken to machine shop and alterations made, then re-tested.

	Head.	Weight.	Rev. per minute.	Horse Power.	Cubic feet.	Per Cent.
Whole Gate, ...	18.01	1025	146.5	68.24	2586.61	.7750
Part Gate, ...	18.15	800	152	55.27	2181.25	.7391
" "	18.25	725	144	47.45	1917.22	.7181
" "	18.27	750	130.5	44.48	1843.14	.6992

A. N. Wolf's Turbine.

Manufactured by Barber & Sons, Allentown. Pa.

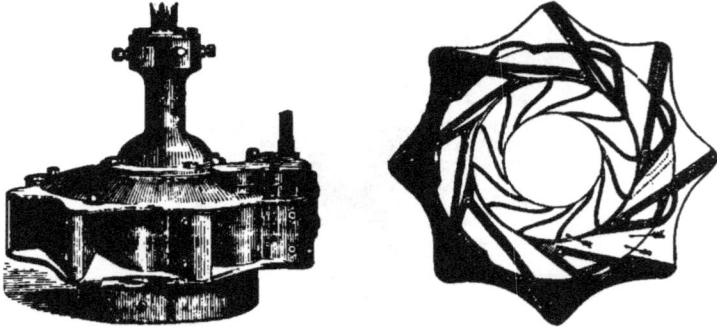

The first Wolf wheels sent to be tested gave exceeding good results and were reported accordingly, which caused manufacturers to order others; as these were sent to be tested they were found to be not only less efficient, but also not well made. The 48-inch wheel reported on next page was ordered for the Newton & Ramage Paper Co. of Holyoke; it was so poorly made that while handling it in order to lower it into testing flume it came apart and the wheel dropped to the bottom; it was sent to machine shop and put into much better condition than when first received. The edges of the buckets were left by the builders square, varying in thickness from one-half to three-fourths of an inch; these were partially rounded, then the wheel was tested, giving the results reported. Mr. Wolf took the wheel out and chipped the buckets to an edge, made it run easier then had it tested again, obtaining the results reported of second 48-inch wheel. In examining the wheel and curb I found the casting to be so thin as to be hardly safe for the pressure of the 24 feet head for which it was ordered; the crown plate or cover was five feet in diameter. Mr. Barber insisted that it was three-fourths of an inch in thickness, but on drilling through, it was found to be but three-eighths; it was rejected.

Test of 24-inch. Multiply revolutions of wheel by 10 to get speed for computing power.

	Head.	Weight.	Rev. per Min.	Horse Power.	Cubic feet.	Per Cent.
Whole Gate, ...	18.21	425	255.3	32.87	1134.57	.8436
Part Gate, ...	18.29	400	240	29.09	1002.07	.8416
" "	18.41	825	247.5	24.37	853.76	.8202
" "	18.55	250	249	18.86	673.63	.8003
" "	18.69	150	234.5	10.65	445.86	.6777

Test of a second 24-inch. Multiply revolutions by 10.

	Head.	Weight.	Rev. per Min.	Horse Power.	Cubic feet.	Per Cent.
Whole Gate, ...	18.45	420	255	32.45	1166.27	.7997
Part Gate, ...	18.50	350	252	26.72	1004.33	.7626
" "	18.55	300	250	22.72	884.27	.7374
" "	18.73	175	251.5	13.33	628.87	.6601
" "	18.83	80	248	6.01	445.39	.3018

Test of a third 24-inch. Multiply revolutions by 15.

	Head.	Weight.	Rev. per Min.	Horse Power.	Cubic feet.	Per Cent.
Whole Gate	18.23	280	253.5	32.26	1164.28	.8078
Part Gate, ...	18.32	220	262.2	26.22	1001.40	.7578
" "	18.41	190	261.5	22.53	891.93	.7275
" "	18.47	150	262.5	17.89	755.83	.6796

Test of a fourth 24-inch.

	Head.	Weight.	Rev. per Min.	Horse Power.	Cubic feet.	Per Cent.
Whole Gate, ...	18.20	300	245	33.40	1183.46	.8206
Part Gate, ...	18.24	270	243	29.82	1093.77	.7910
" "	18.38	220	241	24.10	922.11	.7700
" "	18.45	170	251.3	19.41	791.52	.7033
" "	18.70	75	235.5	8.92	470.67	.5376

Test of a 30-inch. Multiply revolutions by 15.

	Head.	Weight.	Rev. per Min.	Horse Power.	Cubic feet.	Per Cent.
Whole Gate, ...	17.86	500	183.5	41.70	1547.74	.8000
Part Gate, ...	18.01	435	182.5	36.08	1325.98	.8011
" "	18.06	400	180.5	32.81	1233.06	.7814
" "	18 43	230	182 5	19.07	819.71	.6850

Test of a 36-inch. Multiply revolutions by 15.

	Head.	Weight.	Rev. per Min.	Horse Power.	Cubic feet.	Per Cent.
Whole Gate, ...	18.03	825	160.7	60.26	2308.08	.7665
Part Gate, ...	18.02	825	161.5	60.25	22 18.75	.7699
" "	18.12	700	160	50.91	1958.80	.7594
" "	18.27	550	149	37.25	1507.43	.7161
" "	18.49	300	157.2	21.42	983.49	.6228

Test of a 18-inch. Multiply revolutions by 10.

	Head.	Weight.	Rev. per Min.	Horse Power.	Cubic feet.	Per Cent.
Whole Gate, ...	18.38	165	304	15.20	579.07	.7574
Part Gate, ...	18.50	160	287	13.91	517.71	.7702
" "	18.55	135	291.5	11.92	454.00	.7506
" "	18.59	105	267	9.30	357.42	.7425
" "	18.61	80	315	7.63	338.38	.6422

Test of a 48-inch. Sent to machine shop for alterations. Multiply revolutions by 20.

	Head.	Weight.	Rev. per Min.	Horse Power.	Cubic feet.	Per Cent.
Whole Gate, ...	17.47	1525	90	83.18	3618.81	.6982
Part Gate, ...	17.65	1000	121	73.33	3110.36	.7088
" "	18.01	600	111.5	40.54	2127.43	.5615

Test of the 48 a second time. Rejected by intended purchaser.

	Head.	Weight.	Rev. per Min.	Horse Power.	Cubic feet.	Per Cent.
Whole Gate, ...	17 60	1300	117	92 18	3640.50	.7630
Part Gate, ...	17.71	1150	117	81 54	3263.49	.7482
" "	17.80	1050	118	75.09	2953.76	.7562
" "	17.94	875	117.3	62.20	2566.52	.7165
" "	18.17	700	114.3	48.49	2112.31	.6701

Test of a 54-inch. Rejected by intended purchaser. Multiply revolutions by 20

	Head.	Weight.	Rev. per Min.	Horse Power.	Cubic feet.	Per Cent.
Whole Gate, ...	17.24	1700	112.6	116.01	4841.07	.7373
Part Gate, ...	17.42	1600	106.3	103.07	4201.41	.7469
" "	17.73	1000	112	67.87	3107.58	.6686

Clark & Chapman, Turner's Falls, Mass.

COLEMAN WHEEL.

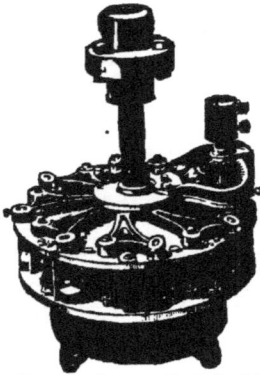

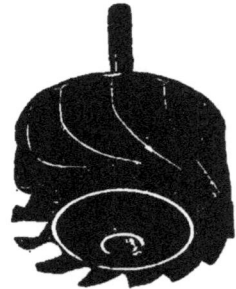

Have had several of the kind tested, and the results given herewith, obtained from the test of a 30-inch, represent the general characteristics of the wheel.

March 31, 1874.	Head.	Weight.	Rev.per minute.	H. P.	Cubic feet.	Per Cent.
Whole Gate,..............	18.45	425	176.8	34.15	1297.40	.7584
Part Gate,.......	18.57	370	173	29.09	1144.55	.7935
" "	18.71	330	174.5	26.18	963.24	.7687
" "	18.66	175	171.5	13.63	687.60	.5696

Wm. F. Mosser & Co., Allentown, Pa.

Two 36-inch wheels. Results below. First wheel had chutes and gate similar to Stout, Mills & Temple curb; wheels downward discharge.

	Head.	Weight.	Rev.per minute.	Horse Power.	Cubic Feet.	Per Cent.
Whole Gate,..............	17.80	770	155.5	54.32	2148.20	.7545
Part Gate,..............	17.87	680	156.5	48.37	1866.51	.7708
" "	17.96	580	157.5	41.52	1611.01	.7610
" "	18.10	460	156	32.62	1331.37	.7195
" "	18.21	300	169	23.00	1047.20	.6709
" "	18.10	220	157.5	15.41	802.86	.5624
Test of Second Wheel, Inside Register Gate.						
Whole Gate,..............	18.04	725	176.5	58.44	2175.31	.7879
Part Gate,..............	18.10	650	164 6	48.64	1916 99	.7415
" "	18.21	520	164.5	38.88	1591.18	.7100
" "	18.34	375	163.5	27.84	1294.82	.6202
" "	18.45	245	165	18.40	1008.06	.5234

"Excelsior," Roland, Benedict & Co., Reading, Pa.

Test of a 42-inch Wheel.

	Head.	Weight	Rev.per minute	H. P.	Cubic feet.	Per Cent.
Whole Gate, ...	18.48	460	149.5	31.26	1136.34	.7880
Part Gate,............. ,....	18.52	300	158.6	21.62	1055.49	.6003
" "	18.64	160	160	11.60	910.64	.3692
" "	18.72	50	164	3.12	785.78	.1123

TYLER FLUME
WETMORE AND WALSH
WHEELS

B. J. BARBER, BALLSTON SPA, N. Y.

[From my Report of 1871.]

Test of a 30 inch.

No. of Test.	Head.	Weight.	Rev. per Minute.	Horse Power.	Weir.	Cubic Feet.	Per Cent.
Whole Gate, 1	18.80	380	198.5	34.28	1.081	1279.46	.756
" 2	18.84	390	196.3	34.80	1.065	1246.49	.760
" 3	18.87	400	195	35.45	1.089	1281.53	.764
" 4	18.88	410	191.5	35.69	1.091	1297.07	.783
" 5	18.60	400	192	34.91	1.083	1282.98	.778
" 6	18.60	410	189	35.25	1.088	1291.77	.778
" 7	18.60	420	186.5	35.63	1.090	1285.30	.783
" 8	18.60	430	181.5	35.47	1.092	1299.15	.777
" 9	18.61	440	181	36.20	1.096	1305.90	.788
" 10	18.60	450	177	36.18	1.098	1309.44	.788
" 11	18.59	445	178	36.04	1.096	1305.90	.785
" 12	18.58	435	181.5	35.89	1.094	1302.30	.785
" 13	18.60	425	185	35.73	1.092	1280.15	.775
" 14	18.61	415	188.5	35.66	1.088	1291.77	.787
" 15	18.61	405	191.5	35.25	1.085	1286.49	.779
HEAD REDUCED.							
Whole Gate, 16	12.32	315	136	19.47	.962	1074.49	.778
" 17	12.32	320	131.5	19.10	.964	1078.83	.760
" 18	12.31	330	126.5	19.27	.968	1085.51	.761
" 19	12.31	310	137.5	19.37	.962	1074.49	.775
" 20	12.32	305	139	19.27	.960	1072.15	.772
Unreliable, 21	12.32	287.5	154.3	20.22	.956	1065.49	.816
HEAD REDUCED.							
Whole Gate, 22	8.91	200	128	11.63	.844	844.00	.781
" 23	8.85	225	114.5	12.01	.855	901.38	.797
" 24	8.85	230	103	11.71	.866	918.85	.762
HEAD REDUCED.							
Whole Gate, 25	6.79	135	119.5	7.33	.766	763.82	.748
" 26	6.77	140	117.5	7.75	.769	768.35	.769
" 27	6.77	150	111.5	7.60	.775	777.41	.764
Part Gate, 28	12.57	225	154.5	15.80	.856	904.97	.736
" 29	12.77	150	154.8	10.55	.739	723.45	.604
" 30	13.02	95	150	6.47	.596	528.59	.497
" 31	12.52	25	161	1.81	.445	380.59	.207

JAMES EMERSON.

HOLYOKE, MASS., September 30, 1872.

Humphrey Turbine.

Manufactured by the Humphrey Machine Co., Keene, N. H.

Of all the turbine builders extant, perhaps excepting J. P. Collins of Norwich, Ct., there is no other probably that can be named, so immensely scientific and so boiling over with theories as is Mr. Humphrey. The tests below will aid the reader to judge whether such theories are practically beneficial. In placing the 21-inch wheel reported below, a Collins' brass bucket wheel was removed, and advantageously so, I believe, it was admitted. The Humphrey wheel has downward and outward discharge, register gate. Tests of three of the wheels for Rawitser & Brother, Stafford Springs, Conn., Nov. 1878. These wheels were manufactured, fitted for their positions, set by the Humphrey Machine Co., and have been in use but a few months. A weir was constructed for each of the wheels. These weirs were of less capacity than desirable, but if there were any errors in measurements through this lack of capacity, such errors would be entirely in favor of the wheels. Each wheel was thoroughly cleaned, previous to its test.

Test of 42-inch wheel, Nov. 13, 1878.

	Head.	Weight.	Rev.	Horse Power.	Weir.	Cubic Feet.	Per-centage
Whole Gate,.......	5.00	375	68	7.72	1.005	1167.34	.7000
"	5.00	400	64	7.75	1.000	1167.34	.7020
Part Gate, ...	5.30	225	62.5	4.26	.755	766.65	.555
"	5.45	325	65	6.40	.870	944.59	.6582

Test of 24-inch wheel, Nov. 15, 1878.

	Head.	Weight.	Rev.	Horse Power.	Weir.	Cubic Feet.	Per-centage
Whole Gate,.......	29.00	135	320	13.09	1.028	589.98	.4051
"	29.00	150	310	14.09	1.030	591.60	.4348
"	29.00	175	291	15.43	1.032	593.26	.4748
"	29.00	200	265	16 06	1.077	629.84	.4655
Part Gate, ...	29.50	150	293	13.31	.921	497.26	.4803
"	29.50	115	270	9.41	.750	269.86	.4566

Test of 21-inch wheel, Nov. 17, 1878.

	Head.	Weight.	Rev.	Horse Power.	Weir.	Cubic Feet.	Per-centage
Whole Gate,.......	47.00	350	379	40.20	1.300	824.72	.5491
"	47.00	300	403	36.63	1.290	815.49	.5060
"	47.00	275	424	35.33	1.284	800.42	.4972
"	47.00	250	435	32 95	1.272	798.99	.4646
"	47.00	225	455	31.02	1.242	773.00	.4520
"	47.00	200	462.5	28.03	1.230	762 76	.4139
Part Gate,.........	47.00	125	446.6	16.91	.994	554.55	.3434
"	47.00	85	416.6	10.73	.852	444.61	.2718

In the mill where the 24-inch wheel is used the main line of shafting is designed to run 100 revolutions per minute; the gears connecting the wheel are one to three, consequently Mr. Humphrey prepared the wheel to run 300 revolutions per minute under 32 feet head.

The main shaft in mill where the 21 inch wheel is used is arranged in connection with the machinery used to run at 160 revolutions per minute. This shaft is connected to the wheel by gears, one to three, consequently, the wheel was prepared to run at a velocity of 480 revolutions per minute. It will be seen, however, by the tests, that Mr. Humphrey was very wild in his calculations for speed.

Mr. Humphrey took a very active interest in the hydrodynamic experiments made by the Holyoke Water Power Co., and promised distinctly, several times, to furnish one of his wheels for trial, but failed to do so.

Stilwell & Bierce Manufacturing Co.,
DAYTON, OHIO.

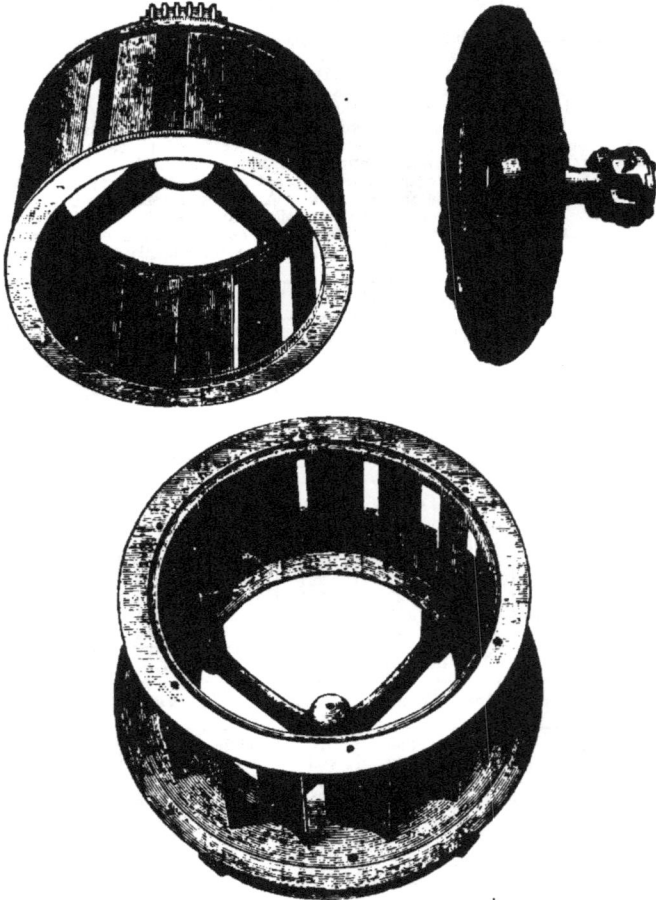

This turbine to be properly classed must be placed with the "Hercules" under the head of the New Departure, established by the production of that wheel.

As may be seen by the cuts, the Victor turbine is very simple in construction, having but few pieces, and those unlikely to get out of order; its inside register gate, so far as my experience goes, works with rapidity and ease, its long, peculiarly shaped buckets may be framed or cast into the rims of the wheel, as may be deemed advisable. Its capacity for its diameter, to be fully realized, must be compared with that of other turbines that were popular but a few years since, when Swain, Houston and Leffel & Co., each claimed to construct wheels of greater capacity for their diameter than those of any other make, and in suppor

Victor Turbine.

of their claims, published tables at least fully up to the capacity of their wheels, the tables of Swain and Houston being computed upon a supposd useful effect of 80 per cent. of the water used; those of the Leffel at 88. Under 18 feet head, the Swain, 15¼ inch, is tabled to give 13½ h. p.; the Houston, 15 inch, 8½; and the Leffel, 15¼ inch, 11.1 h. p.; while the Victor, 15 inch, as may be seen by the test herewith annexed, under 18.34 head, actually gave 29.36 h. p., and a useful effect of .8808 per cent.

There can be no question but what the Victor, with the exception of the Hercules, has taken a position in advance of all other turbines—not because the same efficiency of useful effect may not be obtained by other wheels, but because at the same cost no other wheel can be made to transmit the same amount of power. Instead of acting the part of Mrs. Partington in opposing the inevitable, it will be well for turbine builders to accept the fact and strive to do still better, for the turbine is a long way from being the perfect engine it may be made. The tests will show that Mr. Stilwell has steadily improved, showing concluively that the wild variations of other builders are owing to the lack of settled plans. At part gate the Victor is about as good as the average, but it would be well for Mr. Stilwell to try a thinner shell next his wheel so that the gate opening may be as near as possible to the wheel. The suggestion will be best understood by observing the filling of a bottle, or what is better, the filling of a canal through a small head gate, where the river may be several feet higher than the surface of the canal; yet the power due that difference is used up by passing through the small head gate, or in the inertia of the water in the canal, so that the part gate efficiency of a wheel must be somewhat in proportion to the size of the chamber inside of the gate.

Test of a 25-inch wheel, July 25, 1877.

	Head.	Weight.	Revolutions.	Horse Power.	Cubic ft.	Percentage.
Whole Gate,...	18.07	625	200	56.81	2214.55	.7533
Part Gate,.....	18.04	600	198	54.00	2208.44	.7192
" "	18.13	500	208	47.27	1964.67	.7042

Test of a 20-inch wheel, July 26, 1877.

	Head.	Weight.	Revolutions.	Horse Power.	Cubic ft.	Percentage.
Whole Gate,....	18.33	500	246	37.27	1387.27	.7777
Part Gate,.....	18.41	425	269	34.64	1284.30	.7774
" "	18.43	390	246	29.07	1145.59	.7305
" "	7.97	75	246	5.59	757.93	.4911

Test of a 20-inch wheel, Feb. 21, 1878.

	Head.	Weight.	Revolutions.	Horse Power.	Cubic ft.	Percentage.
Whole Gate,...	18.01	490	266.5	38.76	1362.39	.8363
Part Gate,.....	18.08	415	265	33.32	1242.03	.7853
" "	18.28	310	266	24.98	1014.03	.7134
" "	18.40	240	263	19.12	870.79	.6310
" "	18.58	100	271.5	8.33	602.23	.3941

Test of a 15-inch wheel, March 26, 1878.

	Head.	Weight.	Revolutions.	Horse Power.	Cubic ft.	Percentage.
Whole Gate,.. ..	18.34	300	323	29.36	974	.8705
Part Gate,......	18.10	300	321.5	29.22	970	.8808
" "	18.39	160	326.5	15.83	755	.6035
" "	18.74	100	320	9.09	492	.5220

Test of a 25-inch wheel, Oct. 28, 1878.

	Head.	Weight.	Revolutions.	Horse Power.	Cubic ft.	Percentage.
Whole Gate,....	17.96	700	209	68.62	2356.54	.8584
Part Gate,.....	17.93	650	208	61 45	2237.00	.8112
" "	18.00	450	200	40.90	1792.69	.6710
" "	18.25	350	205	32.61	1567.18	.6036
" "	18.37	175	211.5	16 81	1180.27	.4098

Test of a 30-inch wheel, Oct. 29, 1878.

	Head.	Weight.	Revolutions.	Horse Power.	Cubic ft.	Percentage.
Whole Gate, ...	11.65	800	144.5	52.54	2751.87	.8676
Part Gate, ...	11.78	675	136.5	41.88	2456.38	.7663
" "	11.92	600	142.5	38.87	2335.58	.7392
" "	11.83	450	145	29.65	1996.36	.6648
" "	12.10	300	144.5	19.70	1621.84	.5316

Eclipse Double Turbine, Manufactured by the same Co.

Test of a 30-inch Eclipse wheel.

Head.	Rev. per minute.	H. P.	Cubic feet.	Per Cent.
18 79	184.5	33 85	1253	.7028
18.93	170	31.66	1214	.7280
19.10	173.5	24.44	1026	.6497
19.10	165	18.00	862	.5786
19.18	166.6	12.11	699	.4779

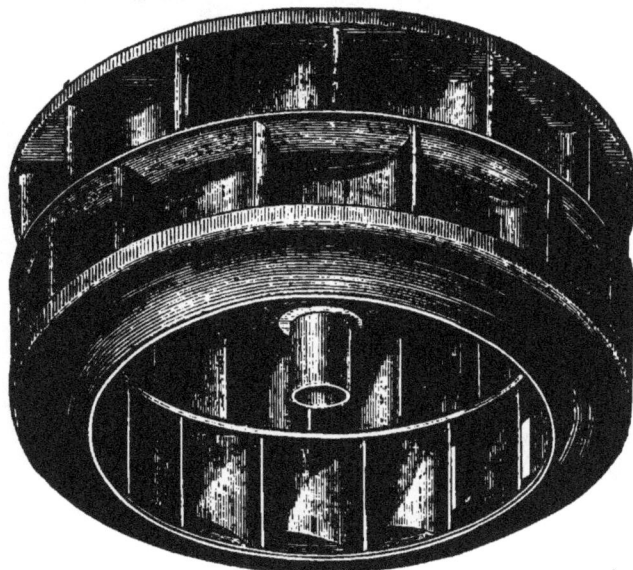

Waldo Whitney, Leominster, Mass.

Wheel downward, and central discharge, similar to the Swain, but with fewer buckets. Inside register gate, the chutes and outer rim, R, being stationary; the thin hoop or gate T, rotating sufficiently to open or close the ports. After my report of the test at the top of next page, Mr. Whitney sent three other wheels for verification, that he had sold with guarantee that they should be as good as the one reported, and he undoubtedly believed them to be so until tested.

First wheel reported, tested January 10, 1873.

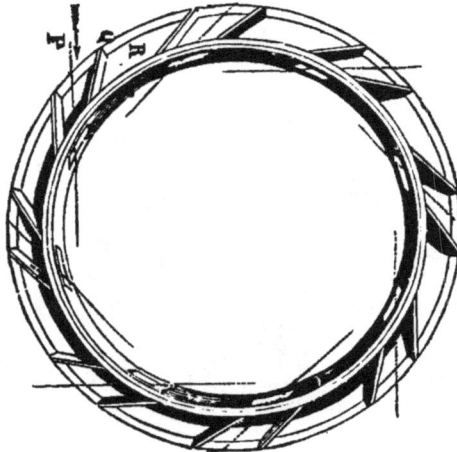

WALDO WHITNEY, LEOMINSTER, MASS.

No. of Test.	Head.	Weight.	Rev. per Minute.	Horse Power.	Weir.	Cubic Feet.	Per Cent.
Whole Gate, 1...	18.48	400	195	35.45	.765	1283.41	.791
" 2...	18.48	410	192	35.78	.768	1291.09	.793
" 3...	18.47	420	188.5	35.99	.771	1298.77	.794
" 4...	18.46	430	186	36.35	.775	1309.03	.796
" 5...	18.46	440	184	36.80	.777	1311 61	.804
" 6...	18.50	450	181	37.00	.779	1316.77	.803
" 7...	18.50	460	178	37.21	.781	1321.93	.805
" 8...	18.50	470	175.5	37.49	.784	1332.23	.805
" 9...	18.51	480	173	37.74	.786	1337.39	.807
" 10...	18.50	490	171	38.09	.790	1347.76	.806
" 11...	18.49	500	168	38.18	.792	1352.96	.807
" 12...	18.49	510	166	38.48	.793	1355.76	.812
" 13...	18.48	520	163	38.52	.796	1371.19	.804
" 14...	18.47	530	162	39.00	.800	1373.78	.813
" 15...	18.48	540	159.5	39.15	.802	1379.00	.813
" 16...	18.46	550	155	38.75	.804	1383.22	.803
" 17...	18.45	560	149	37.92	.806	1388.44	.783
Head Reduced.							
Whole Gate, 19...	12.67	350	131	20.84	.706	1137.50	.765
" 20...	12.63	370	123	20.68	.710	1159.69	.747
" 21...	12.61	390	116	20.56	.718	1166.53	.740
" 22...	12.62	360	128	20.94	.709	1157.22	.759
" 23...	12.63	340	133	20.55	.702	1189.94	.755
" 24...	12.64	330	137	20.55	.699	1120.31	.768
" 25...	12.65	320	141	20.51	.695	1110.51	.772
Head Reduced.							
Whole Gate, 27 ..	6.81	150	111.5	7.60	.554	785.43	.752
" 28...	6.77	160	105	7.61	.558	792.53	.750
" 29...	6.75	170	99	7.65	.564	807.72	.742
" 30...	6.72	180	98.5	8.06	.567	814.66	.779
" 31...	6.70	190	88	7.60	.573	825.68	.727
" 32...	6.66	200	86.5	7.83	.580	841.31	.739
Part Gate,							
7/8 34...	18.51	450	163	33.34	.731	1197.48	.796
3/4 35...	18.74	390	162	28.72	.660	1024.16	.792
1/2 36...	19.20	200	157	14.27	.500	666.34	.590
1/2 37...	19.45	50	211	4.79	.378	460.83	.282
1/4 38...	19.45	50	142	3.22	.310	342.70	.255

Whole Gate.	Head.	Weight.	Rev. per minute.	Horse Power.	Cub. ft.	Per Cent.
2d wheel, June 25, 1873, ...	18.40	450	174	35.59	1345	.762
	18.39	480	163	35.56	1374	.746
	18.39	500	154.5	35.11	1389	.728
3d wheel, July 17, 1873,.....	18.24	450	166.5	34.51	1342	.747
	18.22	480	158.5	34.58	1370	.734
	18.22	440	169	33.80	1332	.738
4th wheel, Aug. 16, 1873, ...	17.91	400	184.3	33.31	1317	.753
	17.90	450	168	34.36	1348	.755
	17.88	480	161	35.11	1371	.759

HOLYOKE MACHINE CO.,

Holyoke, Mass.

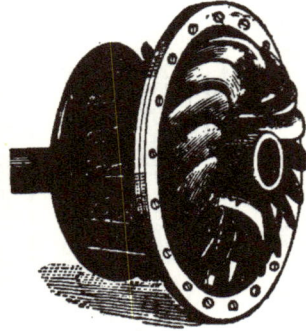

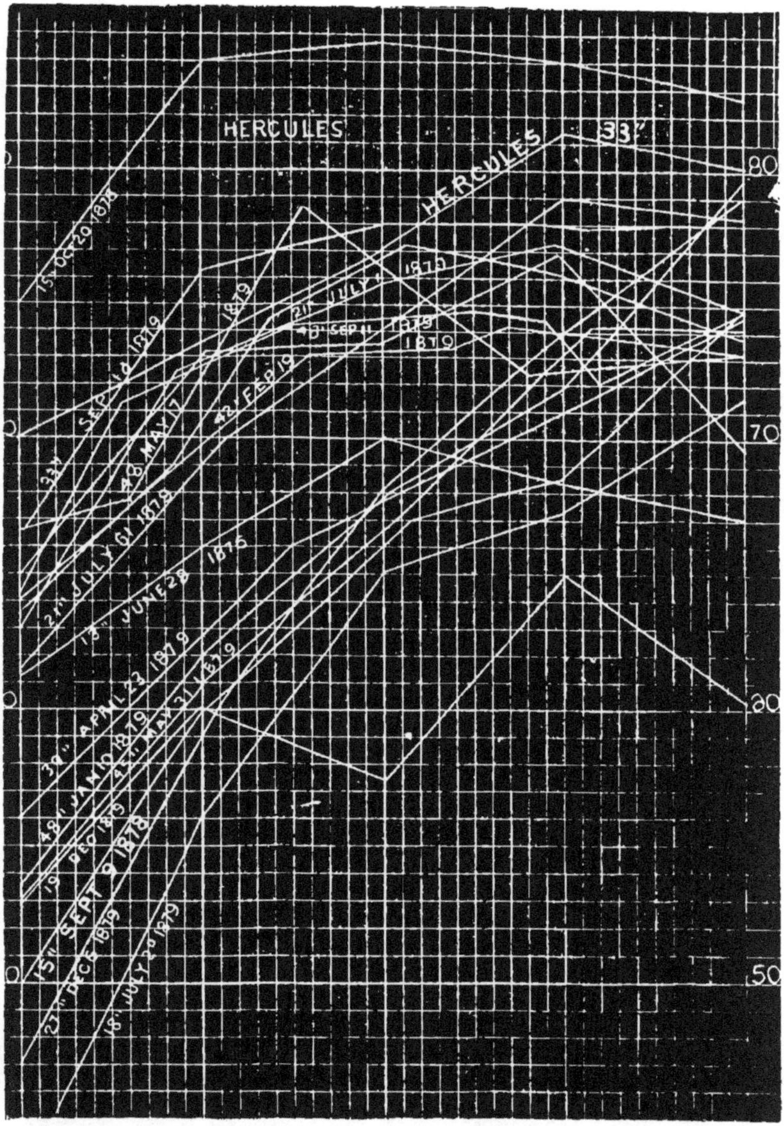
HERCULES
HERCULES
33"
80
70
50
15" SEPT 9 1878
48 MAY 17

HERCULES WHEELS

Holyoke Machine Company, Holyoke, Mass.

This company build the Boyden and Hercules turbines, the latter invented by John B. McCormick, of Brookville, Pa. Of the former, it is unnecessary to say anything here, as its merits are given on several other pages of this work.

THE HERCULES.

In March, 1876, several of the above named wheels, 24 inches in diameter, each differing somewhat from the others, were brought to Holyoke to be tested. All gave remarkable results: one, 87 per cent. useful effect, and a power so extraordinary that the wheel was taken up and examined. A few changes were made, then it was reset and again tested, when the following results were obtained:

Gate Opened		Head	Rev per minute	Horse Power	Cubic Feet	Per Cent
Whole Gate,	. . . .	18.02	217	70.58	2478.60	.8361
" "	. . . .	18.04	206	70.52	2466.04	.8386
Part Gate,	. . . .	18.06	214	70.03	2391.04	.8579
" "	. . . .	18.17	214.5	64.35	2167.29	.8644
" "	. . . .	18.23	213.5	64.05	2083.25	.8922
" "	. . . .	18.26	212	57.81	1944.50	.8612
" "	. . . .	18.34	210	53.45	1820.13	.8470
" "	. . . .	18.38	209.5	48.56	1690.89	.8267
" "	. . . .	18.57	211	32.12	1250.50	.7291

As high useful effect at whole gate had been obtained by several builders, but no such average at all stages of gate opening. In capacity, however, the Hercules took a stand so entirely above that of any turbine ever before produced, that it seemed a good starting point for bringing all builders into harmony for their own and the public good. I immediately opened a correspondence with the best builders of the country, urging the abandonment of inferior wheels and the advantage of uniting upon the plans of the Hercules, paying the patentees a small royalty, and each builder striving to excel. The idea was favorably received. In the meantime the patentees hastily disposed of their right to build for the Western States. This, of course, ended the chance for a union of builders. The contract, however, was soon canceled; then the patentees offered the Holyoke Machine Company certain exclusive rights in their patent. I opposed the negotiations, because I believed then, and continue to believe, that it would be better for all to have a union of the best builders, instead of a continuance of the ruinous competition of the past upon inferior plans, which only serves to hinder the perfection of the best. Turbine building is not sufficiently understood to allow of its being considered a science; it is simply "cut and try." I know of no builder that with certainty can make two turbines that will give the same results, even made from the same pattern. Until that can be done, the manufacturing interest must suffer, unless manufacturers use the greatest care in the selection of turbines. What is almost invariably needed, is a wheel that will economize water at any stage of gate opening, so that either the abundance of the spring and fall months, or the scarcer quantity of summer, may be utilized in full. To do this, turbine builders must be able to produce turbines that will give their best percentage at either one-half, five-eighths, three-fourths, seven-eighths, or whole gate, as may best adapt each for the place where it is to be used. Such a wheel should be good at any stage of gate opening, and when such can be produced with certainty, then turbine building may properly be considered a science and not before, for such wheels are possible. Our rivers and streams are all extremely variable in supply of water—that of the summer months often being less than one-fourth of what it is for the rest of the year; and three-fourths of the larger quantity is almost invariably allowed to run to waste, because wheels of sufficient capacity to utilize the maximum have generally proved to be incapable of transmitting any power from the use of the minimum supply.

I have tested about eighty of these wheels, and, as may be seen by the diagrams, the variations have been great, and there are no good reasons for believing that they will be less so in the future. Many of them are sent to purchasers without flanges on gate. In such case, in my opinion, the Victor would be preferable. That the company desire to do an honorable business, may be seen by the following extract, taken from second edition of this work :

" From our experience we are satisfied the interest of the purchaser requires that wheels should be tested before acceptance, and hereafter we shall furnish tested wheels when desired to do so; and if a purchaser desires to have his wheel tested after it is set in his mill, we will make the test there, if the purchaser will pay the extra expense incurred thereby. And we believe the safest and most economical way to furnish mills with power, is to first ascertain exactly what is required, and we will send an engineer at the expense of the purchaser, to any mill, who will consult with the proprietor, make examinations. using suitable instruments when deemed necessary, ascertain the quantity of water available, etc., etc., and then furnish wheels that we will guarantee to do the work in an economical manner and to give the maximum of power promised by us; but it must be plainly understood that we do not promise to furnish a given amount of power with a less quantity of water than fixed upon at the time of making the examination, or that our wheel or wheels will run the mill if additional machinery is added.

"HOLYOKE MACHINE COMPANY.

" November 20, 1875."

The constant variations of the wheels, and a lack of appreciation of purchasers, caused an abandonment of the the plan suggested in the card, and now the wheels are sent away without any knowledge of their efficiency.

D. P. Blackstone's Wheel, Berlin, Wis.

The Blackstone wheel has been tested in the Elmer curb represented above; in the Leffel; also, the Stout, Mills & Temple.

Test of a 40-inch, in Elmer curb.

	Head.	Weight.	Rev. per minute.	H. P.	Cubic feet.	Per Cent.
Whole Gate,......	17.60	925	158	66.73	2416.89	.8313
Part Gate,................	17.69	800	157	57.57	2184.17	.7895
"	17.88	520	156	36.87	1744.10	.6435
"	18.20	240	156.5	17.07	1261.62	.3940

Another 40-inch, in Leffel curb.

	Head.	Weight.	Rev. per minute.	H. P.	Cubic feet.	Per Cent.
Whole Gate, ...	17.75	1000	172	78.18	3143.27	.7424
Part Gate,.............. , ...	17.87	900	170	69.64	2734.01	.7543
"	18.05	750	168	57.27	2299.64	.7310
"	18.29	450	175	35.79	1621.44	.6394

CHASE WHEELS, FURNISHED BY THE ORANGE TURBINE CO., ORANGE, MASS.

No. of Test. **42-inch Wheel.**	Head.	Weight.	Rev. per Minute.	Horse Power.
Whole Gate, 1	15.17	500	194	44.09
" 2..............	15.17	700	161.5	51.17
" 3..............	15.17	750	158	53.86
" 4..............	15.17	800	149	54.18
" 5..............	15.17	850	143	55.25
" 6..............	15.17	900	135	55.22
" 7..............	15.17	950	128	55.27
" 8..............	15.17	1050	116	55.36
" 9..............	15.17	1100	104	52.00
54-inch Wheel.		.		
Whole Gate, 11..............	15.00	500	148	33.63
" 12..............	15.00	600	141	38.38
" 13..............	15.00	700	138	43.91
" 14..............	15.00	1000	117	53.20
" 15..............	15.00	1000	120.6	54.82
" 16..............	15.00	1050	118	56.32
" 17..............	15.00	1100	117	58.50
" 18..............	15.00	1200	112	61.09
" 19..............	15.00	1500	100	68.18
" 20..............	15.00	1800	90	73.63
" 21..............	15.00	2000	82	74.55

The 54-inch wheels were tabled and were geared to run at 138 revolutions per minute, to give 112 horse power; actual results obtained, 43.91, while the percentage of useful effect from the water used could not have exceeded 25 per cent., but at 82 revolutions it might have reached 35 or 40.

The water in the race below the mill was 30 inches average depth, 29 feet in width, its velocity being so great as to cause it to break white, the fall being at least one foot in a hundred.

TESTS OF A 36-INCH AND OF A 48-INCH TUTTLE WHEEL AT
SMITH & MEADER'S MILL, WATERVILLE, ME.

36-inch. No. of Test.	Rev. per Minute.	Weight in Pounds.	Horse Power.	Head in Feet.	Weir.	Cubic Feet Disch'd.	Per Cent.
Whole Gate, 1...	190.8	425	36.86	15.85	1.438	2746.83	.4479
" 11...	162.6	600	44.32	15.84	1.450	2779.65	.5325
" 12...	161	615	45.01	15.79	1.446	2768.70	.5572
Part Gate, 13...	175	325	25.82	16.58	1.250	2227.83	.3698
" 14...	168	130	9.93	17.22	.963	1489.98	.2066
48-inch.							
Whole Gate, 1...	148.4	900	60.71	14.53	1.634	4135.02	.5243
" 8...	149	1000	63.63	14.47	1.630	4145.97	.5612
" 9...	138.5	1025	64.53	14.45	1.630	4145.97	.5831
" 10...	137	1050	65.38	14.48	1.632	4149.62	.5625
" 11...	130	1100	65.00	14.48	1.630	4145.97	.572
" 12...	125	1150	66.38	14.48	1.632	4149.62	.57?
½ Gate, 13...	161	425	31.11	15.73	1.312	2992.99	.349

TESTS OF A 42-INCH TYLER, AND A 42-INCH REYNOLDS WHEEL,
AT VASSALBORO' WOOLEN MILLS, VASSALBORO', ME.

Tyler. No. of Test.	Rev. per Minute.	Weight in Pounds.	Horse Power.	Head in Feet.	Weir.	Cubic Feet Disch'd.	Per Cent.
Whole Gate, 1...	168.6	700	53.65	27.21	1.264	2609.50	.3999
" 2...	164.4	750	56.05	27.21	1.282	2664.84	.4089
" 3...	160.8	800	58.47	27.21	1.282	2664.84	.4266
" 4...	144.6	1100	72.30	27.21	1.297	2776.10	.5063
Reynolds.							
Part Gate, 1...	139	800	50.54	28	1.210	2436.82	.3918
" 2...	168.5	750	57.44	28	1.210	2436.82	.4453
" 3...	172.5	775	57.74	28	1.210	2436.82	.4477
" 4...	173.5	800	63.09	28	1.210	2436.82	.4892
" 5...	161	850	62.20	28	1.210	2436.82	.4823
" 6...	149.6	900	61 20	28	1.210	2436.82	.4745
Whole Gate, 7...	166	1000	75.45	28	1.297	2776.10	.4683
" 8...	161.6	1100	80.80	28	1.297	2776.10	.5015

This certifies, that on the 8th and 9th of this month I tested two 42-inch water-wheels at the Vassalboro' Woolen Mills, Vassalboro', Me., George Wilkins, Agent.

The first was called a Tyler wheel, though not made or furnished by Mr. Tyler. Regulator speed of wheel 170 revolutions per minute. The test proved that it was run at a velocity much too high to utilize its greatest effectiveness.

Second wheel, a "Reynolds." Testing first with gate open the same as when running all the machinery attached to it, six tests; then the gate was opened in full; with 1100 pounds on the scale beam the wheel ran very unsteadily, so much so that it was considered useless to try it with more weight.

Weir 10 feet in length, sectional area approaching weir 25 feet in width, depth below crest 2.5 feet.

April 15, 1872.   JAMES EMERSON.

Boyden Turbine.

In the purchase of this turbine, more ignorance is displayed than a well-wisher of his race likes to acknowledge lies dormant in the average business man of the times; in purchasing any other kind of turbine the purchaser almost invariably makes inquiries in order to get the best; the Boyden seeker makes no inquiries except, perhaps, as to capacity and cost, supposing all to be alike as to efficiency, whether made by an expert mechanic or the veriest botch. There is no reason to doubt but what at whole gate an outward discharge wheel may be made to give a high useful effect, but every intelligent turbine builder knows that of all wheels the outward discharge is the most difficult to get just right; also, that good part gate results are impossible with such discharge. There are vague rumors of remarkable results obtained by Mr. Boyden, as there are of the Humphrey and every other turbine, but such results are rarely confirmed when the wheels are tested by competent disinterested engineers. Of four Boyden wheels tested in a Connecticut mill, three were found to be giving 46 per cent. useful effect, the fourth gave 47. At Unionville, Conn., Platner & Porter Mf'g Co., a test of one gave 61 per cent. The wheel was built by the Ames Mf'g Co., of Chicopee, and is named in a recent circular of that Company in commendation of that style of turbine. A nice brass bucket wheel, made by the same Company in 1871, 72 inches in diameter, 51 openings, each 7x26 inches in height, 1½ inches in width, rated to discharge 6360 cubic feet of water under 24 feet head, at 98 revolutions per minute, and to give 217 horse power. was tested at the Dwight No. 7 mill, Chicopee, Mass., Nov. 6, 1878. The tests at its geared speed are given below:

	Head in Feet.	Weight.	Revolutions.	Horse Power.	Cubic ft.	Percentage.
Whole Gate, . . .	22.1	3730	97	219.2	7141.70	.7353
Part Gate, ...	23 20	2000	98	118.78	5446.80	.4977
" "	23.50	1000	96.4	58.43	4341.98	.3031

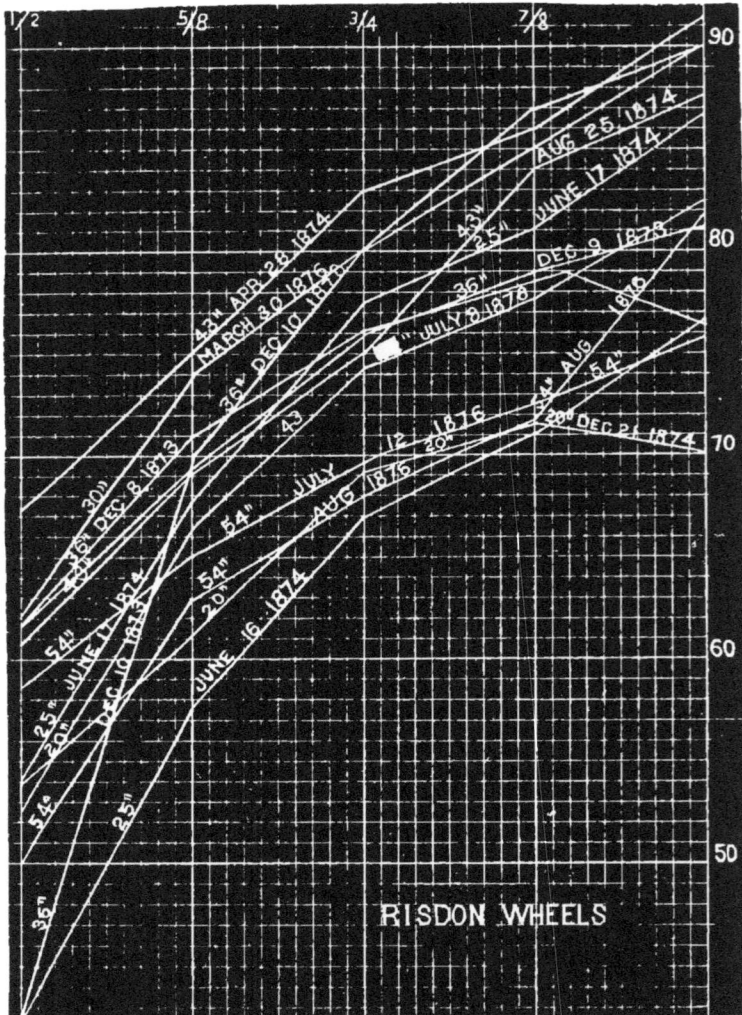

RISDON WHEELS

From the time of the Philadelphia turbine tests in 1859–60, up to the present, Mr. Risdon has continued an almost unbroken series of experiments for the purpose of perfecting the turbine; yet the lines above show decided variations in useful effect. If such is the case with wheels constructed by one so skillful, how must it be with those turned out by machine companys merely as a business, without other supervision than that of the ordinary foreman? The 54-inch, represented above, was put together by the Holyoke Machine Co , though Mr. Risdon furnished plans and core-boxes for forming the buckets. Many more of the Holyoke made wheels were tested than those made wholly by Mr. Risdon. As a general thing the wheels, when first tested, were rather low in efficiency; but, after making alterations suggested by such tests, the results were often very high at whole gate, and the tests proved conclusively that purchasers who accept untested turbines, generally do so at a loss of from ten to twenty per cent. of what they might have with more care.

THOMPSON & HOLCOMB WHEEL.

This certifies that a Water Wheel, 30 inches in diameter, made of cast iron, fly-trap gates—downward discharge,—in form somewhat like the Houston, known as the Thompson Turbine, was sent to the Holyoke Testing Flume by A. P. Holcomb of Silver Creek, N. Y., to be tested. The figures showing the results obtained by me, may be found below. During the test, the scale beam was attached to the brake at a point, which, if revolving, would describe a circle fifteen feet in circumference, consequently the revolutions of the wheel must be multiplied by fifteen to obtain the correct speed.

Length of Weir, ... 6 feet.
Temperature of Water, ... 32° Fah.
Weight of Water, per cubic foot, ... 62.875.
Correction for Leakage, 18 feet head, ... 13.10 cubic feet.
Correction for Leakage, 12 feet head, ... 11.10 cubic feet.
Correction for Leakage, 6 feet head, ... 9.10 cubic feet.

A second trial of the same wheel to ascertain the effect of short extensions added to the outer end of chutes, for the purpose of rounding or flaring them when open. These extensions prevented the gates from being opened quite as wide as without them, consequently less water was discharged. The partial gate at first trial gave best percentage, but owing to a breakage of the gates by the ice at the second trial, no part gate tests could be taken.

The wheel run very steady, was easily regulated, and from its high speed is a favorite with those who have it in use; its gates, like the Leffel and all of that class, would be likely to become leaky.

No. of Test.	Head.	Weight	Rev. per Minute.	Horse Power.	Weir.	Cubic Feet.	Per Cent.
Whole Gate, 2	18.44	390	220.5	39.09	1.226	1547.76	.725
" 3	18.45	400	220	40.00	1.243	1579.39	.726
" 4	18.42	410	221	41.18	1.254	1599.95	.741
" 5	18.40	420	218	41.62	1.252	1614.97	.748
" 6	18.39	430	215	42.02	1.260	1611.20	.762
" 7	18.88	440	211	42.20	1.258	1607.48	.756
" 8	18.38	450	207.5	42.44	1.251	1613.09	.757
" 9	18.35	440	203	42.42	1.260	1611.20	.759
" 10	18.26	450	205.3	41.99	1.200	1611.20	.755
HEAD REDUCED.							
Whole Gate, 13	12.19	260	183	21.62	1.080	1285.96	.730
" 14	12.18	270	178	21.84	1.082	1289.47	.726
" 15	12.18	280	172	21.89	1.085	1294.75	.734
" 16	12.18	290	170	22.40	1.088	1300.08	.745
" 17	12.17	300	161.5	21.95	1.091	1305.30	.731
" 18	12.17	310	156.7	22.08	1.091	1305.30	.735
" 19	12.18	285	169.5	21.98	1.087	1298.27	.725
" 20	12.18	295	164	21.99	1.088	1300.03	.721
HEAD REDUCED.							
Whole Gate, 23	6.60	130	139	8.34	.878	940.26	.711
" 24	6.59	135	136	8.21	.878	948.28	.695
" 25	6.58	140	132.5	8.43	.876	945.08	.717
" 26	6.58	145	128	8.43	.877	946.68	.700

July 12th.

Test of the same wheel before extensions were added.

No. of Test.	Head.	Weight	Rev. per Minute.	Horse Power.	Weir.	Cubic Feet.	Per Cent.
Whole Gate, 28	18.10	460	206	43.07	1.174	1713.41	.787
" 29	6.41	160	103.5	7.53	.800	974.88	.689
Part Gate, 30	18.25	400	203	36.60	1.040	1433.27	.742
" 31	18.38	325	208	30.70	.988	1280.52	.723
" 32	18.44	250	207	23.86	.818	1004.54	.673
" 33	18.50	200	207.5	18.86	.790	880.88	.651
" 34	6.60	125	102	6.26	.669	747.68	.676

JAMES EMERSON.

HOLYOKE, MASS., December 18, 1872

Rodney Hunt Machine Co.,

ORANGE, MASS.

‒•‒

THIS certifies that a WATER WHEEL, thirty inches in diameter, made of cast iron, central and downward discharge, known as the Hunt Double Action Turbine was sent to the HOLYOKE TESTING FLUME by the Rodney Hunt Machine Co., of Orange, Mass., to be tested. The date of each test and the figures showing the exact results obtained by me, may be found on the following pages. During the test the scale beam was attached to the brake at a point; which, if revolving, would describe a circle fifteen feet in circumference, consequently the revolutions of the wheel must be multiplied by fifteen to obtain the correct speed. Data for one minute:

Length of Weir, ...	6 feet.
Temperature of Water, ...	40° Fah.
Weight of Water, per cubic foot, ...	62.373.
Correction for Leakage, 18 feet head, . . .	14.20 cubic feet.
Correction for Leakage, 12 feet head, . . .	12.20 cubic feet.
Correction for Leakage, 6 feet head, . . .	10.20 cubic feet.

No. of Test.	Head.	Weight.	Rev. per Minute.	Horse Power.	Weir.	Cubic Feet.	Per Cent-age.
Whole Gate,	18.34	650	125.5	37.08	1.192	1473.94	.7245
	18.36	525	177.5	42.36	1.179	1460.17	.8388
" 12...	18.35	540	171	41.06	1.180	1461.99	.8187
" 13...	18.35	550	168.5	42.12	1.184	1469.30	.8275
" 14...	18.34	560	166	42.25	1.187	1474.80	.8260
" 15...	18.34	530	176.5	42.52	1.181	1463.82	.8385
" 16...	18.35	520	180.5	42.66	1.178	1458.34	.8433
" 17...	18.36	510	183	42.42	1.173	1449.23	.8425
" 18...	18.36	500	185	42.04	1.170	1443.77	.8409
Part Gate, 19...							
" 20...	18.37	475	190	41.02	1.158	1422.00	.8306
" 21...	18.38	490	186	41.42	1.160	1425.62	.8374
" 22...	18 36	500	183	41.59	1.163	1427.43	.8395
" 23...	18.42	440	182.7	36 54	1.123	1359.02	.7722
" 24...	* 18.40	450	185.5	37.94	1.125	1362.62	.8055
" 25...	18.40	460	176.6	36.92	1.128	1368.59	.7756
" 26...	18.61	250	179	20.34	.932	1030.04	.5613
" 27...	18.60	235	185	19.76	.927	1022.66	.5495
" 28..	18.86	75	176.2	6.01	.727	710.89	.2376
Head Reduced.							
" 30...	12.24	250	173.3	19.69	.983	1117.88	.7630
" 31...	12.20	275	165	20.62	.995	1138.16	.7856
" 32...	12.19	300	157.5	21.47	1.007	1158.55	.804
" 33...	12.17	320	151	21.96	1.018	1187.33	.804
" 34...	12.15	310	145	22.41	1.027	1192.77	.8199
" 35...	12.13	350	141.5	22.51	1.032	1201.37	.817
" 36...	12.13	360	137	22.41	1.036	1208.27	.8089
" 37...	12.13	370	134.5	22.62	1.037	1209.99	.8172
" 38...	12.13	380	129	22.28	1.037	1209.99	.803
" 39...	12.13	390	124	21.98	1.040	1215.17	.7888

Hunt's Double Action Turbine Wheel.

The cut at the left represents the Hunt curb, with the downward and outward discharge wheel which gave the results reported in the second table below; the other cut represents the wheel generally used by the Hunt Machine Co., (the Swain); and the one giving the results reported in the first table below, also, those upon the opposite page.

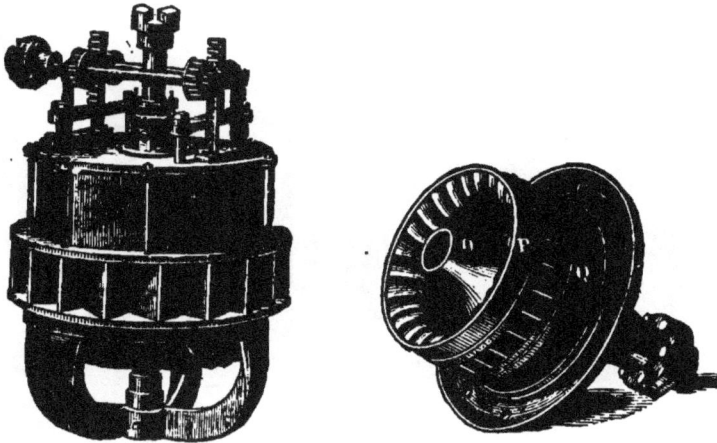

Test of a 48-inch Hunt-Swain wheel.

	Head.	Weight.	Rev. per minute.	H. P.	Cubic feet.	Per Cent.
Whole Gate,..............	11.71	1500	83	56.59	3454.74	.757
Part Gate,..............	11.89	1275	86.5	50.13	3252.31	.684
" "	12.30	850	85.5	33.03	2713.01	.524
" " ,...............	12.61	200	87.7	7.97	1617.18	.254

Test of a Hunt-Flint wheel, downward and outward discharge; see bottom of buckets in curb above.

	Head.	Weight.	Rev. per minute.	H. P.	Cubic feet.	Per Cent.
Whole Gate, ...	18.31	675	176.3	54.09	1732.10	·9050
Part Gate,..............	18.31	575	176.5	46.13	1672.72	.7992
" "	18.33	500	175	39.77	1564.08	.7361
" "	18.53	200	175	15.71	1067.10	.4215

Test of a Hunt wheel, downward discharge, in the same curb as the one above.

	Head.	Weight.	Rev. per Min.	Horse Power.	Cubic feet.	Per Cent.
Whole Gate,..............	18.28	665	180	54.40	1800.73	.8780
Part Gate,..............	18.32	525	178	42.47	1681.87	.7314
" "	18.40	395	178	31.95	1435.63	.6432
" "	18.49	270	181	22.21	1245.48	.5184
" "	18.63	150	188	12.81	870.44	.4082

Gates Curtis, Ogdensburg, N. Y.

CURTIS TURBINE.

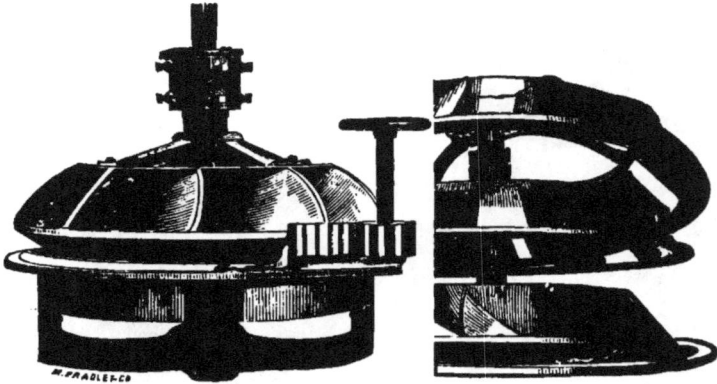

This wheel is diagonal in shape, like the Houston, but has an Inside Register gate.

Test of a 47-inch wheel.

	Head.	W'ht.	Rev.	H. P.	Cubic Feet.	P C.
Whole Gate,	17.71	1150	115	80.15	3041.00	.788
Part Gate,	17.98	1000	115.5	70.00	2648.24	.778
"	18.07	850	111.5	57.44	2345.83	.717
"	18.17	570	116.5	40.24	1846.92	.635
"	18.32	450	107	29.58	1470.12	.582
"	18.32	400	114	27.63	1441.54	.554

Mr. Curtis also makes the wheel with open chutes, omitting gate, allowing the wheel to run at full gate at all times, regulating speed by head in forebay, using a wicket gate between flume and forebay. A 25-inch made in that way, tested at my flume gave the following results

	Head.	W'ht.	Rev.	H. P	Cubic Feet	P.C.
Whole Gate,	18.21	500	220	33.33	1095.93	.8842
Same wheel in another set of chutes.						
Whole Gate,	18.20	465	223.2	31.45	1099.41	8322
chutes stopped,	18.29	400	213	25.81	957.63	.7801
chutes stopped,	18.37	300	214	19.75	816.25	.6973
Same wheel tested in a curb with gate.						
Whole Gate,	18.40	415	224.5	28.23	1017.27	.7984
Part Gate,	18.42	360	211.3	23.05	886.22	.7510
"	18.51	290	209.5	18.41	751.00	.7012
"	18.60	215	213	13 87	615.70	.6413
"	18.68	165	199.2	9.96	491.65	.5742

Humming Bird Wheels.

48-inch wheels, sent by Willis Read, Danbury, Conn.

Through some peculiarity of construction, which, without illustration, is inde-scribable, these wheels keep up a constant humming sound while running; hence their name. Mr. Read was promptly on hand with his wheel, which was tested Sept. 6. From information obtained by the test, he took a new departure and constructed another wheel, which was tested Oct. 15. The results of each may be found below. The workmanship of the wheels would hardly cause manu-facturers to look for machinery in Danbury.

Data below for one minute. Multiply revolutions by 20.

Gate Opened		Head	Weight	Rev per minute	Horse Power	Cubic Feet	Per Cent
Whole Gate.	. . . . .	17.95	1550	000	000		000
" "	. . . . .	18.02	750	103	46.81	2187.30	.6287
" "	. . . . .	18.02	775	102	47.91	2211.71	.6366
" "	. . . . .	18.00	800	100.5	48.72	2218.70	.6473
" "	. . . . .	18.00	825	98.2	49.60	2232.70	.6533
" "	. . . . .	17.98	850	97.5	50.22	2246.73	.6581
" "	. . . . .	17.98	875	96	50.90	2260.77	.6629
" "	. . . . .	17.97	900	95	51.81	2271.33	.6720
" "	. . . . .	17.95	925	93.5	52.42	2306.62	.6702
" "	. . . . .	17.94	950	90	51.82	2338.53	.6537
Part Gate.	. . . . .	18.40	425	98.3	24.03	1210.67	.5734
" "	. . . . .	18.41	400	100	24.24	1196.08	.5823
" "	. . . . .	18.45	400	92	22.30	1089.69	.5872
" "	. . . . .	18.41	420	96.7	24.78	1255.49	.5670
" "	. . . . .	18.40	500	91.5	27.72	1269.56	.6283
" "	. . . . .	18.34	600	88.5	32.18	1398.88	.6640
" "	. . . . .	18.32	600	91	33.09	1472.63	.6493
" "	. . . . .	18.18	700	96	40.72	1827.76	.6488
" "	. . . . .	18.16	750	93.2	42.36	1887.49	.6541
" "	. . . . .	18.13	775	93.5	43.94	1944.34	.6598
" "	. . . . .	18.22	650	95	37.42	1732.81	.6275

Tested October 15.

Gate Opened		Head	Weight	Rev per minute	Horse Power	Cubic Feet	Per Cent
Whole Gate.	. . . . .	17.81	1600	000	000	2642.89	000
" "	. . . . .	17.85	800	107.5	52.12	2474.90	.6246
" "	. . . . .	17.85	850	103	53.06	2485.76	.6331
" "	. . . . .	17.84	900	95.8	52.25	2551.18	.6078
" "	. . . . .	17.83	825	106	53.00	2503.88	.6286
" "	. . . . .	17.83	850	103.3	53.21	2514.77	.6283
" "	. . . . .	17.85	875	99.6	52.78	2554.83	.6127
Part Gate.	. . . . .	18.06	700	97.3	41.27	2042.38	.5923
" "	. . . . .	18.02	675	100.3	41.03	2035.56	.5922
" "	. . . . .	18.04	650	102.6	40.41	2035.56	.5826
" "	. . . . .	18.20	500	104.2	31.57	1683.73	.5454
" "	. . . . .	18.18	525	103.2	32.83	1640.51	.5689
" "	. . . . .	18.32	400	95.6	23.17	1328.28	.5041
" "	. . . . .	18.34	350	103	21.90	1301.28	.4847
" "	. . . . .	18.57	200	93.5	11.94	908.29	.3760
" "	. . . . .	18.44	250	105	11.59	1079.87	.3081
" "	. . . . .	18.06	650	106	41.75	1981.13	.6192
" "	. . . . .	18.20	500	106.8	32.36	1651.78	.5699
" "	. . . . .	18.35	350	106	22.48	1277.42	.5078
Whole Gate.	. . . . .	17.84	850	107.5	55.38	2532.96	.6489

HOUSTON WHEEL.

This certifies, that a Water Wheel 50 inches in diameter, made of cast iron, cast whole, Register gate, known as the Houston Water Wheel, was sent to the Holyoke Testing Flume by O. E. Merrill & Co., Beloit, Wisconsin, to be tested.

No. of Test.	Head.	Weight.	Rev. per Minute.	Horse Power.	Weir.	Cubic Feet.	Per Cent.
Whole Gate, 1...	18.00	1500	112	101.80	1.555	3660.33	.817
" 2...	17.96	1600	106	102.78	1.545	3625.01	.835
" 3...	18.22	1650	106.6	106.60	1.554	3656.85	.848
" 4...	18.07	1700	103.6	106.72	1.569	3710 12	.842
" 5...	18.08	1725	102	106 60	1.573	3720.80	.838
" 6...	18.06	1620	111	108.95	1.565	3695.88	.8635
" 7...	18.04	1630	109	107.68	1.566	3699.44	.8535
" 8...	18.04	1640	110	109.33	1.565	3695.88	.8675
" 9...	18.04	1660	108	108.65	1.570	3713.66	.858
" 10...	18.04	1610	113	110.25	1.560	3678.12	.880
Part Gate, 12...	18.05	1400	113	95.88	1.520	3536.99	.797
" 13...	18.10	1240	117	87.92	1.484	3411.39	.721
" 14...	18.20	1040	119	75.01	1.400	3128.62	.556
" 15...	18.40	800	109	52.85	1.300	2823.82	.568
" 16...	18.58	575	110	88.33	1.126	2240.12	.539
" 17..	18.70	275	112	18.66	.960	1749.84	.302
Second Day,	Flaring	Extensi	ons to	chutes	off.		
Whole Gate, 28...	17.95	1550	109	102.39	1.535	3689.72	.818
" 29...	17.95	1590	105	101.18	1.537	3596.78	.829
" 30...	17.97	1535	113	105.12	1.525	3554.54	.8716
" 31...	17.97	1515	111	101.31	1.525	3554.54	.840
Part Gate, 33...	18.02	1300	116	91.39	1.490	3362.90	.798
" 34...	18.15	1140	115	79.45	1.400	3123.62	.741
" 35...	18.25	1000	110	66.67	1.315	2840.26	.680
" 36...	18.48	800	103	49.94	1.146	2301.59	.621
" 37...	18.62	500	105	31.81	1.030	1952.16	.463
" 38...	18.91	200	102	12.36	.850	1445.74	.240

Previous to the trial of this wheel it had been frozen solid in ice at the bottom of the flume for two weeks; to clear it, crowbars, blocks of wood, axes and other implements were used, some of which entered the wheel with a crash when it first started, probably throwing it out of center, for it required the strength of two men applied to the rim of the brake (six feet in diameter) to turn the wheel when the gate was closed.

E. L. SMALL, URBANA, OHIO.

The results obtained may be found below. The peculiarity of the wheel consists in its gates and buckets, the gates being simply large faucets. The buckets are like shallow boxes, — Mr. Small believing angles better than curves for surfaces.

No. of Test.	Head.	Weight.	Rev. per Minute.	Horse Power.	Weir.	Cubic Feet.	Per Cent.
Whole Gate, 1	18.39	490	181.6	39.62	1.148	1658.40	.688
" 2	18.40	540	170	41.72	1.147	1656.28	.725
" 3	18.29	570	149	38.61	1.147	1656.28	.675
" 4	18.29	580	147	38.75	1.147	1656.28	.677
" 5	18.28	535	169	41.09	1.148	1658.40	.720
" 6	18.28	540	169	41.48	1.148	1658.40	.724
" 7	18.28	545	165	40.88	1.148	1658.40	.714
" 8	18.28	550	165	41.25	1.148	1658.40	.718
Head Reduced. Gates Reversed.							
Whole Gate, 10	11.95	225	156	15.96	.916	1183.55	.505
" 11	12.03	310	136	19.34	.962	1276.34	.617
" 12	12.03	320	134.5	19.56	.962	1276.84	.674
Head Reduced.							
Whole Gate, 14	6.53	165	100	7.50	.776	984.87	.688
" 15	6.55	155	105	7.40	.769	915.44	.654
3-4 Gate, 17	18.52	330	170	25.50	.938	1227.15	.594
" 18	18.51	325	174	25.70	.937	1229.09	.590
" 19	18.52	340	170	26.27	.941	1235.35	.568
1-2 Gate, 21	18.72	130	170.2	11.60	.721	840.00	.301
" 22	18.60	190	171	13.68	.718	823.45	.471
4 Gates cl's'd, 24	18.51	270	170	20.86	.840	1043.90	.572

J. W. UPHAM, WORCESTER, MASS.

Mr. Upham has been in the Water Wheel business for many years, and is known for his sterling integrity. The wheel he now builds is one similar to the Houston Wheel inverted. It has a register gate that works very easily, as it is on the inside at the top and small. The figures below were obtained from trials at my Lowell Flume. The two last sets of figures are given to show the speed at which it may be run, and produce good power.

J. E.

No. of Test.	Head.	Weight.	Rev. per Minute.	Horse Power.	Weir.	Cubic Feet.	Per Cent.
Whole Gate, 1	15.43	250	238.5	27.10	.965	1259.88	.737
" 2	15.45	245	201.5	26.10	.945	1221.02	.732
" 3	15.46	310	193.5	27.27	.987	1295.89	.774
" 4	15.465	310	179.5	26.92	.982	1195.96	.770
" 5	15.42	275	216.5	27.06	.952	1234.58	.752
" 6	15.42	250	200.5	27.09	.944	1219.06	.763
" 7	15.42	300	198	27.00	.940	1211.36	.765
" 8	15.425	310	193.5	27.27	.936	1203.06	.777
" 9	15.42	320	186.5	27.13	.932	1196.96	.778
" 10	15.43	330	176.5	26.47	.928	1188.90	.764
" 11	15.39	300	198	27.00	.940	1211.36	.766
Tests of another wheel of the same kind.							
Whole Gate, 13	15.60	100	300	18.18	1.117	1519.08	.473
" 14	15.49	250	170	25.76	1.024	1175.01	.7508

E. G. Libby, Medford, Mass.

The wheel, illustrated in the Upham report above, was designed by Mr. Libby, who has recently applied the water to the same kind of wheel, but through chutes similar to those of the Hercules. A 25-inch wheel so arranged was tested by me, Aug. 5, 1878, giving the following results:

Head.	Weight.	Rev. per min.	Horse Power	Cubic feet.	Per Cent.
18.23	350	298	47.40	2101.35	.6552
18.35	250	288.5	33.92	1847.54	.5297
18.38	200	293	26.63	1688.39	.4543
18.48	100	309.5	14.06	1393.52	.2890

N. F. BURNHAM, YORK, PENN.

No. of Test.	Head.	Weight.	Rev. per Minute.	Horse Power.	Weir.	Cubic Feet.	Per Cent.
Whole Gate, 1...	18.09	750	156	53.18	1.308	1994.52	.7824
" 2...	18.12	800	151	54.91	1.309	1996.77	.787
" 3...	18.10	810	150	55.22	1.311	2001.27	.808
" 4...	18.10	820	148	55.16	1.312	2003.53	.807
" 5...	18.10	830	147	55.46	1.313	2005.78	.810
" 6...	18.09	840	143.5	52.25	1.316	2012.55	.761
2 chutes stopped with blocks, 7...	18.30	680	146.4	45.25	1.200	1755.97	.747
3 chutes stopped with blocks, 8...	18.30	615	146.4	40.93	1.147	1642.74	.722
4 chutes stopped with blocks, 9...	18.22	500	147.2	33.47	1.062	1464.92	.665
6 chutes stopped with blocks, 10...	18.49	365	147	24.39	.919	1180.30	.579
Whole Gate, 11...	18.11	830	146.4	55.23	1.327	2037.41	.794
Part Gate, 12...	18.20	680	147.4	45.56	1.226	1812.58	.733
Without bl'ks, 13...	18.25	615	145	40.53	1.178	1708.50	.689
Whole Gate, 14...	18.29	500	146	33.18	1.082	1505.95	.639
" 15...	18.37	365	146	24.35	.959	1252.11	.567
Head Reduced.							
Whole Gate, 17...	12.14	450	137	28.02	1.120	1586.21	.772
" 18...	12.15	475	133.5	28.82	1.127	1600.95	.778
" 19...	12.13	500	128.5	29.21	1.134	1615.71	.773
" 20...	12.13	525	123.5	29.47	1.139	1626.29	.792
" 21...	12.11	550	117	29.25	1.143	1634.77	.784
" 22...	12.09	575	115	30.05	1.151	1651.76	.798

·Patent Curbs.

Designed to Economize Water at Part Gate.

This, by W. S. Davis, Warner, N. H., has 16 chutes or gates that open successively, two at a time, tested May, 1871. Wheel a rough imitation of the Swain.

16 chutes open, perc'tge, .6346.

14 chutes open, perc'tge, .4765.

10 chutes open, perc'tge, .3955.

6 chutes open, perc'tge, .2968.

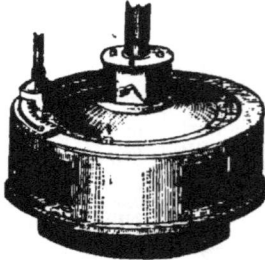

J. T. Case, Bristol, Conn.

National Water Wheel Company.

See Report of Tests for that Company.

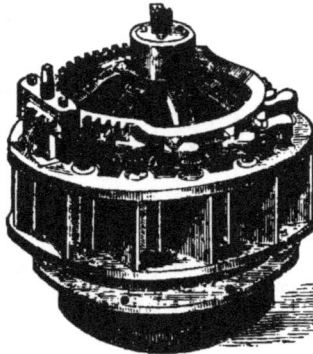

John L. Stowe, Newark, New Jersey.

Test of a 24-inch, April, 1878.

Head.	W'ht.	Rev.	H. P.	Cubic feet.	Per-Cent
18.26	425	217.5	28 01	1005.12	.8075
18.40	345	216	22.58	854.33	.7599
18.55	235	211	15.02	607.23	.7054

The Davis and Case chutes are closed at their outer ends, while the Stowe plan closes them at their inner end.

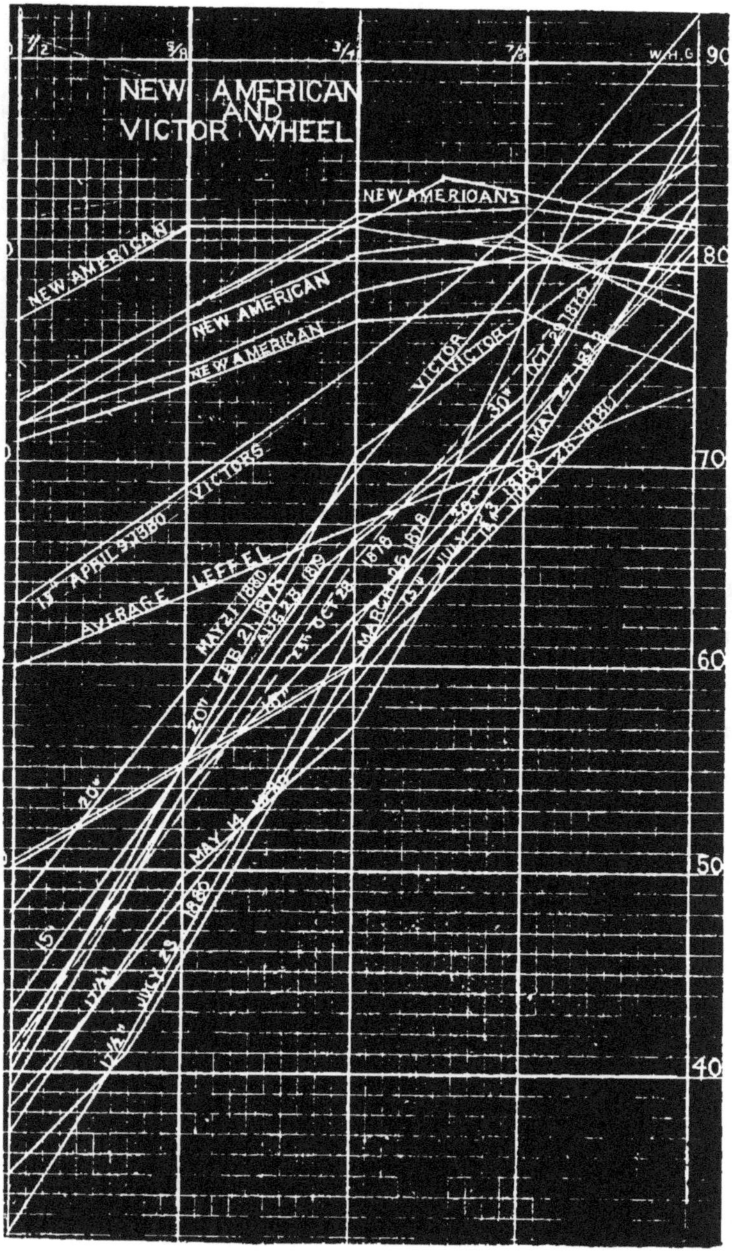
NEW AMERICAN
AND
VICTOR WHEEL
1/2
5/8
3/4
7/8
W.H.G
90
80
70
60
50
40
NEW AMERICANS
NEW AMERICAN
NEW AMERICAN
NEW AMERICAN
VICTOR
VICTOR
VICTORS
AVERAGE LEFFEL

List of Wheels Tested.

Those having a star placed before name are specially reported,

*AMERICAN, Stout, Mills & Temple, Dayton, Ohio. The best of the early wheels.

*ANGELL, Providence, R. I. Double discharge, central and down. Buckets cast separate, then bolted to hub, very apt to shear off. Fly trap gates, very leaky; is steadier, gives more power and higher useful effect with central discharge stopped.

ARROWSMITH, Lockport, N. Y. Central discharge with sheets of steel extending the inner edge of buckets until they met like the sides of a wedge upon the supposition that at part gate the pressure of water would regulate the opening, and produce high percentage at any stage of gate. The plan was a failure. Highest useful effect, 68 per cent.

*BURNHAM, York, Pa. Downward discharge. Outside register gate.

*BOYDEN FOURNEYRON. Made at Chicopee, Holyoke and other places. Outward discharge. Poor at part gate and of small capacity for diameter. Useful effect of those I have tested has varied from 46 to 85 per cent.

BUZZELL, St. Johnsbury, Vt. Scroll. Downward discharge. So arranged that proportionally it gives good part gate results. Highest percentage, 56 per cent.

BASTION, Canton, N. Y. Similar to the Curtis, but I think not manufactured now. Tested one with wicket gate in draft tube below the wheel, which proved the plan to be bad. With register gate, highest useful effect, 70 per cent.

BEE, Lancaster, Mass. Downward discharge. Babbitted in the upper bearing, and became bound while being tested, so that 58 per cent., the highest result obtained, was no indication of what the wheel would have done if it had been in a proper condition.

BRYANT BRO'S., Westchesterfield, Mass. Downward discharge. Gave 65 per cent.

BRYSON TURRETT, Miles Greenwood, Cincinnati, Ohio. Down and central. 75 per cent. Not manufactured now.

BLAKE, Pepperell, Mass. Scroll. Obsolete. 50 per cent.

*BARBER, Ballston Spa, N. Y. 79.29 per cent.

*BLACKSTONE, in Elmer, Leffel and American curbs. See special reports.

BODINE JONVAL, Mount Morris, N. Y. If made at all. 76 per cent.

*BOLLINGER, York, Pa. Central discharge. 70 per cent.

*COX, Ellsworth, N. Y. Double, downward discharge. 70 per cent.

*CASE, National Water Wheel Co., Bristol, Conn. See special report.

*CHASE, Orange, Mass. See report.

CUSHMAN, Hartford, Conn. Scroll. 50 per cent. Discharge up and down.

*COLEMAN, Turner's Falls.

*CURTIS, Ogdensburg, N. Y.

Cook, Lake Village, N. H. Has had several kinds tested, but builds upon a different plan now. Highest useful effect of those tried, .7752 per cent.

Chapman, Clark & Chapman, Turner's Falls, Mass. Highest efficiency,52 per cent.

*Eclipse, Stilwell & Bierce Manf'g Co., Dayton, Ohio.

Grow, Dubuque, Iowa. 69 per cent.

Gillespie, Turner's Falls, Mass. Two wheels upon horizontal shaft. Fourneyron wheels. 54 per cent.

Green, Juda, Wis. 50 per cent.

Geyline, Philadelphia, Pa. Jonval wheels. Telescopic gate below wheels. 56 per cent.

Holman, Adams, N. Y. 47 per cent.

Humming Bird, Willis Read, Danbury, Conn. Two. One central, one downward discharge. 62 per cent.

*Houston, Beloit, Wis. Has had many wheels tested. Useful effect, ranging from .774 to .9006 per cent. Gate works very hard, and is poor at part gate.

*Hercules, Holyoke, Mass. See special report.

*Holyoke Machine Co., Holyoke, Mass. See special report.

*Hunt, Orange, Mass. See special report.

*Humphrey, Humphrey Machine Co., Keene, N. H.

Kindleberger, Cincinnati, Ohio. .6246 per cent.

Knowlton, Saccarappa, Maine. 59 per cent. Abandoned.

Leavitt, Lebanon, N. H. .637 per cent.

Luther, Iowa. Scroll. 70 per cent.

*Leffel, Springfield, Ohio. Have tested many of them. Useful effect varied from 40 to 79 per cent.

*Lucas, Hastings, Minn. See special report.

*Libby, Medford, Mass. See special report.

Lesner, Fultonville, N. Y. Central discharge. Central discharge wheels are behind the age.

*Mullikin, Lansing, Iowa. See special report. The wheel is very poorly made.

*Mosser, Allentown, Penn. See special report.

Mallery, Dryden, N. Y. .769 per cent.

*National, Josiah Buzzby, Crosswicks, N. J. .676 per cent. Complicated gates.

*National, Bristol, Conn. See special report of the Case wheel.

*Perry, Bridgton, Maine. See special report.

Platt, New Brighton, Pa. Two wheels upon a horizontal shaft. .585 per cent.

Raney, New Castle, Penn. Became bound in its stuffing box while being tested, so that the test was no indication of what it would have done if it had been well constructed. Useful effec·, per test, .667 per cent.

*Rindon, Mt. Holly, New Jersey. See special report.

Reynolds, Oswego, N. Y. Scroll. 50 per cent.

Reaser, Milwaukee, Wis. Flutter wheel placed on end between plates; would not run its own weight to speed.

Sherwood, Independence, Iowa. A Fourneyron, 63 per cent., and a downward discharge. .761 per cent.

*Swain, North Chelmsford, Mass. See special report.

*Smith, York, Pa. See special report.

STEVENSON, New York City. Two Jonval wheels placed together, one discharging downward the other upwards, the upper discharge passing into a dome "or vacuum," then downward in an annular tube, as shown in the Fulton & Myers' plan, which is illustrated in the group of perpetual motion inventions.

*SMALL, Urbana, Ohio. See report.

STETSON, Fitchburg, Mass. Central and downward discharge, register gates, not manufactured now. .793 per cent.

*STOWE, Newark, New Jersey.

STAPLES, Boston, Mass. Central discharge, three divisions, with a cylinder gate raised by a screw similar to that of the Hercules; the object of the three divisions of the wheel was to gain high part gate results, as it was supposed that either division would give as high results as the whole combined. Highest results obtained, 77 per cent.

TRULLINGER, Oswego, Oregon. Discharge down and up into a vacuum like Stevenson's. 70 per cent.

TYLER, Claremont, N. H. Old scroll, useful effect ranged from 50 to 67 per cent.

*TYLER. New scroll and flume wheels. See special reports.

TELLER, Fort Plain, N. Y. Wheel in divisions like the Staples and for the same purpose. Useful effect, .645 per cent.

TERRY, Terryville, Ct. Boyden or Fourneyron with two register gates, one inside of chutes, the other outside. 58 per cent. Abandoned.

*TUTTLE, Waterville, Maine. 58 per cent.

TICE, Cincinnati, Ohio. Re-invention of the old Schiele wheel, illustrations of it may be found in Wiesbach's or almost any other work treating of turbines twenty years since.

*THOMPSON, Springfield, Mo., and Silver Creek, N. Y.

*TWITCHELL, Pulaski, N. Y. See under the head of Perpetual Motion.

UPHAM, Worcester, Mass. Central discharge, tried in scroll, also in flume curb. 72 in scroll. 68 per cent. in flume curb. Abandoned.

*UPHAM & LIBBY. See special report.

*VICTOR, Stilwell & Bierce Manfg Co., Dayton, Ohio.

VANDEWATER, Rochester, N. Y. Downward discharge, cylinder gate. .778 per cent. Wheel struck bad in curb while being tested.

WATSON JONVAL, Paterson, N. J. Old. 49 per cent.

*WALSH, Waupaca, Wis.

*WHITNEY, Leominster, Mass. Old plan in flume and scroll curbs abandoned. Percentage of scroll, old wheel, 40 per cent. Flume, 72. For new plan, see special report.

WAGNER, Chicago, Ill. Foolishly complicated in discharge and limited capacity. Highest useful effect, .738 per cent.

WHEELER, Berlin, Mass. Central and downward discharge; but did best every way with central discharge stopped with blocks. Discharged the same quantity of water after blocking central discharge. .745 per cent. Not manufactured now.

*WYNKOOP. See special report.

*WETMORE, Claremont, N. H. See special report.

*WOLF, Allentown, Pa. In taking one of the make apart, a few days since, many small pieces were found that were used for blocking up gate suspension. Such pieces are very liable to get lost and might, with little trouble, be rendered unnecessary, by casting projecting pieces on the surfaces. "Patchwork" is objectionable in turbine building. See special report for efficiency.

THE UNITED RAILWAYS
SAFETY
CAR HEATING COMPANY.

The purpose of this combination is to obtain and control the most perfect devices for the safety, comfort, and convenience of the traveling public, and employees of the roads, also convenience and economy for the companies.

HORACE H. STEVENS, President.   C. H. COLE, Treasurer.

DIRECTORS :

N. J. RUST, President of Lincoln National Bank, Boston.

O. J. LEWIS, Director of Lincoln Bank, Boston.

OAKES A. AMES, Director of Easton National Bank and Lincoln Bank, Boston.

HORACE H. STEVENS, Director of Globe National Bank, Boston, and Chicago & Eastern Illinois R. R.

GEORGE H. BALL, President of the Norwich & Worcester R. R., Director of Globe National Bank, Boston, Chicago & Eastern Illinois R. R., Rutland R. R,, and Peterborough R. R.

JOHN MULLIGAN, President C. R. R. R. Company.

JAMES EMERSON,

The Mechanical Engineer.

May 2, 1892.

Emerson's New System of Car Heating.

My attention was called to the subject of car heating early in 1854. On fast day of that year I wrote to the editor of the *Scientific American*, suggesting a plan of placing a small boiler in each car, connecting it with the locomotive boiler and a system of piping for warming the cars and operating the brakes. A written reply was returned in which it was stated that George Stephenson tried to warm trains from his locomotive but failed. Numerous inventions then in hand prevented me from proceeding in that at that time. Still the subject was kept in mind, and the almost yearly announcements of futile attempts to heat cars from the engine were carefully considered, resulting in the belief that the locomotive boiler could not furnish steam for the purpose.

A boiler in the baggage car was suggested. In 1881 Mr. Mulligan, superintendent of the Connecticut River road, offered me a train to experiment with. A small boiler was placed in the baggage car, the steam from which warmed three cars. The capacity of the boiler proved the practicability of taking the necessary quantity of steam from the locomotive boiler, and a change to that was immediately made.

Mr. George A. Houston was sent by the managers of the Atchison, Topeka & Santa Fe R. R. Co., to examine and report upon the merits of the various systems. The substance of his report is here given:

BELOIT, WIS., March 30, 1887.

MR. W. B. STRONG, Pres. A., T. & S. F. R. R. Co., Boston, Mass.

Dear Sir:—Referring to the matter of warming cars, I have examined several systems now in use and being introduced for warming by steam, viz.: The MARTIN, the SEWALL, the EMERSON, and the GOLD. The C. R. R. R. Co. placed a train at my disposal to test the quantity of steam used for heating, this test made with four cars and during twelve hours. From this result, I am satisfied that cars can be warmed during a northern winter with an average of not to exceed three-fourths horse-power of steam per car. This test was made with the EMERSON system, and I recommend the EMERSON system as the best.

Mr. Houston's report was accepted, and train fitted up.

Atchison, Topeka & Santa Fe Railroad Company.
Topeka, Aug. 7, 1888.

James Emerson, Esq.

Dear Sir:—Your letter to Mr Hilton was handed me by him yesterday. In reply to same I will say that I continued to use the cars you fitted up until late in the spring. They gave entire satisfaction, did not have any trouble with them, whatever. I think all cars fitted up should have coils put under the seats, as they can be heated so much quicker and kept more comfortable. I am, yours very respectfully,
Samuel Black,
Conductor A., T. & S. F. R. R.

There are hardly any of the devices for car heating known that have not been tried upon the C. R. road, yet now in the eleventh year my devices are preferred to those of any and all others, and the experienced ubiquitous drummer in cold weather often expresses satisfaction to get into the pleasantly heated and ventilated cars of that road.

I think that I may justly claim to be the first to produce a successful system for heating cars from the locomotive, and the only one that has produced a complete system for ordinary use and emergencies. My plans are now turned over to the United Railways Company. JAMES EMERSON.

CAR HEATING BY STEAM FROM THE LOCO-MOTIVE.

Undoubtedly the method of the future, because the simplest, safest, cheapest, most comfortable, and convenient; but to obtain the advantages named above, common sense must be used in fitting up the cars for the heating.

In no way can a car or room be so pleasantly heated as by having a steam chamber beneath and the floor perforated with minute openings throughout its entire surface, but as that is not conveniently practicable the next best plan is to distribute the heat in small pipes over as much of the floor space as is practicable, and a liberal supply of the pipe should be placed at the ends of the car near the doors.

All who have traveled in cars where the Martin system is in use know how the feet and legs suffer through the intense heat from those large pipes; if there is any possible danger of scalding passengers by steam escaping from broken pipes it rests entirely in the use of large pipes, for with pipe sufficient for the purpose the steam cannot escape fast enough to create heat.

Steam has no heat unless compressed, and a car has too many openings to allow of compression unless through the use of pipes that no competent master mechanic would allow to be used after a moment's consideration.

For thirty years, experience has proved one and one-fourth inch pipe best for car heating by hot water, and the caliber of that pipe is reduced by the use of " double thick " to about the same as that of the ordinary inch pipe, consequently the two-inch pipe carrying four times the steam contained in the one-inch, the danger from scalding is increased four to one, while its heating capacity is but two to one.

The average maximum heat that can be produced by the hot water system throughout a car is 168°, while the average from steam is at least one-third greater, consequently, as the inch pipe is four-fifths the heating capacity of the inch and a fourth, the inch pipe with steam must exceed the one and one-fourth inch pipe for heating with hot water, leaving no excuse whatever for increasing the danger through the unnecessary use of two-inch pipe; besides the space for piping a car is limited, so the smaller the pipe the better for the space.

The various supply pipes in use are at the best but make-shifts and used at serious loss of steam. The proper place for such pipe seems to be through the buffers, then in direct line between the floor timbers of the car as shown, free from all abrupt turns, also out of the way of repairs below, yet leaving it in the most accessible condition for repair that is possible.

The piping of cars piped and coupled as in this system cannot freeze up as is so common with the other systems of piping, for there are no depressions for the condensation to lodge in.

276

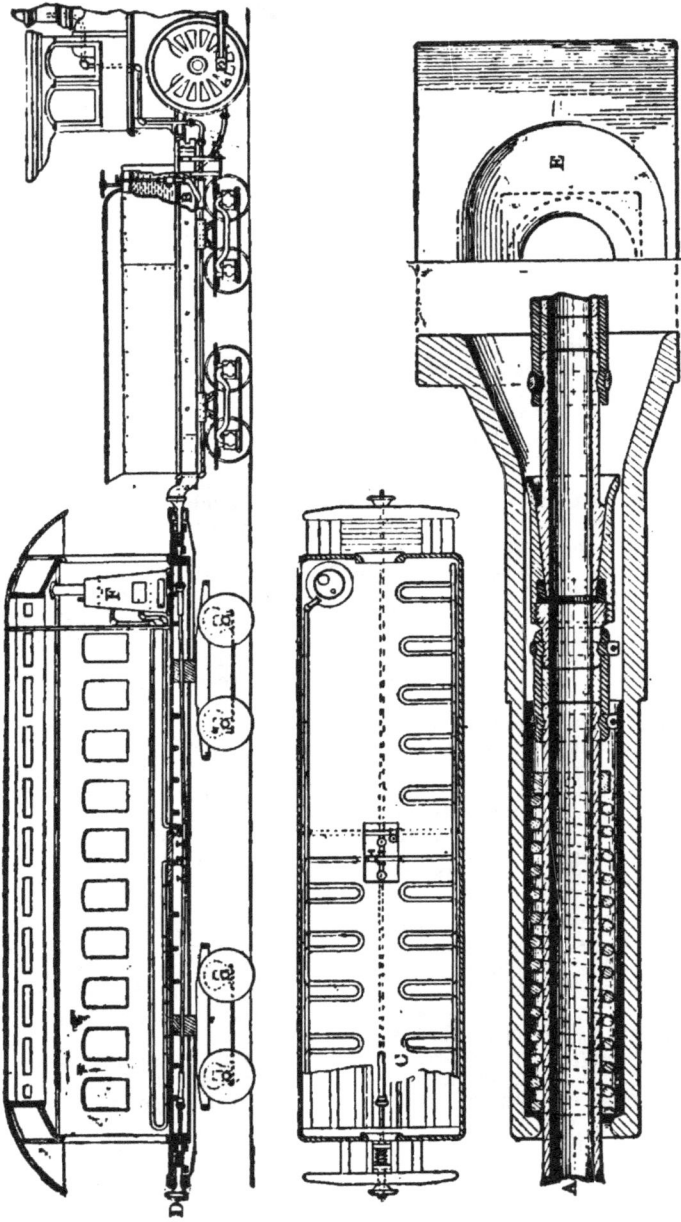

A, supply pipe.   B, injector for filling heater.   C, sliding joint in supply pipe.   D, buffer.   E, end view of buffer.   F, heater.

TO RAILROAD MANAGERS.

Gentlemen, why not save and utilize your Hot Water Heaters ?

The system of piping found best for such after thirty years' experience is far better adapted for rapid and economical car heating than the system of piping employed by Sewall, Martin, or Gold, and at small expense may easily be so arranged that steam from the locomotive may be substituted for the hot water circulation or the hot water circulation restored at will.

The change either way is easily made, without attracting attention, while the train is running.

No trap of any kind is needed, for the temperature is controlled inside of car at any time when in use.

When heating from locomotive the fire is drawn from heater and water from the pipes.

Half an hour before stopping car for night or long detachment from locomotive, open all valves and **blow all condensation from pipes by hot steam from locomotive.**

Leave all valves open until steam again enters pipes for heating.

A few minutes before arriving at a place where a car is to be set off and kept warm, fill pipes with water from the tender, start the fire in heater, and the hot water circulation is at once restored.

This was done at first by taking hot water from the lower part of boiler along through the steam supply pipe A, but that water was so expanded by its intense heat that it required an auxiliary tank above the heater to supply the shrinkage invariably following the filling in that way.

Then an injector placed in a pipe taken from the tender as shown at B was tried and proved perfect, as the steam forcing the water heated it to a desirable temperature for instant use, so that a Pullman or excursion car may use steam or the hot water system at will. As the plan has been in use two years it is past the experimental stage.

All who ride much in cars fitted with hot water heaters know how uncomfortable such cars are in the spring and fall. This is entirely remedied by changing them so as to use steam, so that any sudden change of temperature may be met at once whether of heat or cold, which is impossible with any of the other plans.

An auxiliary heater is necessary on all roads.

A car from a Connecticut River Railroad train is daily taken from Windsor, Vt., to White River Junction by a Central Vt. train. That car stands at the Junction over night without heat, then in the morning it is hitched to a freight train to take early passengers over the road fourteen miles, before steam for heat can be obtained. Sometimes an attempt is made to start a fire in a stove, for the writer early in the winter, during a snow storm, saw the conductor after collecting tickets strike a match and stick it into the stove ; but the match soon went out, and he did the same, leaving us to enjoy the winter weather in full.

Properly fitted cars may be set off with sleeping passengers to wait for morning or to be hitched to freight or branch trains, or as stop over excursion trains, without requiring stationary steam heating facilities or any special arrangements whatever.

SAFETY AUXILIARY CAR HEATER.

An illustration of this heater may be seen upon the opposite page, made with double shells of quarter inch steel plates of such height as to do away with the necessity for separate expansion tank and numerous connecting joints which, accidentally ruptured by derailment, collision, or other causes, allow the burning coals to be thrown around the car.

In this heater there is no coil to be burst by freezing or burned out, as is so commonly the case with the Baker heaters.

As the hot water circulation is only designed to be used in emergencies, such as the absence or disability of the locomotive, stop-over sleeping, or excursion cars set off to be hitched to freight or branch road trains, the heater is so arranged that the fire may be instantly dumped and the burning coals removed from the car as the steam from the locomotive drives the water from the heater and circulating pipes.

Where cars are already fitted for hot water heating, the heaters may be retained, but they are not so convenient, effective, safe, or economical as the one illustrated.

Cars properly piped with this system should never have the ventilators closed, and with very little care the temperature in the car need never vary over two degrees. There should be a thermometer at or near each end of the car.

The usual drip is under the middle of the car, but that may be closed when nearing a station, another opened above the heater, and the train may stand in the station an hour without wetting the floor.

The same process may be followed where a car is to be set off and kept warm by hot water circulation, thus saving the condensation for refilling the heater.

Any car fit to be used can be kept properly warmed and well ventilated by the use of three-fourths of a h. p. of steam in sharp winter weather by the use of this system.

A STRANGE SYSTEM FOR CAR HEATING.

Of all the many wild plans for car heating developed by the demand for a safe substitute for the deadly stove, no other plan can be named so dangerous, extravagant, inconvenient, and uncomfortable as the continuation of the hot water circulation, if the water is to be heated by steam from the locomotive.

In no way can steam be so rapidly condensed as by discharging into water. Then night and day, while at rest, the heat must be kept up by stationary boilers so that at least five times the steam necessary to heat direct is required to heat by such hot water circulation, which is the worst of all systems for meeting sudden changes of temperature, liable at all times in extreme cold weather to freeze up or be unable to keep the cars warm.

Then if a pipe bursts the whole boiler pressure is behind the barrel of boiling water ready, in the old war style of repelling boarders, for boiling the passengers. The danger is so obvious that a jury would hardly excuse a manager on the plea that "he didn't think it was loaded."

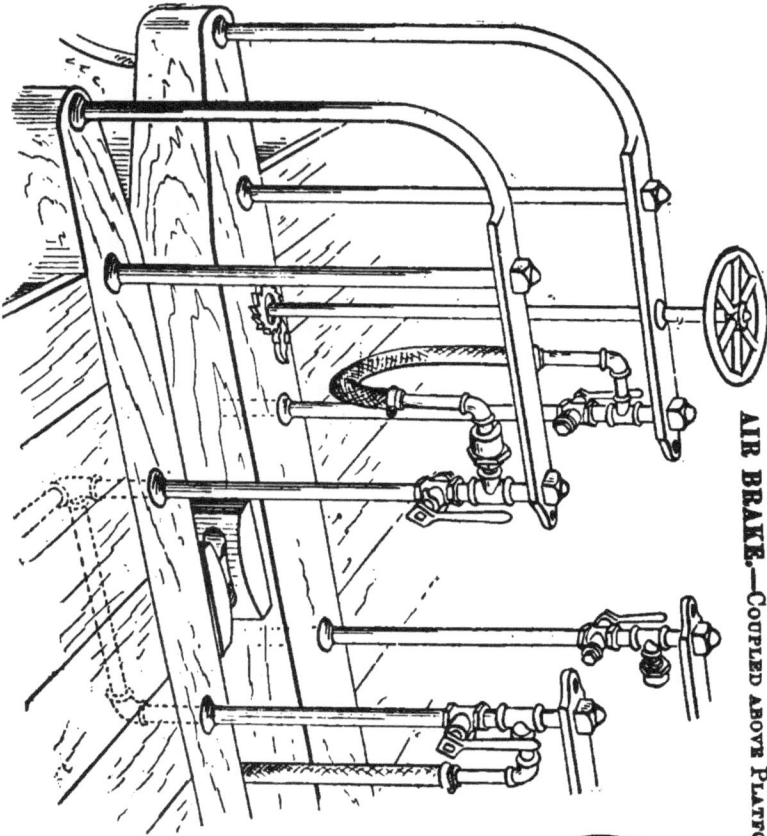

AIR BRAKE.—Coupled above Platform.

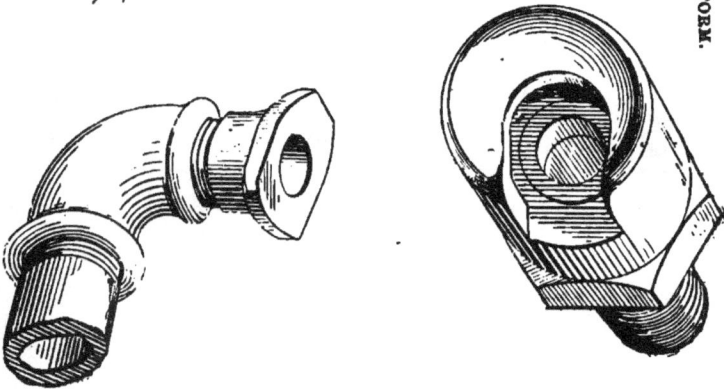

STEAM COUPLINGS FOR RAILROAD CARS.

All couplings similar in style to the illustration cause slow heating, are dangerous, inconvenient, extravagant in the use of steam, expensive, and certainly are not the product of mechanical or experienced reasoning minds.

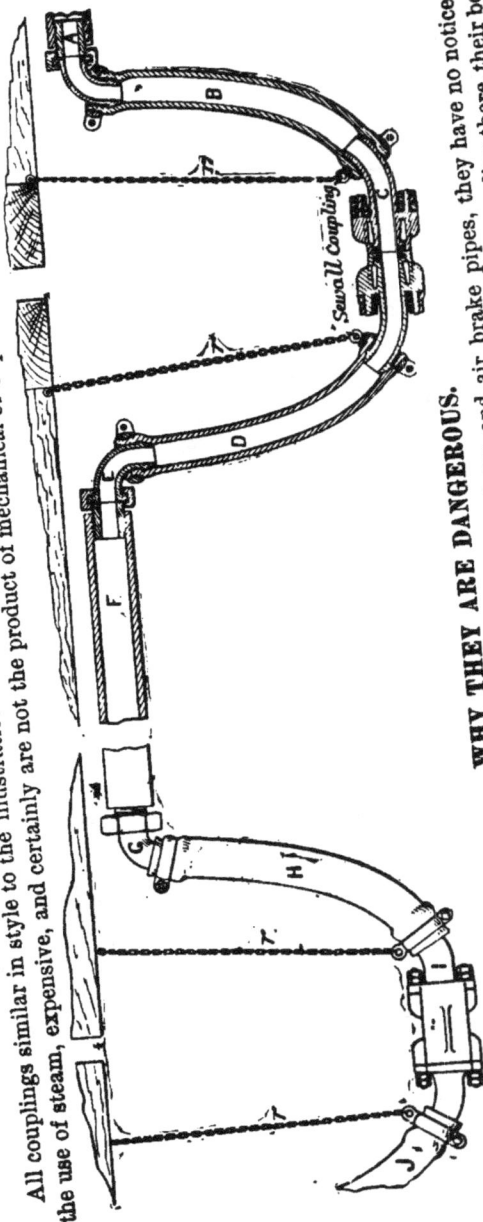

Sewall Coupling

WHY THEY ARE DANGEROUS.

At stations where train hands go under the cars to couple the steam and air brake pipes, they have no notice of the train starting except the striking of the engine bell. Suppose half dozen engines to be standing there, their bells ringing, who is to know which is to start? The danger from such coupling will increase with increase of traffic. The writer has twice been caught in that way and only saved by clinging to the brake rods until the train could be stopped.

UNITED RAILWAYS SAFETY CAR HEATER.

STRONG, DURABLE, NO COILS TO BURN OUT OR BURST BY FREEZING, AS IS SO COMMON WITH THE BAKER HEATER.

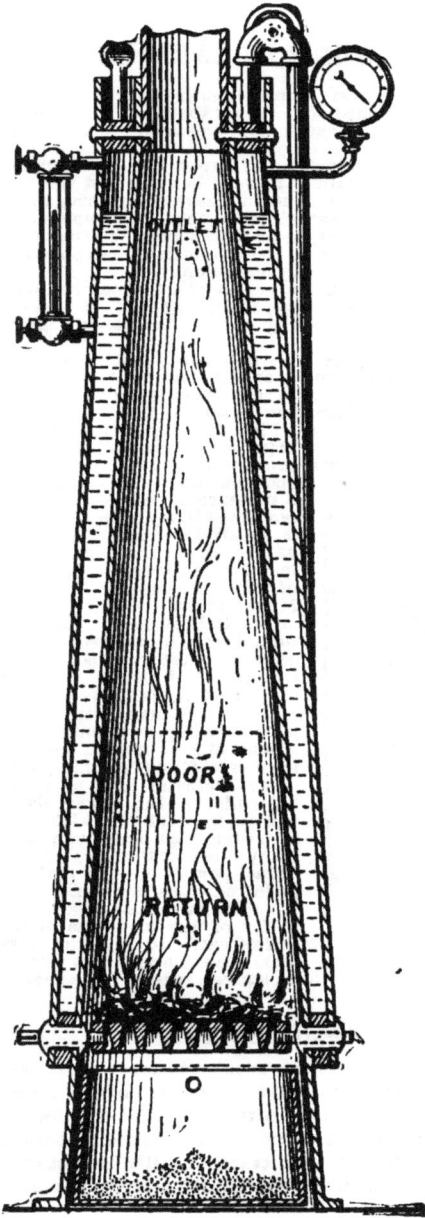

Do railroad managers and the public realize the danger of heating cars by hot water, the water kept hot by steam from the locomotive? The danger is so obvious that a manager in case of general scalding would hardly be excused by a jury, on the plea that "*he didn't think it was loaded.*"

COMPOUND STEAM ENGINES.

That there have been great improvements made in obtaining power from steam during the past third of a century there can be no question, but there fairly may be as to whether such improvements are in any way due to the use of compound engines.

For many years there was quite as strong belief in double turbines, but positive tests proved the fallacy of such beliefs ; and the tests that I have been able to make of compound engines have not shown gain for that method of construction. Twenty years since the test of a compound proved it to be giving far less than expected, and the test of a Westinghouse compound a few months since proved it to be less economical than a simple Buckeye and much less satisfactory in its daily operation.

The marine engine, with its short cylinders, producing rapid rotary motion, may in that way obtain advantage, but it may fairly be questioned whether its increased economy is not owing more to the use of high pressure steam than to triple expansion.

Recently numerous papers have published articles relative to the wonderful efficiency of the Pelton water wheel and that some great English engineer had selected that wheel in preference to that of any other to be used at Niagara Falls in the new plans now under way there,—which may all be true, but as that wheel is simply the old Flutter wheel slightly modified in form, its efficiency can hardly exceed 70 per cent. in useful effect, yet under a head of several hundred feet it may produce an astonishing amount of power to those not acquainted with such matters. So of steam engines working under

Emerson's Drawbar Scale.

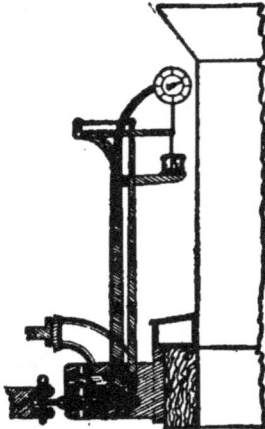

a pressure of 160 pounds instead of the 40 pounds of thirty years since. The locomotive is generally considered an extravagant type of engine, but that idea is founded upon the lack of knowledge of the enormous amount of work the locomotive performs. The White Mountain train, running during the summer on the C. R. R., made up of seven cars all told, going north requires 370 h. p. An ordinary passenger coach upon that road, on straight and level track making local schedule time, requires 50 h. p.

There are many reports of engines that produce a h. p. per hour for each 2, 2½, or 3 pounds of coal burned, but the best result I have ever found was 4.28 pounds per h. p.

The Indicator is of no value whatever in determining the power developed by an engine, in proof of which the tests on the following page are given as but a few of many I have made.

One pound on the dial of the drawbar scale indicates one hundred on the link of the drawbar. This scale is placed in the buffer of the tender as shown, and can be shifted easily to any other tender using the same kind of buffer ; its cost is small and its use might prevent many useless changes, save in the selection of oils and in many other ways.

SUGGESTIONS.

Twenty feet head room for bridges to avoid grade crossings means steep grades and much digging and filling. Why not instead, spread tracks three feet and have eighteen inch walk with rail on side of freight cars?

Prevention is better than cure. The practice of building cars with windows and door outlets that cannot readily be opened for egress will some day result in terrible loss of life; it is the unexpected that astonishes us.

Hartford Engineering Co., Buckeye Twin Engine; Cylinders, 14 inches Diameter, 28-inch Stroke. Simultaneous Trial by Indicator Cards and Power Scale.

WILLIAM A. CHASE,
AGENT HOLYOKE WATER POWER CO.*

Dear Sir:—On Thursday last the trial for power, etc., at the New York Woolen Mills, Connor Brothers, was conducted as follows:—

Ten "sets" were run through the day of eleven hours. The coal was taken from the surface of pile and weighed as used; though not screened, it was much cleaner than the average of the pile.

The weight on Power Scale was taken every fifteen minutes. The boiler pressure was kept at 70 pounds. The driving pulley on engine, 9 feet diameter, with 30-inch double belt, drove 5-feet pulley upon main line. Throwing on and off machinery caused variation of four revolutions of pulley on engine, or from 120 down to 116 per minute.

Mr. Hayes took cards at various times, seemingly with care and skill. The results obtained by the Power Scale, a No. 5, were as follows:—

Divisions of 46 timings gave	1,248 lbs.
Revolutions, 196 per min. cen. force, ...	85
Average net weight for 11 hours, ...	1,163 lbs.
Coal burned in 11 hours, ...	4,955 lbs.
Average power in 11 hours, ...	82.9 H. P.

4955 ÷ 11 = 450.4 ÷ 82.9 = 5.43 lbs. coal per horse-power per hour.

An attempt was made Friday morning to do the work with one cylinder, resulting in a complete failure. Sixty-five horse-power, with 70-pounds boiler pressure, would be all one cylinder could stand steady under. Indicated force, 101.5 horse power.

Respectfully yours,   JAMES EMERSON.

WILLIMANSETT, MASS., Sept. 14, 1884.

E. BLAKE, Needle Works, Chicopee Falls, Mass.
Rated by indicator to use 6.22 horse power.
Maximum possible with every machine in the works running, shown by power scale to be 2.74 horse power, but with the machinery ordinarily in use, 1.24 horse power.
Oct. 21, 1884.

AMOS W. PAGE, Needle Works, Chicopee Falls, Mass.
Rated by indicator to use 7.38 horse power.
Maximum with all machinery in works running, shown by scale to be 3.35 horse power, but with the machinery generally in use 2.49 horse power.   JAMES EMERSON.
Oct. 27, 1884.

TESTING FLUME
HOLYOKE, MASS

F. BOLLES JR SC.

MEASURING PIT, & BRAKE AS APPLIED TO HORIZONTAL SHAFT

HOLYOKE

Hydrodynamic Experiments.

To make the matter generally understood, the following notice is here republished:

HOLYOKE WATER POWER COMPANY,

Holyoke, Mass., April 10, 1879.

NOTICE TO TURBINE BUILDERS AND MANUFACTURERS.

The practice of testing turbines, so common the past ten years, has undoubtedly done much towards bringing the best into use; but there has been one serious defect in the system; that is, the practice has generally been confined to the trial of small wheels, owing to the great expense that would be caused by the tests of large sizes. As it is a matter of vast importance that the best turbine plans should be established beyond chance for doubt, this Company has provided means for a thorough competitive test of the various kinds of turbines that may be offered for trial, and invite Water Power Companies, cities that pump their water supply, and all others interested in the matter, to take part therein. Each builder shall superintend the setting of his wheel—the setting and testing to be done at the expense of the Water Power Company. *Capacity of each wheel to be sufficient to discharge about 5000 cubic feet of water per minute, under 18 feet head. Each wheel will be thoroughly tested from half to whole gate, and, if deemed best, under at least two different heads; also under several feet of back water. At the conclusion of the trial, a full report will be made of the results obtained and of the workmanship, and probable durability of each kind of wheel tried. Turbine builders of this or any other country are invited to furnish wheels, and those proposing to do so should give notice of such intention as soon as possible.

Tests to commence the first day of September next.

HOLYOKE, MASS., June 2, 1879.

---

*Builders who have not got patterns for wheels of so large capacity may enter their largest size, but it is better that all should discharge about the same quantity.

The parties here named have either entered wheels for the trial or have made application for information as to conditions to be observed, &c.

Swain Turbine Co., Lowell, Mass.
Houston Turbine,
    Fales & Jenks, Pawtucket, R. I.
Wolf, Allentown, Pa.
Victor, Stilwell & Bierce M'fg Co.,
    Dayton, Ohio.
Hercules, Holyoke Machine Co.,
    Holyoke, Mass.
Henry Vandewater & Co.,
    Auburn, N. Y.
Willis Reed, Danbury, Ct.
E. Dodge, Spencer, N. Y.
Edward Wemple, Fultonville, N. Y.
Joseph Hough, Mechanics Valley, Pa.

Humphrey Machine Co., Keene, N. H.
S. Sleeper, Mt. Morris, N. Y.
Knowlton & Dolan, Logansport, Ind.
National, Bristol, Conn.
Little Giant, Auburn, N. Y.
T. H. Risdon, Mt. Holly, N. J.
Rodney Hunt Machine Co.,
    Orange, Mass.
W. D. King & Co., Pontiac, Mich.
N. F. Burnham, York, Pa.
Wm. F. Perry, Bridgeton, Maine.
Goldie, McCulloch & Co.,
    Galt, Canada.
Gates Curtis, Ogdensburg, N. Y.

As is often the case in such trials, few of those desirous of taking advantage of the Company's offer were ready at the time named, and, as the notice did not state any time for closing, builders have been tardy in sending their wheels. The ordinary work of the testing flume has been continued during the time, so that the wheels reported are only about one-half the number tested; and any one acquainted with the matter will see that there has been no unnecessary delay in making the report.

The experiments were announced as competitive, meaning, in general utility, economy in the use of water, convenience, cost and durability.

Large turbines were called for, that their discharge might be greater than could be measured in the testing flume of any turbine builder, but this was not insisted upon, as, to have done so, would have limited the competition to a few old builders with full sets of patterns, whose wheels have often been tested and reported. Experience has not yet produced any fact that even hints that any particular size of turbine, small or large, can be made to produce higher results than any other size of the same make. Consequently, builders were allowed to send wheels the most convenient in size for themselves, and it is not known that any one of experience furnished a wheel with the expectation that it would give the highest possible results, but that its general merits should commend it to the public, and that the value of any peculiarity in its construction should be determined.

Competitive turbine tests, in the common meaning of the term, have been useful in the past, as they have enabled those interested in such matters to decide upon the most desirable plans. At the present time, however, such tests can have no public value, because each turbine tested only represents itself in efficiency. Another of the same size and make might and probably would give quite different results, so that should each competitor have a second, third or a tenth wheel tried, his standing would be likely to change with each wheel tested. The Fourneyron, Boyden, Birkinbine and Centennial tests all prove this fact, as they also prove that the builders who have furnished the turbines that have given the highest efficiency reported, have only had a brief popularity, as manufacturers have found other turbines more desirable for business; and it will be evident from the results obtained in these experiments, that builders have taken this fact into consideration and have generally tried to produce turbines economical at any stage of gate opening, rather than to gain the highest possible efficiency at whole gate, where, in practical use, it is rarely used. And in this there has been a decided gain, as there has also in an increased capacity for a given diameter of wheel, noticeable in the Rechard as well as the Hercules and New American.

In considering the comparative merits of the wheels here reported, it should be understood that previous to 1876 turbines of any make for a given diameter generally gave about the same power. There were builders who believed in some mysterious power in *leverage*, who constructed wheels with extended diameter and proportionally small discharge, but these were exceptional; the rule held good, and it will be necessary to take this fact into consideration to realize the improvements in turbines during the past four or five years.

Turbine builders were requested to furnish draft tubes of different sizes with their wheels, that the efficiency of such tubes might be determined; and that the loss in transmission through belts and gears might also be ascertained, several well known gear-making firms were requested to furnish gears for trial.

The experiments have been conducted upon the supposition that their purpose was to ascertain the real utility of the various devices tested under the every-day ordinary conditions to which such plans are subjected in practical use, rather than possibilities in exceptional cases under the most favorable circumstances; and features of known interest developed are recorded in connection with their development. It was expected that the experiments would require much time, and as they were made in the public testing flume, it was necessary that each should be conducted as expeditiously as accuracy would permit; consequently, James Emerson, from his intimate familiarity with such matters and experience in handling wheels, was employed to see that each turbine was set in a manner satisfactory to its builder, and to have a general supervision over the work.

Samuel Webber, Civil Engineer of Manchester, N. H., known in connection with the Centennial tests, was selected to assist in making the experiments, and reports herewith.

Theo. G. Ellis, Civil Engineer of Hartford, Conn., well known through his published works and long employment by the government in river and harbor improvements, was selected by the turbine builders to see that the experiments were skillfully and fairly conducted, whose report is appended.

For the information of the uninitiated, it is proper to state that a turbine, under a given head, does its best at a certain speed. To find this point it is necessary, in testing, to begin with a light weight, run a minute or more, then add weight and repeat until the best point is found; and the test that fixes that point is the speed at which the wheel should be geared to work, and the efficiency at that point is the efficiency of the wheel. The average efficiency from a part to whole gate means when the wheel is running at that speed at any stage of gate opening, and the efficiency at other speeds is to be considered only so far as it shows the loss that will occur through gearing above or below the proper point.

The tests are supposed to be correct and complete in each case as given, but for the information of students or others wishing to work out the data for themselves, the following is given in explanation of the statement at the head of each test: multiply revolutions by 10, 20, &c. It must be understood that during each test the scale beam is attached to the brake at a point which, if revolving, would describe a circle of 10, 15 or 20 feet in circumference. Consequently, the revolutions must be multiplied by the number given, as for example: Of the first New American wheel tested— rev. per minute, 207.5; weight, 675. $207.5 \times 15 = 3112.5 \times 675 = 2100937.5 \div 33000 = 63.66$ h. p.

To make this report really useful, it is issued in size convenient for the pocket.

WM. A. CHASE, Agent.

ENGINEERS' REPORTS.

REPORT OF THEO. G. ELLIS.

HARTFORD, CONN., *September* 13, 1880.

WILLIAM A. CHASE, ESQ.,
· *Agent of the Holyoke Water Power Co.*

SIR: Having been requested to take part in the interesting experiments upon turbines made by your Company in October and November, 1879, at the Holyoke testing flume, I did so with great reluctance as, owing to many professional engagements, I could not give so much time to the subject as its importance seemed to warrant, and could not possibly be at Holyoke at all times during the experiments. I finally, however, agreed to be present at part, at least, of the tests in behalf of the turbine builders, to see that the experiments were fairly conducted as far as lay in my power, and to make such observations as I thought best.

It was understood that the mechanical work of setting the wheels and making the experiments was to be superintended by James Emerson, whose previous experience in the testing of turbines at the same locality eminently fitted him for the task. The flume and apparatus used was mostly, if not entirely, designed and constructed by him, and he was familiar with all its details and capabilities. Whatever may have been his previous published views, it is believed that in the present tests all the turbines presented for trial have received the same careful attention and trial. In some cases the record does not appear to show as full and complete a trial as in others, but there was always some good reason, irrespective of any prejudices for or against that particular wheel, for the apparent limitation of the trial.

Mr. Samuel Webber, civil engineer, of Manchester, N. H., who had superintended the Centennial tests of turbines, was present during the whole of the experiments, and I availed myself of an association with him in overlooking the experiments, so that one of us should be present at every trial, and thus always have a disinterested party to record the readings of the dynamometer and gauges, and the time of the experiment, to serve as a check upon the readings recorded by Mr. Emerson's assistant and taken by him. Mr. Webber was assisted most of the time by Mr. Stockwell Bettes, civil engineer, of Springfield, Mass., who read the gauges and otherwise checked the readings taken and recorded by Mr. Emerson.

All of Mr. Emerson's readings, and such of Mr. Webber's as he desired, were recorded in a book kept for the purpose. These records were kept and all the computations therefrom were made by Miss Charla Adams, who for a long time has been familiar with such experiments and computations as an assistant of Mr. Emerson, and who, I am satisfied from a personal examination of her work, has performed the duty in a careful, accurate and thorough manner.

Experiments upon the following wheels were all witnessed by Mr. Webber, and part of them by myself:

October 10, 1879, Tyler Wheel.
      " 11, " Thompson Wheel.
      " 14, " New American Wheel.
      " 15, " "Humming Bird" Wheel.
      " 16, " Success Wheel.
      " 17, " Two Tait Wheels.
      " 18, " Repeated Test of Tait First Wheel
                         (buckets chipped).
      " 18, " Sherwood Wheel.
      " 21, " Nonesuch Wheel.
      " 22, " Curtis Wheel.
      " 28, " Pair of Curtis Wheels set horizontally.
November 11, " Hercules Wheel.
      " 12, " Hercules Wheel.
      " 13, " Houston Wheel.
      " 14, " Wetmore Wheel.
      " 15, " Monarch Wheel.

The computed volumes of discharge, and the percentage of efficiency of the foregoing wheels, as shown in your Report, the proof of which has been submitted to me, have been carefully examined with a view to determine the relative value of the wheels named, and their respective performances under the different conditions and amounts of water with which they were tested.

In the testing of turbines, it has been the practice to first determine the velocity at which the wheel will give its greatest effect when using all the water that will run through it with the gates or entrance apertures open to their full extent, or at "full gate;" then to diminish the quantity of water to three-quarters and one-half, as nearly as practicable, and to estimate the power of the wheel when running at the same velocity. The experiments at Holyoke were conducted practically in this manner. The best velocity was found for "full gate," and then the amount of water was diminished gradually in successive experiments to the neighborhood of half the quantity, with the wheel running as nearly as might be at the same speed.

This is perhaps the best way to make such tests, everything considered. But it does not in all cases give the exact relative value of the wheels. Some turbines might give a better result at a different velocity when using a less amount of water, and make their average, say, from half to full gate better than by the former method. The difficulty, however, of getting at the exact velocity at which any turbine would give its best results when using different quantities of water, is too great to warrant such determinations in a series of comparative tests such as were made at Holyoke. The same method must be established for all, and the customary one appears to be the fairest, as no other would probably be agreed to by all the turbine builders. In the practical use of turbines for power, it is rarely the case that a wheel is put in of the exact power required. A margin must be left for an excess of power to meet emergencies, and allowance must be made for an increase of machinery, so that a larger wheel is ordinarily purchased than would just suffice to meet present requirements. For this reason, it is not the wheel which gives the highest percentage of efficiency at "full gate" that is really the best wheel. There can be no point fixed at which any wheels should be compared, but it is thought that perhaps "three-quarters gate" is about the average point at which wheels are used, and their comparative efficiency at from one-half to their full power sufficiently represents their real value. It would probably be a better comparative test of wheels to get their best velocity at "three-quarters gate" and run them with the same velocity for greater and less quantities. This would give the real value of the wheel better than the present practice, but it would probably not be generally agreed to. In using the terms "full gate," "half gate," "three-quarters gate," etc., the relative quantity of water is meant. The opening of the wheel gates themselves is not considered. Their construction is often such that opening or closing them a certain proportion does not affect the quantity of water in the same manner. It not unfrequently happens that a slight closing of the gate increases the quantity of water passing through them, so that the gates themselves are deceptive and are no criterion of the amount of water used. The gate opening is sometimes used to deceive the uninitiated in the circulars of unscrupulous turbine builders, calling "half gate" perhaps two-thirds the whole quantity of water, so as to give a higher percentage of efficiency, but the only true standard of comparison is the actual amount of water measured as it leaves the wheel.

The experiments upon the before-named wheels have been carefully plotted with the amounts of water and the percentage of efficiency as co-ordinates, and a mean curve drawn through the points for each wheel. These curves have been all reduced to a uniform horizontal scale for the purpose of comparison, so as to obtain their relative efficiency at all proportions of the whole amount of water from half to full gate. The curves of the eight wheels giving the highest efficiency are shown on the annexed diagram. The horizontal scale shows the parts of the whole quantity of water from half to full gate, and the vertical scale shows the percentage of efficiency at all points corresponding to the amount of water indicated.

The average percentage of efficiency for these eight wheels has been computed for the amount of water from half to three-quarters gate, from half to full gate, and from three-quarters to full gate, as shown in the following table:

TABLE SHOWING AVERAGE PERCENTAGE AT PART GATE.

NAME.	$\frac{1}{2}$ to $\frac{3}{4}$. Per cent.	$\frac{1}{2}$ to full. Per cent.	$\frac{3}{4}$ to full. Per cent.
Hercules, ...	.737	.805	·771
New American, . . .	.732	.795	·763
Success, ...	.708	.786	.747
Tyler, ...	.665	.766	.715
Tait, ...	.680	.744	.712
Thompson, ...	.696	.721	.709
Nonesuch, ...	.619	.712	.666
Houston, ...	.397	.717	.557

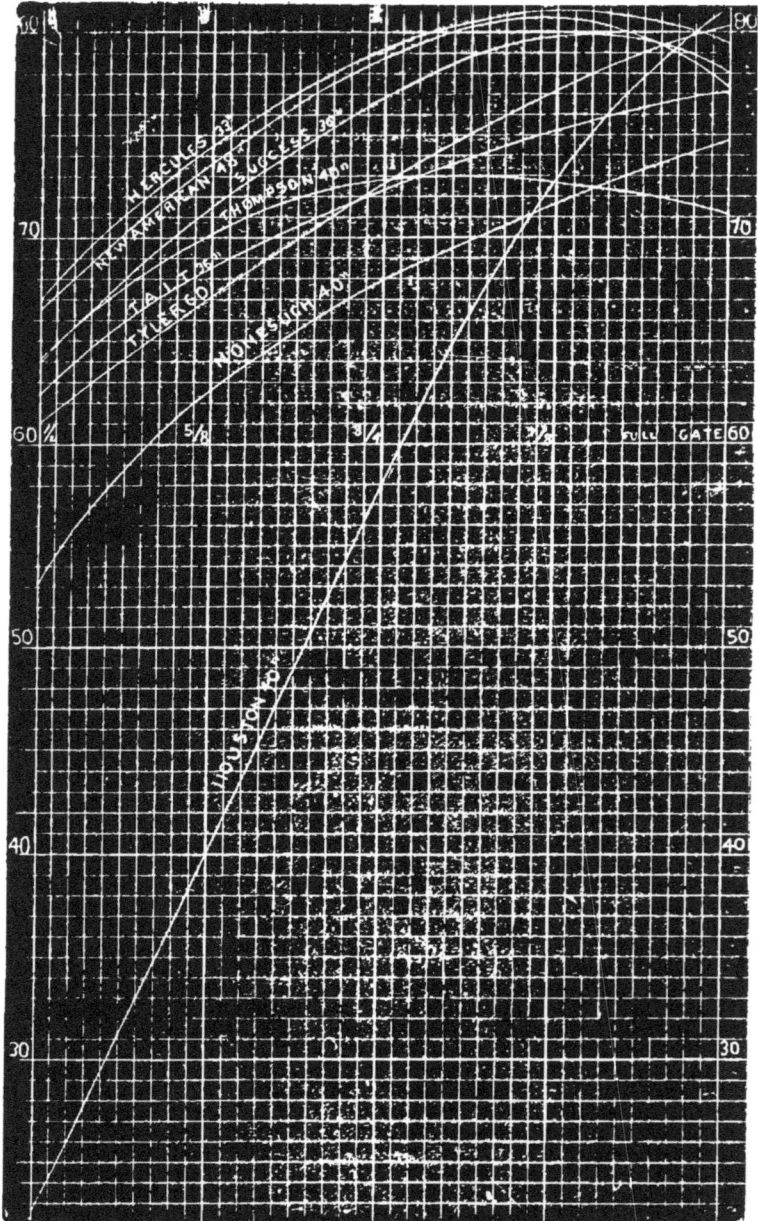

By examining the diagram and the foregoing table, the peculiarities of the several wheels will be readily seen. It will be observed that the Houston turbine, which has the highest percentage of effect at full gate, is really the least efficient at from half to three-quarters, and from half to full gate, of all those shown on the diagram, and is only superior to the Nonesuch at from three-quarters to full gate, and that by a very trifling amount; so that the wheel which apparently has the highest percentage is really the least desirable for actual use. The Thompson turbine, which has the lowest percentage of those shown, at full gate, rises to the sixth place at from one-half to full gate, and to the fourth place at from one-half to three-quarters gate. The Tyler turbine, which has the second highest percentage at full gate, falls to the sixth place at from one-half to three-quarters gate. The Hercules turbine, which stands third only at full gate, takes the first rank at from half to full gate, or any of its subdivisions. The New American turbine, which stands only fifth in the percentage at full gate, is second only to the Hercules at from one-half to full gate or either of its subdivisions, and, indeed, differs from the Hercules very slightly in its useful effect through the whole range shown.

Taking the average useful effect of the wheels shown from one-half to full gate as a measure of their efficiency, their relative value is in the order shown in the table.

Among the turbines tested at about the time of the experiments upon the wheels before named, were two very remarkable ones on account of their very different qualities and performance. These were the Rechard, a statement of which is included in your Report, and the Victor, which was used in the gear experiments, likewise attached to your Report. The first-mentioned has a percentage of useful effect of only 69 at full gate, while the latter has a percentage of 92. At thirteen-sixteenths of full gate, the percentage of efficiency becomes reversed, and below that the Rechard is by far the most effective turbine. From one-half to full gate the efficiency of the Rechard is second only to the Hercules, while for the same range the Victor would come fourth in the list.

Neither Mr. Webber nor myself witnessed the experiments upon these wheels, but they are mentioned to show that a high percentage at full gate is often deceptive and does not always indicate the best wheel for practical use.

In the foregoing Report, with the exception of the last two wheels, only such wheels are considered as were tested in the presence of Mr.

Webber or myself. The list appears to embrace all the really good wheels presented, and gives their efficiency as we saw it. Some of these wheels show a little higher percentage than I have given in some of the other experiments in your Report, particularly the New American, but I have thought best to confine myself to those experiments that were witnessed and verified by the attending engineers.

With the sincere hope that comparative and competitive tests of turbines will be continued, and that thereby the public and users of power will know more fully the qualities of the wheels they purchase, and the useful effect they are likely to derive from them,

I remain, very respectfully yours,

THEO. G. ELLIS, CIVIL ENGINEER.

REPORT OF SAM'L WEBBER.

WM. A. CHASE, Esq.,
Treasurer Holyoke Water Power Co.

DEAR SIR: I was requested by you in October, 1879, to come to Holyoke and be present at a series of competitive tests of turbines, and to see that the measurements were correctly made, and the apparatus in perfect order. I was, accordingly, present the greater part of the time from October 9th to November 15th, and witnessed the tests of the following wheels, viz. :

Oct. 9th and 10th,	The " Tyler" Wheel.		
"	11th,	" " Thompson" Wheel.	
"	14th,	" " New American," being a wheel of the Swain type of bucket, with the case and gates formerly used for the "American Wheel."	
October	15th,	The " Humming Bird " Wheel.	
"	16th,	" " Success " Wheel.	
"	17th,	" " Tait Centennial," 2 wheels.	
"	18th,	" " " " 1st wheel repeated.	
"	"	" " Sherwood " Wheel.	
"	21st,	" " Nonesuch" Wheel, from Clark & Chapman.	
"	22nd,	" " Gates Curtis " Wheel.	
"	27th,	" " pair of wheels on draft tube.	
Nov. 11th and 12th,	" " Hercules " Wheel.		
"	13th,	" " Houston " Wheel.	
"	14th,	" " Wetmore " Wheel.	
"	15th,	" " Monarch " Wheel.	

During all these tests, I verified the measurements of the weir, the revolutions of the wheel, the head of water, and the weight on the steelyard, and in these measurements I was assisted by Mr. Stockwell Bettes; and from the data so obtained I have made up complete calculations of the results.

I have examined the proof sheets sent me by Mr. James Emerson, of his report and calculations of these tests, and have no hesitation in accepting them, as in very many cases we agree exactly, while in no case is there a variation of over 1 per cent., and these differences are mainly due to slight differences in the weir readings, as taken by Mr. Emerson and Mr. Bettes.

I was also present during a portion of the gear and belt tests in April, 1880, and can certify to the correctness of Mr. Emerson's report of those tests, so far as the results then obtained are concerned.

I cannot, however, consider these tests as conclusive, from the fact that the gears were entirely new, and that there was no accurate method of regulating the proper depth to which the gears should be put in contact—a slight change in such depth having shown a great difference in the net power attained.

Neither was there any method for regulating or ascertaining the the tension of the belts.

Nor should I be satisfied to accept the result obtained from the 15-inch Victor wheel as conclusive of the merits of wheels of that make, as from various tests the very small wheels of almost all patterns usually give a higher percentage than the larger ones.

Yours very truly,

SAM'L WEBBER, C. E.

REPORT OF JAMES EMERSON.

WILLIAM A. CHASE,
Agent Water Power Co., Holyoke, Mass.

SIR: Having, in connection with the engineers named, completed the series of turbine and dynamic experiments announced by your Company, the results obtained by myself, with accompanying remarks, are here submitted for your consideration.

In presenting this report, it is a pleasure to recall the interest taken in the experiments, from the beginning to their close, by engineers and experts in such matters. There was hardly a trial of any kind without the presence of such. Mr. Bettes assisted almost invariably; James M. Sickman, C. E. of Holyoke, often examined the arrangements; Prof. Norton, of the Sheffield Scientific School of New Haven, Ct., with members of his class, spent a day in witnessing the tests, and, later, six graduates of his class assisted in testing the 15-inch Victor. Prof. Whittaker, of the Massachusetts Institute of Technology, with some sixteen members of his class, not only witnessed the experiments, but had charge of the apparatus for several hours, and tested the 33-inch Hercules for practice. The Principal of the Holyoke High School, with a large delegation of scholars, both male and female, spent some hours in witnessing the tests, and seemingly with much pleasure. There were also witnesses from very distant places, and some that one would hardly expect would feel an interest in such matters, but they seemed to do so.

JAMES EMERSON.

WILLIMANSETT, MASS., Aug. 1, 1880.

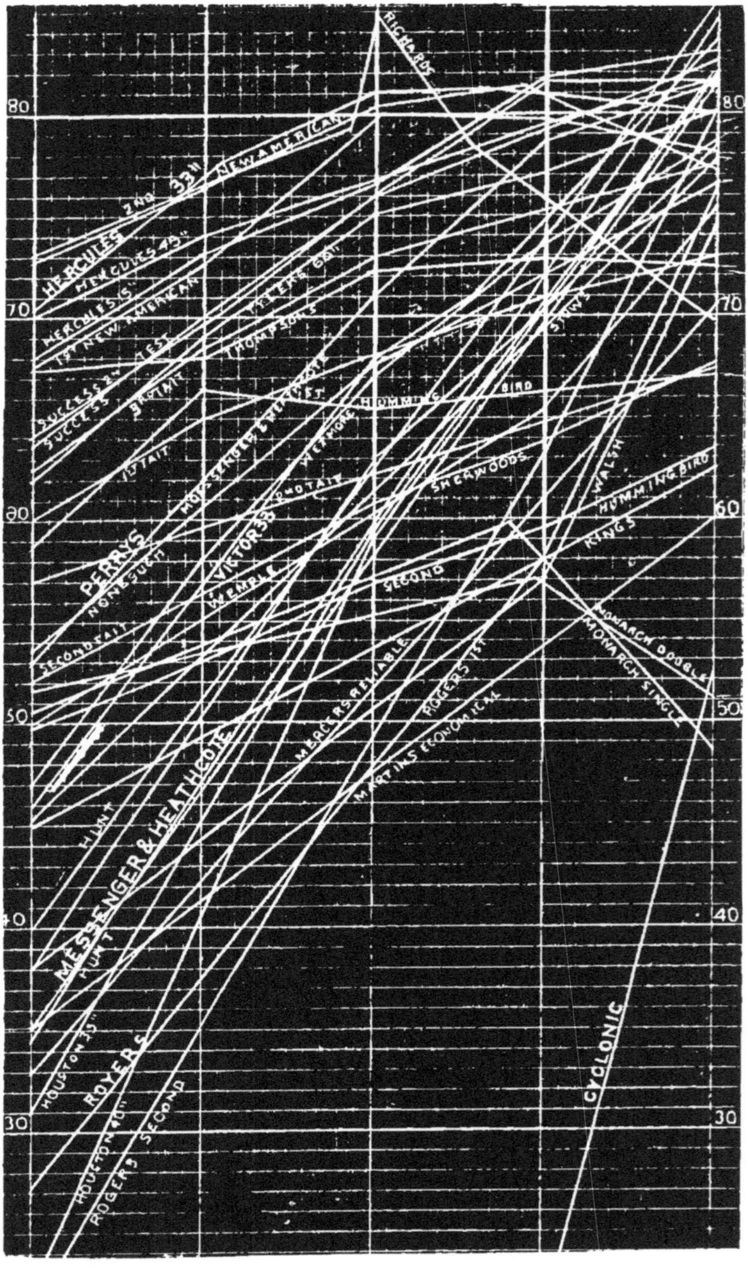
80
70
60
50
40
30
RICHARDS
NEW AMERICAN
HERCULES 2ND 33"
HERCULES
HERCULES 45"
HERCULES 5'
1ST NEW AMERICAN
THOMPSON'S
SUCCESS
1ST TAIT
HUMMING BIRD
PERRYS
NONESUCH
VICTOR 35
WEMPLE
SECOND TAIT
SHERWOODS
WALSH
HUMMING BIRD
KINGS
SECOND
MONARCH DOUBLE
MONARCH SINGLE
ROGERS 1ST
MARTINS ECONOMICAL
HUNT
MESSENGER & HEATHCOTE
HUNT
HOUSTON 35"
ROYERS
HOUSTON 40"
ROGERS SECOND
CYCLONIC

Wemple Wheel.

Sent by Wm. Wemple's Sons, Fultonville, N. Y.

18-inch wheel. Central and downward discharge. Inside register gate.

Data below for one minute. Multiply revolutions by 10. April 17, 1879.

Gate Opened	Head	Weight	Rev per minute	Horse Power	Cubic Feet	Per Cent
Whole Gate.	18.30	300	000	000		
" "	18.24	150	335.3	15.24	623.06	.7265
" "	18.36	160	327.6	15.88	627.43	.7298
" "	18.40	170	319	16.43	640.98	.7375
" "	18.26	180	303.5	16.55	645.48	.7434
". "	18.40	190	296	17.04	648.48	.7561
" ".	18.35	200	280	16.97	651.49	.7516
" "	18.24	210	259.5	16.51	660.44	.7257
" "	18.29	185	293	16.42	648.48	.7329
" "	18.17	195	282	16.66	651.49	.7451
Part Gate.	18.23	150	326	14.81	624.55	.6887
" "	18.20	175	296	15.69	626.04	.7290
" "	18.20	170	294.6	15.17	642.48	.6868
". "	18.21	165	282 5	14.12	599.42	.6850
" ".	18.22	160	290	14.06	596.42	.6851
" "	18.24	140	278	11.79	542.63	.6306
". "	18.24	125	293.5	11.11	525.45	.6137
" "	18.34	100	259	7.85	437.65	.5178
". "	18.34	100	291.5	8.83	447.21	.5700
" "	18.43	75	227.3	5.65	330.80	.4906
" "	18.39	75	300.5	6.83	370.86	.5302
" "	18.41	75	301	6.84	373.48	.5842
" "	18.42	80	291.5	7.07	382 69	.5309
" "	18.48	55	288.5	4.81	299.28	.4722
" "	18.48	50	303	4.59	298.03	.4412

Mr. Wemple not being able to get up a wheel of the size required in time, allowed this to be reported as a representative of the kind.

Tyler Wheel.

30-inch wheel, sent by John Tyler, Claremont, N. H.

This wheel was tested a few days before the time named for the general test, that it might be used.

This wheel was furnished for the purpose of enabling those seeking for such information to compare its power of transmission with those of the same size made by others, as the most of the popular builders have had 30-inch wheels tested. One fact, however, must be taken into consideration in making such comparisons, namely, that while the increase in the sizes of one builder is, say, 6, 12, 18, 24 and 30 h. p., the increase in another make will be 6, 9, 18, 40, 48, 75, &c.; but, in the aggregate, the total power of all the sizes of each builder amount to about the same. The Tyler flume wheel represents very fairly the average capacity of the most popular turbines known previous to 1876, excepting, however, the Boyden, which, for its diameter, is far less in capacity than any of the others.

This particular wheel was made from the same patterns as the one tried at the Centennial tests, and several times at the Holyoke flume. Special pains was taken that it should be an exact duplicate of that one. The curb was the same as the Centennial, yet, as will be seen by those who have the means to make the comparison, the discharge of this wheel was one-sixth greater than the first. Mr. Tyler was so unwilling to accept the results, that he had the wheel taken out, reset, and retested on three successive days, each trial giving the same results.

Data below for one minute. Multiply revolutions by 15. Aug. 1, 1879.

Gate Opened							Head	Weight	Rev per minute	Horse Power	Cubic Feet	Per Cent.
Whole Gate.	.	.	.	.	.	.	18.30	375	218	37.15	1373.63	.7831
"	"	.	.	.	.	.	18.28	385	213.7	37.42	1373.63	.7896
"	"	.	.	.	.	.	18.27	400	209.6	38.10	1373.63	.8045
"	"	.	.	.	.	.	18.27	425	201.6	38.96	1386.77	.8148
"	"	.	.	.	.	.	18.27	440	198.5	39.70	1400.00	.8225
"	"	.	.	.	.	.	18.26	450	194	39.68	1421.11	.8103
"	"	.	.	.	.	.	18.25	475	180	38.86	1445.03	.7809
"	"	.	.	.	.	.	18.28	440	194.5	38.90	1418.46	.7950

Moessinger & Heathecote.

Sent by Moessinger & Heathecote, Glenrock, Pa.

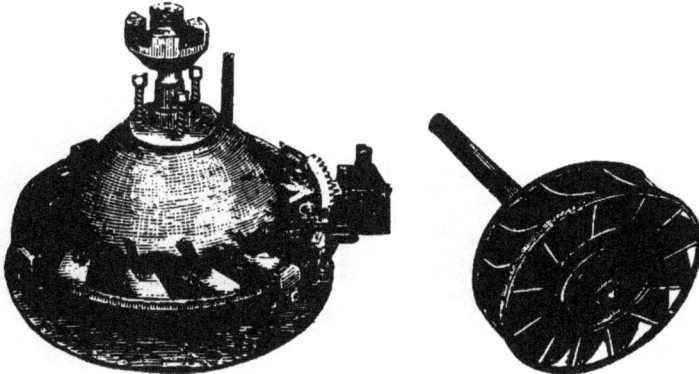

20-inch wheel.

This turbine was a Jonval, with register gate, as represented above.

Data below for one minute. Multiply revolutions by 10. Sept. 3 and 4, 1879.

Gate Opened						Head	Weight	Rev per minute	Horse Power	Cubic Feet	Per Cent
Whole Gate.	.	.	.	.	.	18.40	100	320.5	10.47	511.13	.5894
"	"	.	.	.	.	18.40	110	325	11.16	513.84	.6250
"	"	.	.	.	.	18.40	120	330	12.00	517.92	.6668
"	"	.	.	.	.	18.39	130	323.5	12.74	524.75	.6988
"	"	.	.	.	.	18.39	140	310.5	13.17	531.59	.7133
"	"	.	.	.	.	18.38	150	300	13.63	535.71	.7329
"	"	.	.	.	.	18.38	160	281.6	13.65	541.23	.7265
"	"	.	.	.	.	18.39	170	254.5	13.11	543.99	.6938
"	"	.	.	.	.	18.38	180	230	12.54	545.37	.6623

The wheel bound upon the step during the above trial; and it was taken out of the flume, overhauled, then re-tested, giving the results recorded below.

Whole Gate.						Head	Weight	Rev per minute	Horse Power	Cubic Feet	Per Cent
Whole Gate.	.	.	.	.	.	18.55	150	316.6	14.39	539.08	.7618
"	"	.	.	.	.	18.56	160	307.5	14.90	546.00	.7784
"	"	.	.	.	.	18.53	170	300.5	15.48	551.51	.8016
"	"	.	.	.	.	18.53	180	287.5	15.68	551.51	.8123
"	"	.	.	.	.	18.53	190	270	15.54	555.68	.7990
Part Gate.	.	.	.	.	.	18.71	50	295	4.47	381.41	.3316
"	"	.	.	.	.	18.65	75	296	6.72	432.22	.4414
"	"	.	.	.	.	18.59	100	292	8.85	478.14	.5272
"	"	.	.	.	.	18.55	125	292.5	11.08	517.17	.6114

Victor Turbine.

Stilwell & Bierce Manufacturing Co., Dayton, Ohio.

This wheel is of recent origin; discharges the water used outward, downward and centrally; has a register gate that works easily and opens in full with half a turn of gate rod. It is so designed that its buckets may be made of bronze, if desired. Its discharge in proportion to its diameter is only equaled by that of the Hercules. Price of this 35-inch wheel, $650; weight. 4500 pounds.

Data below for one minute. Multiply revolutions by 20. Sept. 5, 1879.

Gate Opened							Head	Weight	Rev per minute	Horse Power	Cubic Feet	Per Cent
Whole Gate.	.	.	.	.	.	.	16.95	2650	000	000.		.000
"	"	.	.	.	.	.	17.18	1500	147.5	134.09	4994.79	.8289
"	"	.	.	.	.	.	17.10	1550	141.5	132.98	4999.22	.8232
"	"	.	.	.	.	.	17.11	1600	137.5	133.33	5012.56	.8230
"	"	.	.	.	.	.	17.09	1650	131.5	131.50	5025.86	.8121
"	"	.	.	.	.	.	17.07	1700	126 6	130.43	5030.31	.8048
"	"	.	.	.	.	.	17.11	1450	150	131.81	4990.36	.8172
"	"	.	.	.	.	.	17.11	1400	156	132.36	4972.64	.8236
"	"	.	.	.	.	.	17.10	1475	150	134.09	4981.50	.8334
"	"	.	.	.	.	.	17.09	1525	142.3	131.52	4985.93	.8172
Part Gate.	.	.	.	.	.	.	17.14	1475	147.3	131.76	4941.67	.8237
"	"	.	.	.	.	.	17.23	1350	152	124.36	4739.63	.8063
"	"	.	.	.	.	.	17.55	1150	133 3	92.90	3920.79	.7131
"	"	.	.	.	.	.	17.56	1100	136 5	91.00	3892.00	.7050
"	"	.	.	.	.	.	17.59	1050	141	89.72	3855.07	.7006
"	"	.	.	.	.	.	17.58	1000	145	87.87	3777.48	.6989
"	"	.	.	.	.	.	17.66	900	149.5	81.54	3619.75	.6754
"	"	.	.	.	.	.	18.00	575	144.2	50.25	2726.05	.5421
"	"	.	.	.	.	.	18.07	500	149.3	45.24	2616.35	.5066

Walsh Double Turbine.

Sent by B. E. Sanford, Sheboygan Falls, Wisconsin.

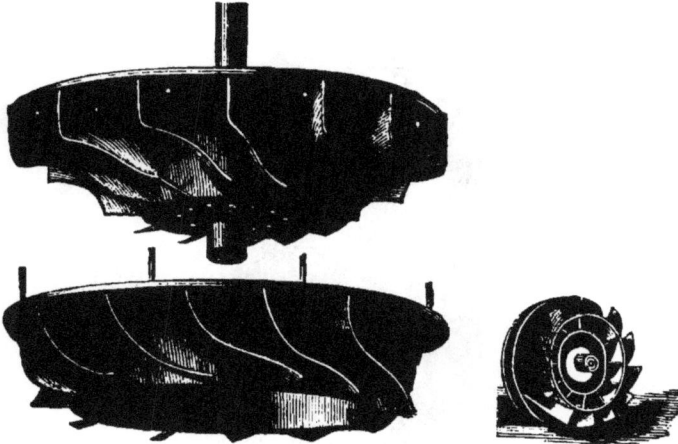

48-inch wheel.

The two wheels represented above were placed together forming one with divided discharge, as represented in the small wheel at the right. The curb had cylinder gate without flange.

Data below for one minute. Multiply revolutions by 20. Sept. 8, 1879.

Gate Opened						Head	Weight	Rev per minute	Horse Power	Cubic Feet	Per Cent
Whole Gate.	.	.	.	.	.	17.29	2525	000	000		000
"	"	.	.	.	.	17.44	1250	129	97.72	4110.90	.7216
"	"	.	.	.	.	17.42	1350	123.5	101.04	4157.13	.7386
"	"	.	.	.	.	17.38	1500	114	103.63	4224.63	.7473
"	"	.	.	.	.	17.37	1600	106.3	103.07	4228.88	.7429
"	"	.	.	.	.	17.38	1650	101	101.00	4228.88	.7292
"	"	.	.	.	.	17.39	1550	109.5	102.86	4219.42	.7423
"	"	.	.	.	.	17.39	1575	107	102.13	4224.63	.7359
"	"	.	.	.	.	17.38	1525	111	102.59	4224.63	.7397
"	"	.	.	.	.	17.38	1475	114.5	102.35	4211.96	.7402
Part Gate.	.	.	.	.	.	17.45	1350	106	86.72	4027.25	.6534
"	"	.	.	.	.	17.45	1300	110.5	87.06	4023.08	.6565
"	"	.	.	.	.	17.45	1275	113	87.31	4019.00	.6591
"	"	.	.	.	.	17.45	1250	114.5	87.73	4019.00	.6623
"	"	.	.	.	.	17.61	1000	114	69.09	3632.61	.5718
"	"	.	.	.	.	17.81	750	113.6	51.63	3038.93	.5051
"	"	.	.	.	.	17.60	900	122	66.54	3600.82	.5558
"	"	.	.	.	.	17.60	950	120	69.09	3608.89	.5759
"	"	.	.	.	.	17.73	850	112.5	57.95	3279.00	.5278
"	"	.	.	.	.	17.98	675	110.6	45.24	2550.32	.5223
"	"	.	.	.	.	17.98	625	115.5	43.75	2550.32	.5170
"	"	.	.	.	.	18.09	450	120	32.72	2311.00	.4143
"	"	.	.	.	.	18.09	500	113.5	34.39	2311.00	.4356
"	"	.	.	.	.	18.24	250	114.5	17.34	1827.74	.2755
"	"	.	.	.	.	18.23	350	125	26.51	2136.79	.3603
"	"	.	.	.	.	18.23	400	116	28.15	2140.23	.3820

King's Turbine.

Sent by A. S. King, Pontiac, Michigan.

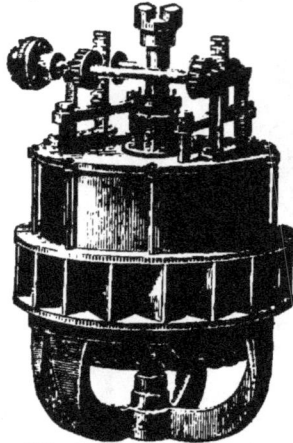

Wheel, 30 inches diameter.

This turbine was a central discharge, constructed with a thick crown plate that could be raised or lowered on the buckets, so that the wheel itself could be changed in depth from ten inch openings to zero—so constructed with the expectation of getting the highest percentage for the water used, whether the wheel was opened two or ten inches. There was no separate gate, the crown plate shutting down to the bottom rim of wheel, thus forming gate in itself.

· Data below for one minute. Multiply revolutions by 15. Sept. 20, 1879.

Gate Opened.						Head	Weight	Rev per minute	Horse Power	Cubic Feet	Per Cent
Whole Gate.	.	.	.	.	.	17.93	920	000	00.00	1866.47	.0000
"	"	.	.	.	.	17.86	450	185	37.84	1969.34	.5695
"	"	.	.	.	.	17.85	500	175	39.77	1987.04	.5937
"	"	.	.	.	.	17.85	550	165	41.25	1989.02	.6120
"	"	.	.	.	.	17.85	600	154	42.00	2037.82	.6256
"	"	.	.	.	.	17.80	650	141.6	41.83	2061.80	.6064
"	"	.	.	.	.	17.80	700	130.5	41.55	2091.88	.5908
"	"	.	.	.	.	17.82	575	157	41.03	2025.86	.6018
"	"	.	.	.	.	17.82	590	154	41.30	2043.80	.6004
"	"	.	.	.	.	17.82	610	150.5	41.73	2043.80	.6067
"	"	.	.	.	.	17.81	625	146.2	41.53	2055.80	.6006
Part Gate.	.	.	.	.	.	17.88	550	152.5	38.12	1936.83	.5829
"	"	.	.	.	.	17.87	525	158.6	37.81	1922.12	.5829
"	"	.	.	.	.	17.93	450	163.8	33.50	1782.48	.5550
"	"	.	.	.	.	17.93	480	156	34.03	1796.88	.5592
"	"	.	.	.	.	17.93	500	152	34.54	1802.64	.5658
"	"	.	.	.	.	18.01	450	150.2	29.01	1640.36	.5199
"	"	.	.	.	.	18.02	425	158.5	30.62	1629.14	.5523
"	"	.	.	.	.	18.10	350	164.5	26.17	1455.45	.5260
"	"	.	.	.	.	18.12	375	155.5	26.51	1455.45	.6322
"	"	.	.	.	.	18.24	300	154	21.00	1222.65	.4998
"	"	.	.	.	.	18.34	250	139.5	15.85	1048.53	.4364
"	"	.	.	.	.	18.35	200	173.5	15.77	1041.16	.4370
"	"	.	.	.	.	18.35	225	159.2	16.28	1043.62	.4501
"	"	.	.	.	.	18.48	140	155	9.86	808.00	.3496
"	"	.	.	.	.	18.57	90	150	3.74	652 00	.1635

Tyler Wheel.

60-inch wheel, sent by John Tyler, Claremont, N. H.

In furnishing wheels for an open comparative trial, Mr. Tyler took a course alike creditable to his manhood and sense of fair dealing. He knew perfectly well that recent improvements in turbines had greatly increased their capacity, without a corresponding increase in cost, and that his wheels would have to contend against such improvements.

This turbine weighed about six tons; price, $1,000. By comparing its cost, capacity of transmission, and general efficiency with the Hercules, Victor or New American, its relative value may be approximated.

It will be noticed that after partially closing the gate, the discharge was greater than with the gate opened in full—a rather curious feature, though the same may be observed in the test of the Monarch, the second test of the Success, and others.

Data below for one minute. Multiply revolutions by 20. Oct. 8, 1879.

Gate Opened						Head	Weight	Rev per minute	Horse Power	Cubic Feet	Per Cent.
Whole Gate.	.	.	.	.	.	16.94	1950	102.5	121.13	4730.67	.7996
" "	.	.	.	.	.	16.94	2000	100.5	121.81	4743.84	.8027
" "	.	.	.	.	.	16.94	2050	98	121.75	4774.60	.7970
Part Gate.	.	.	.	.	.	16.88	2000	98.6	119.51	4809.82	.7775
" "	.	.	.	.	.	16.88	1900	106	122.06	4809.82	.7959
" "	.	.	.	.	.	17.13	1800	97	105.81	4251.42	.7692
" "	.	.	.	.	.	17.15	1750	99	105.00	4192.08	.7733
" "	.	.	.	.	.	17.18	1700	100.7	103.72	4162.50	.7679
" "	.	.	.	.	.	17.27	1500	105	95.45	3878.45	.7545
" "	.	.	.	.	.	17.28	1500	105	95.45	3870.19	.7557
" "	.	.	.	.	.	17.28	1550	102	95.81	3890.86	.7544
" "	.	.	.	.	.	17.64	1200	98.2	71.41	3137.24	.6832
" "	.	.	.	.	.	17.65	1150	100.5	70.04	3040.77	.6809
" "	.	.	.	.	.	17.67	1100	104	69.33	3106.27	.6687
" "	.	.	.	.	.	17.85	950	98.3	56.62	2619.96	.6414
" "	.	.	.	.	.	17.85	850	102	52.54	2510.49	.6208
Full Gate.	.	.	.	.	.	16.85	2000	100.3	121.57	4757.01	.8030

Thompson Wheel.

Sent by Thompson Iron Works, Union City, Pa.

40-inch wheel, diagonal in shape, like the Houston.

Data below for one minute. Multiply revolutions by 20. Oct. 11, 1879.

Gate Opened							Head	Weight	Rev per minute	Horse Power	Cubic Feet	Per Cent
Whole Gate.	.	.	.	.	.	.	17.66	1800	000	000	2958.73	000
"	"	.	.	.	.	.	17.49	900	139.6	76.14	3302.55	.6982
"	"	.	.	.	.	.	17.47	1000	128	77.57	3314.46	.7092
"	"	.	.	.	.	.	17.48	1100	117	78.00	3334.34	.7085
"	"	.	.	.	.	.	17.50	1200	102.5	74.54	3330.38	.6771
"	"	.	.	.	.	.	17.50	1075	120.5	78.50	3334.34	.7122
"	"	.	.	.	.	.	17.50	1125	115	78.40	3342.30	.7096
"	"	.	.	.	.	.	17.48	1100	118.6	79.06	3346.28	.7155
"	"	.	.	.	.	.	17.48	1100	119.2	79.46	3346.28	.7192
"	"	.	.	.	.	.	17.50	1125	117.3	79.97	3342.30	.7239
Part Gate.	.	.	.	.	.	.	17.58	1100	111	74.00	3133.18	.7113
"	"	.	.	.	.	.	17.58	1100	112	74.66	3133.18	.7176
"	"	.	.	.	.	.	17.56	1050	119.7	76.17	3114.64	.7529
"	"	.	.	.	.	.	17.72	950	118	67.94	2783.48	.7292
"	"	.	.	.	.	.	17.94	800	120	58.18	2380.93	.7212
"	"	.	.	.	.	.	18.16	600	126.5	46.00	1963.12	.6832
"	"	.	.	.	.	.	18.15	700	114.5	48.57	2004.19	.7069
"	"	.	.	.	.	.	18.10	675	117.5	48.06	1993.90	.7052
"	"	.	.	.	.	.	18.23	500	118.5	35.90	1545.00	.6748
"	"	.	.	.	.	.	18.27	475	123	35.41	1538.57	.6671

Perry's Improved Turbine.
Sent by Wm. F. Perry, Bridgton, Me.

Downward discharge. Register gate. 36-inch wheel.
Data below for one minute. Multiply revolutions by 15. Oct. 13, 1879.

Gate Opened							Head	Weight	Rev per minute	Horse Power	Cubic Feet	Per Cent
Whole Gate.	.	.	.	.	.	.	17.98	600	201	54.81	2108.42	.7655
"	"	.	.	.	.	.	17.96	650	191.5	56.58	2120.50	.7866
"	"	.	.	.	.	.	17.96	700	182	57.90	2135.60	.7992
"	"	.	.	.	.	.	17.95	750	173	58.97	2138.64	.8133
"	"	.	.	.	.	.	17.95	800	161.5	58.72	2144.70	.8075
"	"	.	.	.	.	.	17.95	725	177	58.32	2129.56	.8077
"	"	.	.	.	.	.	17.95	740	175	58.86	2132.58	.8142
"	"	.	.	.	.	.	17.95	760	170.5	58.90	2138.64	.8124
"	"	.	.	.	.	.	17 94	775	167.5	59.01	2147.73	.8109
Part Gate.	.	.	.	.	.	.	18.06	700	157	49.95	1894.69	.7727
"	"	.	.	.	.	.	18.06	690	160.5	50.33	1894.69	.7786
"	"	.	.	.	.	.	18.06	675	165	50.62	1891.77	.7844
"	"	.	.	.	.	.	18.06	665	167.5	50.63	1888.85	.7859
"	"	.	.	.	.	.	18.17	550	155	38.75	1592.69	7090
"	"	.	.	.	.	.	18.16	550	161	40.25	1628 89	.7204
"	"	.	.	.	.	.	18.13	565	173.5	44.56	1719.00	.7569
"	"	.	.	.	.	.	18.10	575	174.5	45.60	1761.74	.7571
"	"	.	.	.	.	.	18.10	595	169.7	45.89	1761.74	.7620
"	"	.	.	.	.	.	18.22	465	173.5	36.67	1513.82	.7043
"	"	.	.	.	.	.	18.22	485	169	34.22	1532.00	.6490
"	"	.	.	.	.	.	18.47	350	146	23.22	1144.80	.5414
"	"	.	.	.	.	.	18.42	270	165	22.25	1072.42	.5964
"	"	.	.	.	.	.	18.39	300	158	21.72	1099.79	.5694
"	"	.	.	.	.	.	18.40	290	160	21.09	1079.92	.5610
"	"	.	.	.	.	.	18.42	250	169	19.20	1069.90	.5158
"	"	.	.	.	.	.	18.37	340	157.5	21.31	1158.96	.5318
"	"	.	.	.	.	.	18.30	375	167	28.46	1298.89	.6339

Reynold's Champion Wheel.

24-inch wheel, sent by Bloomer & Co., Ellenville, N. Y.

Downward discharge. Register gate.

Data below for one minute. Multiply revolutions by 10. Oct. 13, 1879.

Gate Opened							Head	Weight	Rev per minute	Horse power	Cubic feet	Per Cent.
Whole Gate.	.	.	.	.	.	.	18.34	550	000	000	1010.64	000
"	"	.	.	.	.	.	18.32	275	313	26.08	1047.06	.7198
"	"	.	.	.	.	.	18.30	300	304	27.63	1059.28	.7564
"	"	.	.	.	.	.	18.30	325	290.5	28.48	1071.55	.7689
"	"	.	.	.	.	.	18.30	350	276.7	29.37	1083.39	.7857
"	"	.	.	.	.	.	18.29	375	260	29.54	1091.20	.7837
"	"	.	.	.	.	.	18.29	400	243.7	29.50	1103.63	.7649
"	"	.	.	.	.	.	18.28	365	266 5	29.47	1088.75	.7836
"	"	.	.	.	.	.	18.28	385	251.5	29.34	1096.20	.7753
Part Gate.	.	.	.	.	.	.	18.27	375	260	29.54	1088.75	.7862
"	"	.	.	.	.	.	18.28	350	273	28.95	1083.39	.7755
"	"	.	.	.	.	.	18.28	375	257.5	29.26	1091.20	.7767
"	"	.	.	.	.	.	18.28	350	268	28.42	1083.85	.7594
"	"	.	.	.	.	.	18.28	365	262.5	29.03	1086.30	.7742
"	"	.	.	.	.	.	18.27	350	263	27.89	1083.39	.7492
"	"	.	.	.	.	.	18.28	315	282.5	26.96	1071.55	.7287
"	"	.	.	.	.	.	18.27	335	272.5	27.66	1071.55	.7479
"	"	.	.	.	.	.	18.27	350	262	27.78	1071.55	.7512
"	"	.	.	.	.	.	18.34	300	261	23.72	950.79	.7202
"	"	.	.	.	.	.	18.40	275	236	19.66	848.08	.6670
"	"	.	.	.	.	.	18.41	250	256.5	19.43	834.35	.6697
"	"	.	.	.	.	.	18.41	245	260.5	19.34	834.35	.6666
"	"	.	.	.	.	.	18.46	200	252.5	15.30	746.66	.5877
"	"	.	.	.	.	.	18.46	190	260	14.99	737.83	.5827
"	"	.	.	.	.	.	18.54	125	243	9.20	584.11	.4498
"	"	.	.	.	.	.	18.54	110	260	8.66	573.81	.4310
"	"	.	.	.	.	.	18.60	90	260	7.09	521.12	.3873

New American Wheel.

48-inch wheel, sent by Stout, Mills & Temple, Dayton, Ohio.

This turbine has the same curb in form as the well-known American Turbine, made by that company; but the wheel is downward discharge—very similar in form and plan to the Swain.

Data below for one minute. Multiply revolutions by 20.   Oct. 14, 1879.

Gate Opened.		Head	Weight	Rev per minute	Horse Power	Cubic Feet	Per Cent.
Whole Gate.	. . . . .	16.45	2935	000	000	5397.95	000
" "	. . . . .	16.33	2000	110.5	133.94	5603.83	.7749
" "	. . . . .	16.32	2050	108.3	134.55	5608.43	.7783
" "	. . . . .	16.30	2100	104.6	133.12	5594.63	.7727
" "	. . . . .	16.29	2150	101.5	129.48	5590.23	.7516
" "	. . . . .	16.32	2025	109	133.77	5590.03	.7763
" "	. . . . .	16.32	2075	105.5	132.67	5603.83	.7679
Part Gate.	. . . . .	16.40	2050	107	132.93	5484.60	.7824
" "	. . . . .	16.40	2025	108	132.54	5475.46	.7814
" "	. . . . .	16.38	2075	106.3	133.68	5484.60	.7879
" "	. . . . .	16.49	2000	108.5	131.51	5280.09	.7996
" "	. . . . .	16.43	2025	106.8	131.07	5271.00	.8013
" "	. . . . .	16.52	1975	109.1	130.58	5257.46	.7961
" "	. . . . .	16.69	1900	109	125.51	4984.41	.7989
" "	. . . . .	16.88	1800	106	115.63	4546.23	.7978
" "	. . . . .	16.90	1700	111.7	115.05	4477.28	.8051
" "	. . . . .	16.89	1750	108.3	114.86	4511.71	.7962
" "	. . . . .	16.87	1775	107.3	115.73	4529.00	.8019
" "	. . . . .	17.16	1500	108.8	98.91	3966.32	.7694
" "	. . . . .	17.15	1525	108.2	100.00	3962.18	.7792
" "	. . . . .	17.17	1475	110.3	98.54	3937.34	.7717
" "	. . . . .	17.43	1175	111.3	79.25	3336.22	.7216
" "	. . . . .	17.44	1200	109.4	79.56	3348.02	.7214
" "	. . . . .	17.60	1050	106.3	67.64	2969.24	.6853
" "	. . . . .	17.67	1000	103.3	62.60	2829.91	.6627
" "	. . . . .	17.68	975	106	62.63	2818.70	.6638
" "	. . . . .	17.69	950	108.3	62.35	2774.00	.6728
Whole Gate.	. . . . .	16.31	2050	106.3	132.07	5567.06	.7701

Success Wheel.

36-inch wheel, sent by S. M. Smith, York, Pa.

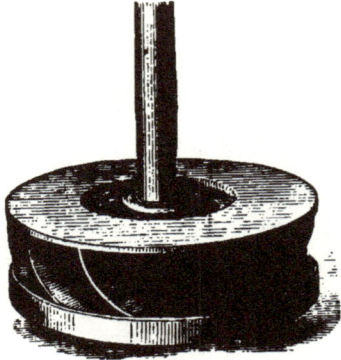

Called the Improved Success, very fragile in construction.

Data below for one minute. Multiply revolutions by 15. Oct. 16, 1879.

Gate Opened							Head	Weight	Rev per minute	Horse Power	Cubic Feet	Per Cent
Whole Gate.	.	.	.	.	.	.	17.99	1350	000	000	2243.17	0000
"	"						17.93	675	191.5	58.75	2380.46	.7287
"	"						17.90	750	185	62.76	2433.92	.7627
"	"						17.89	800	178.2	64.80	2437.50	.7867
"	"						17.87	850	170.5	65.88	2484.14	.7857
"	"						17.87	900	162.5	66.47	2494.94	.7893
"	"						17.86	950	153.3	66.19	2523.82	.7774
"	"						17.85	1000	145.5	66.13	2523.82	.7773
"	"						17.86	875	165	65.62	2491.34	.7809
"	"						17.85	900	161.5	66.06	2502.15	.7829
"	"						17.85	925	157	66.07	2512.98	.7798
Part Gate.	.	.	.	.	.	.	17.96	800	163.2	59.34	2197.96	.7959
"	"						17.98	825	159.2	59.70	2208.37	.7961
"	"						18.19	550	163.5	40.87	1649.44	.7212
"	"						18.19	575	159.2	41.60	1653.80	.7321
"	"						18.09	650	165	48.75	1876.29	.7604
"	"						18.26	450	171.5	35.07	1477.05	.6884
"	"						18.26	475	165	35.62	1486.77	.6947
"	"						18.25	500	160	36.36	1499.10	.7037
"	"						18.25	525	154.5	36.86	1511.46	.7076
"	"						18.37	375	159.5	27.18	1223.47	.6403
"	"						18.37	375	157.5	26.84	1217.67	.6352
"	"						18.34	375	162.5	27.69	1258.46	.6351

Second test of the same wheel, the buckets having been chipped and other changes made.

Gate Opened							Head	Weight	Rev per minute	Horse Power	Cubic Feet	Per Cent
Whole Gate.	.	.	.	.	.	.	17.78	900	164.1	67.13	2482.66	.8051
Part Gate.	.	.	.	.	.	.	17.80	800	179	65.09	2410.98	.8031
"	"						17.80	800	178	64.72	2378.93	.8091
.	"	"					17.88	800	166	60.36	2168.65	.8241
"	"						17.76	875	167.5	66.61	2464.68	.8051
"	"						17.74	925	161.2	67.61	2482.66	.8126
"	"						17.75	900	165	67.50	2493.46	.8076

Nonesuch Wheel.

40-inch wheel, sent by A. S. Clark, Turners Falls, Mass.

The designer sends the following description:

The wheel consists of downward discharge buckets, enclosed by bell-shaped cylinders. The one forming the hub of the wheel has the concave surface next to the buckets. The other forms the flange or band which encloses the lower or reacting parts of the buckets, and has the convex surface next to them, or larger end downward. By this construction, the lower parts of the buckets are expanded on their outer extremity, which gives a very easy discharge. The curb of the wheel has a short draft tube in which is the step on which the wheel revolves. The water enters the wheel at the side and above the outer flange, through a system of straight chutes, within which is a cylinder gate having on the lower edge fins or blades, which extend into the chutes. The downward pressure on these blades and the weight of the gate is counterbalanced by an upward pressure on an external sectional flange near the top of the gate, and within the dome in which the gate rises to open. By this means the gate opens easy under pressure. The wheel is constructed on the theory that water should not be changed in direction horizontally after leaving the chutes, but take a downward direction only, as the wheel absorbs the power of the moving water.

This wheel was very deep, like the Hercules; conical in shape, 40 inches in diameter at the top and 48 at the bottom, which turned outward like the Risdon—hardly distinguishable in outward appearance of curb from the Hercules.

Data below for one minute. Multiply revolutions by 20. Oct. 21, 1879.

Gate Opened							Head	Weight	Rev per minute	Horse Power	Cubic Feet	Per Cent
Whole Gate.	.	.	.	.	.	.	17.37	2100	000	000	3999.93	000
" "	.	.	.	.	.	.	17.14	1100	157.6	105.06	4449.00	.7294
" "	.	.	.	.	.	.	17.15	1200	149	108.36	4453.29	.7512
" "	.	.	.	.	.	.	17.12	1300	139.2	109.67	4461.89	.7600
" "	.	.	.	.	.	.	17.15	1400	131	111.15	4470.51	.7676
" "	.	.	.	.	.	.	17.13	1500	118.5	107.74	4470.51	.7448
" "	.	.	.	.	.	.	17.12	1600	105.5	102.30	4449.00	.7112
" "	.	.	.	.	.	.	17.11	1375	130.5	108.75	4444.38	.7571
Part Gate.	.	.	.	.	.	.	17.12	1425	125.5	108.75	4444.38	.7503
" "	.	.	.	.	.	.	17.12	1400	128.2	108.77	4449.00	.7562
" "	.	.	.	.	.	.	17.19	1400	120	101.82	4256.59	.7368
" "	.	.	.	.	.	.	17.19	1350	125.5	102.68	4239.56	.7461
" "	.	.	.	.	.	.	17.42	1300	106	83.51	3711.41	.6839
" "	.	.	.	.	.	.	17.38	1300	114	89.81	3859 00	.7090
" "	.	.	.	.	.	.	17.38	1200	124.7	90.70	3838.38	.7206
" "	.	.	.	.	.	.	17.48	1100	125	83.33	3865.58	.7079
" "	.	.	.	.	.	.	17.65	950	125	71.96	3197.18	.6750
" "	.	.	.	.	.	.	17.75	800	130	63.03	2935.63	.6404
" "	.	.	.	.	.	.	17.75	850	123.5	63.62	2928.04	.6481
" "	.	.	.	.	.	.	17.84	700	127.5	54.09	2666.18	.6021
" "	.	.	.	.	.	.	17.92	675	123	50.31	2512.67	.6053
" "	.	.	.	.	.	.	17.92	650	126.5	49.83	2501.81	.5885
" "	.	.	.	.	.	.	18.04	500	131	39.69	2224.52	.5237
" "	.	.	.	.	.	.	18.03	550	122.5	40.83	2221.03	.5398
" "	.	.	.	.	.	.	17.09	1400	126	106.91	2449.00	.7443

Tait Wheel.

Sent by Thomas Tait, Rochester, N. Y.

36-inch wheel.

This wheel discharged downward. It had thick cast iron buckets, left square at the edge, between the hoop and crown plate.

Data below for one minute. Multiply revolutions by 15. Oct. 17, 1879.

Gate Opened							Head	Weight	Rev per minute	Horse Power	Cubic Feet	Per Cent
Whole Gate.		.	.	.	.	.	18.25	1125	000	000	1685.09	000
"	"	.	.	.	.	.	18.27	550	156.5	39.12	1614.60	.7022
"	"	.	.	.	.	.	18.26	560	154	39.20	1618 97	.7021
"	"	.	.	.	.	.	18.25	570	152.5	39.51	1622.16	.7066
"	"	.	.	.	.	.	18.25	580	151.5	39.94	1627.35	.7119
"	"	.	.	.	.	.	18.25	590	150	40.22	1633.73	.7142
"	"	.	.	.	.	.	18.25	600	147.5	40.22	1640.13	.7109
"	"	.	.	.	.	.	18.25	610	146.2	40.50	1643.33	.7149
"	"	.	.	.	.	.	18.24	620	144.7	40.77	1643.33	.7202
"	"	.	.	.	.	.	18.24	630	142.5	40.81	1656.15	.7153
"	"	.	.	.	.	.	18.24	650	139	41.06	1665.78	.7154
"	"	.	.	.	.	.	18.22	700	133	42.31	1694.78	.7271
"	"	.	.	.	.	.	18.21	750	125	42.61	1720.68	.7200
"	"	.	.	.	.	.	18.20	800	113	41.09	1749.96	.683
Part Gate.		.	.	.	.	.	18.30	500	158.3	35.97	1485.61	.7005
"	"	.	.	.	.	.	18.30	515	156	36.82	1494.94	.7126
"	"	.	.	.	.	.	18.29	530	152.5	36.73	1507.41	.7053
"	"	.	.	.	.	.	18.29	545	149.7	37.11	1516.78	.6921
"	"	.	.	.	.	.	18.33	500	152	34.54	1420.78	.7022
"	"	.	.	.	.	.	18.32	515	149.2	34.92	1426.92	.7074
"	"	.	.	.	.	.	18.37	450	153.5	31.40	1293.67	.6996
"	"	.	.	.	.	.	18.37	480	147.5	32.18	1311.62	.7071
"	"	.	.	.	.	.	18.42	430	146	28.53	1196.28	.6854
"	"	.	.	.	.	.	18.43	400	152.5	27.72	1173.01	.6788
"	"	.	.	.	.	.	18.49	350	145	23.06	1011.04	.6530
"	"	.	.	.	.	.	18.50	320	155	22.54	994.39	.6487
"	"	.	.	.	.	.	18.56	250	157	17.84	848.19	.6000
"	"	.	.	.	.	.	18.56	270	147.5	18.10	848.19	.6080

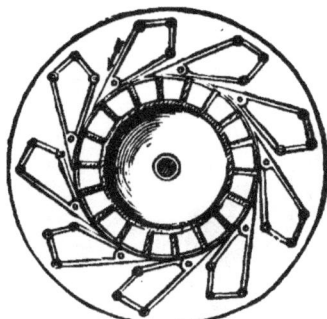

TAIT (*Continued*).

Another wheel, similar to the first, but the edge of the buckets had been finished "quarter round." It was tested in the same curb as the first.

Gate Opened	Head	Weight	Rev per minute	Horse Power	Cubic Feet	Per Cent
Whole Gate.	18.18	550	158.3	39.57	1779.38	.6476
" "	18.17	575	155	40.51	1782.66	.6620
" "	18.17	600	150	40.90	1818.84	.6553
" "	18.16	625	148	42.04	1838.66	.6667
" "	18.16	650	144	42.54	1845.26	.6722
" "	18.15	675	140.5	43.11	1871.83	.6718
" "	18.14	700	136	43.27	1888.48	.6687
" "	18.13	725	133.5	43.99	1905.18	.6743
" "	18.13	750	129.5	44.14	1911.87	.6742
" "	18.13	775	125	44.03	1918.57	.6701
" "	18.11	800	121.5	44.18	1938.71	.6663
Part Gate.	18.18	600	146.2	39.87	1802.36	.6592
" "	18.19	575	150	39.20	1749.96	.6520
" "	18.25	470	160.5	34.28	1601.88	.6208
" "	18.24	500	155	35.22	1615.78	.6327
" "	18.23	530	150.2	36.18	1637.00	.6419
" "	18.28	470	151.5	32.36	1504.29	.6231
" "	18.35	410	151.7	28.27	1341.69	.6079
" "	18.43	350	149.8	23.83	1170.11	.5850
" "	18.50	265	150.2	18.09	972.31	.5324
" "	18.55	240	143.5	15 65	850.80	.5250
" "	18.52	270	145	17.79	934.03	.5445
" "	18.53	250	150	17.04	924.88	.5264

Second test of the No. 1 Tait wheel, the buckets having been "chipped" back three-eighths of an inch, and edges rounded on front side, so as to leave them sharp on back side, between the hoop and crown plate.

Gate Opened	Head	Weight	Rev per minute	Horse Power	Cubic Feet	Per Cent
Whole Gate.	18 29	700	144.5	45.98	1710.11	.7787
" "	18.27	725	140	46.13	1719.83	.7772
" "	18 32	675	148.7	45.63	1697.18	.7771
" "	18.31	650	151.5	44.76	1687.50	.7670
Part Gate.	18.34	650	144.7	42.75	1610.69	.7663
" "	18.36	600	153.5	41.86	1579 00	.7645
" "	18 39	600	145.5	39.68	1509.92	.7566
" "	18.40	580	149	39.28	1500.57	.7533
" "	18 45	525	150	35.79	1380.58	.7440
" "	18.50	475	146.2	31.56	1248.86	.7230
" "	18.50	450	152	31.09	1240.00	.7176
" "	18.57	350	157.2	25.01	1161.43	.6717
" "	18 57	375	150.5	25.65	1067.07	.6853
" "	18 66	300	141 5	19.29	869.71	.6293
" "	18.67	270	151.2	18.55	843.22	.6298
Full Gate.	18.28	750	135	46.02	1742.56	.7648

Hercules Wheel.

Holyoke Machine Co., Holyoke, Mass.

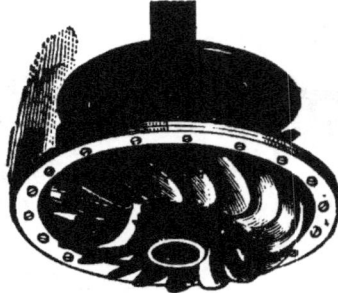

33-inch wheel; weight, 4,000 pounds; price, $550.
Tested Nov. 4.

Data below for one minute. Multiply revolutions by 20.

Gate Opened							Head	Weight	Rev per minute	Horse Power	Cubic Feet	Per Cent
Whole Gate.	.	.	.	.	.	.	16.87	2500	000	000	5137.95	.000
"	"	.	.	.	.	.	17.12	1000	184.5	111.81	4608.44	.7503
"	"	.	.	.	.	.	17.09	1100	177.2	118.13	4664.96	.7844
"	"	.	.	.	.	.	17.06	1200	166.4	121.02	4691.06	.8004
"	"	.	.	.	.	.	17.07	1300	154.6	121.80	4721.69	.8000
"	"	.	.	.	.	.	17.04	1400	142.6	121.00	4743.56	.7925
"	"	.	.	.	.	.	17.02	1500	130	118.18	4765.41	.7716
"	"	.	.	.	.	.	17.05	1250	160.5	121.59	4708.58	.8017
"	"	.	.	.	.	.	17.04	1350	148.5	121.50	4721.69	.7995
"	"	.	.	.	.	.	17.04	1275	157	121.32	4704.15	.8013
"	"	.	.	.	.	.	17.04	1300	154.5	121.72	4708.58	.8031
"	"	.	.	.	.	.	17.04	1325	150.2	120.61	4712.95	.7951
Part Gate.	.	.	.	.	.	.	17.13	1300	148	116.60	4513.21	.7984
"	"	.	.	.	.	.	17.14	1250	154.5	117.04	4500.34	.8033
"	"	.	.	.	.	.	17.25	1200	153.5	111.63	4206.06	.8147
"	"	.	.	.	.	.	17.23	1175	156	111.09	4197.56	.8133
"	"	.	.	.	.	.	17.28	1100	158.5	105.66	4038.24	.8016
"	"	.	.	.	.	.	17.26	1150	155	108.63	4059.11	.8209
"	"	.	.	.	.	.	17.25	1100	155	103.33	3926.08	.8079
"	"	.	.	.	.	.	17.42	1050	156	99.27	3696.75	.8162
"	"	.	.	.	.	.	17.40	1050	150	95.45	3527.48	.8214
"	"	.	.	.	.	.	17.38	1000	153.5	93.03	3507.49	.8079
"	"	.	.	.	.	.	17.39	1000	145	87.87	3368.55	.7924
"	"	.	.	.	.	.	17.21	900	155	84.55	3270.35	.7953
"	"	.	.	.	.	.	17.33	850	151.6	78.09	3095.86	.7705
"	"	.	.	.	.	.	17.35	800	150	72.72	2905.45	.7638
"	"	.	.	.	.	.	17.35	750	152.5	69.31	2860.30	.7396
"	"	.	.	.	.	.	17.37	730	155	66.57	2841.55	.7355
"	"	.	.	.	.	.	17.51	650	157	61.84	2626.87	.7166
"	"	.	.	.	.	.	17.50	675	153	62.59	2656.17	.7129
"	"	.	.	.	.	.	17.34	600	150	54.54	2428.14	.6859
"	"	.	.	.	.	.	17.40	580	152.5	53.65	2381.87	.6848
"	"	.	.	.	.	.	17.40	550	155.5	51.84	2360.60	.6683

Hercules Wheel.

Holyoke Machine Co., Holyoke, Mass.

33-inch wheel; weight, 4,000 pounds; price, $550.

Tested Nov. 11.

Same wheel as tested Nov. 4th. It stood in the flume during the interval, and received some hard knocks during the time from a gang of mill-wrights who were fitting up the nearly horizontal draft tube illustrated further along in report. Previous to this second trial the step was taken out, examined and quite likely "trued up," after which operation the wheel became unsteady in motion and difficult to control by brake, though quite the reverse during the first test.

Data below for one minute. Multiply revolutions by 20.

Gate Opened	Head	Weight	Rev per minute	Horse Power	Cubic Feet	Per Cent,
Whole Gate.	16.68	1275	152	117.42	4661.56	.7995
" "	16.68	1300	148.7	117.15	4652.86	.7991
" "	16.67	1250	154.5	117.04	4652.86	.7989
" "	16.65	1250	151.6	114.85	4639.84	.7879
" "	16.70	1200	159.2	115.78	4631.16	.7926
Part Gate.	17.00	1150	154	107.33	4178.34	.8000
" "	17.00	1175	148	105.39	4174.14	.7510
" "	17.08	1100	156	104.00	3990.56	.8080
" "	17.09	1125	154	105.00	4015.36	.8101
" "	17.28	1025	154.5	95.98	3675.20	·8001
" "	17.28	1050	153	97.36	3679.24	.8110
" "	17.37	1025	147	91.31	3514.60	.7919
" "	17.40	1000	150.2	91.03	3486.74	.7944
" "	17.54	900	155	84.54	3297.41	.7738
" "	17.67	950	151	86.93	3375.90	.7756
" "	17.74	850	156	80.36	3099.87	.7755
" "	17.23	1300	156 7	123.46	4766.24	.7961
" "	17.23	1350	151	123.54	4740.01	.8009
" "	17.68	1050	156.7	99.72	3707.58	.8054
" "	17.68	1100	150	100.00	3736.00	.8015
" "	17.99	800	150	72.72	3015.83	.7100
" "	18.10	650	155	61.06	2566.63	.6958
" "	18.08	675	151.5	61.97	2502.00	.7000
" "	18.10	600	155.5	56.54	2469.49	.7012
" "	18.10	640	151	58.56	2483.81	.6897
" "	18.08	650	150.2	59.70	2473.07	.7070
" "	18.09	675	147.5	60.40	2505.33	.7068

Hercules Wheel.

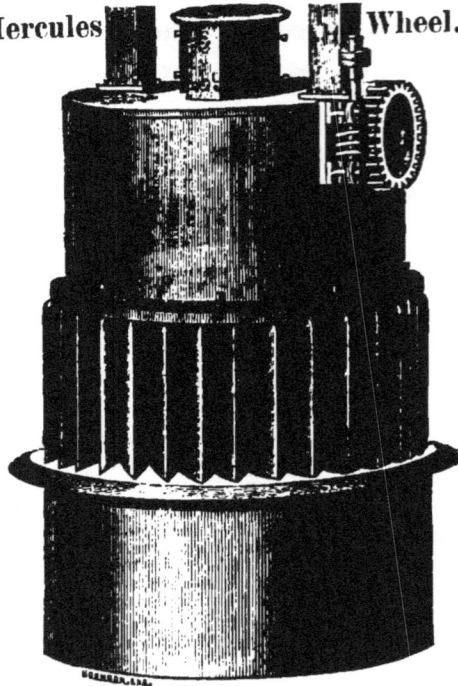

A third test, Nov. 12, step having again been taken out and examined.
Data below for one minute.   Multiply revolutions by 20.

Gate Opened	Head	Weight	Rev per minute	Horse Power	Cubic Feet	Per Cent
Whole Gate.	17.28	1300	155.3	122.35	4732.11	.7923
" "	17.27	1325	152.5	122.46	4710.31	.7984
" "	17.26	1350	150.5	123.13	4723.04	.7996
" "	17.28	1375	146.5	122.08	4745.22	.7892
Part Gate.	17.35	1300	150	118.18	4510.79	.7996
" "	17.36	1250	157	118.94	4493.89	.8071
" "	17.42	1250	147.7	111.89	4238.33	.8024
" "	17.43	1200	154.5	112.37	4217.74	.8091
" "	17.43	1225	152	112.85	4217.74	.8128
" "	17.55	1150	149.6	104.26	3913.03	.8038
" "	17.55	1125	152.5	103.97	3899.56	.8044
" "	17.57	1050	153.5	97.68	3700.64	.7954
" "	17.58	1050	146	92.90	3539.76	.7904
" "	17.59	1000	151	91.51	3527.79	.7809
" "	17.62	975	154	91.00	3490.89	.7812
" "	17.28	950	145	83.48	3322.19	.7700
" "	17.30	900	150.5	82.00	3275.29	.7671
" "	17.25	850	148	76.24	3070.62	.7620
" "	17.25	825	151.2	75.60	3062.99	.7576
" "	17.01	750	147.7	67.13	2832.72	.7377
" "	16.99	725	151	66.34	2814.09	.7346
" "	17.20	650	153	60.27	2604.47	.7123
" "	17.32	600	153.5	55.81	2478.28	.6881
" "	17.37	600	150.5	54.72	2442 58	.6674

Houston Wheel.

40-inch wheel, sent by Fales & Jenks Machine Co., Pawtucket, R. I.

Gate worked **very** hard.
Data below for one minute. Multiply revolutions by 15. Nov. 14, 1879.

Gate Opened		Head	Weight	Rev per minute	Horse Power	Cubic Feet	Per Cent
Whole Gate.	. . . . .	16.41	2125	000	000	2786.04	000
" "	. . . . .	16.56	1050	143.6	68.53	2697.24	.8123
" "	. . . . .	16.57	1075	140	68.41	2704.60	.8083
" "	. . . . .	16.56	1100	137	68.50	2697.24	.8120
" "	. . . . .	16.56	1125	133.5	68.27	2704.60	.8069
" "	. . . . .	16.54	1025	146.7	68.55	2682.52	.8179
Part Gate.	. . . . .	16.62	700	138.5	44.06	2145.70	.6541
" "	. . . . .	16.57	650	149.5	44.17	2176.65	.6484
" "	. . . . .	16.53	675	147.5	45.25	2200.82	.6585
" "	. . . . .	16.53	695	146	46.12	2200.82	.6712
" "	. . . . .	16.59	900	147.5	60.34	2522.32	.7634
" "	. . . . .	16.59	925	144.5	60.75	2522.32	.7688
" "	. . . . .	16.88	800	143	52.00	2315.85	.7043
" "	. . . . .	16.70	790	141	50.63	2298.30	.6984
" "	. . . . .	16.68	770	145	50.75	2284.30	.7053
" "	. . . . .	16.60	600	142	38.72	2050.22	.5743
" "	. . . . .	16.54	575	144.5	37.76	2033.38	.5943
" "	. . . . .	16.94	450	142	26.01	1820.95	.4464
" "	. . . . .	16.87	425	145.5	24.10	1807.91	.4876
" "	. . . . .	17.00	350	133.5	21.23	1448.80	.4564
" "	. . . . .	17.25	150	145	9.88	1346.51	.2252

Wetmore.

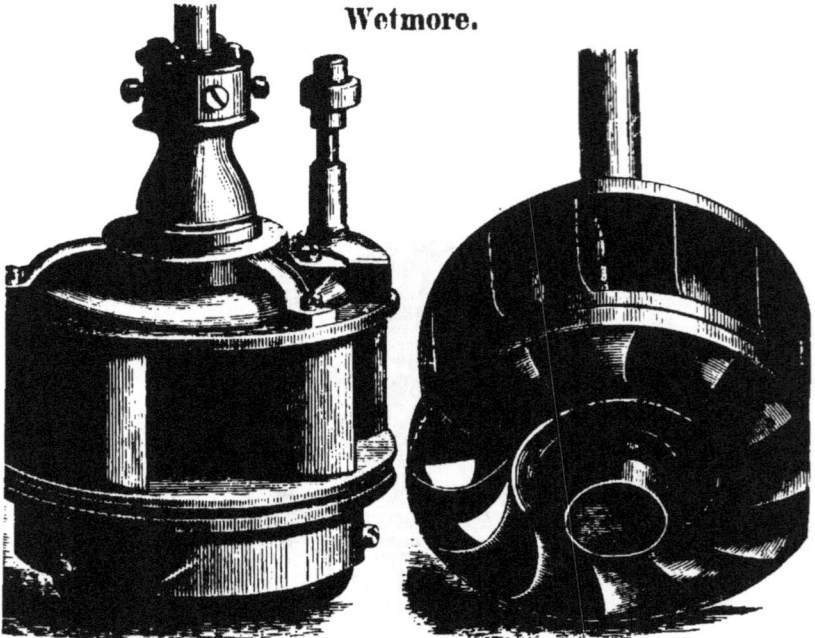

To the Engineers making Hydro-Dynamic Experiments for Water Power Co.,
Holyoke, Mass.

GENTLEMEN: The wheel which we had tested by you was an experimental
one, differing somewhat from the others heretofore tested, and from what we
furnish our customers. The results you obtained did not warrant us in continuing its manufacture, so it has been abandoned, and we have returned to our
original plans represented above.

Respectfully,

SULLIVAN MACHINE CO.

Nov. 14, 1879.   C. B. RICE, Treas.

Gate Opened						Head	Weight	Rev per minute	Horse Power	Cubic Feet	Per Cent
Whole Gate.	.	.	.	.	.	18.38	350	250	39.77	1508.19	.7596
"	"	.	.	.	.	18.38	375	237	40.36	1511.29	.7692
"	"	.	.	.	.	18.38	400	224	40.72	1511.29	.7762
"	"	.	.	.	.	18.38	425	210.5	40.66	1505.09	.7781
"	"	.	.	.	.	18.38	450	196	40.09	1502.00	.7689
"	"	.	.	.	.	18.39	390	227.5	40.32	1502.00	.7727
"	"	.	.	.	.	18.38	410	217	40.44	1498.89	.7772
Part Gate.	.	.	.	.	.	18.44	350	223 5	35.51	1361.62	.7488
"	"	.	.	.	.	18.20	300	199	27.13	1144.66	.6894
"	"	.	.	.	.	18.21	275	216	27.00	1138.96	.6892
"	"	.	.	.	.	18.21	260	223.5	26.41	1138.96	.6741
"	"	.	.	.	.	18.39	200	204	18.54	917.75	.5818
"	"	.	.	.	.	18.39	175	225	17.89	917.75	.5613
"	"	.	.	.	.	18.53	125	222	12.61	761.30	.4733
"	"	.	.	.	..	18.70	275	225	28.12	1176.21	.6769

Houston Wheel.

35-inch wheel, sent by one who had purchased the wheel.

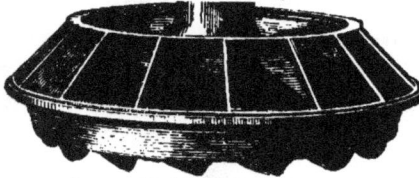

Data below for one minute.   Multiply revolutions by 15.   Nov. 28, 1879.

Gate Opened.	Head	Weight	Rev per minute	Horse Power	Cubic Feet	Per Cent
Whole Gate.	14.07	550	165.8	41.45	1944.61	.8022
" "	14.05	600	155	42.27	1944.61	.8192
" "	14.04	625	149.2	42.38	1946.63	.8166
" "	14.05	650	143	42.25	1956.67	.8129
" "	14.01	675	138	42.30	1964.81	.8135
Part Gate.	14.11	625	146.2	41.53	1918.46	.8121
" "	13.62	600	135.5	36.95	1812.81	.7925
" "	13.66	575	142	37.11	1818.75	.7907
" "	14.15	500	136.5	31.02	1635.19	.7099
" "	14.29	450	149.2	30.51	1623.66	.6960
" "	13.85	250	136	15.45	1223.47	.4827
" "	13.68	225	146.5	14.98	1202.14	.3869
" "	14.58	120	139	7.58	922.81	·2983
" "	14.45	120	142	7.74	939.37	.3019
" "	14 28	120	148.5	8.10	964.36	.3114

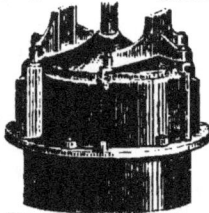

Sherwood Wheel.

20-inch wheel.

Downward discharge, similar to the Risdon, with plain cylinder gate; had been in use two years; was sent for the purpose of ascertaining the efficiency of the plan.

Data below for one minute.   Multiply revolutions by 10.   Oct. 7, 1879.

Gate Opened	Head	Weight	Rev per minute	Horse Power	Cubic Feet	Per Cent
Whole Gate.	18.31	260	248.2	19.55	835.13	.6769
" "	18.32	270	242.2	19.81	848.37	.6748
" "	18.32	280	230	19.51	848 37	.6647
" "	18.31	250	259.2	19.63	848.37	.6692
Part Gate.	18.36	260	248	19.53	805.01	.6996
" "	18.41	260	212	16.70	746.59	.6432
" "	18.43	225	255	17.38	756.00	.6606
" "	18.43	245	231.5	17.18	754.42	.6543
" "	18.43	235	243	17.30	754.42	.6588
" "	18.43	230	251	17.49	754.42	.6508

Royer Wheel.

24-inch wheel, sent by R. R. Royer, Ephrata, Pa.

Downward discharge, having plain cylinder gate.

Data below for one minute. Multiply revolutions by 10. Dec. 5, 1879.

Gate opened							Head	Weight	Rev per minute	Horse Power	Cubic Feet	Per Cent.
Whole Gate.	.	.	.	.	.	.	18.05	225	277.5	18.92	829.98	.6686
"	"	.	.	.	.	.	18.02	250	261	19.77	840.29	.6913
"	"	.	.	.	.	.	18.00	275	249.5	20.80	854.88	.7158
"	"	.	.	.	.	.	17.98	300	238	21.63	867.91	.7339
"	"	.	.	.	.	.	17.97	325	227.5	22.41	877.72	.7184
"	"	.	.	.	.	.	17.95	350	213.7	22.66	889.19	.7517
"	"	.	.	.	.	.	17.93	375	198.5	22.49	897.41	.7400
"	"	.	.	.	.	.	17.93	400	181.2	21.96	905.66	.7161
"	"	.	.	.	.	.	17.95	340	220.5	22.71	885.91	.7562
"	"	.	.	.	.	.	17.95	360	206.5	22.52	890.83	.7456
Gate closed	4 turns.		.	.			17.96	325	228	22.45	845.91	.7471
"	"	4	"		.	.	17.96	340	207.5	21.37	885.91	.7595
"	"	8	"		.	.	17.95	325	216.7	21.34	897.41	.7014
"	"	8	"		.	.	17.96	315	223	21.28	895.96	.7003
"	"	12	"		.	.	17.96	300	215	19.54	884.27	.6514
"	"	12	"		.	.	17.97	290	222	19.50	879.35	.6519
"	"	16	"		.	.	18.03	250	211	15.98	812.90	.5772
"	"	16	"		.	.	18.02	235	223	15.88	798.48	.5843
"	"	20	"		.	.	18.15	175	214	11.35	683.14	.4847
"	"	20	"		.	.	18.15	165	222.5	11.25	679.54	.4828
"	"	24	"		.	.	18.30	75	226	5.14	529.35	.2745

Monarch Wheel.

Sent by Albred & Koellsch, Randleman Mf'g Co., High Point, N. C.

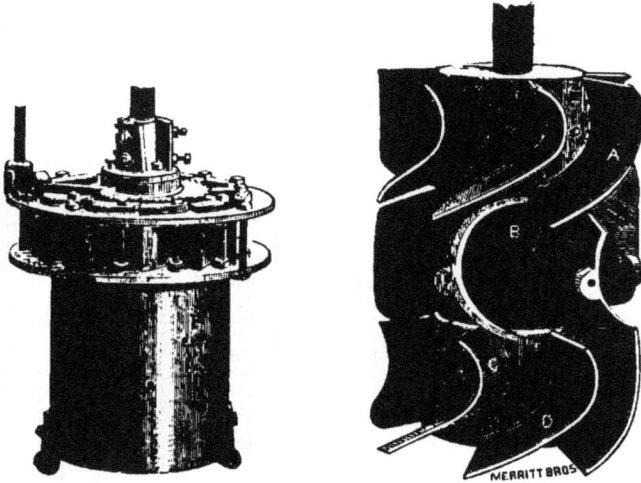

Three wheels, placed one above the other, the middle wheel being loose on shaft, but being bolted firmly to the curb—arranged in this manner that it might act as chutes to the lower wheel. Chutes and gates to upper wheel similar to the Leffel, but so very leaky as to be anything but creditable to the workmanship.

HIGH POINT, N. C., August 15, 1879.

W. A. CHASE, ESQ.,

Dear Sir: I have a turbine water wheel, finished; size, sixteen inches—a new invention, which has not been tested except by myself. It will use the *water twice*, and *increases the power one-quarter over any wheel known.* My 16-inch wheel run over *eight horse* power, under *nine* foot *head*, with *34 square inches* discharge. As the test is open to all wheels, I would be pleased to send on my wheel to *you*, under such *rule* and *regulations as you desire*, for a test with other wheels.

Very respectfully,

H. L. KOELLSCH.

The letter of Mr. Koellsch is given as the best means of introducing his device and ideas; also, as a sample of hundreds of other letters received of the same tenor.

During the past few years many patents have been issued for devices known to be perfectly worthless by those acquainted with the subjects to which they belong. Particularly has this been the case in turbine plans. It is hardly possible to conceive of a device, no matter how absurd, that has not been tried in the hopes of circumventing nature in its claim for friction and waste, or, what is more generally the case, hoping to achieve "perpetual motion" through a double use of the same fall of water. Boyden's "Diffuser," or the "Double Turbines" of Wynkoop, Leffel, or any other make, have proved equally fallacious. The highest results have been obtained from the single, simple plans. As the most effective means of presenting this fact to Mr. Koellsch, the Monarch was first tested in the combined form designed. The results may be seen in the first table below. Then the lower wheel C and chutes B were removed and the wheel A alone tested; results obtained in the lowest table. Whenever the efficiency of a single turbine is increased by the addition of a second wheel or diffuser beneath, it may safely be concluded that the upper wheel is defective.

Data below for one minute. Multiply revolutions by 10. Nov. 15, 1879.

Gate Opened					Head	Weight	Rev per minute	Horse Power	Cubic Feet	Per Cent
Whole Gate.					18.48	75	265.5	6.09	420.92	.4145
"	"				18.49	85	250	6.40	423.66	.4326
"	"				18.49	95	288.3	6.86	427.76	.4592
"	"				18.51	105	223	7.06	429.14	.4708
"	"				18.51	120	207.5	7.54	429.14	.5025
"	"				18.51	130	172	7.56	429.14	.5007
"	"				18.51	140	182	7.72	430.50	.5135
"	"				18.51	150	166	7.54	433.24	.4978
Gate closed 5 turns.					18.51	130	194.5	7.66	430.50	.5095
"	"	10	"		18.51	135	187	7.65	430.50	.5087
"	"	"	"		18.52	135	187	7.65	429.14	.5096
"	"	15	"		18.53	135	193.7	7.92	418.19	.5411
"	"	"	"		18.52	140	187	7.90	415.47	.5436
"	"	"	"		18.55	130	194.5	7.66	383.11	.5706
"	"	"	"		18.56	140	179.5	7.61	376.44	.5767
"	"	"	"		18.67	75	152.6	3.46	242.80	.4135
"	"	"	"		18.65	95	184	5.29	292.24	.5139

After the above tests were made, the lower wheel and set of chutes were removed.

Test of upper wheel A.

Gate Opened					Head	Weight	Rev per minute	Horse Power	Cubic Feet	Per Cent
Whole Gate.					18.32	130	267	9.00	602.24	.4319
"	"				18.33	140	229	9.71	594.81	.4405
"	"				18.34	150	184.5	8.38	587.40	.4119
Gate closed 5 turns.					18.34	130	270.3	10.64	605.22	.4847
"	"	"	"		18.35	140	232.5	9.86	600.75	.4775
"	"	10	"		18.35	130	277.5	10.93	605.22	.5210
"	"	"	"		18.37	140	232.5	9.86	590.36	.4813
"	"	15	"		18.37	130	293·5	11.56	578.54	.5750
"	"	"	"		18.38	140	247.5	10.50	565.33	.5350
"	"	18	"		18.39	130	290.5	11.44	539.11	.5969
"	"	"	"		18.37	140	239	10.17	524.68	.5587
"	"	21	"		13.46	110	267	5.90	429.14	.5942
"	"	"	"		18.47	105	235	7.47	404.62	.5292
"	"	22½	"		18.65	75	263.3	5.98	317.65	.5344
"	"	24	"		18.62	50	215	3.25	242.80	.3806

New American Wheel.

48-inch wheel, sent by Stout, Mills & Temple, Dayton, Ohio.

Chutes and gates complete.   Gates cut away.

Another turbine of the same size, but of increased discharge, made after the test of the one recorded upon the opposite page. The capacity of this wheel is double that of the old 48-inch American with central discharge.

Data below for one minute. Multiply revolutions by 20. Jan. 3, 1880.

		Head	Weight	Rev per minute	Horse Power	Cubic Feet	Per Cent.
Whole Gate.	. . . . . .	13.36	1650	109	109.00	5823.77	.7418
" "	. . . . .	13.22	1800	104.6	114.10	5922.79	.7715
" "	. . . . .	13.09	1900	96.5	111.12	6016.59	.7471
" "	. . . . .	12.92	2000	90.5	109.69	6030.76	.7454
" "	. . . . .	13.20	1750	105	111.36	5857.38	.7626
" "	. . . . .	13.21	1775	105	112.05	5862.04	.7723
" "	. . . . .	13.10	1825	100	110.60	5876.05	.7608
" "	. . . . .	13.06	1850	99	111.00	5885.39	.7647
" "	. . . . .	13.11	1800	101	110.18	5871.38	.7578
Part Gate.	. . . . .	13.47	1700	101.5	104.57	5685.60	.7231
" "	. . . . .	13.48	1750	106.5	112.95	5722.61	.7752
" "	. . . . .	13.70	1700	108.5	111.78	5574.99	.7749
" "	. . . . .	12.45	1750	96.5	102.34	5414.85	.7855
" "	. . . . .	12.62	1700	97	99.93	5278.71	.7943
" "	. . . . .	13.22	1500	112.3	102.09	5031.86	.8126
" "	. . . . .	13.20	1550	107.5	100.98	5054.15	.8014
" "	. . . . .	12.95	1700	100	103.03	5211.04	.8083
" "	. . . . .	13.10	1350	106	86.72	4462.82	.7853
" "	. . . . .	13.34	1400	102	86.54	4351.51	.7893
" "	. . . . .	13.17	1450	101	88.75	4441.35	.8034
" "	. . . . .	14.40	1200	107.8	78.40	3823.42	.7716
" "	. . . . .	13.07	1150	99	69.00	3637.22	.7685
	. . . . .	13.08	850	10.2	52.84	2963.38	.7177

Retest of the same, having cut the wings A of gates off. This change was made for the purpose of ascertaining whether those wings had an injurious effect upon the efficiency of the wheel when the gates were opened in full.

		Head	Weight	Rev per minute	Horse Power	Cubic Feet	Per Cent.
Whole Gate.	. . . . .	13.02	1750	102.5	108.63	5829.43	.7450
" "	. . . . .	13.19	1800	102	111.27	5852.72	.7631
" "	. . . . .	13.20	1850	100	112.12	5862.04	.7671
" "	. . . . .	13.31	1700	106.5	109.72	5806.15	.7517
" "	. . . . .	13.17	1900	98	112.84	5876.05	.7721
Part Gate.	. . . . .	14.15	1700	107	110.24	5383.01	.7663
" "	. . . . .	13.14	1700	98.5	101.48	5233.57	.7814
Whole Gate.	. . . . .	13.10	2000	92	111.51	5946.20	.7577

Hercules Wheel.

Holyoke Machine Co., Holyoke, Mass.

48-inch wheel.

Data below for one minute. Multiply revolutions by 20. Jan. 10, 1880.

Gate Opened				Head	Weight	Rev per minute	Horse Power	Cubic Feet	Per Cent.
Whole Gate 20¼ inches.			.·	11 32	4400	000	000	8151.87	0000
"	"	"	"	11.04	2600	84.5	133.15	8041.76	.7941
"	"	"	"	11.47	2700	82.5	135.00	8122.32	.7767
"	"	"	"	11.40	2800	78.3	132.87	8151.87	.7669
"	"	"	"	11.45	2900	75.2	132.17	8180.04	.7472
"	"	.·	"	11.78	2600	80.5	141.03	8122.32	.7604
"	"	"	"	11.74	2650	93.2	149.68	8145.96	.8286
"	"	"	"	11.74	2550	94.5	146.04	8111.38	.8119
"	"	"	"	11.68	2750	81.5	135.83	8157.78	.7647
Gate open 19 inches.			.	11.13	2750	78.3	130.50	7892.70	.7866
"	"	18	"	11.70	2650	81.5	130.89	7750.30	.7643
"	"	17	"	11.49	2600	78.5	123.69	7456.50	.7644
"	"	17	"	11.54	2550	80	123.63	7433.75	.7631
"	"	17	"	11.62	2500	83.5	126.51	7439.67	.7748
"	"	16	"	11.20	2300	81.8	114.02	7000.30	.7700
"	"	15	"	11.45	2100	87	110.72	6721.00	.7619
"	"	15	"	11.41	2200	83.2	110.93	6764.60	.7610
"	"	14	"	11.62	2100	85.7	109.07	6493.63	.7654
"	.·	14	"	11.68	2150	85	110.75	6541.40	.7674
"	"	13	"	11.69	2100	83	105.63	6236.89	.7672
"	"	12	"	11.60	1900	82.5	98.2	5841.73	.7659
"	"	11	"	11.25	1750	80	84.84	5352.88	.7459
"	"	11	"	11.47	1750	81	85.90	5368.15	.7387
"	"	10	"	11.00	1500	82	74.54	4852.87	.7394
"	"	9	"	11.70	1300	88.5	69.72	4463.90	.7068
"	"	9	"	11.77	1350	85.5	69.95	4536.06	.6937
"	"	9	"	11.70	1400	84	71.27	4574.69	.7050

Hercules Wheel.

Holyoke Machine Co., Holyoke, Mass.

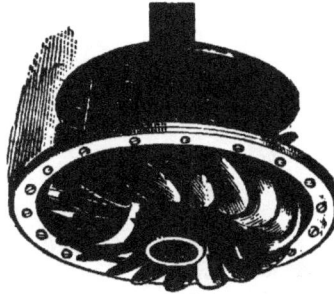

15-inch wheel.

Data below for one minute. Multiply revolutions by 10.   March 5, 1880.

Gate Opened.							Head	Weight	Rev per minute	Horse Power	Cubic Feet	Per Cent.
Whole Gate,	.	.	.	.	.	.	17.90	575			1172.22	
"	"	.	.	.	.	.	17.94	250	375.5	28.41	1048.13	.8000
"	"	.	.	.	.	.	17.92	275	356.5	29.71	1058.33	.8294
"	"	.	.	.	.	.	17.90	300	327.5	29.77	1068.55	.8241
"	"	.	.	.	.	.	17.90	325	298	29.65	1080.51	.8121
"	"	.	.	.	.	.	17.86	275	356	29.66	1061.73	.8281
Part Gate.		.	.	.	.	.	17.88	275	345	28.75	1036.27	.8215
"	"	.	.	.	.	.	17.86	270	346	28.30	1029.51	.8148
"	"	.	.	.	.	.	17.87	265	352	28.26	1022.76	.8185
"	"	.	.	.	.	.	17.86	260	356.5	28.28	1019.31	.8226
"	"	.	.	.	.	.	17.86	250	375	28.39	1017.71	.8269
"	"	.	.	.	.	.	17.86	250	355.5	26.93	987.54	.8084
"	"	.	.	.	.	.	17.85	250	346	26.21	962.58	.8076
"	"	.	.	.	.	.	17.90	245	347.5	25.79	954.30	.7993
"	"	.	.	.	.	.	17.89	240	352	25.60	951.00	.7966
"	"	.	.	.	.	.	17.85	235	356.5	25.38	946.05	.7957
"	"	.	.	.	.	.	17.90	235	345	24.56	918.11	.7912
"	"	.	.	.	.	.	17.91	230	353	24.60	911.57	.7978
"	"	.	.	.	.	.	17.88	225	357.5	24.37	905.04	.7974
"	"	.	.	.	.	.	17.95	225	343.5	23.42	880.68	.7848
"	"	.	.	.	.	.	17.96	210	356	22.65	866.15	.7708
"	"	.	.	.	.	.	17.97	200	351.6	22.30	827.14	.7586
"	"	.	.	.	.	.	18.00	195	357.5	21.12	811.85	.7652
"	"	.	.	.	.	.	18.01	195	343	20.26	796.06	.7429
"	"	.	.	.	.	.	17.98	185	354	19.84	777.22	.7517
"	"	.	.	.	.	.	18.02	180	360.5	19.06	770.98	.7493
"	"	.	.	.	.	.	18.04	175	348.5	18.48	738.38	.7346
"	"	.	.	.	.	.	18.06	170	354	18.23	730.68	.7314
"	"	.	.	.	.	.	18.05	165	362.5	18.12	727.61	.7303
"	"	.	.	.	.	.	18.07	160	350	16.96	689.46	.7207
"	"	.	.	.	.	.	18.08	155	354.5	16.65	684.92	.7118
"	"	.	.	.	.	.	18.10	145	353.5	15.53	641.44	.7082
"	"	.	.	.	.	.	18.10	140	355	15.06	659.34	.6681
"	"	.	.	.	.	.	18.15	130	348.5	13.72	587.09	.6817

Royer Wheel.

24-inch wheel, sent by R. R. Royer, Ephrata, Pa.

After the test of the first wheel, Mr. Royer returned home and prepared the one here reported.

Data below for one minute. Multiply revolutions by 10. March 9, 1880.

Gate Opened							Head	Weight	Rev per minute	Horse Power	Cubic Feet	Per Cent
Whole Gate.		.	.	.	.	.	17.86	655	000	000	978.04	000
"	"	.	.	.	.	.	17.79	325	235.7	23.03	890.33	.7698
"	"	.	.	.	.	.	17.76	350	227	24.07	904.95	.7920
"	"	.	.	.	.	.	17.74	375	212.5	24.14	918.00	.7848
"	"	.	.	.	.	.	17.70	400	198.5	24.06	929.47	.7744
"	"	.	.	.	.	.	17.69	425	185.5	23 89	931.97	.7673
"	"	.	.	.	.	.	17.70	340	229.5	23.64	901.69	.7843
"	"	.	.	.	.	.	17.70	350	226.5	24.02	904.95	.7939
"	"	.	.	.	.	.	17.69	360	222.5	24.27	908.21	.8000
"	"	.	.	.	.	.	17.73	370	217.5	24.38	913.11	.7974
Part Gate.		.	.	.	.	.	17.73	370	215	24.10	918.00	.7889
"	"	.	.	.	.	.	17.77	370	207.6	23.27	922.91	.7512
"	"	.	.	.	.	.	17.80	360	215	23.45	922.91	.7733
"	"	.	.	.	.	.	17.82	350	219.5	23.28	920.64	.7514
"	"	.	.	.	.	.	17.84	350	206	21.84	909.84	.7124
"	"	.	.	.	.	.	17.86	330	219.5	21.95	904.95	.7191
"	"	.	.	.	.	.	17.90	300	215.7	19.60	880.61	.6434
"	"	.	.	.	.	.	17.92	290	223	19.59	869.32	.6668
"	"	.	.	.	.	.	17.95	250	225	17.01	815.06	.6299
"	"	.	.	.	.	.	17.97	270	212 5	17.38	818.18	.6258
"	"	.	.	.	.	.	17.98	250	187	14.16	769.47	.5418
"	"	.	.	.	.	.	18.01	225	207.5	14.14	750.83	.5536
"	"	.	.	.	.	.	18.02	215	216	14.07	746.19	.5541
"	"	.	.	.	.	.	18 09	150	216	9.81	640.16	.4485
"	"	.	.	.	.	.	18.13	145	224	9.84	632.75	.4541
"	"	.	.	.	.	.	18.17	125	201	7.61	574.35	.3861
"	"	.	.	.	.	.	18.17	115	209	7.28	567.16	.3740
"	"	.	.	.	.	.	18.17	105	217	6.60	562.85	.3417
"	"	.	.	.	.	.	18.25	75	191	4.34	478.59	.2630
"	"	.	.	.	.	.	18.25	60	210	3.82	467.64	.2370
"	"	.	.	.	.	.	18.27	50	222	3.37	462.18	.2113

Cyclonic Turbine.

More than ordinary pains was taken to obtain a decisive trial of this device, not from any belief in its superior efficiency, but because *cyclonic* minds, filled with *vorticose* ideas, are far more abundant than is generally realized, not only with the illiterate but quite as plentifully with the educated, the turbine user as well as builder. The cyclone, the whirlpool and centrifugal force have been harped upon in connection with turbine building since the conception of that business,—Uriah A. Boyden and the author of the cyclonic alike trying to profit thereby, to gain something from nothing. It should be plain to any level headed person that to produce a centrifugal force of one hundred pounds, a somewhat greater force must be expended to do it. Were the reverse the case, then "perpetual motion" would not only be possible, but would be very philosophical. The following explanation and description is by the author :

The laws that govern the action of this wheel, as its name implies, is copied from Nature, and is founded on the principles and laws that govern the rotary motion of the Cyclone—the great motor engine of our atmosphere. It is a well known fact in meteorology, that all storms, from the smallest whirlwind to the most extended cyclone, are translated along their course in a rapid vorticose motion, revolving around its axis, which is the point of lowest barometer. Immediately the vapor ladened air rushing along the earth's surface from points of high barometer, rise in spirals till they reach the cooler currents of the upper atmosphere, and there rapidly condense into clouds and rain, setting free the latent heat produced by condensation and greatly expanding the surrounding atmosphere and correspondingly increasing the point of low barometer. This rapid rotary motion calls into play the centripetal and centrifugal forces, and they, acting almost equally in opposite directions, and on both sides of the whirling air, it escapes spirally upwards with the power of both forces combined. It is the upward, twisting vorticose motion that makes the tornado the most destructive engine that comes within our experience, and as nature ever follows the line of least resistance, so it must be the most perfect and powerful mechanical contrivance with which we are acquainted—air and water in motion being governed by the same laws, with the exception that air is compressible and elastic

In order to meet the differences, I have made the upper part of my wheel a large air chamber, then, as the water comes up into the wheel, instead of striking an iron plate, it strikes a column of confined air, and by the force of elasticity, it is thrown back upon the wheel without loss of power and escapes horizontally at the perimeter of the wheel—thus doing away with most of the impact and friction which seems to be a necessity to most other wheels. The claims that I have got allowed are, first, the air chamber, which is described as spherical, surmounting the wheel; second, a scroll shaped flume, with a central aperture through the top plate corresponding to one in the lower section of the wheel. The water enters the flume and is made to assume a vortical or cyclonic motion before it reaches the wheel, so that the wheel does not have to expend the power in changing a direct motion of the water column to a rotary or spiral one, but it gains in power from the application of the cyclonic motion, which the water has gained in passing through the flume, so that the wheel gets not only the head pressure but that due to the acquired centrifugal motion. The value of this wheel seems to be, first, in rapid whirlpool motion before it touches the wheel, and consequently does not have to perform that labor; second, its great velocity of revolution; third, the water coming in at the center and flowing outward makes the most of centrifugal force, which force is additional to head pressure, and will increase in proportion to the square of its velocity; fourth, a small wheel will do as much work as others two or three sizes larger, because the pressure, being greater, will discharge more water through the same vent with corresponding power.

24-inch wheel; six outlets, each 2⅜ inches square,

Data below for one minute. Multiply revolutions by 10. March 10, 1880.

Gate Opened						Head	Weight	Rev per minute	Horse Power	Cubic Feet	Per Cent
Whole Gate.	.	.	.	.	.	17.10	200	000	0.00	307.20	000
"	"	.	.	.	.	16.98	50	380	5.75	459.87	.3900
"	"	.	.	.	.	17.03	75	299	6.70	428.46	.4861
"	"	.	.	.	.	17.02	100	204	6.18	392.29	.4901
"	"	.	.	.	.	17.00	60	352	6.40	443.41	.4495
"	"	.	.	.	.	17.00	70	321.5	6.81	437.96	.4842
"	"	.	.	.	.	17.00	80	288.5	6.99	425.77	.5113
"	"	.	.	.	.	17.00	90	252.5	6.88	409.61	.5231
"	"	.	.	.	.	17.00	85	266.5	6.86	406.93	.5250
"	"	.	.	.	.	17.00	85	247	6.36	384.36	.5153
"	"	.	.	.	.	17.00	75	290	6.59	402.92	.5094
"	"	.	.	.	.	17.00	65	323.5	6.37	420.36	.4719
"	"	.	.	.	.	17.00	50	195	2.95	268.93	.3416

Another test of same wheel, the outlets being enlarged to 2⅞ inches square.

Whole Gate.						Head	Weight	Rev per minute	Horse Power	Cubic Feet	Per Cent
Whole Gate.	.	.	.	.	.	16.93	230	000	0.00	416.31	000
"	"	.	.	.	.	16.80	75	322.5	7.32	564.50	.4086
"	"	.	.	.	.	16.80	85	289.5	7.15	555.81	.4224
"	"	.	.	.	.	16.83	95	262	7.54	522.86	.4537
"	"	.	.	.	.	16.84	105	235.5	7.49	508.69	.4630
"	"	.	.	.	.	16.80	100	246.5	7.46	515.76	.4558
"	"	.	.	.	.	16.82	90	278	7.58	534.23	.4466

Hunt Wheel.

Sent by R. Hunt Machine Co., Orange, Mass.

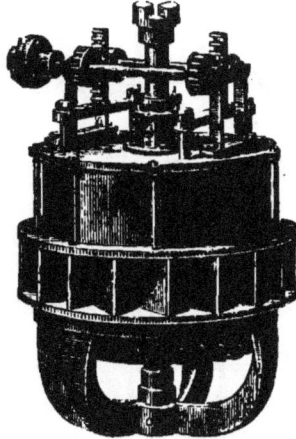

36-inch wheel. Downward and outward discharge.

Data below for one minute. Multiply revolutions by 15. May 19, 1880.

Gate Opened										Head	Weight	Rev per minute	Horse Power	Cubic Feet	Per Cent
Whole Gate.	.	.	.	.	.	.				17.73	1610	000	000	2431.00	000
"	"	.	.	.	.	.	.			17.75	800	179	65.09	2399.36	.8091
"	"	.	.	.	.	.	.			17.78	820	176.5	65.78	2396 20	.8174
"	"	.	.	.	.	.	.			17.71	840	174.5	66.62	2402.52	.8290
"	"	.	.	.	.	.	.			17.74	860	170.5	68.65	2418.32	.8226
"	"	.	.	.	.	.	.			17.73	880	169	67.60	2424.65	.8324
"	"	.	.	.	.	.	.			17.69	900	166.5	68.11	2434.16	.8374
"	"	.	.	.	.	.	.			17.69	920	164	68.58	2450.02	.8376
"	"	.	.	.	.	.	.			17.68	940	161.5	69.00	2459.55	.8401
"	"	.	.	.	.	.	.			17.69	960	160	69.81	2469.09	.8461
"	"	.	.	.	.	.	.			17.70	980	156	69.49	2481.94	.8374
"	"	.	.	.	.	.	.			17.67	1000	154.5	70.22	2497.79	.8425
"	"	.	.	.	.	.	.			17.66	1025	150.5	70.31	2507.37	.8382
"	"	.	.	.	.	.	.			17.67	1050	148	70.63	2513.77	.8419
"	"	.	.	.	.	.	.			17.68	1100	139.5	69.75	2513.77	.8309
Part Gate.	.	.	.	.	.	.				18.34	200	148.6	13.50	1167.94	.3337
"	"	.	.	.	.	.	.			18.20	300	165	25.53	1473.79	.5639
"	"	.	.	.	.	.	.			18.20	350	152.5	24.26	1498.56	.4709
"	"	.	.	.	.	.	.			18.07	500	151.5	34.43	1781.38	.5663
"	"	.	.	.	.	.	.			17.94	700	143.3	45.56	2044.46	.6576
"	"	.	.	.	.	.	.			17.90	700	150	47.72	2086.79	.6763
"	"	.	.	.	.	.	.			17.82	850	141.2	54.65	2260.65	.7169
"	"	.	.	.	.	.	.			17.79	850	146	56.40	2298.97	.7304
"	"	.	.	.	.	.	.			17.79	850	152.5	59.13	2352.15	.7481

Hunt Wheel.

Sent by R. Hunt Machine Co., Orange, Mass.

36-inch wheel.   Downward discharge.

This wheel was made from the same patterns as the one upon the opposite page, but the hoop extended down to the bottom of the bucket, completely closing the outward discharge. The shaft of each wheel was extra heavy, or of large diameter, as they were made to work under high heads.

Data below for one minute.   Multiply revolutions by 15.   May 20, 1880.

Gate opened							Head	Weight	Rev per minute	Horse Power	Cubic Feet	Per Cent
Whole Gate.	.	.	.	.	.	.	17.18	1700	000	000	2568.33	.000
"	"	.	.	.	.	.	17.65	850	165	63.75	2283.40	.7998
"	"	.	.	.	.	.	17.65	875	162.5	64.63	2305.21	.8410
"	"	.	.	.	.	.	17.61	900	159	65.04	2317.69	.8437
"	"	.	.	.	.	.	17.58	925	155.5	65.38	2320.82	.8484
"	"	.	.	.	.	.	17.60	950	150.5	64.97	2336.47	.8363
"	"	.	.	.	.	.	17.58	975	149.5	66.25	2342.74	.8516
"	"	.	.	.	.	.	17.58	1000	145.5	66.10	2355.29	.8452
Part Gate.	.	.	.	.	.	.	18.27	200	174	15.81	1147.42	.3993
"	"	.	.	.	.	.	18.27	250	162 5	17.55	1204.10	.4223
"	"	.	.	.	.	.	18.26	290	153	20.16	1235.34	.4732
"	"	.	.	.	.	.	18.15	400	160	29.09	1498.56	.5662
"	"	.	.	.	.	.	18.15	450	150	30.68	1528.98	.5854
"	"	.	.	.	.	.	17.99	700	134	42.63	1880.66	.6670
"	"	.	.	.	.	.	17.98	600	154.5	42.13	1851.22	.6701
"	"	.	.	.	.	.	17.96	650	144	39.51	1874.78	.6212
"	"	.	.	.	.	.	17.86	750	157.5	51.64	2089.82	.7325
"	"	.	.	.	.	.	17.68	875	149.6	59.50	2267 89	.7874

Mercer's Reliable Turbine.

24-inch wheel, sent by Mercer & Stinman, Lancaster, Pa.

Downward discharge. Outside register gate.

Data below for one minute. Multiply revolutions by 10. May 29, 1880.

Gate Opened			Head	Weight	Rev per minute	Horse Power	Cubic feet	Per Cent.
Whole Gate.		.	18.28	775	000	000	1028.77	000
"	"	.	18.24	350	211.6	22.42	998.95	.6514
"	"	.	18.31	360	209.5	22.85	1001.43	.6449
"	"	.	18.31	370	206.7	23.17	1001.43	.6690
"	"		18.30	380	204.5	23.54	1003.91	.6784
"	"		18.30	390	201.5	23.81	1006.39	.6845
"	"		18.29	400	200	24.24	1008.87	.6954
"	"		18.29	410	198.5	24.66	1011.35	.7058
"	"		18.28	420	193.5	24.62	1013.87	.7033
"	"		18.29	430	189	24.62	1016.32	.7012
Part Gate.		.	18.22	375	195	22.15	961.56	.6694
"	"	. . .	18.28	375	192.5	21.87	947.34	.6686
"	"	. . . .	18.36	325	198.7	19.56	896.45	.6292
"	"	. . . .	18.34	275	201.5	16.79	839.22	.5775
"	"	. . . .	18.25	250	210.5	15.94	836.86	.5525
"	"	. . . .	18.50	190	207	11.91	730.07	.4668
"	"	. . . .	18.52	175	212	11.24	723.23	.4443
"	"	. . . .	18.57	150	225	10.22	698.31	.4173
"	"	. . . .	18.71	115	197	6.86	555.72	.3302
"	"	. . . .	18.71	100	212.5	6.44	551.46	.3304

Rechard Wheel.

24-inch wheel, sent by George F. Baugher, York, Pa.

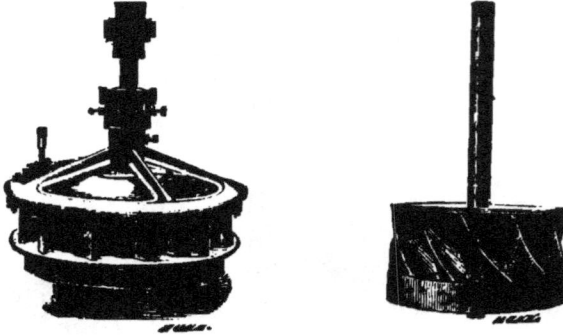

Turbine building, like the other arts, started with low beginnings, how far back it is impossible to determine. Water wheels, working upon verticle shafts, were used centuries since. The tub wheel, with buckets made of wood, and shaped substantially like those of the Jonval wheel, were the earliest in my recollection, though the impact, flutter, undershot, breast and overshot were also common at that time—all of which were objectionable under certain conditions. Fourneyron, Jonval, Parker, Boyden, and many others, attempted to produce wheels free from such objections, but, in doing so, overlooked the essential feature necessary to make their efforts successful.

In supplying a mill with motive power, a surplus for emergencies is absolutely necessary. The plans of the builders alluded to were generally capable of producing wheels reasonably efficient, when working with the maximum supply of water that would pass through their openings. Half that quantity would hardly turn the wheel to speed. Consequently, with such wheels in use upon our variable streams, it was necessary to have them so small that, during nine months of each year, from half to three-fourths of the water would run to waste over the dam, or the works must stand idle through the dry months—a fact that prejudiced manufacturers to such an extent that breast or overshot wheels have been displaced with reluctance.

Mr. Rechard, like a few other recent builders, has worked upon a different plan, as may be seen by an examination of the tabulated results below, or in the diagram connected with this report, instead of striving for high results at full gate, where a turbine is seldom used unless during back water, when the quantity used is of no account. He has so arranged chutes and buckets as to gain his eighty-five per cent. at three-fourths gate, or at the point at which the wheel is most likely to be used, instead of from thirty to sixty per cent. that would be realized by the use of the Fourneyron or any of the early whole gate turbines. Wheels equal to the one tested of this make are far superior in efficiency to any breast or overshot wheel that can be produced, no matter what the head may be; and such wheels enable the user to get the full benefit of his stream, either in its highest or lowest supply.

The results below show this wheel to be the most economical in the use of water at about three-fourths discharge; and Mr. Baugher takes the very novel course of tabling the capacity of his wheels at that point, thus insuring the purchaser not only the full power represented in the table, but a surplus for emergencies.

Data below for one minute. Multiply revolutions by 15. June 8, 1880.

Gate Opened							Head	Weight	Rev per minute	Horse Power	Cubic Feet	Per Cent.
Whole Gate.	.	.	.	.	.	.	18.03	660	000	000	1749.33	000
"	"	.	.	.	.	.	18.05	300	280.5	38.25	1669 66	.6719
"	"	.	.	.	.	.	18.05	325	269	39.73	1678.15	.6944
"	"	.	.	.	.	.	18.05	350	252	40.09	1686.64	.6972
"	"	.	.	.	.	.	18.04	375	234	39.88	1703.67	.6870
"	"	.	.	.	.	.	18.00	400	215	39.00	1712.20	.6700
"	"	.	.	.	.	.	18.04	340	257	39.71	1689.48	.6898
"	"	.	.	.	.	.	18.02	360	244.5	40.00	1692.31	.6943
"	"	.	.	.	.	.	18 02	370	235.5	39.61	1695 15	.6865
Part Gate.	.	.	.	.	.	.	18.18	325	247	36.48	1356.80	.7829
"	"	.	.	.	.	.	18.22	300	255	34.77	1288.13	.7844
"	"	.	.	.	.	.	18.21	325	236.5	34.93	1293.38	.7851
"	"	.	.	.	.	.	18.26	300	255.5	34.84	1192.19	.8481
"	"	.	.	.	.	.	18.28	275	253	31.62	1153.92	.7937
"	"	.	.	.	.	.	18.34	250	253 5	28.80	1070.96	.7763
"	"	.	.	.	.	.	18.38	225	253.5	25.92	977.65	.7639
"	"	.	.	.	.	.	18.39	225	248	25.21	960.73	.7555
"	"	.	.	.	.	.	18.41	225	244.5	25.01	946.29	.7600
"	"	.	.	.	.	.	18.46	185	251.5	21.15	861.59	.7123
"	"	.	.	.	.	.	18.46	200	240	21.81	856.26	.7305
"	"	.	.	.	.	.	18.52	150	250	17.04	728 26	.6689
"	"	.	.	.	.	.	18.51	150	247.5	16.87	721 57	.6687
"	"	.	.	.	.	.	18.55	150	233	15.88	606.43	.6800
"	"	.	.	.	.	.	18.62	100	243.5	11.06	555.61	.5662
"	"	.	.	.	.	.	18.61	80	257	9.34	534 91	.4968
"	"	.	.	.	.	.	18.67	80	224	8.14	452.20	.5104

The Economical Turbine.

24-inch wheel, sent by S. Martin, York, Pa.

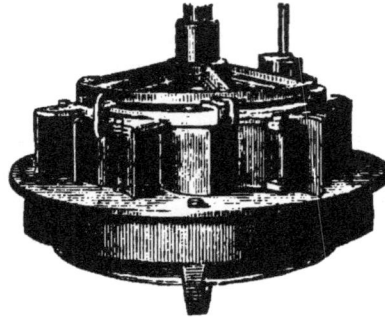

This turbine consisted of an upper plain downward discharge wheel above one of an outward discharge. The builder declined to have a test made of the upper wheel alone.

During this test, the area of aperture was 102 square inches.

Data below for one minute. Multiply revolutions by 10. June 15, 1880.

Gate Opened	Head	Weight	Rev per minute	Horse Power	Cubic feet	Per Cent
Whole Gate.	18.29	485	000	0000	752.36	000
" "	18.26	200	244.5	14.81	788.68	.5446
" "	18.25	210	234	14.89	787.09	.5489
" "	18.25	220	225	15.00	785.50	.5527
" "	18.27	230	215.5	15.02	783.91	.5552
" "	18.27	240	206	14.98	780.74	.5560
" "	18.27	250	194.2	14.71	777.58	.5482
Part Gate.	18.58	50	234.3	3.55	371.10	.2726
" "	18.55	70	205.5	4.35	368.46	.3369
" "	18.53	100	203	6.15	415.23	.4232
" "	18.51	95	210	6.05	424.94	.4072
" "	18.49	125	201.5	7.63	478.63	.4766
" "	18.48	120	209	7.60	481.43	.4522
" "	18.44	140	209.5	8.88	524.00	.4866
" "	18.41	160	210	10.18	570.30	.5134
" "	18.41	170	197.5	10.17	573.24	.5102
" "	18.41	165	205.5	10.27	571.77	.5154
" "	18.39	190	204.7	11.78	623.57	.5438
" "	18.34	210	198.5	12.63	656 70	.5552
" "	18 35	205	204.5	12.70	659.73	.5554

Second test of same wheel, area of aperture being reduced to 72 square inches.

Gate Opened	Head	Weight	Rev per minute	Horse Power	Cubic feet	Per Cent
Whole Gate.	18.31	200	231.2	13 12	685.66	.5908
" "	18.31	210	222.5	13.10	681.07	.6007
" "	18.32	220	212.5	11.37	682.60	.5195
" "	18.33	230	204	11.86	682.60	.5498
" "	18.33	250	184	9.75	684.13	.5872
Part Gate. . . .	18.33	215	201.5	13.12	638.58	.5934
" "	18.34	210	206	13.10	634.07	.5964
" "	18.39	175	214.5	11.37	590.87	.5541
" "	18.40	190	206	11.86	582.04	.5864
" "	18.44	150	214.5	9.75	521.08	.5372
" "	18.46	160	206	9.98	510.49	.5607
" "	18.52	125	194	7.34	431.57	.4862
" "	18.54	115	207	7.21	424.74	.4848
" "	18.61	65	196	3.86	321.72	.3413
" "	18.61	60	202	3.67	320.44	.3558

Stowe Wheel.

24-inch wheel, sent by E. W. Roff, Newark, N. J.

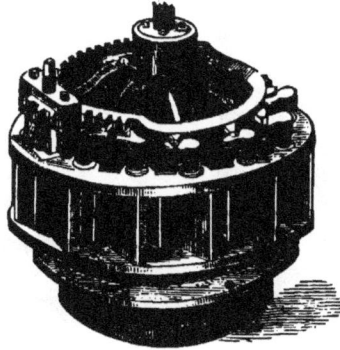

The claim for merit in this combination is upon the arrangement of gates, which open two at a time, up to sixteen in all. The plan of closing a part of the chutes or buckets of a turbine, for the purpose of using the water economically with a partial supply or at "part gate," has been tried by all of our noted turbine builders, and is still a favorite idea with amateurs or inexperienced persons interested in such matters. Walter S. Davis, of Warner, N. H., patented a plan nearly identical with that of the Stowe about 1870. J. B. Case, of Bristol, Ct., also, at about the same time, patented a plan the same in principle, though differing in detail.

Data below for one minute. Multiply revolutions by 10. June 17, 1880.

			Head	Weight	Rev per minute	Horse Power	Cubic Feet	Per Cent.
16 Gates Opened.		. . .	17.85	755	000	0000	1429.48	000
"	"	" . . .	18.00	340	276	28.43	1137.77	.7350
"	"	" . . .	18.08	350	274	29.06	1139.52	.7466
"	"	" . . .	18.05	360	269	29.31	1150.00	.7484
"	"	" . . .	18.04	370	267	29.90	1158.76	.7574
"	"	" . . .	18.04	380	265	30.51	1167.54	.7669
"	"	" . . .	18.05	390	260	30.72	1176.33	.7556
"	"	" . . .	18.05	400	255.5	30.97	1188.66	.7639
"	"	" . . .	18.02	410	253.5	31.49	1195.73	.7739
"	"	" . . .	18.01	420	.247.5	31.50	1202.81	.7700
"	"	" . . .	18.02	430	245	31.92	1213.44	.7729
"	"	" . . .	18.01	450	238	32.45	1224.06	.7795
"	"	" . . .	17.97	475	231	33.25	1256.05	.7800
"	"	" . . .	17.95	500	221	33.45	1265.21	.7799
"	"	" . . .	17.94	550	202	33.66	1292.21	.7678
10	"	" . . .	18.22	300	249.3	22.66	942.03	.6989
"	"	" . . .	18.17	325	238	23.43	968.68	.7047
8	"	" . . .	18.33	225	244	16.63	728.55	.6592
"	"	" . . .	18.31	235	240	17.09	744.09	.6642
"	"	.' . . .	18.31	245	233.5	17.33	753.45	.6650
6	"	" . . .	18.46	175	233	12.35	561.98	.6303
"	"	" . . .	18.43	165	236	11.80	563.43	.6017
4	"	" . . .	18.60	100	224	6.78	363.48	.5309
"	"	" . . .	18.60	90	231	6.30	362.17	.4952
"	"	" . . .	18.61	85	234.2	6.03	359.57	.4770
2	"	" . . .	18.18	50	210	3.18	203.66	.4546

Hard Working Gate.

Risdon Wheel.

Gate.

To ascertain the comparative efficiency of a plain cylinder gate at different stages of gate opening, the following experiments were made: A 36-inch Risdon turbine was selected for the purpose. It was one of the best, and from the same patterns the 90 per cent. wheels reported of that make were made. The gate hoisting rods and geared levers were changed to the plan to be seen upon the Hunt wheel reported upon another page. As the gate raised to open, it worked the other side up from what it is illustrated here and the four hoisting rods were connected to what is represented as the bottom, running up, and in no way obstructing the chutes. In this condition the wheel was carefully tested.

Data below for one minute. Multiply revolutions by 15.

Gate Opened	Head	Weight	Rev per minute	Horse Power	Cubic Feet	Per Cent
Whole Gate.	18.19	700	170.6	54.28	1966.76	.8033
" "	18.19	725	166	54.90	1972.76	.8070
" "	18.19	750	161.3	54.98	1970.76	.8120
" "	18.18	775	155.6	54.80	1970.76	.8099
" "	18.18	800	150.3	54.65	1979.76	.8039
" "	18.18	740	163	54.82	1966.76	.8118
" "	18.18	760	158.2	54.65	1953.76	.8110
Part Gate.	18.19	760	152.1	52.54	1901.16	.8044
" "	18.20	745	155	52.48	1901.16	.8030
" "	18.22	725	159.5	52.50	1898.19	.8045
" "	18.25	725	151	49.76	1824.45	.7914
" "	18.27	700	156.5	49.79	1818.59	.7934
" "	18.27	675	162.5	49.85	1818.59	.7944
" "	18.30	665	154.5	46.70	1728.38	.7817
" "	18.31	645	158	46.32	1722.60	.7775
" "	18.32	625	163.7	46.50	1713.96	.7840
" "	18.37	600	154.5	42.13	1608.25	.7550
" "	18.38	585	158.2	42.06	1602.59	.7560
" "	18.38	570	162.5	42.10	1605.42	.7654
" "	18.42	525	155.5	37.10	1476.72	.7222
" "	18.42	510	158.8	36.81	1476.72	.7164
" "	18.44	495	162.8	36.63	1479.49	.7180
" "	18.49	450	152.5	31.19	1326.84	.6730
" "	18.49	435	155.5	30.74	1324.15	.6709
" "	18.50	415	161.2	30.40	1321.47	.6583
" "	18.57	350	156	24.81	1160.69	.6095
" "	18.57	340	159.5	24.65	1160.09	.6055
" "	18.57	330	160.5	24.07	1155.52	.5939
" "	18.62	300	143.7	19.59	983.26	.5666

Easier Working Gate.

Risdon Wheel.

Gate.

Retest of the same wheel, the flange of the gate having been cut away about half the length of the chutes, as represented above.

Data below for one minute. Multiply revolutions by 15.

Gate Opened	Head	Weight	Rev per minute	Horse Power	Cubic Feet	Per Cent
Whole Gate.	18.21	700	171	54.40	1962.74	.8058
" "	18.20	725	165.7	54.60	1965.53	.8080
" "	18.21	750	162	55.22	1968.52	.8149
" "	18.21	775	155.2	54.67	1965.53	.8086
" "	18.21	800	149.7	54.43	1977.49	.8003
" "	18.21	740	162.5	54.65	1971.51	.8058
" "	18.20	760	158.5	54.75	1983.48	.8031
Part Gate.	18.23	760	151.5	52.33	1911.94	.7950
" "	18.22	745	154.3	52.25	1908.98	.7952
" "	18.23	725	159.5	52.56	1906.02	.8008
" "	18.25	725	150.5	49.59	1835.31	.7838
" "	18.26	700	155.3	49.41	1829.39	.7832
" "	18.26	675	160.5	49.24	1826.48	.7817
" "	18.30	665	151.3	45.73	1727.80	.7656
" "	18.29	645	156	45.73	1719.16	.7700
" "	18.31	625	160.6	45.62	1716.28	.7685
" "	18.35	600	151	41.18	1613.65	.7363
" "	18.34	585	155.2	41.26	1605.18	.7420
" "	18.35	570	158	40.91	1605.18	.7352
" "	18.34	555	163	41.12	1601.56	.7413
" "	18.40	510	152.6	35.37	1482.40	.6865
" "	18.40	495	158.7	35.70	1479.64	.6943
" "	18.40	480	163	35.56	1476.88	.6928
" "	18.46	430	154.6	30.11	1335.51	.6466
" "	18.46	415	158.5	29.89	1330.15	.6401
" "	18.47	400	163.5	29.72	1327.47	.6403
" "	18.53	340	155.6	24.04	1169.65	.5739
" "	18.54	325	160	23.63	1167.07	.5781
" "	18.53	315	163	23.33	1164.49	.5726
" "	18.60	260	155	18.31	1000.06	.5211

Easy Working Gate.

Risdon 36-inch wheel. Gate.

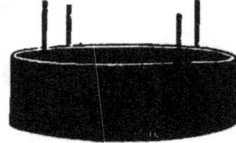

A third test of the same wheel, the flange of the gate having been cut entirely away, leaving a plain cylinder gate.

Data below for one minute. Multiply revolutions by 15.

Gate Opened						Head	Weight	Rev per minute	Horse Power	Cubic Feet	Per Cent
Whole Gate.	•	•	•	•	•	18.23	700	177	56.31	2120.25	.7713
"	"	•	•	•	•	18.24	725	171.2	56.48	2139.46	.7662
"	"	•	•	•	•	18.25	750	165.6	56.75	2139.46	.7695
"	"	•	•	•	•	18.24	775	160.5	56.54	2136.39	.7681
"	"	•	•	•	•	18.22	800	154.5	56.18	2136.39	.7643
"	"	•	•	•	•	18.23	740	168	56.50	2139.46	.7609
"	"	•	•	•	•	18.23	760	163.7	56.41	2136.39	.7609
Part Gate.	•	•	•	•	•	18.24	760	156.5	54.06	2096.53	.7484
"	"	•	•	•	•	18.26	745	161.5	54.68	2090.41	.7585
"	"	•	•	•	•	18.24	725	165.6	54.27	2081.25	.7569
"	"	•	•	•	•	18.26	725	155	51.07	2023.49	.7317
"	"	•	•	•	•	18.27	700	161.5	51.38	2017.43	.7380
"	"	•	•	•	•	18.26	675	166.6	51.12	2017.43	.7347
"	"	•	•	•	•	18.28	665	157	47.45	1942.20	.7076
"	"	•	•	•	•	18·30	645	158.5	46.46	1939.21	.6932
"	"	•	•	•	•	18.28	625	166	47.17	1923.23	.7103
"	"	•	•	•	•	18.30	600	155 5	42.40	1850.02	.6631
"	"	•	•	•	•	18.31	585	158.5	42.14	1850.02	.6586
"	"	•	•	•	•	18.31	570	163	42.23	1844.12	.6622
"	"	•	•	•	•	18.31	555	166.2	41.92	1838.22	.6594
"	"	•	•	•	•	18 36	510	157.7	36.55	1727.17	.6101
"	"	•	•	•	•	18.38	495	164	36.90	1727.17	.6154
"	"	•	•	•	•	18.37	480	167.5	36.54	1727.17	.6097
"	"	•	•	•	•	18.42	430	158	31.18	1601.01	.5600
"	"	•	•	•	•	18.42	415	162.5	30.65	1595.34	.5323
"	"	•	•	•	•	18.42	400	167.5	30.45	1595.34	.5486
"	"	•	•	•	•	18.48	340	160	24.72	1458.15	.4856
"	"	•	•	•	•	18.47	325	165	24.37	1455.39	.4800
"	"	•	•	•	•	18.48	315	167.6	23.96	1455.39	.4717
"	"	•	•	•	•	18.55	260	156.6	18.51	1289.50	.4097
"	"	•	•	•	•	18.54	245	162.5	18.09	1284.16	.4023
"	"	•	•	•	•	18.54	230	167.2	17.48	1281.49	.3895
"	"	•	•	•	•	18.62	150	164	11.18	1083	.2935
"	"	•	•	•	•	18.64	140	166	10.56	1077.89	.2782

EXPERIMENTS

WITH

Gears, Belts and Draft Tubes.

[These experiments occupied the time from March 18 to
April 23 inclusive.]

In presenting these results, it is not pretended that they exhaust
the subjects, for such is far from being the case, as every change
made, no matter how slight, caused a change in the rate of transmission. The best results obtained are given, while the conditions
under which they were obtained were certainly quite as favorable
as gears and shafting are likely to be placed in mills. The great
loss in transmission through the spur gears was entirely unexpected,
and the experiment was repeated at intervals, during several weeks,
with substantially the same results at each repetition, and it would
seem desirable to make a more exhaustive trial by trying a greater
variety of gears of different make and relative proportion, and particularly of gears made from the same patterns, but of different
brands of iron. There must be some discoverable cause why one
gear will run without perceptible wear for years, when another, put
in to replace it, cuts out in a day or two. So of water wheel steps,
where two wheels, seemingly alike, placed in the same pit, with one
the step lasts for years, while the other requires a new one monthly.
Is there not some property in the iron that causes such different
effects? At any rate, it is hardly worth while to spend time, brains
and money in efforts to produce turbines and other engines of the
highest efficiency, unless corresponding efforts are made to transmit
a reasonable proportion of such efficiency.

To find the loss of power in transmission through gears, and the
loss by use of draft tubes, the highest efficiency in each case must
be compared with that of the 15-inch Victor wheel reported upon
the next page.

Victor Turbine.

.5 inches in diameter. Price, $250.

This wheel was in use several weeks to make the following gear, draft tube and belt experiments. The results below show the efficiency of the wheel. Data for one minute. Multiply revolutions by 10.

Gate Opened	Head	Weight	Rev per minute	Horse Power	Cubic Feet	Per Cent
Whole Gate.	17.98	310	323.2	30.36	981.15	.9111
" "	17.97	320	300.5	29.13	981.15	.8747
" "	17.97	290	348.5	30.62	974.47	.9258
" "	17.98	290	347.5	30.53	972.80	.9242
" "	18.00	280	355	30.12	969.47	.9139
" "	17.98	300	337.3	30.66	977.81	.9234
Part Gate.	17.99	300	331	30.09	972.80	.9102
" "	17.99	290	345	30.31	972.80	.9174
" "	18.00	300	334.5	30.40	972.80	.9191
" "	17.99	290	334	29.35	971.13	.8896
" "	17.99	275	339	28.25	962.82	.8634
" "	18.02	260	331.5	26.11	901.88	.8506
" "	18.03	250	338.5	25.64	897.00	.8394
" "	18.09	230	331.2	23.08	820.67	.8231
" "	18.09	225	339.5	23.14	808.53	.8376
" "	18.20	175	339	17.97	695.06	.7538
" "	18.38	105	334	10.62	482.59	.6345
" "	18.41	95	340	9.78	460.56	.6108

Re-test of the wheel some weeks later, several alterations having been made.

Whole Gate.	17.94	285	352	30.40	981.46	.9141

The results obtained from a 23-inch Boyden wheel, price $500, tested in the same place and under precisely the same conditions is here given. The Boyden wheel, however, had a sort of flanged gate specially fitted for the trial. With the ordinary gate, the results are shown in the lowest table. Made at Ames Works.

Best Whole Gate. . . .	18.16	195	263.5	15.21	553.15	.8364
Part Gate. . . .	18.14	155	263	12.35	477.27	.7551
" "	18.29	75	264	6.00	325.49	.5336
Whole Gate.	18.25	195	257.5	15.21	545.79	.8084
Part Gate.	18.33	75	259.5	5.87	380.63	.4973

Draft Tube in Backwater.

Experiment to determine whether a draft tube causes a loss of efficiency during backwater.

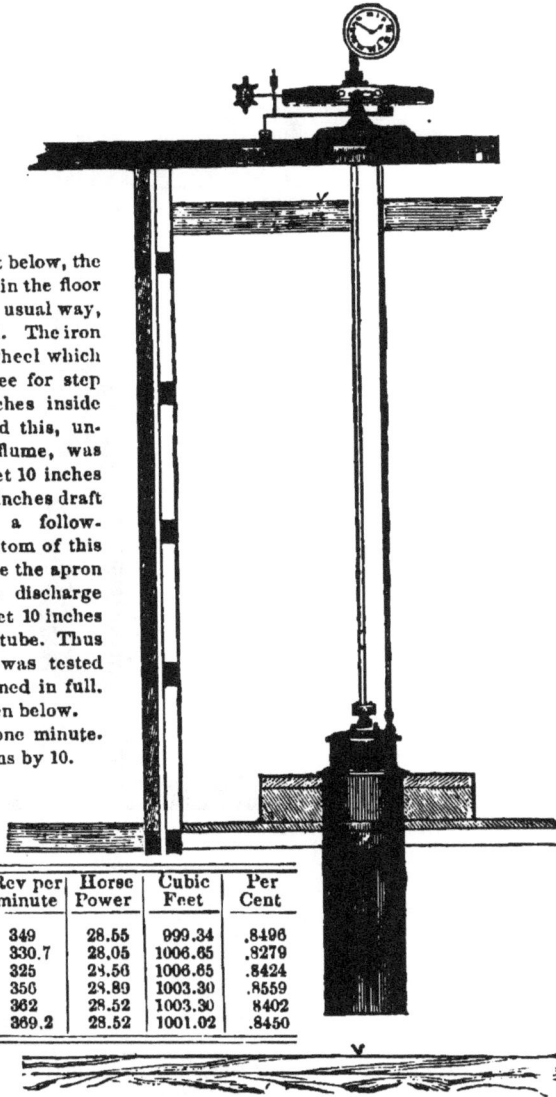

To make the test below, the wheel was placed in the floor of the flume in the usual way, under the full head. The iron draft tube of the wheel which held the bridge-tree for step was about 21 inches inside diameter. Around this, underneath floor of flume, was placed a piece 6 feet 10 inches in length of the 23 inches draft tube described on a following page. The bottom of this was 22 inches above the apron of wheel pit, the discharge being through 6 feet 10 inches of submerged draft tube. Thus placed, the wheel was tested with the gate opened in full. Results may be seen below.

Data below for one minute. Multiply revolutions by 10.

Head	Weight	Rev per minute	Horse Power	Cubic Feet	Per Cent
17.80	270	349	28.55	999.34	.8496
17.80	280	330.7	28.05	1006.65	.8279
17.81	290	325	28.56	1006.65	.8424
17.81	265	356	28.89	1003.30	.8559
17.83	260	362	28.52	1003.30	8402
17.84	255	369.2	28.52	1001.02	.8450

Draft Tube Experiments.

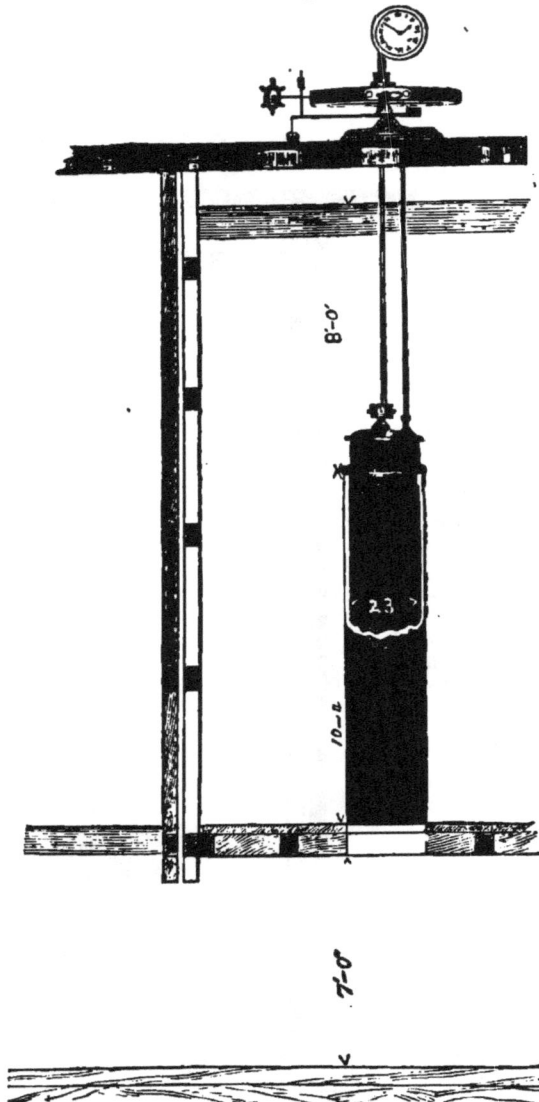

In preparing for these tests, the wheel was placed 10 feet above the flume floor upon the top of a draft tube 23 inches inside diameter, 10 feet 4 inches in length. Results on opposite page.

DEPARTMENT OF THE INTERIOR, UNITED STATES PATENT OFFICE,
Washington, D. C., June 17th, 1880.

SIR: In reply to your letter of 14th inst., you are informed that the records of this office show that the first patent granted for "Draft Tube for Water Wheels" was issued June 28th, 1840, No. 1658. It appears to have been the invention of Zebulon and Austin Parker of Licking Co., Ohio. The patent was issued to Zebulon Parker and R. McKilby, administrator of Austin Parker, deceased.

Respectfully yours,

F. A. SEELEY, *Chief Clerk.*

JAMES EMERSON, Willimansett, Mass.

Tests of 15 inch wheel placed as shown on opposite page.

The wheel was far less steady during this trial than when placed at the bottom of the flume. As the tube was surrounded by 8 feet of water, of course there was no leakage of air.

Gate Opened		Head	Weight	Rev per minute	Horse Power	Cubic Feet	Per Cent
Whole Gate.	. . . . .	17.73	285	322	27.80	959.51	.8651
" "	. . . . .	17.77	295	307.5	27.48	957.85	.8548
" "	. . . . .	17.77	305	291.5	26.94	961.18	.8352
" "	. . . . .	17.79	270	345	28.22	954.52	.8799
" "	. . . . .	17.78	275	336.2	28.02	954.52	.8741
" "	. . . . .	17.79	280	326	27.66	961.18	.8369
" "	. . . . .	17.80	275	338.5	28.20	957.85	.8737

Test of the same, the lower end of draft tube being unsubmerged.

Whole Gate.	. . . . .	17.80	100	266	8.06		

Second test of the same draft tube taken several days later.

Whole Gate.	. . . . .	17.91	200	349.3	21.17	869.88	.7194
" "	. . . . .	17.87	230	356.2	24.82	937.09	.7848
" "	. . . . .	17.82	250	362	27.42	957.00	.8512
" "	. . . . .	17.81	260	354.5	27.93	967.12	.8585
" "	. . . . .	17.81	270	341.7	27.95	974.77	.8523
" "	. . . . .	17.79	280	325.5	27.62	983.75	.8356
Part Gate.	. . . . .	17.96	200	328.5	19.90	789.75	.7428
" "	. . . . .	17.95	190	340.7	19.61	783.43	.7383
" "	. . . . .	18.11	100	356.6	10.80	550.66	.5733
" "	. . . . .	18.11	110	345	11.50	549.22	.6122

The wheel was more difficult to control with brake than during the first trial. It took a long time to clear the tube of air. Quite a number of tests were taken before anything like the power due the head could be obtained, though they were not recorded.

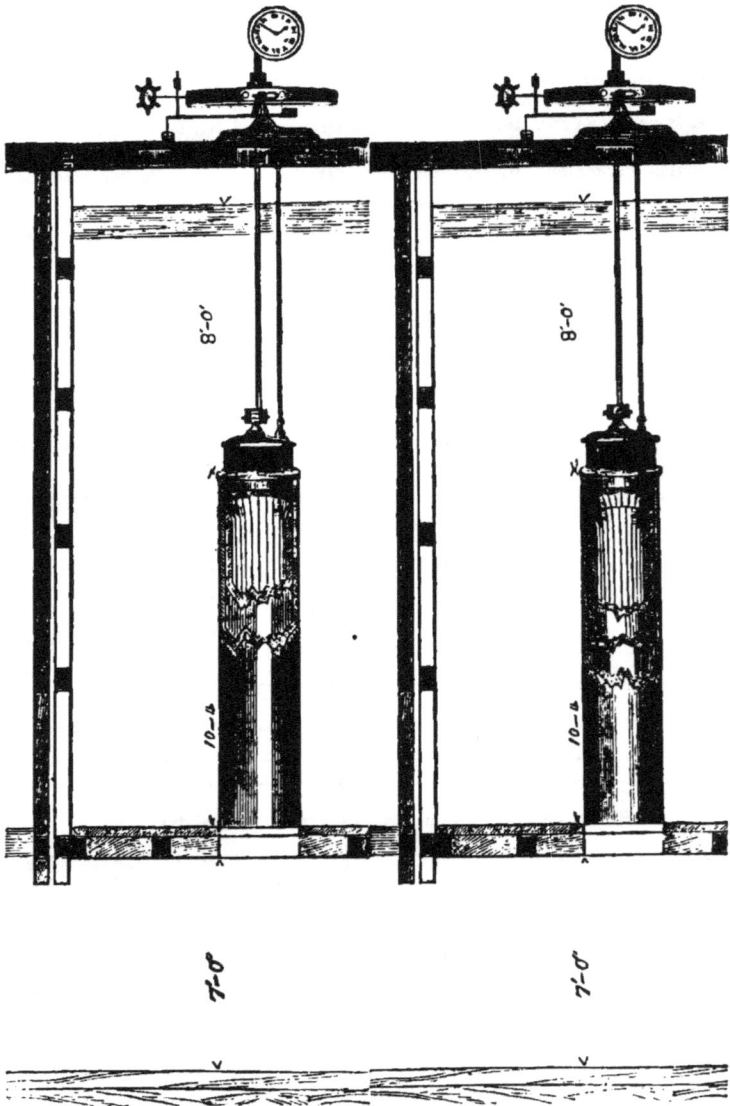

Reduced Draft Tube.

Test with 19-inch draft tube.

During this test the wheel was placed at the top of the before mentioned 23-inch draft tube, that having been diminished in diameter by the insertion of a lining 2 inches in thickness, leaving the inside diameter of tube 19 inches in the clear, and 10 feet 4 inches in length as before; and, as before, about 8 feet of the head above the wheel.

Gate Opened	Head	Weight	Rev per minute	Horse Power	Cubic Feet	Per Cent.
Whole Gate.	17.90	250	354	26.81	959.52	.8264
" "	17.88	260	337.5	26.57	961.18	.8185
" "	17.89	270	324.2	26.52	964.50	.8137
" "	17.89	280	309	26.21	966.17	.8029
" "	17.88	240	365	26.54	957.86	.8204

Test of the above arrangement the lower end of tube being unsubmerged.

Gate Opened	Head	Weight	Rev per minute	Horse Power	Cubic Feet	Per Cent.
Whole Gate.		130	369	14.53		
" "		140	367	15.57		

Draft Tube Again Reduced.

Test with 15-inch draft tube.

Continuation of the same arrangement of tubes as before, another lining having been inserted, leaving inside diameter of tube 15 inches; length, 10 feet 4 inches, as before.

Gate Opened	Head	Weight	Rev per minute	Horse Power	Cubic Feet	Per Cent.
Whole Gate.	17.88	200	376	22.81	890.78	.7584
" "	17.87	225	336	22.90	898.97	.7546
" "	17.85	250	296	22.42	905.53	.7339
" "	17.86	240	310.6	22.58	905.53	.7391
" "	17.86	230	324.5	22.61	902.25	.7429
" "	17.86	220	339.5	22.63	898.97	.7462
" "	17.86	210	355.5	22.62	894.05	.7500
" "	18.08	125	338.5	12.82	591.92	.6343
Gate open two-thirds. . .	18.24	70	322.5	6.84	415.42	.4779
" " one-half. . .	18.24	65	329.2	6.48	415.42	.4527
" " one-half. . .	18.24	60	338	6.14	411.37	.4333

Test with the lower end of draft tube unsubmerged.

Gate Opened	Head	Weight	Rev per minute	Horse Power	Cubic Feet	Per Cent.
Whole Gate.		200	365	22.30		
" "		225	323	22.02		

Elevation of Testing Flume and Draft Tube.

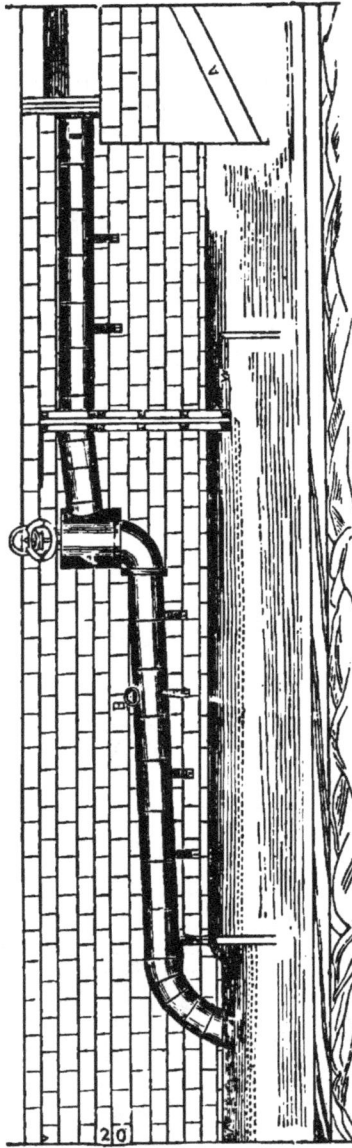

The Draft tube represented above along the side of flume and over the measuring pit, was so constructed that the water passing through it might be discharged below the weir, in order to allow a continued use of the wheel, with which the experiments were made, to add to the power used in the Whiting Paper Mill, near by. As may be seen, the water enters the round iron trunk above the flume. This trunk is four feet in diameter and about fifty feet in length to the wheel case, A. From the wheel case, the draft tube four feet in diameter, descending one foot in forty, carries the discharge over the weir, a distance of about fifty feet from the wheel. A 27-inch Hercules wheel, having a plain unflanged cylinder gate, was first tested in the ordinary way in the testing flume—the wheel standing in the opening of the floor, marked W. The results may be seen in the upper table on the opposite page. In the same place, with twenty feet head, the wheel would give 104 h. p., and make about 193 revolutions per minute. After the test in the flume, the wheel was placed in the curb, A, and the brake was applied at the top of shaft fitted for the crown gear. The results given in the lower table on opposite page show the efficiency of that style of draft tube.

The Hercules.

Test of wheel in flume in the ordinary way.
Data below for one minute. Multiply revolutions by 15. Dec. 6, 1879.

Gate opened			Head	Weight	Rev per minute	Horse Power	Cubic Feet.	Per Cent
Whole Gate.	. . . . .		17.09	1000	177.3	80.59	3264.21	.7648
"	"	. . . . . .	17.13	1050	167.5	79.94	3288.37	·7514
"	"	. . . . . .	17.02	1100	157.5	78.75	3288.37	.7450
"	"	. . . . . .	17.15	950	190	82.04	3240.09	.7817
"	"	. . . . . .	17.16	900	199	81.41	3205.72	.7835
"	"	. . . . . .	17.16	975	182.5	80.89	3233.21	.7719
"	"	. . . . .	17.16	925	194	81.56	3216.02	.7824

Test of wheel for power after it was placed in the wheel case, A, and previous to its being geared to the machinery in the mill near by.

Gate opened			Head	Weight	Rev per minute	Horse Power	Cubic Feet.	Per Cent
Whole Gate.	. . . . .		20.00	900	173	70.77	. . .	. . .
"	"	. . . . . .	20.00	850	184	71.09	. . .	. . .
"	"	. . . . .	20.00	800	188	68.36	. . .	. . .

View of Testing Flume, Horizontal Wheels and Draft Tube.

Curtis Wheel.

Sent by Gates Curtis, Ogdensburg, N. Y.

The results in the table below were obtained from the test of a 35-inch wheel upon upright shaft in the usual way. The inside register gate had been left out, so the chutes were open in full and the water was applied by the head gates of testing flume.

Data below for one minute. Multiply revolutions by 15. Oct. 22, 1879.

Gate Opened	Head	Weight	Rev per minute	Horse Power	Cubic Feet	Per Cent
Whole Gate.	17.69	2225	000	000	3115.69	.0000
" "	17.77	900	193	78.95	2905.30	.8096
" "	17.77	1000	179	81.36	2935.63	.8258
" "	17.75	1100	165.5	82.75	2950.83	.8365
" "	17.74	1200	152	82.90	2981.29	.8300
" "	17.73	1300	138.5	81.82	3004.21	.8133
" "	17.75	1050	171.2	81.70	2954.63	.8248
" "	17.74	1150	159	83.11	2906.05	.8363
" "	17.80	100	273.5	12.43	2574.47	.0000

After the above test, the same wheel, with a left-hand mate of the same supposed efficiency, was fixed upon a horizontal shaft, then placed in the flume at the top of a square draft tube ten feet in height, as shown on the opposite page. The draft tube and fittings were furnished by Mr. Curtis, and upon the same scale that he had furnished for other wheels of the kind for mills. The dotted lines in bulk-head show the application of the brake for testing. The same may perhaps be more clearly seen in the illustration of Measuring Pit in the first part of this report.

Data below for one minute. Multiply revolutions by 20.

Gate Opened	Head	Weight	Rev per minute	Horse Power	Cubic Feet	Per Cent
Whole Gate.	16.26	1500	141	128.18	5794.09	.7204
" "	16.38	1400	150.5	127.70	5779.76	.7141
" "	16.37	1450	145	127.42	5788.43	.6940
" "	16.37	1500	140.6	127.81	5817.38	.7089
" "	16.40	1350	151	126.81	5761.51	.7105
" "	16.39	1300	159.2	125.43	5738.27	.7161

New American Wheel.

30-inch wheel, sent by Stout, Mills & Temple, Dayton, Ohio.

Tests made to ascertain whether flaring the ordinary draft tube of a turbine at the bottom adds to its efficiency. During this trial the water, in passing through the wheel, made a constant rumbling or humming sound, whether the wheel was running or held stationary by the brake.

Data below for one minute. Multiply revolutions by 15. July 2, 1880.

	Head	Weight	Rev per minute	Horse Power	Cubic Feet	Per Cent.
Whole Gate 10½in. or 72 turns	17.75	1100	000	000	2504.00	000
" "	17.75	650	209.3	61.83	2389.50	.7735
" "	17.74	700	199	63.31	2424.32	.7793
" "	17.74	750	186.5	63.57	2478.42	.7655
" "	17.73	800	175.5	63.81	2539.27	.7523
" "	17.71	850	163	62.72	2574.71	.7284
" "	17.77	675	207.5	63.66	2430.66	.7803
" "	17.73	725	194	63.93	2478.42	.7703
Part Gate.	17.84	650	203	59.98	2267.22	.7852
" "	17.81	675	202	61.97	2354.82	.7823
" "	17.85	675	201.6	61.85	2329.69	.7874
" "	17.87	650	206.5	61.02	2261.00	.7997
" "	17.89	625	207.5	58.94	2202.14	.7922
" "	17.91	600	208.5	53.86	2113.18	.7954
" "	17.91	610	206.5	57.25	2128.44	.7951
" "	17.90	610	205.5	56.97	2116.23	.7962
" "	17.90	600	207.5	56.89	2091.86	.8044
" "	17.94	580	207.5	54.70	2013.23	.8017
" "	17.92	600	202	55.09	2031.30	.8012
" "	17.97	550	207.5	51.87	1914.68	.7980
" "	17.96	550	205	51.25	1896.82	.7963
" "	18.04	500	207	47.04	1759.40	.7861
" "	18.01	525	202	48.20	1794.24	.7904
" "	18.04	500	205	46.57	1727.66	.7912
" "	18.11	430	209.5	40.97	1561.16	.7673
" "	18.10	440	206.5	41.30	1577.00	.7661
" "	18.10	450	205.5	42.34	1589.21	.7793
" "	18.09	475	199	42.96	1613.56	.7792
" "	18.17	390	208	36.84	1443.91	.7434
" "	18.16	400	205.5	37.36	1427.51	.7631
" "	18.24	335	206	31.36	1234.86	.7372

The average efficiency from half to whole gate, .779

New American Wheel.

30-inch wheel.

Retest of the wheel after slightly reducing its diameter, as it was found to have touched the curb during the former trial. As may be seen, this change raised the whole gate efficiency at the expense of that of the part gate.

Data below for one minute. Multiply revolutions by 15. July 7, 1880.

Gate Opened	Head	Weight	Rev per minute	Horse Power	Cubic Feet	Per Cent.
Whole Gate 10¼ in or 72 turns	17.67	675	210.4	64 55	2487.94	.7774
" " " " "	17.64	685	208.5	64.92	2494.32	.7812
" " " " "	17.65	700	203	64.59	2507.07	.7727
" " " " "	17.63	715	201.5	65.46	2513.46	.7821
Gate open 8 1-16in. or 59 turns	17.67	700	200.5	63.79	2405.55	.7964
" " " " "	17.71	675	203	62.28	2349.00	.7926
" " " " "	17.72	650	209.5	61.90	2323.99	.7958
" " 7⅜ " 55 "	17.75	650	202.5	59.83	2286.62	.7804
" " " " "	17.74	640	208.5	60.66	2268.01	.7983
" " " " "	17.72	675	200	61.36	2299.06	.7976
" " 7 " 51 "	17.77	650	201.2	59.44	2212.41	.7990
" " " " "	17.77	630	207	59.27	2200.12	.8027
" " 6½ " 47 "	17.81	625	204	57.95	2141.97	.8042
" " " " "	17.80	645	198	55.02	2151.12	.7608
" " " " "	17.81	600	206.7	56.37	2135.87	.7845
" " 5 15-16 " 43 "	17.85	575	205	53.58	2018.06	.7875
" " " " "	17.85	560	208	52.94	2009.08	.7816
" " " " "	17.84	600	201	54.81	2036.06	.7989
" " 5 5-16 " 39 "	17.92	525	206	49.15	1890.44	.7681
" " " " "	17.90	545	204	50.53	1899.26	.7869
" " 4¾ " 35 "	17.97	500	203	46.13	1765.26	.7700
" " " " "	17.96	490	204	45.43	1759.50	.7611
" " " " "	17.96	480	207	45.16	1750.86	.7604
" " 4½ " 31 "	18.03	465	197	41.63	1639.76	.7482
" " " " "	18.04	450	203	41.52	1631.30	.7471
" " 3½ " 27 "	18.11	390	206	36.51	1453.89	.7358
" " " " "	18.12	400	203	36.90	1473.00	.7319
" " 2 15-16 " 23 "	18.17	300	220	30.00	1251.00	.6988
" " " " "	18.16	335	285.5	31.44	1292.99	.7089
" " " " "	18.16	350	200.5	31.90	1293.00	.7192

Average, .771

New American Wheel.

30-inch Wheel.

Retest of the same wheel after changing the flaring for a straight draft tube. The gate openings were the same through the three trials. The 10½ inches at whole gate means the extreme swing of gate, the openings at outer end of chutes being 7½ inches only; but the gate had to move the distance named to clear the openings. The averages are found by adding the thirty tests of each trial together and dividing by that number.

Data below for one minute. Multiply revolutions by 15. July 8, 1880.

	Head	Weight	Rev per minute	Horse Power	Cubic Feet	Per Cent.
Whole Gate 10½in. or 72 turns	17.73	675	209.8	64.37	2491.16	.7715
" " " " " " "	17.70	700	205.2	65.29	2500.72	.7809
" " " " " " "	17.68	725	198.5	65.41	2513.48	.7794
" " " " " " "	17.67	750	192.5	65.92	2545.04	.7759
Gate open 59 turns. " "	17.77	675	203.5	62.43	2355.41	.7900
" " 55 " " "	17.79	625	212	60.22	2252.75	.7950
" " " " " "	17.79	650	206.5	60.71	2258.93	.8000
" " " " " "	17.78	675	200	61.36	2289.92	.7979
" " 51 " " "	17.82	650	202	59.68	2225.00	.7969
" " " " " "	17.81	625	207.5	58.94	2194.25	.7984
" " 47 " " "	17.85	600	208.5	56.86	2136.19	.7895
" " " " " "	17.85	625	203.5	57.41	2145.33	.7937
" " " " " "	17.85	650	197	58.20	2154.48	.8012
" " 43 " " "	17.90	575	208	54.36	2015.49	.7978
" " " " " "	17.90	600	201.5	54.95	2151.48	.7923
" " 39 " " "	17.96	525	208	47.63	1885.29	.7760
" " " " " "	17.96	550	206.5	51.62	1899.79	.8009
" " " " " "	17.94	575	200	52.27	1952.93	.7900
" " 35 " " "	18.01	475	211	45.56	1751.52	.7647
" " " " " "	18.00	500	206	46.00	1786.09	.7574
" " " " " "	18.00	525	201	47.96	1806.35	.7810
" " 31 " " "	18.06	425	211.5	40.85	1599.41	.7488
" " " " " "	18.05	450	207.5	42.44	1632.08	.7629
" " " " " "	18.04	475	198	42.75	1666.00	.7531
" " 27 " " "	18.12	375	213.2	36.34	1454.85	.7299
" " " " " "	18.11	400	207.5	37.72	1484.90	.7426
" " " " " "	18.09	425	200	38.63	1509.60	.7489
" " 23 " " "	18.17	325	216.5	31.97	1308.32	.7119
" " " " " "	18.18	350	208	33.09	1328.49	.7253
" " " " " "	18.17	375	200	34.09	1363.11	.7287

Average, .774

Experiments with Gears.

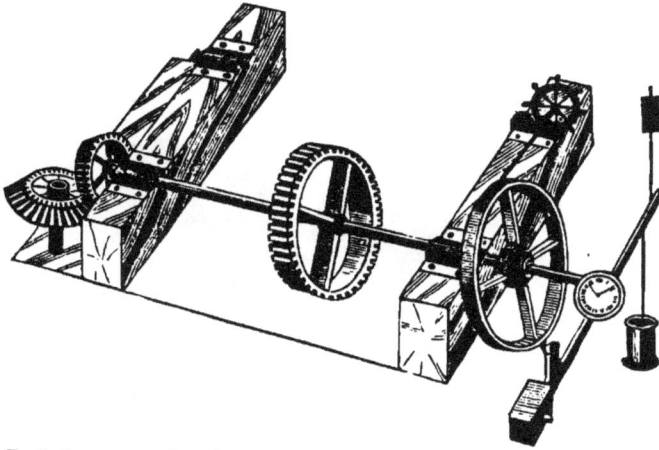

Test of gears continued, the arrangement of gears named on previous page being reversed, or the small gear having 26 teeth being on turbine shaft, that of 46 teeth on "Jack Shaft"—gears being worked without lubrication of any kind. Data below for one minute. Multiply revolutions by 10.

Head	Weight	Rev per minute	Horse Power	Cubic Feet	Per Cent
18.10	210	267.5	17.02	882.34	.5642
18.08	220	261.5	17.43	893.69	.5712
18.07	230	256	17.84	896.95	.5826
18.07	240	253	18.40	898.58	.6000
18.06	250	248.5	18.82	901.84	.6117
18.05	260	245	19.30	910.00	.6221
18.03	270	241	19.71	923.10	.6270
18.03	280	239.2	20.95	926.38	.6650
18.04	290	237.5	20.87	926.38	.6613
18.03	300	230.5	20.95	928.02	.6628

Test of above named arrangement of gears, the gears being well oiled.

Head	Weight	Rev per minute	Horse Power	Cubic Feet	Per Cent
17.83	350	229	24.28	902.45	.7989
17.81	400	221	26.78	937.06	.8494
17.78	425	213	27.43	962.00	.8490
17.77	450	204	27.81	968.68	.8555
17.76	475	196	28.21	972.00	.8653
17.75	500	187	28.33	978.71	.8634
17.74	525	173.5	27.60	798.71	.8416

Verification of the same arrangement of gears taken several days later.

Head	Weight	Rev per minute	Horse Power	Cubic Feet	Per Cent
18.02	475	197.5	28.42	963.56	.8665
18.03	525	176.5	29.07	969.63	.8500
18.02	512.5	180.6	28.04	971.31	.8482
18.05	500	187.8	28.45	978.59	.8571

During the above tests, the teeth of the gears ran rather close together, though perfectly free and were correctly placed according to the opinion of experts in such matters. They were separated about 1-16 in. more, then gave the results below.

Head	Weight	Rev per minute	Horse Power	Cubic Feet	Per Cent
18.02	500	191.5	29.01	972.67	.8762
17.99	510	187.5	28.97	979.33	.8706
18.00	520	184.2	29.02	981.00	.8700

Experiments with Gears.

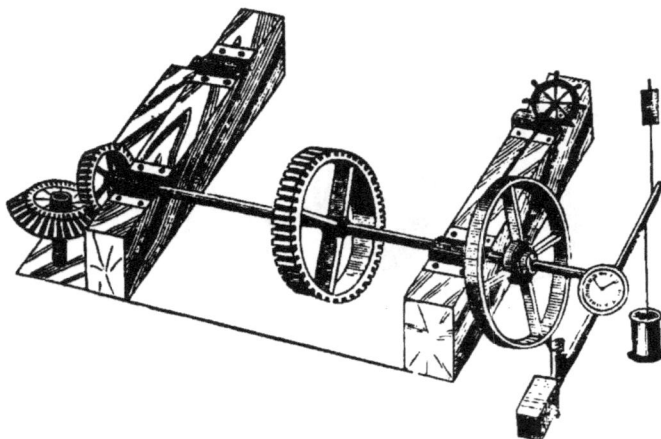

Tests made for the purpose of ascertaining the loss of power in transmission through gears. To make these the brake, as shown above, was placed upon one end of a horizontal shaft, representing "Jack Shaft," the other end being connected to the turbine shaft in the usual way by bevel gears. These gears, shafts and fittings were generously furnished for the purpose by the Messrs. Poole & Hunt, of Baltimore, Md. Other gear makers were applied to but none of them seemed willing to submit their gears to such trial. Plain cast gears with unfinished surfaces were furnished. The workmanship of the gears, shafts and boxes was pronounced by experts to be excellent and superior to the average work of the kind furnished in this vicinity. The form of the teeth of the gears was invariably approved. With every change of gears, experts were called in to examine their position and condition. During these experiments the largest gear, which had 46 teeth, was used upon the turbine shaft as crown gear, while the smallest, which had 26 teeth, was on the horizontal or "Jack Shaft." The bearings were kept well oiled, but, as it is a common idea with gear makers that the teeth of gears *roll* together so that they work just as easy when dry as when well lubricated, the first trial was made with dry gears. The table below shows results.

Data below for one minute. Multiply revolutions by 10.

Head	Weight	Rev per minute	Horse Power	Cubic Feet	Per Cent
17.96	150	487.5	22.14	1012.14	.6449
17.98	125	611.5	23.16	997.02	.6840
17.96	135	564	23.07	997.02	.6821
17.96	130	565	22 25	993.67	.6601
17.96	120	6 2	22. 5	995.31	.6599

Test through same gears, the gears being thoroughly lubricated.

Head	Weight	Rev per minute	Horse Power	Cubic Feet	Per Cent
18.04	150	646	29 36	961.93	.8957
18.04	160	606	29.38	966.94	.8913
18.04	170	558	28.74	978.66	.8619
18.03	180	506	27.60	976.14	.8303
18.05	165	584	29.20	975.31	.8779

Experiments with Gears.

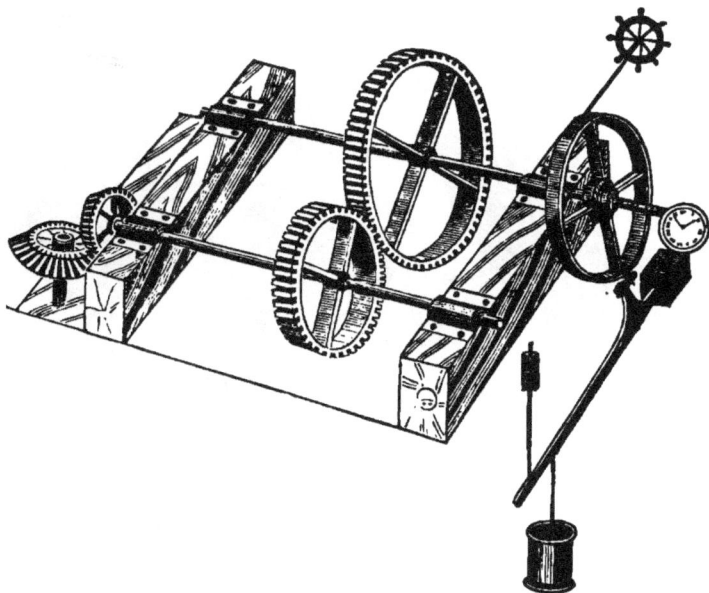

Test of gears continued, a second horizontal shaft being added to the previous arrangement described on foregoing page. This shaft, representing the main line of shafting through a mill, was connected to the "Jack Shaft" by a pair of spur gears—the large one, about 27 inches diameter, 1½ inches pitch. 5-inch face, having 49 teeth, was secured upon the second horizontal shaft or main line, and was driven by a gear on "Jack Shaft," same face and pitch as the above, and about 16½ inches diameter, having 30 teeth. The brake was placed upon the end of second line, the power of wheel being transmitted through the two pairs of gears, as represented above.

Data below for one minute. Multiply revolutions by 10.

	Head	Weight	Rev per minute	Horse Power	Cubic Feet	Per Cent
	17.94	300	165.5	15.16	841.18	.5319
	17.91	350	158.5	16.81	857.35	.5796
	17 90	400	151	18.30	870.33	.6234
	17.87	500	139	21.06	906.32	.6884
	17.85	550	133.5	22.25	934.34	.7064
	17.84	600	126.5	23.00	947.68	.7202
	17.90	625	125	23.65	939.37	.7443
	17.90	675	118	24.13	956.01	.7465
	17.94	650	120	23.63	966.03	.7219
	17.85	700	109	23.12	964.36	.7112

The gears were thoroughly lubricated with a mixture, used for the same purpose in a mill near by, probably composed of tallow and tar.

Experiments with Gears.

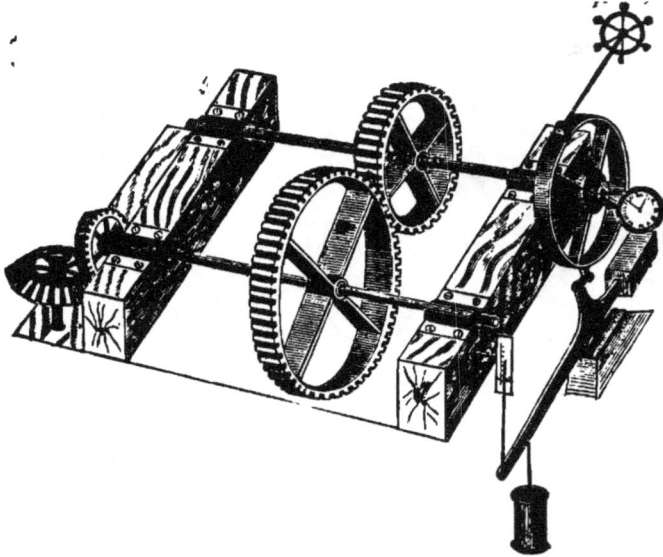

Continuation of the combined spur and bevel gear experiments, the spur gears having been changed, the one having 49 teeth being placed upon the "Jack Shaft" and working into the one having 30 teeth on second horizontal shaft upon which the brake was placed—the small bevel gear being continued as crown gear through all these tests.

Data below for one minute.   Multiply revolutions by 10.

Head	Weight	Rev per minute	Horse Power	Cubic Feet	Per Cent
17.86	270	310	25.36	972.87	.7727
17.86	265	317.5	25.49	977.28	.7731
17.84	260	323.5	25.52	971.21	.7798
17.84	280	301.5	25.58	978.96	.7755
17.84	290	277.5	24.47	985.61	.7380
17.84	250	326.7	24.75	961.18	.7641

Verification test, taken several days later.

18,03	275	305.1	25 42	980.22	.7614

Another test of the same arrangement after being taken down, then reset.

17.66	285	278.5	24.05	972.68	.7409
17.67	275	286.1	23.84	962.70	.7419
17.69	265	304	24.41	964.36	.7576
17.78	270	297.5	24.34	971.00	.7504
17.82	275	296.5	24.41	974.33	.7442
17.86	270	300.5	24.58	971.00	.7504

Belt Experiments.

To prepare for the experiments to determine the loss of power in transmission through belts, the wheel was raised in flume sufficiently to bring top of shaft above upper bearing, to give room for placing a 30-inch pulley thereon; this was done by adding another 10-inch platform to the first.

The wheel itself was first tested by placing the brake on the wheel shaft in the usual way. That it did not repeat the efficiency shown previously, was due to alterations made in the conditions. First, the step was altered somewhat in form, then the wheel was placed considerably above the floor of the flume for the purpose named above, and the difference in the head probably effected it; but the conditions, however, continued the same through the belt tests.

Wheel Test.

Data below for one minute. Multiply revolutions of wheel by 10.

Gate Opened	Head	Weight	Rev per minute	Horse Power	Cubic Feet	Per Cent
Whole Gate.	12.03	150	292.5	13.29	735.43	.7954
" "	12.00	170	278.5	14.34	740.05	.8550
" "	11.97	185	270	15.13	760.16	.8804
" "	11.95	200	239	14.48	772.61	.8308
" "	11.94	195	247	14.59	772.61	.8375
" "	11.95	190	262.5	15.11	771.06	.8682
" "	11.96	180	271	14.78	763.27	.8559

Quarter-Turn Belt.

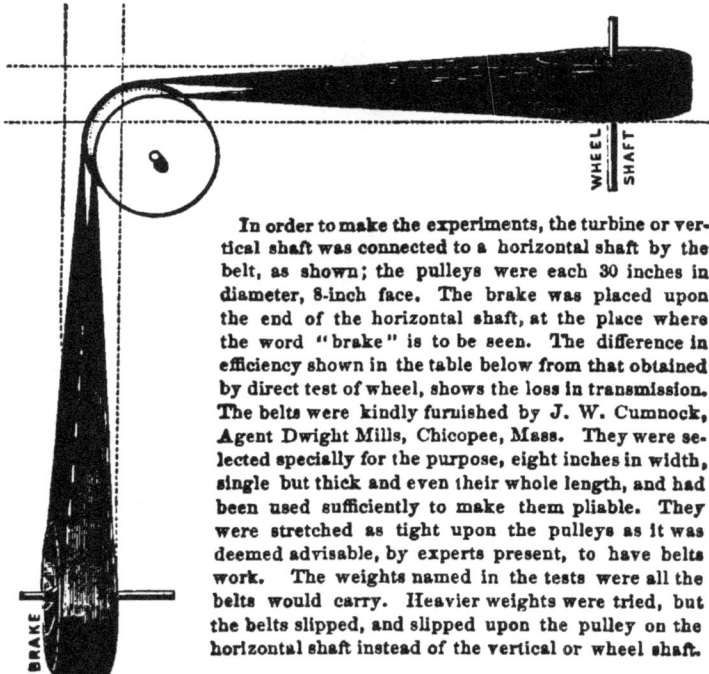

In order to make the experiments, the turbine or vertical shaft was connected to a horizontal shaft by the belt, as shown; the pulleys were each 30 inches in diameter, 8-inch face. The brake was placed upon the end of the horizontal shaft, at the place where the word "brake" is to be seen. The difference in efficiency shown in the table below from that obtained by direct test of wheel, shows the loss in transmission. The belts were kindly furnished by J. W. Cumnock, Agent Dwight Mills, Chicopee, Mass. They were selected specially for the purpose, eight inches in width, single but thick and even their whole length, and had been used sufficiently to make them pliable. They were stretched as tight upon the pulleys as it was deemed advisable, by experts present, to have belts work. The weights named in the tests were all the belts would carry. Heavier weights were tried, but the belts slipped, and slipped upon the pulley on the horizontal shaft instead of the vertical or wheel shaft.

Whole length of belt, 46 feet.

Data below for one minute. Multiply revolutions by 10.

Gate Opened							Head	Weight	Rev per minute	Horse Power	Cubic Feet	Per Cent
Whole Gate.	.	.	.	.	.	.	12.40	125	308	11.47	794.52	.6134
"	"	.	.	.	.	.	12.42	135	279.5	11.43	787.66	.6185
"	"	.	.	.	.	.	12.35	145	256	11.24	787.66	.6116
"	"	.	.	.	.	.	12.28	155	236.5	11.11	803.96	.5957
"	"	.	.	.	.	.	12.30	120	300	10.91	783.68	.5992
"	"	.	.	.	.	.	12.27	130	285.8	11.25	786.24	.6158

Quarter-Twist Belt.

Pulley, 30 inches in diameter; 8-inch face.

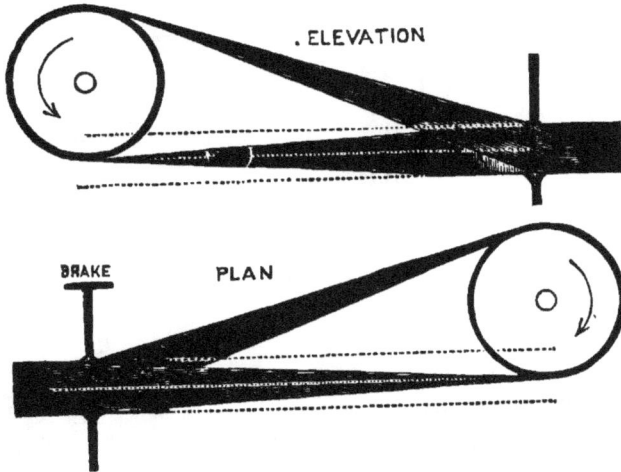

.ELEVATION

BRAKE   PLAN

Whole length of belt, about 85 feet,

Gate Opened							Head	Weight	Rev per minute	Horse Power	Cubic Feet	Per Cent
Whole Gate.	.	.	.	.	.	.	12.13	100	349.6	10.59	724.86	.6376
"	"	.	.	.	.	.	12.00	125	319.5	12.10	761.71	.7009
"	"	.	.	.	.	.	11.98	135	305	12.47	767.73	.7177
"	"	.	.	.	.	.	11.96	145	295	12.96	775.73	.7396
"	"	.	.	.	.	.	11.95	155	231.5	13.22	782.00	.7490
"	"	.	.	.	.	.	11.94	165	268	13.40	783.54	.7584
"	"	.	.	.	.	.	11.95	175	252	13.30	783.54	.7521

Open Belt.

Whole length of this belt about 36 feet.

Data for one minute. Multiply revolutions by 10.

Gate Opened						Head	Weight	Rev per minute	Horse Power	Cubic Feet	Per Cent
Whole Gate.	.	.	.	.	.	12.28	150	324.2	14 73	769.49	.8253
"	"	.	.	.	.	12.11	175	286.2	15.17	785.11	.8447
"	"	.	.	.	.	12.05	190	260.5	14.99	789.81	.8359
"	"	.	.	.	.	11.99	180	273.5	14.92	788.24	.8359
"	"	.	.	.	.	11.98	185	261.5	14.66	788.24	.8220
"	"	.	.	.	.	11.97	170	273.5	14.08	780.36	.7980
"	"	.	.	.	.	11.96	165	289.2	14.47	788.24	.8126

Cross Belt.

Pulleys the same and in the same position as when tried with open belt.

Gate Opened						Head	Weight	Rev per minute	Horse Power	Cubic Feet	Per Cent
Whole Gate.	.	.	.	.	.	12.03	150	311	14.13	774.17	.8032
"	"	.	.	.	.	11.99	160	291	14.10	778.85	7993
"	"	.	.	.	.	11.97	170	271.5	13.98	783.54	.7891
"	"	.	.	.	.	11.96	180	251.5	13.71	788.24	.7700
"	"	.	.	.	.	11.99	140	317	13.45	769.49	.7719

PHENOMENAL TURBINES.

It is here necessary to utter a caution against the selection of a turbine from any make because one of the kind has been reported as giving remarkable results. Mr. Boyden reported in an exceptional case high efficiency, yet the builders of that wheel refuse to guarantee above 75 per cent.; and tests prove many of them to be below that. Stevenson's wheel was reported above 90 per cent. at the Birkinbine, Philadelphia, tests; yet it would now be difficult to find a Stevenson wheel in use. The Risdon, reported so high at the Centennial tests, is little talked of now. The Hercules, reported as giving the highest average results ever obtained from a turbine, often gives less, though remarkably efficient, if care is used in the selection, as may be seen by the results obtained in the trials here reported. The first Wolf wheel tested gave it a reputation that was soon lost by subsequent tests of other, and particularly larger wheels. The same is true of the Walsh. As to the small Victor reported in connection with the gear, belt and draft tube experiments, probably not one in a thousand of that make would repeat those results. Two of the same size since tested, under the same conditions, sent to fill orders, did not reach 80 per cent. in either case. At whole gate the Victor stands unequaled in efficiency, but care is necessary in selection with that as with any other make of turbine, and particularly in the selection of the larger sizes which have not proved so efficient as the smaller ones.

Caution is also necessary in the consideration of the " part gate " claims, published in circulars. It can be a matter of little importance to mill owners whether the gate is one-fourth, one-half, three-fourths or wholly opened, if the same quantity of water is discharged in either case. Many of the turbine gates may be closed one-half without diminishing the discharge materially. The gate of a 36-inch Swain turbine raised four inches as its maximum. With one inch of that opening, it discharged three-fifths of the whole quantity. In a circular before me, the builder states that his wheel gives 70 per cent. at three-eighths gate. The report of the test is published therewith. The whole-gate discharge is 2300 cubic feet per minute, while the discharge at his three-eighths gate is over 1600 cubic feet per minute.

MYTHS OF PREHISTORIC TIMES OF UNKNOWN AND UNKNOWABLE ORIGIN,

THE CREATION. THE FALL. THE EXPULSION. THE LAST JUDGMENT,

Certainly myths of a cruel and low state of civilization, when the few lived upon and ruled the many with an iron hand, which, unhappily, through superstitions ignorance is still continued and must be until the priest is relegated to the oblivion of the megatherium and pterodactyl; Ireland nor the countries of South America will ever have stable or peaceful governments while priestcraft dominates. A constant struggle in this country is going on for more power to the hierarchic influence, but history proves beyond dispute the fact that as that influence decreases the elevation of the masses increases.

In Roman Catholic countries lewd books and pictures may be found in profusion, and in such pictures the male figure invariably is a priest. For the general prevalence of such belief there must be cause. While writing this article, happening to cast my eyes over a daily paper, an article in large type met my gaze giving an account of a Massachusetts minister following one of the fair ones of his flock so persistently that her husband stoned him away as he would have stoned a hog from his garden. One can hardly look over a paper without meeting something of the kind. Ministers and priests should be looked upon as sharp business men. Priest-craft as a trade has paved its cruel pathway with fire and blood; its unscrupulous lust for power has been unending, its abnegation of trade interest only upon compulsion. It is a curse to humanity and should end. A gospel shop is a place of business and should be taxed.

Lot and His Lovely Daughters.

GEN. XIX.

* 8 Behold now, I have two daughters which have not known man; let me, I pray you, bring them out unto you, and do ye to them as *is* good in your eyes: only unto these men do nothing; for therefore came they under the shadow of my roof.

31 And the first-born said unto the younger, Our father *is* old, and *there is* not a man in the earth to come in unto us after the manner of all the earth:

32 Come, let us make our father drink wine, and we will lie with him, that we may preserve seed of our father.

33 And they made their father drink wine that night: and the first-born went in, and lay with her father; and he perceived not when she lay down, nor when she arose.

34 And it came to pass on the morrow, that the first-born said unto the younger, Behold, I lay yesternight with my father: let us make him drink wine this night also, and go thou in, *and* lie with him, that we may preserve seed of our father.

35 And they made their father drink wine that night also: and the younger arose and lay with him; and he perceived not when she lay down, nor when she arose.

36 Thus were both the daughters of Lot with child by their father.

Mothers, look at the three! Can you desire your children to accept such filth as from the Creator of this beautiful world? The hard-shelled Baptist that opened his prayer with "Oh, Thou great and diabolical God!" was not much off his base.

* Do spiritualists do anything like that?

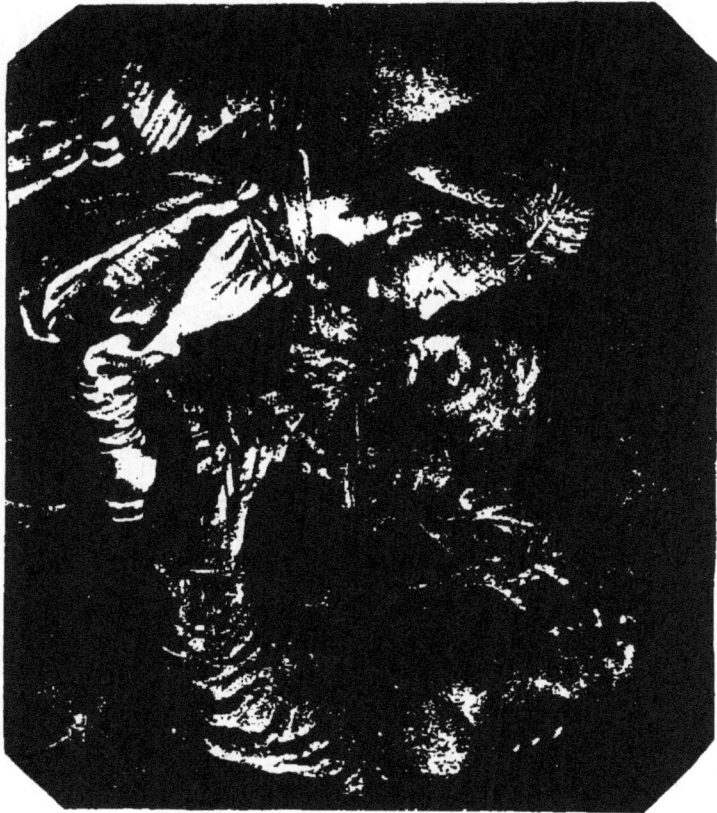

THE FEARFUL PRESENCE OF GOD UPON THE MOUNT.

EXODUS XIX

AN ANGEL, DESTROYETH THE ASSYRIAN HOST.

2 CHRONICLES XXXII.

BEARS DESTROY THE CHILDREN THAT MOCKED ELISHA

2 KINGS

THE MIDIANITES SPOILED, AND BALAAM SLAIN.

NUMBERS XXXI

15 And Moses said unto them, Have ye saved all the women alive?

16 Behold, these caused the children of Israel, through the counsel of Balaam, to commit trespass against the Lord in the matter of Peor, and there was a plague among the congregation of the Lord.

17 Now therefore kill every male among the little ones, and kill every woman that hath known man by lying with him.

18 But all the women-children, that have not known men by lying with them, keep alive for yourself.

35 And thirty and two thousand persons in all, of women that had not known man by lying with him.

40 And the persons were sixteen thousand, of which the Lord's tribute was thirty and two persons.

41 And Moses gave the tribute, *which was* the Lord's heave-offering, unto Eleazaar the priest, as the Lord commanded Moses.

The priest was there.

DANIEL'S VISION OF THE FOUR BEASTS.
DANIEL VII. 3.

And four great beasts came up from the sea, diverse one from another.

> "Vice is a monster of so frightful mien,
> As to be hated needs but to be seen;
> Yet seen too oft, familiar with her face,
> We first endure, then pity, then embrace."

And to prevent such familiarity the Bible should be banished from our schools.

Holbein's Bible illustrations so much admired in the sixteenth century that they were painted upon the walls of buildings of the streets; if so exhibited to-day our clerical Pandarus, Anthony Comstock, would have the author arrested for obscenity and profanity. The illustrations of the Creation, Expulsion, Jewish, and Puritan ideal of the Creator, with the vision above, are from that lovely work and are published as the readiest means of displaying the cream of Bible ideal. If we should hear of "Times, times, and a half times," from an author to-day, we should look upon it as the maudlin utterance of a lunatic or inebriate, and such visions as the above or the beast with its seven heads and ten horns of Revelations as the effect of nightmare or delirium tremens.

The Bible represents the opinions of the writers of its times and is as much out of place to-day in family or school as would be the writings of Apuleius, Boccaccio, Rabelais, Fielding, or Smollett, yet either may be very useful to the student. Blind, ignorant prejudice and idolatry only can account for the continuance of the former in schools; the woman that reads the passages illustrated (by no means the worst that can be found therein), and then desires its continued use as the word of God, must be a human monstrosity and certainly unfit for motherhood; but she presents a terrible example of the effect of early instruction in religious superstition.

THE TRIAL OF ABRAHAM'S FAITH.
GENESIS XXII. 10.

JUDGES I. 6.

URIAH SLAIN BY DAVID'S CONTRIVANCE.
2 SAMUEL XI. 14.

An all-wise God obliged to experiment to determine his creatures' faith. Is it to be understood that the modern imitators of old David, who murder husbands in order to get their wives, are also men after God's own heart? If not, why continue to cultivate such belief? The book of Mormon fiercely condemns polygamy, so that Brigham Young had to go to the Bible of Christianity for authority for that practice. I have the statement over his own signature.

THE LOVE OF THE CHURCH UNTO CHRIST.

SONG OF SOLOMON 1.
13 A bundle of myrrh is my well-beloved unto me: he shall lie all night betwixt my breasts.

And she smote twice upon his neck with all her might, and she took away his head from him. JUDITH XIII. 8.

JUDITH CUTTETH OFF THE HEAD OF HOLOFERNES.

ABISHAG CHERISHETH DAVID.

1 KINGS I. 3.

What mother finding that her young daughter had cuddled her *feller* all night between her breasts would believe it was only from love to Christ? Would a good mother encourage emulation of the tigress Judith, or the lecherous Abishag? A continued reverence for such ideas denotes barbarism, superstition, ignorance, and selfishness, certainly no thought of universal good for all.

THE POPE OR GOD.

The Virgin Mary as intercessor.

Tetzel selling through tickets via the Virgin Mother. Peculiar motherhood, but not patentable from lack of novelty.

THE MAHOMETAN'S IDEAL HEREAFTER.

The above ideal seems on the first thought to offer a chance for the seventy thousand unmarried women of Massachusetts to get even, but more matured reflection causes the thought to arise that if supplied as liberally to each saint as indicated by the illustration, he would soon regret his success in traveling the hair bridge and wish that he had taken up his abode with the more open countenances represented in the other place.

Mr. POSTMASTER GENERAL:

If espionage and discrimination are to be practiced at all, is this a proper book to be carried in the mails?

If so, then please define where *sacred license* ends and *profane obscenity* begins,

OH, YE HOLY HYPOCRITES,

that shriek so loudly for suppression of the Louisiana Lottery and efforts of the masses for freedom from your cruel bondage; yet smile so benignly upon church fairs, bucket shops, dealings in "futures," "corners" in the necessaries of life, "tariff for the rich," "coal combines," etc., etc.

"The Bible."

My mother, like most New England mothers of her time, firmly believed in the infallibility of the Bible and insisted that her children should study its contents, so that its stories from my earliest childhood have been familiar to me, for I cannot remember back when I could not read any story that I could get hold of. I can well remember how the story of "Susanna and the Elders" was given to me to read as a reward for some slight assistance in her many duties, the dear soul not thinking that the unbiased mind of a child might see the rascality of the priest as well as the smartness of the Daniel.

The reliability of the Bible stories have often been fiercely disputed in my hearing, yet without causing me to think the authors guilty of intentional misrepresentation, but I do believe that through ignorance or fraudulent piety what were beautiful allegories have been given meanings very different from their original purposes.

Allegory has ever been common with primitive people. Many beautiful ones have been handed down through Homer, Hesiod, and others, not as original with them or their times but as fragmentary traditions of a much earlier people, so probably of the Bible stories.

It was common with the writers of Plato's time to commence a story with "Away back in the dark ages." Many of those writers mention dates of ten to twenty thousand years previous, and in the works of one I cannot recall it is stated that the Babylonians claim to have authentic records reaching back four hundred and seventy thousand years, certainly sufficient time for the production of myths.

We have stories of the Cyclops, Polyphemus, Perseus, and Andromeda, Penelope, Æneas, Anchises, the beautiful story by Apuleius of Psyche, her envious sisters, Cupid and his mother Venus, and a thousand others all coeval with the Bible myths and quite likely different versions from the same originals.

It is not difficult for a person of ordinary imagination to perceive how easy it would be to construct an allegory from the meaning of the names of individuals as given in the appendix of all complete editions of the Bible. "David dancing naked before the Lord," as there explained, can hardly be looked upon as a kingly performance, but allegorically it might mean much; so of Jonah and the fish story, Samson and Delilah, etc.

It is my sincere belief that if we could have the true meaning of the Bible stories, we should at least have common sense and often very applicable parables, instead of which they have been so distorted that only fanaticism can make their application perceptible. For instance to pretend that the salacious rhapsodies of Solomon's songs refer to the love of the Church for Christ, puts the love of the Church on a very low plane to say the least, and makes a large draft upon the credulity of the unbiased mind and certainly is expecting too much to suppose that children uninstructed will ever look upon such reading in that way. The allegory of the Witch of Endor as given by W. H. C., coincides so entirely with my idea of Bible stories, that I herewith give it space.

THE WITCH OF ENDOR.

The definitions of words as understood by the ancients is necessary to be learned before it is possible to understand this beautiful allegory.

"Saul," in the Hebrew, means death, or hell, or the grave, or winter, or demanded, or sepulcher, or lent, or ditch; for every noun and verb in that jargon, erroneously called a "language," had a great variety of significations, often self-contradictory. Winter was the beggar, the asker, the receiver.

"David" means the lover, the beloved, the giver, the summer, etc.

"Samuel" means heard of God, or asked of God, or earth at the vernal equinox, where Samuel died and was buried, where the Jewish ecclesiastic year always began and does to this day, the civil year beginning at the autumnal equinox.

"Endor" means fount of the dwelling place, or the last summer constellation, or Virgo, the virgin.

Winter ended at the vernal equinox, and it was there that summer began. Saul, or winter, arrives there and finds David with the Philistines (those that dwell in villages, or summer constellations) gathered to meet him, "and he was afraid." He wanted a fortune teller to advise him, but he had "put away those that had familiar spirits, and the wizards, out of the land." That is, Virgo had set the previous year, just as Aries, the harbinger of summer, rose in the east with the sun. But now, at the vernal equinox, where winter must end, Virgo was visible; for the first point of this constellation is distant from Aries 150 and the last point 180 degrees.

"Saul disguised himself." This is a very pretty conception on the part of the author, for winter moderates as the sun approaches the vernal equinox, about March 21, and is not at all like the winter in January. So it is no wonder the old woman of Endor did not know him. But when Aries rose with the sun she knew the end of winter was at hand; that is, knew Saul, which means the five winter months, or the brethren of the rich man in hell. Saul asked her what she saw, and she replied: "I saw gods [Elohim in the Hebrew, and the very word which is translated God, as the God of the Bible] ascending out of the earth."

At the vernal equinox the sun enters Aries, and the two together, sun and Aries, are Elohim in the plural number, or "gods," for *im*, added to the singular, forms the plural in Hebrew; thus *cherub*, a bull; *cherubim*, bulls. Therefore, as Virgo was setting in the west she saw the "gods," sun and Aries, rising out of the earth, or Ramah, where Samuel was buried.

During winter the earth may be said to be "dead," but is revived at each coming spring.

So Virgo raised Samuel from the dead, for, as she sets in the west, up comes the sun and Aries in the east, the signal for the death of Saul, or end of winter. Saul complained to Samuel that the Lord had departed from him; that is, the cold, the spirit of winter; even Jack Frost would not answer when he called. The earth in spring putting on her beautiful garments of green, now informs winter that its last hour is at hand. Once more the battle has been fought between heat and cold, light and darkness, and once more cold and darkness have been conquered.

"Then Saul fell straightway all along on the earth, and was sore afraid because of the words of Samuel; and there was no strength in him [of course not, for cold is the strength of winter]; for he had eaten no bread all the day, nor all the night." I. Sam. xxviii., 20.

The supply of provisions for the winter was often exhausted before the sun reached Pisces, the fishes, when the people lived on fish for just forty days before the sun reached the vernal equinox, or Aries, the "Lamb of God that takes away the sins of the world"; not the sins of the people: but the evils of winter. Here was the origin of Lent, or abstaining from meat and living on fish.

All the ancient mythologies abound with allegories descriptive of the changes from summer to winter, and winter to summer. Vishnu had a thousand names, and it may be summer and winter had equally as many; but whether more or less, the prominent idea seemed to be that all those names for summer meant heat and light, while those for winter meant cold and darkness. Twice each year these opposing elements made war upon each other, the decisive battles being fought at the two equinoxes. Light always conquered at the vernal equinox, only to be defeated by darkness six months later at the autumnal equinox. "More light!" was the agonized cry of those in the bonds of darkness, or "outer darkness," weeping and gnashing their teeth because they had no food to gnash. True, the sun is darkened during winter by reason of the clouds and storms, but its "fire is never quenched," and the fire of the sun is the only fire that time does not quench.

Samuel anointed Saul king of winter, well knowing that Saul would be dethroned by the king of summer when the sun reached the spring equinox. David, a mere youth, was chosen king of summer. He was sent to Saul on an ass (the sun while transiting through Cancer, a summer constellation, passes the two asses, "whereon no man ever sat"). Leaving Cancer, the sun transits through Leo, the lion (Hercules), passing a conjunction of Ursa Major, the bear, when both the lion and the bear are invisible, being metaphorically slain. David boasts of these victories, and prepares to meet Goliath (passage, revolution, heap, discovery), the spirit of summer, which can be "laid" only by winter. Therefore, he takes "five smooth stones," symbolical of the five winter months, from the brook, or by metonymy, the zodiac, and kills this giant.

Saul was so delighted with the valor of the beardless youth (the crops were not yet ready for harvest) that he gave him his daughter.

This is very ingenious, depicting the strategies of war. David plays the courtier to Saul, yet means to overcome him in the end; Saul professes to love David, but is jealous of him, and gave him Michal (complete) as a snare; that is, leaving Leo, the sun comes to Virgo, which "completes" the summer. Not much gift about it, however, for Virgo was a summer constellation and belonged to David, king of summer. The strife between Saul and David was descriptive of the struggle between cold and heat. David is conquered when the sun leaves Virgo, and must now flee before Saul till the end of winter, when Saul falls "all along on the earth."

W. H. C.

God has just called the servant from a $5,000 to a $10,000 salary.

Oh, Mr. Congress, do put God in the Constitution and stop Sunday cars! If the multitudes may have excursions instead of sermons, how are we to live?

The Master.

Oh, Brother Malbadge, Theodore Parker preached long sermons but none left, while half my hearers have gone before I get hell painted red.

Well, Brother Bulton, old chestnuts are stale; be sensational; a few lies about Spiritualism will take with our hearers and the ignorant generally.

The Servant.

CHRISTIANITY.

Where and when this atrocious delusion originated God alone knows; but certainly far back in the days of superstition when man made his God in his own image and was not likely to make him of a high standard ; yet it is hard to conceive of the natural man creating a God who would find it necessary to repair his own blunder by the murder of his son. The idea is so abhorrent and unnatural that it is only continued by debasing the minds of children in their earliest infancy with its inhumanity. That the most of the ideas were common in India a thousand years before their recorded promulgation in Palestine is well known; that the Jesus of Nazareth was a myth or a man matters little; as a God he was certainly a failure, for there is no recorded evidence of his superiority to what any real good man ought to be, while many of his represented sayings and acts would be objected to by all right minded persons.

What the human race has suffered, and still continues to suffer, in consequence of the old time preaching of cruel hell fire and brimstone terrors, it will never be possible to compute. The deep and lasting injuries wrought by the relentlessly steady inculcation of these most woful of dogmas can never be compensated for in untold generations. Think of the murderous wars between different people; of the reckless dismemberment of empires; of the barbarous sacrifice of innocent and unoffending lives; and, not least of all, of the insanity caused by these events and the tenets that were their undeniable cause; and then say, if it be possible, that the world has in the whole course of its experience undergone equal paroxysms of torture and wretchedness from any other cause, or because of any combination of circumstances whatever.

Tertullian (A. D. 200) held that the "Books of Moses" were "not only all truth, but that all truth was contained in them." Consequently every attempt to promulgate knowledge was met by horrible persecution. Cyril's mob—of many monks—seized Hypatia and dragged her from her carriage one morning, as she was riding to her academy, stripped her of her clothing, then cut her body into pieces, scraped the flesh from her bones with shells, and burned her, piecemeal.

In 529 the Christian emperor Justinian suppressed the schools of philosophy of Athens, and the night of "the dark ages" closed down on what was then known as the Christian world; the night of a thousand years, in which the church ruled both temporally and spiritually; a church that claims to be the light of the world; and yet this period was the darkest that history has known.

Think of Copernicus, Giordano Bruno, Savonarola, Servetus, Joan of Arc, and thousands of other victims of the damnable delusion. Think of the Inquisition and how quick it would be re-established could either the Roman Catholic or Calvinist have complete control. Think of the thirty-five thousand diseased natives of the Sandwich islands, all that remain of the four hundred thousand after being subjected to the Christianizing process named below. "A steamer recently left her European port for the Congo country, now exciting such unmeasured sympathy on account of its paganism and want of modesty in dress, with a cargo of 60,000 gallons of rum, 720 gallons of gin, 460 tons of gunpowder, and twelve missionaries!"

For fifteen hundred years Christianity has held undisputed sway, and to-day every man is looked upon as a thief. Corruption in our government is openly talked of, free passes are readily accepted by our legislators who well know at the time that much of the legislation will be relative to the business of those from whom the passes are received. A car conductor is not allowed to take a five cent ticket unless tied to a bell punch; a clerk in a store must be checked and counter-checked; if you ask for butter you get grease ; if you ask for first quality cheese you receive anything but that; if you require medicine it is adulterated; if you vote, you must do so through a process that implies that rascality is general; in short, that society is rotten to the core. And this state of affairs exists, say the shallow-minded, because there is not enough of Christianity ; an assertion easily disproved by turning to the description of its most flourishing days as described by Boccaccio, Rabelais, or any other early writer.

No crime or wrong can be named that has not been tolerated by Christianity. Whenever it has been found profitable, lying and deception have been cardinal principles. An age that could have produced the works of

Herodotus must have had contemporaneous historians. Why were their works destroyed; did they tell too much ?

I believe that the proselyting Christian of to-day is a far more injurious citizen than the rum seller, because he begins his pernicious work with the infancy of the individual, which is seldom the case with the rum seller. The Young Men's Christian Association I sincerely believe to be very injurious to the intelligence and morals of the world. I well know how the statement will be received, but it will keep and can afford to wait.

Constantine is called the first Christian emperor, and Henry the Eighth, without much straining of the facts, might be considered the last, and it would be difficult to decide which of the two was the greater scoundrel; but to the latter much of our religious freedom is due, which proves that selfishness is one of the improving powers of the world, yet in itself is not admirable.

Holyoke for its size contains probably more professing Christians and dogs (no reflection upon the dogs) than any other place that can be named, yet in no other place have I ever known such strenuous efforts to be made to prevent the laboring classes from procuring homes of their own. And now in all seriousness

What Good Has Christianity Ever Done ?

The teachings, nominally of Jesus of Nazareth, were like those of our Spiritualists of to-day, and for the purpose of substituting a living religion for that of the dead belief then as now popular. With those teachings went the inspirations and manifestations now so common.

"By their works ye shall know them." The works referred to are ignored by all of our popular churches.

Every phase of mediumship practiced now, was practiced then by the Christians, and now by the pretended Christians ridiculed. New gospels were produced in abundance then as are the spiritual wonders now, and this continued up to the council of Nice, and the organization of the Christian Church, when inspiration and angel visits ceased and Christianity like a dead world, our moon, became dead, having neither life, light, nor warmth therein, but instead was fitted out with an impossible and incomprehensible God of three in one, the idea of which could only have originated from the ancient Phallic worship that certainly should cause any modest, intelligent woman to hesitate before professing a belief therein, at any rate, Miss Abby A. Judson, born in India, where the Phallic worship is likely to be understood, has abandoned the religion of her father, the once well known missionary, Adoniram Judson, and taken up with Spiritualism as the living religion of to-day.

Can an instance be named where Christianity has made a people better? The victims of a single battle field have exceeded all the sacrificial victims that would have been required in a thousand years. Think of the battles fought to prove Christians to be cannibals and vampires, worse in fact, for they claim to eat the flesh and suck the blood of their God.

Is the Christian's oath in court or his note in bank preferred to that of the unbeliever ? Is he a better neighbor or citizen? Is it possible that a noble mind can desire to benefit through the sufferings of another ? Can belief in vicarious atonement produce noble people ?

Are there any countries upon the earth where such strong bank vaults are required as among Christians, or where crime is more common ?

The absurd pretensions of exceptional goodness among Christians are so patent that, were it not for the perversion of the minds of infants and the ignorant, the belief would die out in a generation. That its creeds and dogmas are unbelieved in by the leaders is easy to demonstrate. A half century since the judgment day and its near approach was preached by all; Miller fixed a time and the churches at once scouted the whole belief. A half century since a belief in spirits and their overlooking our actions was generally latent in the Christian mind, but at the advent of the Rochester knockings in practical demonstration of the belief the clergy went mad in opposition.

For ages evolution has been the principal gun of the belief, namely, that by sniveling and professing belief in the saving blood of Christ the most atrocious murderer would instantly evolute from the gallows, drop to a reserved seat at the right hand of the Father; but the moment evolution was taught as the means of physical progression, a howl from the Christian leaders at once went forth, in protest.

I believe the time will come when it will be acknowledged that nine-tenths of all the crime and suffering of mankind is due to what are called the three learned professions, and efforts of those professors to retain the control of the masses for personal benefit, regardless of the benefit of humanity. Religion and medicine are now on the defense, the jargon called "the law" is rotten and top-heavy, and bound to tumble. The human body is the acme of mechanism, offering the broadest field for skill as a mechanic for the physician. The empty churches so much wailed about are not so much due to science as to the rule or ruin policy of the clergy and their Partingtonian attempts to block the wheels of progress, and sensational attempts of materialistic preachers to attract notice. If Talmage declaims on Mount Calvary, Parkhurst airs his Phallic belief in a brothel, for the purpose of notoriety; but truth requires no such expedients for its support, and it may fairly be questioned whether the intelligent readers of the Parkhurst report would not have been better satisfied with the results had he been sent to the Island to keep the brothel-house keeper company, and whether the city would not have been purer for so doing.

During the dozen years in which I made a business of testing turbines, the gauges, weights and revolutions were called as each change of weight was made. These were recorded at the time by an assistant in a book made up of printed blank forms, duplicates of which were copied and furnished as certificate of result to each party having a wheel tested. The experiments were public, and any one sufficiently interested could examine gauges to make sure they were correctly called. The books were subjected to rough usage, and when I gave up the business, they were anything but ornamental; still, they contained a complete record of all the tests made, and, as a means of continuing their usefulness, they were presented to the Engineering Department of Yale, through Prof. Norton, who has always been prompt in attendance to witness any experiments in hydrodynamics from which there was a possibility of gaining information likely to benefit his department.

SHEFFIELD SCIENTIFIC SCHOOL,
New Haven, Nov. 30, 1880.

MR. JAMES EMERSON.

Dear Sir: The Governing Board of the Sheffield Scientific School, composed of the Professors of the School, at a meeting held last evening, instructed me to convey to you their thanks for your very acceptable gift of three volumes of Notes of your Experimental Tests of Turbines. In doing this, I wish to add my personal thanks in consideration of the advantage I shall derive from them personally, and to assure you that they will undoubtedly prove of great service to the Engineering Department of the School, and will add materially to its means of instruction relative to hydraulic motors. They have been deposited in the Library of the Department, and will be held for ready consultation by the students in engineering.

Very truly yours,

W. A. NORTON.

Propelling Screws.

Is there no better plan than the one so common with government engineers of placing the screws in a ship, then lashing the ship to a wharf, fire up and run the engines day after day, to ascertain how fast the screws can be driven, and how fast the ship ought to run, if all the decimals can be depended upon; would it be better, cheaper and far more decisive to take a screw, place it in a frame representing the stern of a ship, but with freedom to move forward; from the after end of the screw have a weight attached, so that if the screw move forward it would have to raise a known weight, then by belt or other means, using a dynamometer in transmission, drive the screw to any speed desired; by such means accuracy could be attained, and the most perfect screw for the purpose could be found at comparatively small expense. The propeller screw partakes too much of the turbine nature to allow of its lines being positively determined by mathematical calculations; at least, the best form might be ascertained with perfect accuracy in the manner suggested above.

Christian Materialist Preacher and Sympathizing Hearers.

Services will open by singing:—

Kindle a flame of sacred love
In these cold hearts of ours.

Come, Holy Spirit, Heavenly Dove,
With all thy quick'ning powers.

Let us pray!

Oh! thou great and fearful Jehovah, look down upon us this morning and pardon our transgression, for if we are weak in works we are strong in faith. We ask this not for our own but for Christ's sake, in the name of the Father, Son and Holy Ghost. Amen.

My friends, my purpose this morning is to caution you against taking up with new gods or that latest work of the devil called Spiritualism.

We the *Elect* of course know that spirits were common in the days of the good old fathers, Abraham, Isaac, and Jacob, and equally well we know that there are no such airy nothings now. Phew! we might as well believe in ghosts!

AN OHIO IDEA.

In one of the western counties of Ohio a petition is being circulated asking Governor Hoadley to pardon a young man sent to the penitentiary for robbing a prominent Free Thinker. The plea is that he should not have been convicted because the victim is a "wicked and perverse infidel." It is peculiarly an Ohio idea that a man who does not profess religion has no rights, and that it is an act of Christian charity to pick his pocket or set fire to his barn. Probably an Ohio office holder would think it a virtue to steal from the government on the same principle.

"The nearer the church the further from God," is an old and a trite saying, but ideas are changing, and we may hope for improvement.

TAXATION OF ALL PROPERTY.

If taxation is right at all, there should be no exception. Church property, usually occupying the best localities, certainly should not be exempt, nor should owners of unimproved land, contiguous to growing cities or towns, be allowed to continue to hold such land at a mere nominal rate of taxation, while others are ready to take it at far higher valuation. Let every owner be his own assessor, but with the understanding that any purchaser may take it at the assessed rate. Of course some provisions may be made to prevent a homestead from being unjustly taken.

Let all property be without the protection of law that has not paid for such protection by its taxation. There is no need for many of our officials.

PROHIBITION.

There is an old saying that most of the unhappiness of life comes to us through the efforts of weak but well meaning persons trying to direct our lives instead of causing general improvement by perfecting their own. Particularly is this the case with the priestly order, and has been so from the beginning of history. The Rev. Mr. Miner, the great advocate of prohibition in this state, must well know that his life has been spent in indoctrinating the minds of his hearers with a superstition that cannot be sustained by evidence, yet he is ready to assume the Creator's place and manage mankind. Prohibition interferes with the rights of all, and with very doubtful effects. Two gallons of liquor, beer included, would more than cover all that I have ever drank, yet I do not believe in prohibition, nor would I vote for license, for to me it would seem wrong to dignify a disreputable business by legal recognition; but as a large portion of crime, poverty, and misery is caused by the traffic, I would have all places where it is carried on taxed at such a rate that the owners would refuse to rent for the purpose.

Belt Transmission.

Of all guess work, there is none more unreliable than that of computing the power transmitted by the width of belt. First, the kind, quality and condition of the belt is to be considered; then the size, distance and position of pulley; whether their surfaces are wood, metal, or covered with leather; whether one is much larger than the other, and whether the belt is running vertically, horizontally, open or crossed; or, what is worse, is running edge up, on pulleys on vertical shafts; whether it is tight or loose; whether it is made of leather or other material, also whether single or double. In testing with lever dynamometer, the speed of belt is determined. A single leather belt, under ordinary conditions, running 1,000 feet per minute, will transmit a h. p. for each inch or width, but the matter is one of the greatest uncertainty

Weight of a Cubic foot of Pure Water at Different Temperatures.

Degrees.	Weight.	Degrees.	Weight.	Degrees.	Weight.	Degrees.	Weight.
32	62.375	45	62.378	59	62.336	73	62.249
33	62.377	46	62.376	60	62.331	74	62.242
34	62.378	47	62 375	61	62.326	75	62.234
35	62.379	48	62.373	62	62.321	76	62.225
36	62.380	49	62.371	63	62.316	77	62.217
37	62.381	50	62.368	64	62.310	78	62.208
38	62.381	51	62.365	65	62.304	79	62.199
39(max)	62.382	52	62.363	66	62.298	80	62.190
39.38	62.382	53	62.359	67	62.292	81	62.181
40	62.382	54	62.356	68	62.285	82	62.172
41	62.381	55	62.352	69	62.278	83	62.162
42	62.381	56	62.349	70	62.272	84	62.152
43	62.380	57	62.345	71	62.264	85	62.142
44	62.379	58	62.340	72	62.257	86	62.132

Table of Inches and Sixteenths Reduced to Decimals of a Foot.

SIXTEENTHS.	0	1	2	3	4	5	6	7	8	9	10	11
$\frac{0}{16}$	.000	.083	.167	.250	.333	.417	.500	.583	.667	.750	.833	.917
$\frac{1}{16}$	.005	.089	.172	.255	.339	.422	.505	.589	.672	.755	.839	.922
$\frac{2}{16}$	.010	.094	.177	.260	.344	.427	.510	.594	.677	.760	.844	.927
$\frac{3}{16}$	.016	.099	.182	.266	.349	.432	.516	.599	.682	.766	.849	.932
$\frac{4}{16}$	.021	.104	.187	.271	.354	.437	.521	.604	.687	.771	.854	.937
$\frac{5}{16}$	.026	.109	.193	.276	.359	.443	.526	.609	.693	.776	.859	.943
$\frac{6}{16}$	.031	.115	.198	.281	.365	.448	.531	.615	.698	.781	.865	.948
$\frac{7}{16}$	.036	.120	.203	.286	.370	.453	.536	.620	.703	.786	.870	.953
$\frac{8}{16}$	.042	.125	.208	.292	.375	.458	.542	.625	.708	.792	.875	.958
$\frac{9}{16}$	.047	.130	.214	.297	.380	.464	.547	.630	.714	.797	.880	.964
$\frac{10}{16}$	.052	.135	.219	.302	.385	.469	.552	.635	.719	.802	.885	.969
$\frac{11}{16}$	.057	.141	.224	.307	.391	.474	.557	.641	.724	.807	.891	.974
$\frac{12}{16}$	.062	.146	.229	.312	.396	.479	.562	.646	.729	.812	.896	.979
$\frac{13}{16}$	.068	.151	.234	.318	.401	.484	.568	.651	.734	.818	.901	.984
$\frac{14}{16}$	.073	.156	.240	.323	.406	.490	.573	.656	.740	.823	.906	.990
$\frac{15}{16}$	.078	.161	.24˙	.328	.411	.495	.578	.661	.745	.828	.911	.995

The Emerson Weir Tables,

For weirs with end contractions, were computed for me by Miss Charla A. Adams, some 20,000 quantities; these have done much towards reducing the cost of water wheel tests and water measurements, at the same time producing far greater accuracy.

These were computed by the Francis formula, from zero up. The experiments upon which that formula was prepared were not extended below a depth of .500 of a foot, but it is often necessary to use it at a much less depth; and experience proves it to be sufficiently accurate for all practical purposes.

The computations are per minute. If the weir is properly constructed there is no need of correction, if not properly constructed a correction is mere guess-work or conjecture.

The Francis tables for the one foot weir are calculated for weir without contraction; consequently, by using those in connection with the others, by adding to or subtracting from, the quantity flowing over a weir of any length may readily be found.

Depth on Weir.	LENGTH OF THE WEIR.									
Feet.	2 Feet.	3 Feet.	4 Feet.	6 Feet.	7 Feet.	8 Feet.	10 Feet.	12 Feet.	16 Feet.	20 Feet.
.001	.013	.019	.025	.038	.044	.051	.063	.076	.101	.127
.002	.045	.067	.064	.134	.154	.179	.224	.268	.358	.475
.003	.077	.115	.103	.230	.264	.307	.385	.460	.615	.823
.004	.109	.163	.142	.326	.374	.435	.547	.653	.872	1.171
.005	.141	.212	.281	.424	.494	.565	.709	.847	1.130	1.410
.006	.194	.289	.384	.542	.674	.775	.966	1.157	1.543	1.927
.007	.247	.366	.487	.661	.854	.981	1.223	1.467	1.956	2.444
.008	.301	.443	.591	.780	1.034	1.188	1.481	1.777	2.369	2.967
.009	.355	.521	.695	.899	1.214	1.375	1.739	2.087	2.782	3.478
.010	.409	.599	.799	1.018	1.397	1.598	1.997	2.397	3.196	3.515
.011	.46	.74	.93	1.24	1.87	1.86	2.33	2.79	3.71	4.666
.012	.52	.83	1.06	1.47	2.10	2.21	2.67	3.19	4.22	5.33
.013	.59	.92	1.19	1.71	2.34	2.40	3.01	3.59	4.74	6.00
.014	.66	1.01	1.32	1.95	2.57	2.67	3.35	3.99	5.25	6.67
.015	.73	1.10	1.46	2.20	2.84	2.94	3.69	4.40	5.87	7.34
.016	.81	1.21	1.62	2.43	3.06	3.25	3.94	4.87	6.50	8.13
.017	.89	1.33	1.78	2.66	3.28	3.56	4.45	5.34	7.13	8.92
.018	.97	1.45	1.94	2.90	3.50	3.87	4.84	5.81	7.76	9.97
.019	1.05	1.57	2.10	3.14	3.72	4.19	5.24	6.28	8.40	10.50
.020	1.13	1.69	2.27	3.38	3.95	4.51	5.64	6.76	9.04	11.30
.021	1.22	1.82	2.44	3.65	4.26	4.87	6.09	7.29	9.75	12.19
.022	1.31	1.95	2.61	3.92	4.57	5.23	6.54	7.83	10.47	13.09
.023	1.40	2.08	2.78	4.19	4.88	5.59	6.99	8.37	11.19	13.99
.024	1.49	2.22	2.95	4.46	5.20	5.95	7.44	8.91	11.91	14.89
.025	1.58	2.36	3.12	4.73	5.52	6.31	7.89	9.45	12.63	15.79
.026	1.67	2.51	3.32	5.02	5.86	6.70	8.38	10.05	13.42	16.78
.027	1.77	2.66	3.52	5.32	6.20	7.10	8.88	10.65	14.21	17.77
.028	1.87	2.81	3.72	5.62	6.55	7.50	9.38	11.25	15.00	18.76
.029	1.97	2.96	3.93	5.92	6.90	7.90	9.88	11.85	15.80	19.75
.030	2.07	3.11	4.14	6.22	7.25	8.30	10.38	12.46	16.60	20.75
.031	2.17	3.27	4.37	6.52	7.63	8.74	10.91	13.10	17.46	21.83
.032	2.28	3.43	4.60	6.82	8.01	9.18	11.45	13.74	18.32	22.91
.033	2.39	3.59	4.84	7.13	8.39	9.62	11.99	14.39	19.18	23.99
.034	2.50	3.75	5.08	7.43	8.77	10.07	12.53	15.04	20.05	25.17
.035	2.61	3.91	5.22	7.84	9.15	10.52	13.07	15.69	20.92	26.15

Depth on Weir.	LENGTH OF THE WEIR.									
Feet.	2 Feet.	3 Feet.	4 Feet.	6 Feet.	7 Feet.	8 Feet.	10 Feet.	12 Feet.	16 Feet.	20 Feet.
.036	2.72	4.08	5.45	8.18	9.55	10.97	13.64	16.38	21.84	27.31
.037	2.83	4.25	5.68	8.53	9.95	11.42	14.21	17.07	22.77	28.47
.038	2.95	4.42	5.91	8.88	10.35	11.87	14.78	17.76	23.71	29.63
.039	3.07	4.60	6.14	9.23	10.76	11.32	15.35	18.46	24.64	30.79
.040	3.19	4.78	6.38	9.58	11.17	12.77	15.93	19.16	25.56	31.95
.041	3.31	4.96	6.62	9.94	11.60	13.26	16.55	19.90	26.54	33.18
.042	3.43	5.14	6.86	10.31	12.03	13.75	17.17	20.64	27.53	34.41
.043	3.55	5.32	7.11	10.68	12.47	14.25	17.80	21.38	28.52	35.65
.044	3.67	5.51	7.36	11.05	12.91	14.75	18.43	22.12	29.51	36.89
.045	3.80	5.70	7.61	11.42	13.35	15.25	19.06	22.88	30.50	38.13
.046	3.93	5.89	7.87	11.81	13.80	15.76	19.71	23.66	31.54	39.43
.047	4.06	6.08	8.13	12.20	14.25	16.28	20.36	24.44	32.58	40.73
.048	4.19	6.27	8.39	12.59	14.70	16.80	21.01	25.22	33.62	42.03
.049	4.32	6.47	8.66	12.98	15.15	17.32	21.66	26.00	34.67	43.34
.050	4.45	6.67	8.93	13.38	15.69	17.84	22.32	26.78	35.72	44.65
.051	4.58	6.87	9.20	13.79	16.09	18.39	23.00	27.60	36.81	46.02
.052	4.71	7.07	9.47	14.20	16.57	18.94	23.68	28.42	37.90	47.39
.053	4.84	7.28	9.74	14.61	17.05	19.49	24.36	29.24	39.00	48.76
.054	4.99	7.49	10.01	15.02	17.53	20.04	25.05	30.06	40.10	50.14
.055	5.13	7.70	10.28	15.43	18.01	20.59	25.74	30.90	41.20	51.52
.056	5.27	7.91	10.56	15.86	18.51	21.16	26.45	31.76	42.35	52.95
.057	5.41	8.12	10.84	16.29	19.01	21.73	27.17	32.62	43.50	54.38
.058	5.55	8.33	11.12	16.72	19.51	22.30	27.89	33.48	44.71	55.81
.059	5.69	8.55	11.41	17.15	20.01	22.87	28.61	34.34	45.86	57.24
.060	5.84	8.77	11.71	17.58	20.52	23.45	29.33	35.20	46.95	58.69
.061	5.98	8.99	12.00	18.02	21.04	24.04	30.07	36.09	48.16	60.18
.062	6.13	9.11	12.30	18.46	21.56	24.64	30.81	36.99	49.37	61.68
.063	6.28	9.33	12.60	18.91	22.08	25.24	31.56	37.89	50.58	63.18
.064	6.43	9.56	12.90	19.36	22.61	25.84	32.31	38.79	51.79	64.68
.065	6.58	9.89	13.20	19.81	23.14	26.44	33.06	39.69	53.00	66.18
.066	6.73	10.12	13.50	20.27	23.68	27.06	33.83	40.62	54.23	67.73
.067	6.88	10.35	13.81	20.74	24.22	27.62	34.61	41.55	55.46	69.28
.068	7.03	10.58	14.12	21.21	24.76	28.24	35.39	42.48	56.69	70.83
.069	7.19	10.81	14.43	21.68	25.30	28.86	36.17	43.41	57.92	72.39
.070	7.35	11.04	14.74	22.15	25.85	29.55	36.95	44.35	59.15	73.95
.071	7.51	11.28	15.06	22.63	26.41	30.19	37.75	45.31	60.46	75.55
.072	7.67	11.52	15.38	23.11	26.97	30.83	38.55	46.28	61.77	77.15
.073	7.83	11.76	15.71	23.59	27.53	31.47	39.36	47.25	63.08	78.75
.074	7.99	11.98	16.03	24.07	28.10	32.12	40.17	48.22	64.38	80.36
.075	8.15	12.25	16.35	24.56	28.67	32.77	40.98	49.19	65.60	81.97
.076	8.31	12.49	16.68	25.05	29.25	33.43	41.81	50.18	66.93	83.64
.077	8.47	12.74	17.03	25.55	29.83	34.09	42.64	51.18	68.26	85.31
.078	8.63	12.99	17.38	26.05	30.41	34.75	43.47	52.18	69.59	86.99
.079	8.80	13.24	17.73	26.55	30.99	35.42	44.30	53.18	70.92	88.67
.080	8.97	13.49	18.01	27.05	31.57	36.09	45.14	54.18	72.26	90.35
.081	9.14	13.74	18.35	27.56	32.17	36.77	45.99	55.20	73.63	92.06
.082	9.31	13.99	18.69	28.07	32.77	37.45	46.85	56.23	75.00	93.78
.083	9.48	14.24	19.03	28.59	33.37	38.14	47.71	57.26	76.37	95.50
.084	9.65	14.49	19.37	29.01	33.97	38.83	48.57	58.29	77.74	97.22
.085	9.82	14.75	19.72	29.62	34.58	39.52	49.43	59.32	79.13	98.94
.086	9.99	15.01	20.07	30.15	35.19	40.22	50.31	60.38	80.54	100.71
.087	10.16	15.27	20.42	30.68	35.81	40.92	51.19	61.44	81.95	102.48
.088	10.33	15.54	20.77	31.21	36.43	41.63	52.07	62.50	83.87	104.28
.089	10.51	15.81	21.12	31.74	37.05	42.34	52.95	63.56	84.79	106.02
.090	10.69	16.08	21.48	32.27	37.67	43.05	53.84	64.63	86.21	107.80
.091	10.87	16.35	21.84	32.81	38.30	43.77	54.75	65.72	87.66	109.62
.092	11.05	16.62	22.20	33.35	38.93	44.50	55.66	66.81	89.11	111.44
.093	11.23	16.89	22.56	33.89	39.57	45.23	56.57	67.90	90.57	113.26

Depth on Weir.	LENGTH OF THE WEIR.									
Feet.	2 Feet.	3 Feet.	4 Feet.	6 Feet.	7 Feet.	8 Feet.	10 Feet.	12 Feet.	16 Feet.	20 Feet.
.094	11.41	17.16	22.92	34.44	40.21	45.96	57.48	68.99	92.03	115.08
.095	11.58	17.44	23.29	34.99	40.85	46.69	58.39	70.09	93.49	116.90
.096	11.76	17.77	23.66	35.55	41.50	47.43	59.32	71.21	94.98	118.78
.097	11.94	17.99	24.03	36.11	42.15	48.17	60.25	72.33	96.47	120.66
.098	12.13	18.27	24.40	36.67	42.80	48.92	61.18	73.45	97.96	122.54
.099	12.32	18.55	24.77	37.23	42.45	49.67	62.12	74.57	99.46	124.42
.100	12.51	18.83	25.14	37.78	44.11	50.42	63.06	75.69	100.96	126.30
.101	12.69	19.11	25.52	38.35	44.77	51.18	64.01	76.84	102.49	128.20
.102	12.88	19.39	25.90	38.92	45.43	51.94	64.96	77.99	104.02	130.10
.103	13.07	19.67	26.28	39.49	46.10	52.70	65.91	79.14	105.55	132.00
.104	13.26	19.96	26.66	40.06	46.77	53.47	66.87	80.29	107.08	133.90
.105	13.45	20.25	27.04	40.64	47.44	54.24	67.83	81.44	108.62	135.81
.106	13.64	20.53	27.43	41.21	48.12	55.02	68.81	82.61	110.17	137.77
.107	13.83	20.81	27.82	41.79	48.80	55.80	69.79	83.78	111.73	139.73
.108	14.02	21.10	28.21	42.38	49.49	56.58	70.77	84.95	113.29	141.69
.109	14.21	21.39	28.60	42.97	50.18	57.36	71.75	85.12	114.85	143.65
.110	14.41	21.71	29.00	43.57	50.87	58.15	72.73	87.30	116.41	145.62
.111	14.60	22.00	29.39	44.17	51.57	58.95	73.73	88.50	118.07	147.62
.112	14.80	22.30	29.78	44.77	52.27	59.75	74.73	89.70	119.67	149.63
.113	15.00	22.60	30.11	45.37	52.97	60.55	75.73	90.90	121.27	151.64
.114	15.20	22.90	30.58	45.97	53.67	61.35	76.73	92.11	122.88	153.65
.115	15.40	23.20	30.98	46.57	54.37	62.15	77.74	93.32	124.49	155.66
.116	15.60	23.50	31.39	47.18	55.08	62.96	78.76	94.55	126.13	157.71
.117	15.80	23.80	31.80	47.79	55.79	63.78	79.78	95.78	127.77	159.76
.118	16.00	24.10	32.21	48.40	56.50	64.50	80.80	97.01	129.41	161.81
.119	16.20	24.41	32.62	49.01	57.22	65.32	81.82	98.25	131.05	163.86
.120	16.41	24.72	33.02	49.63	57.94	66.24	82.85	99.49	132.69	165.91
.121	16.61	25.03	33.43	50.25	58.67	67.07	83.89	100.74	134.36	168.00
.122	16.81	25.34	33.84	50.87	59.40	67.90	84.95	102.09	136.03	170.09
.123	17.01	25.65	34.26	51.49	60.17	68.74	85.98	103.34	137.70	172.18
.124	17.22	25.97	34.68	52.11	60.90	69.58	87.03	104.59	139.38	174.27
.125	17.43	26.27	35.09	52.75	61.59	70.42	88.08	105.74	141.06	176.38
.126	17.64	26.58	35.51	53.38	62.33	71.27	89.14	107.02	142.76	178.51
.127	17.85	26.89	35.93	54.02	63.07	72.13	90.20	108.30	144.47	180.64
.128	18.06	27.21	36.36	54.66	63.81	72.99	91.27	109.58	146.18	182.78
.129	18.27	27.53	36.79	55.30	64.55	73.85	92.34	110.86	147.89	184.91
.130	18.48	27.85	37.22	55.94	65.30	74.71	93.41	112.14	149.60	187.06
.131	18.69	28.17	37.65	56.59	66.06	75.57	94.49	113.44	151.34	189.23
.132	18.90	28.49	38.06	57.24	66.82	76.44	95.57	114.70	153.08	191.40
.133	19.11	28.81	38.49	57.89	67.58	77.31	96.66	116.04	154.82	193.58
.134	19.33	29.13	38.92	58.54	68.34	78.18	97.75	117.35	156.56	195.76
.135	19.55	29.46	39.37	59.19	69.11	79.05	98.84	118.66	158.30	197.94
.136	19.76	29.78	39.81	59.85	69.88	79.92	99.94	119.98	160.07	200.15
.137	19.97	30.11	40.25	60.51	70.65	80.79	101.04	121.31	161.84	202.37
.138	20.19	30.44	40.69	61.17	71.42	81.67	102.14	122.64	163.61	204.59
.139	20.41	30.77	41.13	61.84	72.19	82.55	103.25	123.97	165.38	206.81
.140	20.63	31.10	41.57	62.51	72.97	83.43	104.37	125.30	167.16	209.03
.141	20.85	31.43	42.01	63.18	73.75	84.33	105.49	126.65	168.96	211.28
.142	21.07	31.76	42.45	63.85	74.53	85.23	106.61	128.00	170.76	213.53
.143	21.29	32.09	42.90	64.52	75.32	86.13	107.74	129.35	172.57	215.79
.144	21.52	32.43	43.35	65.19	76.11	87.03	108.87	131.70	174.37	218.05
.145	21.74	32.77	43.80	65.87	76.90	87.93	110.00	132.06	176.19	220.31
.146	21.96	33.11	44.25	66.55	77.70	88.84	111.14	133.43	178.02	222.60
.147	22.18	33.45	44.71	67.23	78.50	89.75	112.28	134.80	179.85	224.90
.148	22.40	33.79	45.17	67.91	79.30	90.67	113.42	136.18	181.69	227.20
.149	22.63	34.13	45.63	68.60	80.10	91.59	114.57	137.56	183.53	229.50
.150	22.86	34.47	46.09	69.29	80.90	92.51	115.72	138.94	185.37	231.80
.151	23.08	34.81	46.55	69.98	81.71	93.43	116.88	140.33	187.23	234.13

Depth on Weir.	LENGTH OF THE WEIR.									
Feet.	2 Feet.	3 Feet.	4 Feet.	6 Feet.	7 Feet.	8 Feet.	10 Feet.	12 Feet.	16 Feet.	20 Feet.
.152	23.31	35.15	47.01	70.67	82.52	94.36	118.04	141.73	189.09	236.46
.153	23.54	35.50	47.47	71.37	83.33	95.29	119.21	143.17	190.96	238.79
.154	23.77	35.85	47.93	72.07	84.14	96.22	120.38	144.57	192.83	241.13
.155	24.00	36.20	48.39	72.77	84.96	97.15	121.55	145.93	194.70	243.47
.156	24.23	36.55	48.85	73.47	85.78	98.09	122.73	147.35	196.59	245.86
.157	24.46	36.90	49.32	74.18	86.61	99.04	123.91	148.77	199.48	248.25
.158	24.69	37.25	49.79	74.89	87.44	99.99	125.09	150.19	201.38	250.64
.159	24.93	37.60	50.26	75.60	88.27	100.94	126.27	151.61	203.28	253.04
.160	25.17	37.96	50.73	76.31	89.10	101.89	127.46	153.03	204.18	255.34
.161	25.40	38.31	51.20	77.02	89.93	102.85	128.66	154.47	206.10	257.75
.162	25.63	38.66	51.68	77.74	90.77	103.81	129.86	155.91	208.03	260.17
.163	25.86	39.01	52.16	78.46	91.61	104.77	131.06	157.35	209.96	262.59
.164	26.10	39.37	52.64	79.18	92.45	105.73	132.26	158.80	211.89	265.01
.165	26.34	39.73	53.12	79.90	93.29	106.69	133.47	160.25	213.82	267.38
.166	26.57	40.09	53.60	80.63	94.14	107.66	134.69	161.71	215.76	269.82
.167	26.81	40.45	54.08	81.36	94.99	108.63	135.91	163.17	217.70	272.26
.168	27.05	40.81	54.56	82.09	95.84	109.60	137.13	164.64	219.65	274.71
.169	27.29	41.17	55.05	82.82	96.69	110.58	138.35	166.11	221.60	277.16
.170	27.53	41.53	55.54	83.55	97.55	111.56	139.57	167.58	223.65	279.61
.171	27.77	41.88	56.03	84.28	98.41	112.54	140.85	169.06	225.62	282.09
.172	28.01	42.23	56.53	85.02	99.27	113.53	142.09	170.54	227.59	284.57
.173	28.25	42.59	57.02	85.76	100.14	114.52	143.33	172.03	229.56	287.05
.174	28.49	42.95	57.51	86.50	101.01	115.51	144.57	173.52	231.54	289.54
.175	28.74	43.31	57.99	87.24	101.88	116.50	145.76	175.01	233.52	292.03
.176	28.98	43.68	58.48	87.99	102.75	117.50	147.01	176.49	235.54	294.54
.177	29.22	44.05	58.98	88.74	103.62	118.50	148.26	177.98	237.56	297.06
.178	29.47	44.43	59.48	89.49	104.50	119.50	149.51	179.47	239.58	299.58
.179	29.72	44.81	59.98	90.24	105.38	120.50	150.77	180.96	241.60	302.10
.180	29.97	45.19	60.48	91.00	106.26	121.51	152.03	182.55	243.62	304.62
.181	30.21	45.57	60.98	91.76	107.14	122.52	153.31	184.07	245.65	307.17
.182	30.45	45.95	61.48	92.52	108.03	123.53	154.59	185.60	247.68	309.72
.183	30.70	46.33	61.98	93.28	108.92	124.54	155.88	187.13	249.71	312.27
.184	30.95	46.71	62.49	94.04	109.81	125.55	157.17	188.66	251.74	314.82
.185	31.20	47.10	63.00	94.80	110.70	126.57	158.43	190.19	253.78	317.38
.186	31.45	47.48	63.51	95.57	111.60	127.60	159.71	191.74	255.85	319.96
.187	31.70	47.86	64.02	96.34	112.50	128.63	160.99	193.29	257.92	322.55
.188	31.95	48.24	64.53	97.11	113.40	129.67	162.27	194.84	259.99	325.14
.189	32.21	48.62	65.04	97.88	114.30	130.71	163.55	196.39	262.06	327.73
.190	32.47	49.01	65.56	98.65	115.20	131.75	164.84	197.94	264.13	330.32
.191	32.72	49.39	66.07	99.43	116.11	132.79	166.14	199.51	266.22	332.93
.192	32.97	49.77	66.58	100.21	117.02	133.83	167.45	201.08	268.31	335.55
.193	33.22	50.16	67.10	100.99	117.93	134.87	168.76	202.65	270.40	338.17
.194	33.47	50.55	67.62	101.77	118.84	135.92	170.07	204.22	272.50	340.79
.195	33.73	50.94	68.14	102.56	119.76	136.97	171.38	205.79	274.60	343.41
.196	33.98	51.33	68.66	103.35	120.68	138.02	172.70	207.37	276.72	346.06
.197	34.24	51.72	69.18	104.14	121.60	139.07	174.02	208.96	278.84	348.72
.198	34.50	52.11	69.70	104.93	122.52	140.13	175.34	210.55	280.96	351.38
.199	34.76	52.50	70.23	105.72	123.45	141.19	176.66	212.14	283.08	354.04
.200	35.02	52.89	70.76	106.51	124.38	142.25	177.99	213.73	285.22	356.70
.201	35.28	53.28	71.29	107.31	125.31	143.32	179.33	215.34	287.36	359.38
.202	35.54	53.67	71.82	107.81	126.24	144.39	180.67	216.95	289.51	362.07
.203	35.80	54.07	72.35	108.61	127.17	145.46	182.01	218.56	291.66	364.76
.204	36.06	54.47	72.88	109.41	128.11	146.53	183.35	220.17	293.81	367.45
.205	36.33	54.87	73.42	110.51	129.05	147.60	184.69	221.78	295.96	370.14
.206	36.59	55.27	73.95	111.32	130.00	148.68	186.04	223.40	298.13	372.86
.207	36.85	55.67	74.48	112.13	130.94	149.76	187.39	225.03	300.30	375.58
.208	37.11	56.07	75.02	112.94	131.90	150.84	188.75	226.66	302.47	378.30
.209	37.37	56.47	75.56	113.75	132.85	151.92	190.11	228.29	304.65	381.02

Depth on Weir.				LENGTH OF THE WEIR.						
Feet.	2 Feet.	3 Feet.	4 Feet.	6 Feet.	7 Feet.	8 Feet.	10 Feet.	12 Feet.	16 Feet.	20 Feet.
.210	37.64	56.88	76.10	114.56	133.81	153.01	191.47	229.92	306.83	383.75
.211	37.90	57.28	76.64	115.37	134.76	154.10	192.84	231.56	309.03	386.30
.212	38.17	57.68	77.18	116.19	135.72	155.19	194.21	233.21	311.23	389.05
.213	38.44	58.08	77.72	117.01	136.68	156.29	195.58	234.86	313.43	391.80
.214	38.71	58.48	78.26	117.83	137.64	157.39	196.95	236.51	315.63	394.55
.215	38.98	58.89	78.81	118.65	138.60	158.49	198.33	238.16	317.84	397.51
.216	39.24	59.30	79.36	119.48	139.56	159.59	199.71	239.82	320.06	400.29
.217	39.51	59.71	79.91	120.31	140.52	160.69	201.09	241.49	322.28	403.07
.218	39.78	60.12	80.46	121.14	141.48	161.70	202.48	243.16	324.57	405.86
.219	40.05	60.53	81.01	121.97	142.45	162.81	203.87	244.83	326.80	408.65
.220	40.32	60.94	81.56	122.80	143.41	164.03	205.26	246.50	328.97	411.44
.221	40.59	61.35	82.11	123.63	144.38	165.15	206.66	248.18	331.22	414.23
.222	40.86	61.76	82.66	124.47	145.36	166.27	208.07	249.86	333.47	417.02
.223	41.13	62.17	83.21	125.31	146.34	167.39	209.48	251.55	335.72	419.82
.224	41.40	62.59	83.17	126.15	147.32	168.50	210.89	253.24	337.97	422.62
.225	41.68	63.01	84.33	126.99	148.30	169.63	212.30	254.93	340.22	425.52
.226	41.95	63.41	84.89	127.83	149.29	170.76	213.71	256.63	342.49	428.36
.227	42.22	63.82	85.45	128.67	150.28	171.89	215.12	258.33	344.76	431.21
.228	42.50	64.23	86.01	129.52	151.27	173.02	216.53	260.03	347.04	434.06
.229	42.78	64.65	86.57	130.37	152.26	174.16	217.95	261.73	349.32	436.91
.230	43.06	65.08	87.14	131.22	153.26	175.30	219.37	263.45	351.60	439.76
.231	43.33	65.50	87.70	132.07	154.26	176.44	220.80	265.17	353.90	442.65
.232	43.61	65.92	88.27	132.92	155.26	177.58	222.23	266.89	356.20	445.53
.233	43.89	66.35	88.84	133.78	156.26	178.72	223.67	268.61	358.50	448.41
.234	44.17	66.78	89.41	134.64	157.26	179.87	225.11	270.33	360.80	451.29
.235	44.45	67.21	89.98	135.50	158.26	181.02	226.55	272.06	363.11	454.15
.236	44.73	67.63	90.55	136.38	159.27	182.17	227.99	273.80	365.43	457.06
.237	45.01	68.05	91.12	137.22	160.28	183.36	229.44	275.54	367.73	459.97
.238	45.29	68.48	91.69	138.08	161.29	184.48	230.89	277.28	370.08	462.88
.239	45.57	68.91	92.26	138.95	162.30	185.64	232.34	279.02	372.43	465.79
.240	45.85	69.34	92.84	139.82	163.31	186.80	233.79	280.77	374.74	468.70
.241	46.13	69.77	93.41	140.71	164.33	187.96	235.24	282.52	377.09	471.65
.242	46.41	70.20	93.99	141.58	165.35	189.13	236.69	284.28	379.44	474.60
.243	46.69	70.63	94.58	143.31	166.37	190.30	237.14	286.04	381.79	477.55
.244	46.98	71.06	95.17	144.18	167.39	191.47	238.59	287.80	384.14	480.51
.245	47.27	71.49	95.73	144.19	168.42	192.64	241.05	289.56	386.49	483.40
.246	47.55	71.92	96.33	145.07	169.45	193.82	242.54	291.33	388.86	486.37
.247	47.83	72.35	96.91	145.95	170.48	195.00	244.03	293.11	391.23	489.34
.248	48.12	72.79	97.49	146.83	171.51	196.18	245.52	294.89	393.60	492.31
.249	48.41	73.24	98.07	147.71	172.54	197.36	247.01	296.67	395.97	495.28
.250	48.70	73.67	98.65	148.60	173.58	198.55	248.50	298.45	398.35	498.25
.251	48.98	74.11	99.24	149.49	174.62	199.74	249.99	300.24	401.74	501.25
.252	49.27	74.55	99.83	150.38	175.66	200.93	251.48	302.03	404.14	504.25
.253	49.51	74.99	100.42	151.27	176.70	202.12	252.97	303.82	406.53	507.25
.254	49.85	75.43	101.01	152.16	177.74	203.31	254.46	305.62	408.92	510.25
.255	50.14	75.87	101.60	153.06	178.78	204.51	255.95	307.42	410.33	513.25
.256	50.43	76.31	102.19	153.95	179.83	205.71	257.46	308.23	412.75	516.27
.257	50.72	76.75	102.78	154.85	180.88	206.91	258.97	310.04	415.17	519.30
.258	51.01	77.19	103.38	155.75	181.93	208.11	260.48	311.85	417.59	522.33
.259	51.30	77.63	103.98	156.65	182.98	209.32	262.02	313.66	420.01	525.36
.260	51.60	78.08	104.58	157.55	184.04	210.53	263.51	316.48	422.44	528.39
.261	51.89	78.52	105.18	158.45	185.10	211.74	265.03	318.31	424.88	531.44
.262	52.18	78.97	105.78	159.36	186.16	212.95	266.55	320.14	427.32	534.50
.263	52.47	79.42	106.38	160.27	187.22	214.16	268.07	321.97	429.76	537.56
.264	52.76	79.87	107.18	161.18	188.28	215.38	269.59	323.80	432.20	540.62
.265	53.06	80.32	107.58	162.09	189.35	216.60	271.11	325.63	434.65	543.68
.266	53.35	80.77	108.18	163.00	190.42	217.82	272.68	327.47	437.11	546.76
.267	53.64	81.22	108.78	163.92	191.49	219.04	274.22	329.31	439.58	549.84

Depth on Weir.	LENGTH OF THE WEIR.									
Feet.	2 Feet.	3 Feet.	4 Feet.	6 Feet.	7 Feet.	8 Feet.	10 Feet.	12 Feet.	16 Feet.	20 Feet.
.268	53.94	81.67	109.39	164.84	192.56	220.27	275.76	331.16	442.05	552.93
.269	54.24	82.12	110.00	165.76	193.63	221.50	277.30	333.01	444.52	556.02
.270	54.54	82.57	110.61	166.68	194.70	222.73	278.80	334.86	446.99	559.11
.271	54.84	83.02	111.22	167.60	195.78	223.96	280.34	336.72	449.47	562.22
.272	55.14	83.47	111.83	168.52	196.86	225.20	281.89	338.58	451.95	565.33
.273	55.44	83.93	112.44	169.44	197.94	226.44	283.44	340.44	454.43	568.44
.274	55.74	84.39	113.05	170.37	199.02	227.68	284.99	342.31	456.91	571.56
.275	56.04	84.85	113.67	171.30	200.11	228.92	286.54	344.18	459.43	574.68
.276	56.34	85.31	114.28	172.23	201.20	230.26	288.10	346.06	461.94	577.84
.277	56.64	85.77	114.89	173.16	202.29	231.51	289.66	347.94	464.45	581.00
.278	56.94	86.23	115.51	174.09	203.38	232.76	291.22	349.82	466.97	584.16
.279	57.24	86.69	116.13	175.02	204.47	234.01	292.79	351.71	469.49	587.33
.280	57.54	87.15	116.75	175.96	205.56	235.16	294.36	353.60	472.01	590.40
.281	57.84	87.61	117.37	176.90	206.66	236.42	295.94	355.49	474.54	593.57
.282	58.14	88.07	117.99	177.84	207.76	237.68	297.52	357.38	477.07	596.74
.283	58.44	88.53	118.61	178.78	208.86	238.94	299.10	359.27	479.60	599.91
.284	58.75	88.99	119.23	179.72	209.96	240.20	300.68	361.16	481.93	603.08
.285	59.06	89.46	119.86	180.66	211.06	241.46	302.26	363.06	484.66	606.25
.286	59.36	89.92	120.48	181.61	212.17	242.73	303.85	364.96	487.21	609.45
.287	59.66	90.38	121.11	182.56	213.28	244.00	305.44	366.87	489.76	612.65
.288	59.97	90.85	121.74	183.51	214.39	245.27	307.03	368.78	492.31	615.85
.289	60.28	91.32	122.37	184.46	215.50	246.54	308.62	370.69	494.87	619.05
.290	60.59	91.79	123.00	185.41	216.61	247.81	310.22	372.60	497.43	622.25
.291	60.89	92.26	123.63	186.36	217.72	249.09	311.82	374.51	500.00	625.47
.292	61.20	92.73	124.26	187.31	218.84	250.37	313.42	376.42	502.58	628.69
.293	61.51	93.20	124.89	188.27	219.96	251.65	315.02	378.34	505.16	631.91
.294	61.82	93.67	125.52	189.23	221.08	252.93	316.63	380.26	507.74	635.14
.295	62.13	94.15	126.16	190.19	222.20	254.22	318.24	382.18	510.32	638.37
.296	62.44	94.62	126.79	191.15	223.32	255.51	319.86	384.14	512.92	641.62
.297	62.75	95.09	127.43	192.11	224.45	256.80	321.48	386.10	515.52	644.87
.298	63.06	95.56	128.07	193.07	225.58	258.09	323.10	388.06	518.53	648.12
.299	63.47	96.04	128.71	194.04	226.71	259.38	324.72	390.02	520.72	651.38
.300	63.69	96.52	129.35	195.01	227.84	260.67	326.34	392.00	523.32	654.64
.301	64.00	96.99	129.99	195.98	228.97	261.97	327.97	393.95	525.94	657.92
.302	64.31	97.47	130.63	196.95	230.11	263.27	329.60	395.90	528.58	661.20
.303	64.62	97.95	131.26	197.92	231.25	264.57	331.23	397.86	531.18	664.48
.304	64.93	98.43	131.91	198.99	232.39	265.87	332.86	399.82	533.80	667.76
.305	65.25	98.91	132.57	199.87	233.53	267.18	334.49	401.78	536.42	671.04
.306	65.56	99.39	133.21	200.85	234.67	268.49	336.13	403.76	539.06	674.34
.307	65.87	99.87	133.86	201.83	235.81	269.80	337.77	405.74	541.70	677.64
.308	66.19	100.35	134.51	202.81	236.96	271.11	339.42	407.72	544.34	680.94
.309	66.51	100.83	135.16	203.79	238.11	272.42	341.07	409.70	546.98	684.24
.310	66.83	101.31	135.81	204.77	239.26	273.73	342.72	411.69	549.63	687.57
.311	67.14	101.79	136.46	205.75	240.41	275.06	344.37	413.68	552.29	690.86
.312	67.46	102.27	137.11	206.74	241.56	276.38	346.03	415.67	554.95	694.19
.313	67.78	102.76	137.76	207.73	242.72	277.71	347.69	417.66	557.61	697.52
.314	68.10	103.25	138.41	208.72	243.88	279.04	349.35	419.65	560.27	700.75
.315	68.42	103.74	139.07	209.71	245.04	280.36	351.01	421.65	562.94	704.18
.316	68.74	104.23	139.72	210.70	246.20	281.69	352.68	423.65	565.62	707.55
.317	69.06	104.72	140.38	211.69	247.36	283.02	354.35	425.65	568.30	710.92
.318	69.38	105.21	141.04	212.69	248.52	284.35	356.02	427.66	570.98	714.29
.319	69.70	105.70	141.70	213.69	249.68	285.69	357.69	429.67	573.66	717.66
.320	70.02	106.19	142.36	214.69	250.85	287.03	359.36	431.69	576.36	721.04
.321	70.34	106.68	143.02	215.69	252.02	288.37	361.04	433.71	579.06	724.42
.322	70.66	107.17	143.68	216.69	253.19	289.71	362.72	435.73	581.76	727.80
.323	70.98	107.66	144.34	217.69	254.36	291.05	364.40	437.75	584.47	731.19
.324	71.30	107.85	145.00	218.70	255.54	292.39	366.09	439.78	587.18	734.58
.325	71.63	108.65	145.67	219.71	256.72	293.74	367.78	441.82	589.89	737.97

Depth on Weir				LENGTH OF THE WEIR.						
Feet.	2 Feet.	3 Feet.	4 Feet.	6 Feet.	7 Feet.	8 Feet.	10 Feet.	12 Feet.	16 Feet.	20 Feet.
.326	71.95	109.14	146.33	220.72	257.90	295.09	369.47	443.86	592.61	741.37
.327	72.27	109.63	147.00	221.73	259.08	296.44	371.11	445.90	595.33	744.77
.328	72.59	110.13	147.67	222.74	260.26	297.79	372.87	447.94	598.06	748.18
.329	72.92	110.63	148.34	223.75	261.44	299.15	374.57	449.98	600.79	751.59
.330	73.25	111.13	149.01	224.76	262.63	300.51	376.27	452.02	603.52	755.09
.331	73.57	111.63	149.68	225.77	263.82	301.87	377.97	454.07	606.26	758.44
.332	73.89	112.13	150.35	226.78	265.01	303.23	378.68	456.12	609.00	761.88
.333	74.22	112.63	151.02	227.80	266.20	304.60	380.39	458.18	611.75	765.32
.334	74.55	113.13	151.69	228.82	267.39	305.97	382.10	460.28	614.50	768.76
.335	74.88	113.63	152.37	229.84	268.59	307.34	384.81	462.30	617.25	772.21
.336	75.20	114.13	153.04	230.86	269.78	308.71	386.53	464.36	620.01	775.66
.337	75.53	114.63	153.71	231.88	270.98	310.08	388.25	466.42	622.77	779.12
.338	75.86	115.13	154.39	232.91	272.08	311.45	389.97	468.49	625.52	782.58
.339	76.19	115.63	155.07	233.94	273.28	312.82	391.69	470.56	628.31	786.04
.340	76.52	116.14	155.75	234.97	274.58	314.19	393.42	472.63	631.08	789.50
.341	76.85	116.64	156.43	236.00	275.78	315.57	395.15	474.71	633.86	792.99
.342	77.18	117.14	157.11	237.03	276.99	316.95	396.88	476.79	636.64	796.48
.343	77.51	117.65	157.79	238.06	278.20	318.33	398.61	478.88	639.42	799.97
.344	77.84	118.16	158.47	239.09	279.41	319.72	400.34	480.97	642.20	803.46
.345	78.18	118.67	159.16	240.13	280.62	321.11	402.08	483.06	644.99	806.96
.346	78.51	119.18	159.84	241.17	281.83	322.50	403.82	485.15	647.79	810.46
.347	78.84	119.69	160.52	242.21	283.04	323.89	405.56	487.25	650.59	813.96
.348	79.17	120.20	161.20	243.25	284.26	325.28	407.31	489.35	653.39	817.47
.349	79.50	120.71	161.90	244.29	285.48	326.68	409.06	491.45	656.19	820.98
.350	79.84	121.22	162.59	245.33	286.70	328.08	410.81	493.55	659.00	824.49
.351	80.17	121.73	163.28	246.37	287.92	329.48	412.57	495.66	661.83	828.03
.352	80.50	122.24	163.97	247.41	289.14	330.88	414.33	497.77	664.66	831.57
.353	80.83	122.76	164.66	248.46	290.36	332.28	416.09	499.89	667.49	835.11
.354	81.17	123.28	165.35	249.51	291.59	334.28	417.85	501.01	670.33	838.66
.355	81.51	123.79	166.04	250.56	292.82	335.08	419.61	504.13	673.17	842.21
.356	81.84	124.30	166.73	251.61	294.05	336.49	421.38	506.25	676.01	845.77
.357	82.18	124.81	167.42	252.66	295.28	337.90	423.15	508.38	678.86	849.33
.358	82.52	125.32	168.12	253.71	296.51	339.31	424.92	510.50	681.71	852.89
.359	82.86	125.84	168.82	254.77	297.75	340.73	426.69	512.63	684.56	856.46
.360	83.20	126.36	169.52	255.83	298.99	342.15	428.46	514.77	687.41	860.03
.361	83.54	126.88	170.22	256.89	300.23	343.57	430.24	516.91	690.27	863.61
.362	83.88	127.40	170.92	257.95	301.47	344.99	432.02	519.05	693.13	867.19
.363	84.22	127.92	171.62	259.01	302.71	346.11	433.80	521.19	695.99	870.78
.364	84.56	128.44	172.32	260.07	303.95	347.83	435.58	523.34	698.85	874.37
.365	84.90	128.96	173.02	261.14	305.20	349.26	437.38	525.49	701.72	877.96
.366	85.24	129.48	173.72	262.20	306.45	350.68	439.17	527.64	704.60	881.57
.367	85.58	130.00	174.42	263.27	307.70	352.11	440.96	529.80	707.48	885.18
.368	85.92	130.52	175.13	264.34	308.95	353.54	442.75	531.96	710.37	888.80
.369	86.26	131.05	175.84	265.41	310.20	354.97	444.55	534.12	713.26	892.42
.370	86.60	131.58	176.54	266.48	311.45	356.41	446.35	536.28	716.15	896.04
.371	86.94	132.10	177.25	267.55	312.71	357.85	448.15	538.45	719.05	899.67
.372	87.28	132.62	177.96	268.62	313.97	359.29	449.95	540.62	721.95	903.30
.373	87.62	133.14	178.67	269.69	315.23	360.73	451.76	542.80	724.86	906.93
.374	87.97	133.67	179.38	270.77	316.49	362.17	453.57	544.98	727.77	910.56
.375	88.32	134.20	180.09	271.85	317.76	363.62	455.38	547.16	730.68	914.20
.376	88.66	134.73	180.80	272.93	319.02	365.06	457.19	549.34	733.60	917.86
.377	89.00	135.26	181.51	274.01	320.28	366.51	459.01	551.52	736.52	921.52
.378	89.34	135.79	182.22	275.09	321.54	367.96	460.83	553.70	739.44	896.04
.379	89.69	136.32	182.93	276.17	322.80	369.41	462.65	555.89	742.36	928.84
.380	90.04	136.85	183.65	277.26	324.06	370.86	464.47	558.08	745.29	932.50
.381	90.39	137.38	184.36	278.34	325.33	372.32	466.27	560.28	748.23	936.18
.382	90.74	137.91	185.08	279.43	326.60	373.78	468.10	562.48	751.17	939.86
.383	91.09	138.44	185.80	280.52	327.87	375.24	469.93	564.68	754.11	943.54

Depth on Weir.	LENGTH OF THE WEIR.									
Feet.	2 Feet.	3 Feet.	4 Feet.	6 Feet.	7 Feet.	8 Feet.	10 Feet.	12 Feet.	16 Feet.	20 Feet.
.384	91.45	138.97	186.52	281.61	329.15	376.70	471.76	568.88	757.05	947.23
.385	91.81	139.51	187.24	282.70	330.43	378.16	473.61	569.08	760.00	950.92
.386	92.15	140.04	187.96	283.79	331.71	379.62	475.45	571.29	762.96	954.62
.387	92.49	140.57	188.68	284.88	332.99	381.09	477.30	573.51	765.92	958.32
.388	92.83	141.11	189.40	285.98	334.27	382.56	479.15	575.73	768.88	962.02
.389	93.18	141.65	190.12	287.08	335.55	384.03	481.00	577.95	771.84	965.71
.390	93.53	142.19	190.83	288.18	336.84	385.50	482.85	580.17	774.81	969.44
.391	93.88	142.73	191.57	289.28	338.12	386.97	484.70	582.39	777.78	973.17
.392	94.23	143.27	192.39	290.38	339.41	388.45	486.55	584.61	780.76	976.90
.393	94.58	143.81	193.12	291.48	340.70	389.93	488.40	586.84	783.74	980.63
.394	94.93	144.35	193.85	292.58	341.99	391.41	490.25	589.07	786.72	984.37
.395	95.28	144.89	194.48	293.69	343.28	392.89	492.10	591.30	789.70	988.11
.396	95.63	145.43	195.21	294.79	344.58	394.37	493.96	593.54	792.69	991.92
.397	95.98	145.97	195.94	295.90	345.88	395.86	495.82	595.78	795.69	995.67
.398	96.33	146.51	196.67	297.01	347.18	397.35	497.68	598.02	798.69	999.42
.399	96.68	147.06	197.40	298.12	348.48	398.84	499.54	600.26	801.69	1003.17
.400	97.04	147.61	198.14	299.23	349.78	400.33	501.41	602.51	804.69	1006.87
.401	97.39	148.15	198.83	300.34	351.08	401.82	503.28	604.76	807.70	1010.64
.402	97.74	148.69	199.56	301.45	352.39	403.31	505.15	607.02	810.71	1014.42
.403	98.10	149.23	200.29	302.57	353.70	404.80	507.02	609.28	813.72	1018.20
.404	98.54	149.77	201.03	303.69	355.01	406.30	508.91	611.54	816.74	1021.98
.405	98.82	150.32	201.81	304.81	356.32	407.80	510.79	613.80	819.77	1025.76
.406	99.17	150.86	202.55	305.93	357.63	409.30	512.67	616.14	822.80	1029.56
.407	99.52	151.40	203.29	307.05	358.94	410.80	514.56	618.40	825.83	1033.36
.408	99.88	151.95	204.03	308.17	360.75	412.30	516.45	620.66	828.86	1047.16
.409	100.24	152.50	204.77	309.29	361.56	413.81	518.34	622.92	831.90	1050.96
.410	100.60	153.05	205.51	310.42	362.87	415.32	520.23	625.13	834.94	1044.76
.411	100.96	153.60	206.25	311.54	364.18	416.83	522.12	627.41	837.99	1048.58
.412	101.32	154.15	206.99	312.67	365.49	418.34	524.02	629.69	841.04	1052.40
.413	101.68	154.70	207.73	313.80	366.80	419.85	525.92	631.97	844.09	1056.22
.414	102.04	155.25	208.48	314.93	368.11	421.37	527.82	634.26	847.15	1060.05
.415	102.40	155.81	209.23	316.06	369.42	422.89	529.72	636.55	850.21	1063.88
.416	102.76	156.36	209.97	317.19	370.76	424.41	531.63	638.84	853.28	1067.72
.417	103.12	156.91	210.72	318.32	372.10	425.93	533.54	641.14	856.35	1071.56
.418	103.48	157.47	211.47	319.45	373.44	427.45	535.45	643.44	859.42	1075.40
.419	103.84	158.03	212.22	320.59	374.78	428.97	537.36	645.74	862.50	1079.25
.420	104.20	158.58	212.97	321.73	376.12	430.50	539.27	648.04	865.58	1083.10
.421	104.56	159.13	213.72	322.87	377.45	432.03	541.19	650.34	868.66	1086.97
.422	104.92	159.70	214.47	324.01	378.78	433.56	543.11	652.65	871.74	1090.84
.423	105.28	160.29	215.22	325.16	380.12	435.09	545.03	654.96	874.82	1094.71
.424	105.64	160.88	215.97	326.30	381.46	436.62	546.05	657.17	877.90	1098.58
.425	106.01	161.32	216.73	327.44	382.80	438.15	548.88	659.58	880.99	1102.45
.426	106.37	161.93	217.48	328.58	384.16	439.69	550.81	661.90	884.10	1106.34
.427	106.73	162.49	218.23	329.73	385.50	441.23	552.75	664.22	887.21	1110.23
.428	107.10	163.05	218.99	330.88	386.84	442.77	554.69	666.54	890.32	1114.12
.429	107.46	163.61	219.75	332.03	388.19	444.31	556.63	668.87	893.43	1118.01
.430	107.83	164.17	220.51	333.18	389.52	445.86	558.53	671.20	896.55	1121.90
.431	108.19	164.73	221.27	334.33	390.87	447.40	560.47	673.54	899.67	1125.81
.432	108.55	165.29	222.03	335.48	392.25	448.95	562.41	675.88	902.79	1129.72
.433	108.92	165.85	222.79	336.63	393.60	450.50	564.35	678.22	905.92	1133.63
.434	109.29	166.41	223.55	337.79	394.95	452.05	566.29	680.56	909.65	1137.55
.435	109.66	166.98	224.31	338.95	396.27	453.60	568.24	682.90	912.18	1141.47
.436	110.02	167.54	225.07	340.11	397.62	455.16	570.19	685.25	915.32	1145.40
.437	110.39	168.10	225.83	341.27	398.98	456.72	572.14	687.61	918.46	1149.33
.438	110.76	168.67	226.59	342.43	400.34	458.28	574.09	689.97	921.60	1153.27
.439	111.13	169.24	227.36	343.59	401.70	459.84	576.05	692.33	924.74	1157.21
.440	111.50	169.81	228.13	344.75	403.06	461.40	578.01	694.64	927.89	1161.15
.441	111.86	170.37	228.89	345.91	404.43	462.96	579.97	697.00	931.05	1165.10

Depth on Weir.				LENGTH OF THE WEIR.						
Feet.	2 Feet.	3 Feet.	4 Feet.	6 Feet.	7 Feet.	8 Feet.	10 Feet.	12 Feet.	16 Feet.	20 Feet.
.442	112.23	170.94	229.66	347.08	405.80	464.52	581.93	699.36	934.21	1169.06
.443	112.60	171.51	230.43	348.25	407.17	466.08	583.89	701.68	937.37	1173.02
.444	112.97	172.08	231.20	349.42	408.54	467.64	585.86	704.04	940.53	1176.98
.445	113.34	172.65	231.97	350.59	409.91	469.21	587.83	706.45	943.70	1180.94
.446	113.71	173.22	232.74	351.76	411.28	470.78	589.80	708.83	946.87	1184.92
.447	114.08	173.79	233.51	352.93	412.65	472.35	591.77	711.21	950.04	1188.90
.448	114.45	174.36	234.28	354.10	414.02	473.92	593.74	713.59	953.22	1192.88
.449	114.82	175.13	235.05	355.28	415.39	475.49	595.72	715.97	956.40	1196.86
.450	115.20	175.51	235.83	356.46	416.77	477.07	597.70	718.35	959.58	1200.84
.451	115.57	176.08	236.61	357.63	418.15	478.75	599.68	720.73	962.77	1204.84
.452	115.94	176.65	237.38	358.81	419.53	480.33	601.67	723.11	965.96	1208.84
.453	116.31	177.22	238.16	359.99	420.91	481.91	603.66	725.49	969.16	1212.84
.454	116.68	177.80	238.94	361.17	422.29	483.50	605.65	727.88	972.36	1216.84
.455	117.06	178.38	239.71	362.35	423.67	484.99	607.64	730.27	975.56	1220.85
.456	117.43	178.95	240.50	363.53	425.06	486.58	609.63	732.67	978.77	1224.87
.457	117.80	179.53	241.30	364.71	426.46	488.17	611.62	735.07	981.98	1228.90
.458	118.18	180.11	242.10	365.90	427.86	489.76	613.61	737.47	985.19	1232.93
.459	118.56	180.69	242.90	367.09	429.26	491.35	615.61	739.88	988.40	1236.96
.460	118.94	181.27	243.60	368.28	430.61	492.94	617.61	742.29	991.62	1240.99
.461	119.31	181.85	244.38	369.47	432.00	494.54	619.61	744.70	994.85	1245.03
.462	119.68	182.43	245.16	370.66	433.40	496.14	621.61	747.10	998.08	1249.07
.463	120.06	183.01	245.94	371.85	434.80	497.74	623.62	749.51	1001.31	1253.11
.464	120.44	183.59	246.72	373.04	436.20	499.34	625.63	751.92	1004.54	1257.16
.465	120.82	184.17	247.50	374.23	437.60	500.94	627.64	754.35	1007.77	1261.19
.466	121.19	184.75	248.29	375.42	439.06	502.54	629.66	756.77	1011.01	1265.25
.467	121.57	185.33	249.08	376.62	440.40	504.14	631.68	759.29	1014.25	1269.31
.468	121.95	185.91	249.87	377.82	441.80	505.75	633.70	761.72	1017.50	1273.38
.469	122.33	186.49	250.66	379.02	442.20	507.36	635.72	764.15	1020.75	1277.45
.470	122.71	187.08	251.46	380.22	444.60	508.97	637.75	766.49	1024.00	1281.52
.471	123.08	187.66	252.25	381.42	446.00	510.57	639.77	768.93	1027.26	1285.61
.472	123.47	188.24	253.04	382.62	447.41	512.21	641.79	771.37	1030.52	1289.70
.473	123.85	188.83	253.83	383.83	448.81	513.85	643.82	773.81	1033.78	1293.79
.474	124.23	189.42	254.62	385.04	449.62	515.49	645.85	776.25	1037.04	1297.88
.475	124.60	190.01	255.42	386.24	451.64	517.17	647.88	778.71	1040.30	1301.96
.476	124.98	190.59	256.21	387.45	453.05	518.83	649.91	781.16	1043.58	1306.06
.477	125.36	191.18	257.00	388.66	454.47	520.49	651.94	783.61	1046.86	1310.17
.478	125.74	191.77	257.80	389.87	455.89	522.16	653.98	786.06	1050.15	1314.28
.479	126.12	192.36	258.60	391.08	457.31	523.83	656.02	788.51	1053.44	1318.39
.480	126.51	192.95	259.40	392.29	458.73	525.18	658.06	790.96	1056.73	1322.50
.481	126.89	193.54	260.19	393.50	460.15	526.81	660.10	793.43	1060.02	1326.63
.482	127.27	194.13	260.99	394.71	461.58	528.44	662.15	795.90	1063.31	1330.76
.483	127.65	194.72	261.79	395.93	463.01	530.07	664.20	798.37	1066.60	1334.89
.484	128.05	195.13	262.59	397.15	464.44	531.70	666.25	800.84	1069.90	1339.03
.485	128.42	195.90	263.39	398.37	465.87	533.33	668.30	803.31	1073.20	1343.17
.486	128.80	196.49	264.19	399.59	467.29	534.97	670.36	805.78	1076.51	1347.32
.487	129.18	197.08	264.99	400.81	469.71	536.61	672.42	808.26	1079.82	1351.47
.488	129.56	197.68	265.79	402.03	471.14	538.25	674.48	810.74	1083.13	1355.62
.489	129.95	198.28	266.59	403.25	472.57	539.89	676.54	813.22	1086.45	1359.77
.490	130.34	198.88	267.41	404.47	473.00	541.53	678.60	815.70	1089.70	1363.92
.491	130.72	199.47	268.21	405.69	474.44	543.17	680.66	818.18	1093.04	1368.09
.492	131.10	200.06	269.01	406.92	475.88	544.81	682.73	820.66	1096.39	1372.26
.493	131.49	200.66	269.82	408.15	477.32	546.45	684.80	823.14	1099.74	1376.43
.494	131.88	201.26	270.63	409.38	478.76	548.10	686.87	825.62	1103.09	1380.60
.495	132.27	201.86	271.44	410.61	480.20	549.75	688.94	828.10	1106.44	1384.78
.496	132.66	202.46	272.25	411.84	481.64	551.41	691.02	830.60	1109.92	1388.97
.497	133.05	203.06	273.06	413.07	483.08	553.07	693.10	833.10	1113.40	1393.16
.498	133.44	203.66	273.87	414.30	484.52	554.73	695.18	835.60	1116.88	1397.35
.499	133.83	204.26	274.68	415.54	485.97	556.39	697.26	838.11	1120.36	1401.54

QUANTITIES OF WATER, IN CUBIC FEET PER MINUTE, FLOWING OVER WEIRS OF DIFFERENT LENGTHS, WITH VARYING DEPTHS OF WATER.

Depth on Weir.	LENGTH OF THE WEIR.									
Feet.	2 Feet.	3 Feet.	4 Feet.	6 Feet.	7 Feet.	8 Feet.	10 Feet.	12 Feet.	16 Feet.	20 Feet.
.500	134.22	204.86	275.50	416.78	487.42	558.06	699.34	840.62	1123.18	1405.74
.501	134.61	205.46	276.31	418.01	488.87	559.72	701.42	843.13	1126.54	1409.95
.502	135.00	206.06	277.12	419.25	490.32	561.38	703.51	845.64	1129.90	1414.16
.503	135.39	206.66	277.93	420.49	491.77	563.04	705.60	848.15	1133.26	1418.37
.504	135.78	207.26	278.74	421.73	493.22	564.71	707.69	850.66	1136.62	1422.58
.505	136.16	207.86	279.57	422.97	494.67	566.38	709.78	853.18	1139.99	1426.80
.506	136.55	208.46	280.38	424.21	496.13	568.05	711.87	855.70	1143.37	1431.03
.507	136.94	209.06	281.20	425.45	497.59	569.72	713.97	858.22	1146.75	1435.27
.508	137.33	209.67	282.02	426.70	499.05	571.39	716.07	860.75	1150.13	1439.51
.509	137.72	210.28	282.84	427.95	500.57	573.06	718.17	863.28	1153.51	1443.75
.510	138.12	210.89	283.66	429.20	501.97	574.74	720.27	865.81	1156.89	1447.94
.511	138.51	211.50	211.71	430.45	503.43	576.41	722.38	868.34	1160.28	1453.20
.512	138.90	212.11	212.53	431.70	504.89	578.09	724.49	870.88	1163.67	1458.46
.513	139.29	212.71	213.35	432.95	506.35	579.77	726.60	873.42	1167.07	1462.72
.514	139.68	213.32	214.17	434.20	507.82	581.45	728.71	875.98	1170.47	1466.46
.515	140.08	213.92	287.76	435.45	509.29	583.13	730.82	878.50	1173.87	1469.24
.516	140.47	214.53	288.58	436.70	510.76	584.81	732.93	881.05	1177.28	1474.51
.517	140.86	215.14	289.40	437.95	512.23	586.40	735.05	883.60	1180.69	1478.78
.518	141.25	215.75	290.23	439.21	513.70	588.09	737.17	886.15	1184.11	1483.06
.519	141.64	216.36	291.06	440.47	515.17	589.78	739.29	888.70	1187.53	1487.34
.520	142.05	216.97	291.89	441.73	516.65	591.57	741.41	891.25	1190.94	1490.62
.521	142.45	217.58	292.71	442.99	518.13	593.26	743.54	893.81	1194.36	1494.91
.522	142.85	218.19	293.54	444.25	519.61	594.95	745.67	896.37	1197.79	1499.20
.523	143.25	218.80	294.37	432.95	521.09	596.65	747.80	898.93	1201.22	1503.49
.524	143.65	219.41	295.20	434.20	522.57	598.35	749.93	901.49	1204.65	1507.79
.525	144.03	220.03	296.03	448.04	524.05	600.05	752.06	904.06	1208.08	1512.09
.526	144.43	220.64	296.86	449.30	525.53	601.75	754.19	906.63	1211.52	1516.40
.527	144.83	221.25	297.69	450.57	527.01	603.45	756.33	909.20	1214.96	1520.71
.528	145.23	221.86	298.52	451.84	528.49	605.15	758.47	911.77	1218.40	1525.03
.529	145.60	222.47	299.36	453.11	429.98	606.85	760.61	914.34	1221.85	1529.35
.530	146.01	223.10	300.20	454.38	531.47	608.56	762.75	916.93	1225.30	1533.67
.531	146.41	223.72	301.03	455.65	532.96	610.27	764.89	919.51	1228.76	1538.00
.532	146.81	224.34	301.86	456.92	534.45	611.98	767.04	922.09	1232.12	1542.33
.533	147.21	224.96	302.70	458.19	535.94	613.69	769.19	924.68	1235.58	1546.66
.534	147.61	225.58	303.54	459.47	537.43	615.40	771.34	927.27	1239.04	1551.00
.535	148.01	226.19	304.38	460.75	538.93	617.12	773.49	929.86	1242.60	1555.34
.536	148.41	226.81	305.21	462.02	540.43	618.83	775.64	932.45	1246.07	1559.69
.537	148.81	227.43	306.05	463.30	541.93	620.55	777.80	935.05	1249.54	1564.04
.538	149.21	228.05	306.89	464.58	543.43	622.27	779.96	937.65	1253.02	1568.40
.539	149.61	228.67	307.73	465.86	544.93	623.99	782.12	940.25	1256.50	1572.76
.540	150.01	229.29	308.57	467.14	546.43	625.71	784.28	942.85	1259.98	1577.12
.541	150.41	229.91	309.41	468.42	547.93	627.43	786.44	945.45	1263.47	1581.49
.542	150.81	230.53	310.25	469.70	549.43	629.15	788.61	948.06	1266.96	1585.86
.543	151.21	231.15	311.09	470.98	550.93	630.88	790.78	950.67	1270.45	1590.23
.544	151.61	231.77	311.94	472.27	552.44	632.61	792.95	953.22	1273.94	1594.61
.545	152.01	232.40	312.79	473.56	553.95	634.34	795.12	955.89	1277.44	1598.99
.546	152.41	233.03	313.63	475.85	555.46	636.07	797.29	958.51	1280.95	1603.38
.547	152.81	233.66	314.47	477.14	556.97	637.80	799.46	961.13	1284.46	1607.77
.548	153.21	234.29	315.32	478.43	558.48	639.53	801.64	963.75	1287.97	1612.17
.549	153.61	234.92	316.27	479.72	559.99	641.27	803.82	966.37	1291.48	1616.57
.550	154.03	235.53	317.02	480.01	561.51	643.01	806.00	969.00	1294.99	1620.97
.551	154.43	236.16	317.87	481.30	563.02	644.75	808.18	971.63	1298.51	1625.38
.552	154.83	236.79	318.72	482.59	564.54	646.49	810.36	974.26	1302.03	1629.71
.553	155.23	237.42	319.57	483.89	566.06	648.23	812.54	976.89	1305.55	1634.12
.554	155.63	238.05	320.42	485.19	567.58	649.97	814.72	979.52	1309.07	1638.54

Depth on Weir.	LENGTH OF THE WEIR.									
Feet.	2 Feet.	3 Feet.	4 Feet.	6 Feet.	7 Feet.	8 Feet.	10 Feet.	12 Feet.	16 Feet.	20 Feet.
.555	156.05	238.66	321.27	486.49	569.10	651.71	816.93	982.16	1312.60	1643.04
.556	156.46	239.29	322.12	487.79	570.62	653.46	819.12	984.80	1316.13	1647.47
.557	156.87	239.92	322.97	489.09	572.14	655.25	821.32	987.48	1319.67	1651.90
.558	157.28	240.55	323.82	490.39	573.67	657.00	823.52	990.12	1323.21	1656.33
.559	157.69	241.18	324.68	491.69	575.20	658.75	825.72	992.76	1326.74	1660.77
.560	158.08	241.81	325.54	493.00	576.73	660.46	827.92	995.37	1330.29	1665.21
.561	158.49	242.44	326.39	494.30	578.26	662.21	830.12	998.02	1333.84	1669.66
.562	158.00	243.07	327.24	495.60	579.79	663.96	832.32	1000.67	1337.39	1674.11
.563	158.41	243.70	328.10	496.91	581.32	665.72	834.52	1003.33	1340.94	1678.56
.564	158.82	244.33	328.96	498.22	582.85	667.48	836.73	1005.99	1344.50	1683.02
.565	160.12	244.97	329.82	499.53	584.38	669.24	838.94	1008.65	1348.06	1687.48
.566	160.53	245.61	330.68	500.84	585.91	671.00	841.15	1011.31	1351.63	1691.95
.567	160.94	246.25	331.54	502.15	587.45	672.76	843.36	1013.97	1355.20	1696.42
.568	161.35	246.89	332.40	503.46	588.99	674.52	845.58	1016.54	1358.77	1700.89
.569	161.76	247.54	333.26	504.77	590.53	676.28	847.80	1019.21	1362.27	1705.36
.570	162.16	248.19	334.13	506.09	592.07	678.05	850.02	1021.98	1365.91	1709.84
.571	162.57	248.82	334.99	507.40	593.61	679.82	852.24	1024.65	1369.49	1714.33
.572	162.98	249.45	335.85	508.72	595.15	681.59	854.46	1027.33	1373.07	1718.82
.573	163.39	250.08	336.71	510.04	596.69	683.36	856.68	1030.01	1376.66	1723.31
.574	163.80	250.71	337.68	511.36	598.24	685.13	858.91	1032.69	1380.25	1727.80
.575	164.21	251.33	338.45	512.68	599.79	686.91	861.14	1035.37	1383.84	1732.30
.576	164.42	251.93	339.31	514.00	601.34	688.68	863.57	1038.06	1387.44	1736.81
.577	164.83	252.53	340.17	515.32	602.89	690.46	865.60	1040.75	1391.04	1741.32
.578	165.24	253.13	341.04	516.64	604.44	692.24	867.83	1043.44	1394.64	1745.83
.579	165.65	253.73	341.91	517.96	605.99	694.02	870.07	1046.13	1398.24	1750.34
.580	166.27	254.53	342.78	519.29	607.54	695.80	872.31	1048.82	1401.84	1754.86
.581	166.69	255.17	343.65	520.61	609.09	697.58	874.55	1051.52	1405.45	1759.29
.582	167.11	255.81	344.52	521.94	610.65	699.36	876.79	1054.22	1409.06	1764.92
.583	167.53	256.45	345.39	523.27	612.21	701.15	879.03	1056.92	1412.67	1769.45
.584	167.95	257.09	346.26	524.60	613.77	702.94	881.27	1059.62	1416.28	1773.98
.585	168.34	257.74	347.13	525.93	615.33	704.73	883.52	1062.32	1419.91	1777.51
.586	168.76	258.38	348.00	527.26	616.89	706.52	885.77	1064.03	1423.54	1782.05
.587	169.18	259.02	348.87	528.59	618.45	708.31	888.02	1066.74	1427.17	1786.60
.588	169.60	259.66	349.74	529.92	620.01	710.10	890.27	1069.45	1430.80	1791.15
.589	070.02	260.31	350.62	531.26	621.87	711.89	892.52	1072.16	1434.43	1795.70
.590	170.41	260.96	351.50	532.60	623.14	713.69	894.78	1075.88	1438.07	1800.25
.591	170.83	261.60	352.37	533.93	624.71	715.49	897.04	1078.60	1441.71	1804.81
.592	171.25	262.24	353.25	535.27	626.28	717.29	899.30	1081.32	1445.35	1809.38
.593	171.66	262.89	354.13	536.61	627.85	719.09	901.56	1084.04	1448.99	1814.95
.594	172.08	263.54	355.01	537.95	629.42	720.89	903.82	1086.76	1452.63	1819.52
.595	172.49	264.19	355.89	539.29	630.99	722.69	906.09	1089.49	1456.29	1823.09
.596	172.91	264.83	356.77	540.63	632.56	724.49	908.36	1092.22	1459.95	1827.67
.597	173.33	265.48	357.65	541.97	634.10	726.30	910.63	1094.95	1463.61	1832.26
.598	173.75	266.13	358.53	543.31	635.68	728.11	912.90	1097.68	1467.27	1836.85
.599	174.17	266.78	359.41	544.66	637.26	729.92	915.17	1100.42	1470.93	1841.44
.600	174.57	267.43	360.29	546.01	638.87	731.73	917.44	1103.16	1474.60	1846.03
.601	174.99	268.08	361.17	547.36	640.45	733.54	919.72	1105.80	1478.27	1850.63
.602	175.41	268.73	362.05	548.71	642.03	735.35	922.00	1108.54	1481.94	1855.23
.603	175.83	269.38	362.93	550.06	643.61	737.16	924.28	1111.28	1485.61	1859.84
.604	176.26	270.03	363.82	551.41	645.19	738.98	926.56	1114.03	1489.29	1864.45
.605	176.67	270.69	364.71	552.76	646.78	740.80	928.84	1116.88	1492.97	1869.06
.606	177.09	271.34	365.59	554.11	648.36	742.62	931.12	1119.63	1496.66	1873.68
.607	177.51	271.99	366.48	555.46	649.95	744.44	933.41	1122.38	1500.35	1878.30
.608	177.94	272.64	367.37	556.81	651.54	746.26	935.70	1125.14	1504.04	1882.92
.609	178.37	273.30	368.26	558.17	653.13	748.08	937.99	1127.90	1507.73	1887.55
.610	178.77	273.96	369.15	559.53	654.72	749.91	940.28	1130.66	1511.42	1892.18
.611	179.19	274.61	370.04	560.88	656.31	751.73	942.57	1133.42	1515.12	1896.82
.612	179.61	275.26	370.93	562.24	658.03	753.56	944.87	1136.19	1518.82	1901.46

Depth on Weir.	LENGTH OF THE WEIR.									
Feet.	2 Feet.	3 Feet.	4 Feet.	6 Feet.	7 Feet.	8 Feet.	10 Feet.	12 Feet.	16 Feet.	20 Feet.
.613	180.04	275.91	371.82	563.60	659.75	755.39	947.17	1138.96	1522.53	1906.10
.614	180.47	276.57	372.71	564.96	661.34	757.22	949.47	1141.73	1526.24	1910.75
.615	180.87	277.23	373.60	566.32	662.68	759.05	951.77	1144.50	1529.95	1915.40
.616	181.30	277.89	374.49	507.68	664.28	760.88	954.07	1147.27	1533.67	1920.06
617	181.73	278.53	375.38	569.04	665.88	762.71	936.28	1150.05	1537.39	1924.72
.618	182.16	279.21	376.27	570.41	667.48	764.55	958.69	1152.83	1541.11	1929.78
.619	182.59	279.87	377.16	571.78	669.08	766.39	961.60	1155.61	1544.83	1934.04
.620	182.99	280.53	378.07	573.15	670.68	768.23	963.31	1158.39	1548.55	1938.71
.921	183.42	281.19	378.96	574.51	672.28	770.07	965.62	1161.17	1552.28	1943.39
.622	183.85	281.85	379.85	575.88	673.79	771.91	967.93	1163.96	1556.01	1948.07
623	184.28	282.51	380.75	577.25	675.40	773.75	970.24	1166.75	1559.74	1952.45
.624	184.71	283.17	381.65	578.63	677.01	775.59	972.55	1169.54	1563.48	1957.43
.625	185.10	283.83	382.55	579.99	678.72	777.44	974.88	1172.33	1567.22	1962.11
626	185.53	284.49	383.46	581.36	680.33	779.29	977.20	1175.12	1570.96	1966.76
627	185.96	285.15	384.37	582.73	681.94	781.14	979.52	1177.92	1574.71	1971.42
628	186.39	285.81	385.28	584.11	683.55	782.99	981.84	1180.72	1578.46	1976.08
629	186.82	286.47	386.18	585.49	685.16	784.84	984.17	1183.52	1582.21	1970.74
630	187.23	287.14	387.05	586.87	686.78	786.69	986.50	1186.32	1585.96	1985.60
.631	187.66	287.80	387.95	588.25	688.39	788.54	988.83	1189.13	1589.72	1990.31
.632	188.09	288.46	388.85	589.63	690.01	790.39	991.16	1191.94	1593.48	1995.02
.633	188.52	289.12	389.75	591.01	691.63	792.25	993.49	1194.75	1597.24	1999.74
.634	188.95	289.79	390.65	592.39	693.25	794.11	995.83	1197.56	1601.01	2004.46
.635	189.36	290.46	391.56	593.77	694.87	795.97	998.17	1200.37	1604.78	2009.18
336	189.79	291.12	392.36	595.15	696.49	797.83	1000.51	1203.19	1606.97	2013.19
337	190.22	291.78	393.27	596.53	698.11	799.69	1002.85	1206.01	1610.37	2018.64
.638	190.65	292.45	394.18	597.91	699.73	801.55	1005.19	1208.83	1614.17	2023.38
639	191.08	293.12	395.09	599.30	701.36	803.42	1007.53	1211.65	1617.97	2028.12
.640	191.50	293.80	396.10	600.69	702.99	805.29	1009.88	1214.48	1623.67	2032.86
641	191.93	294.46	397.00	602.08	704.62	807.16	1012.23	1217.31	1627.46	2037.61
.642	192.36	295.12	397.91	603.47	706.25	809.03	1014.58	1220.14	1631.25	2042.36
.643	192.79	295.79	398.82	604.86	707.88	810.90	1016.93	1222.97	1635.04	2047.11
.644	193.22	296.46	399.73	606.25	709.51	812.77	1019.28	1225.80	1638.83	2051.86
.645	193.65	297.14	400.64	607.64	711.14	814.64	1021.64	1228.63	1642.63	2056.62
.646	194.08	297.81	401.55	609.03	712.77	816.51	1023.99	1231.47	1646.43	2061.39
.647	194.51	298.48	402.46	610.42	714.40	818.38	1026.35	1234.31	1650.23	2066.16
.648	194.94	299.15	403.37	611.82	716.04	820.26	1028.71	1237.21	1654.04	2070.93
.649	195.37	299.82	404.29	613.22	717.68	822.14	1031.07	1240.05	1657.85	2075.70
.650	195.80	300.50	405.21	614.62	719.32	824.02	1033.43	1242.84	1661.66	2080.48
.651	196.23	301.17	406.12	616.01	720.96	825.90	1035.80	1245.69	1665.48	2085.26
.652	196.66	301.84	407.03	617.41	722.60	827.78	1038.17	1248.54	1669.30	2090.05
.653	197.09	302.51	407.94	618.81	724.24	829.66	1040.54	1251.39	1673.12	2094.84
.654	197.52	303.19	408.86	620.21	725.88	831.55	1042.91	1254.24	1676.94	2099.63
.655	197.95	303.87	409.78	621.61	727.53	833.44	1045.28	1257.10	1680.76	2104.42
.656	198.38	304.54	410.60	623.01	729.17	835.33	1047.65	1259.96	1684.59	2109.22
.657	198.81	305.21	411.52	624.41	730.82	837.22	1050.02	1262.82	1688.42	2114.03
.658	199.24	305.88	412.44	625.82	732.47	839.11	1052.40	1265.68	1692.26	2118.84
.659	199.67	306.56	413.36	627.23	734.12	841.00	1054.78	1268.55	1696.10	2123.65
.660	200.12	307.25	414.38	628.64	735.77	842.90	1057.16	1271.42	1699.94	2128.46
.661	200.55	307.92	415.30	630.05	737.42	844.79	1059.55	1274.29	1703.79	2133.28
.662	200.98	308.59	416.22	631.46	739.07	846.69	1061.95	1277.16	1707.64	2138.10
.663	201.41	309.27	417.14	632.87	740.72	848.59	1064.35	1280.03	1711.49	2142.92
.664	201.84	309.95	418.07	634.28	742.38	850.49	1066.75	1282.91	1715.34	2147.75
.665	202.29	310.64	418.99	635.69	744.04	852.39	1069.09	1285.79	1719.19	2152.58
.666	202.72	311.32	419.91	637.10	745.70	854.31	1071.48	1288.67	1723.05	2157.42
.667	203.15	312.00	420.83	638.51	747.36	856.23	1073.87	1291.55	1726.91	2162.26
.668	.. ....	312.68	421.75	639.92	749.02	858.15	1076.26	1294.43	1730.71	2167.10
.669	......	313.36	422.68	641.34	750.68	860.08	1078.66	1297.32	1734.63	2171.95
.670	.. ....	314.04	423.61	642.76	752.34	861.91	1081.06	1300.21	1738.50	2176.80

Depth on Weir.	LENGTH OF THE WEIR.								
Feet.	3 Feet.	4 Feet.	6 Feet.	7 Feet.	8 Feet.	10 Feet.	12 Feet.	16 Feet.	20 Feet.
.671	314.72	424.53	644.18	754.00	863.82	1083.46	1303.10	1742.37	2181.66
.672	315.40	425.46	645.60	755.66	865.73	1085.86	1305.99	1746.25	2186.52
.673	316.08	426.39	647.02	757.32	867.64	1088.26	1308.88	1750.13	2191.38
.674	316.76	427.32	648.44	758.99	869.55	1090.66	1311.78	1754.01	2196.24
.675	317.45	428.25	649.86	760.66	871.47	1093.07	1314.68	1757.89	2201.10
.676	318.13	429.28	651.28	762.33	873.38	1095.48	1317.58	1761.78	2205.97
.677	318.81	430.21	652.70	764.00	875.30	1097.89	1320.48	1765.67	2210.85
.678	319.50	431.14	654.12	765.67	877.22	1100.30	1323.38	1769.56	2215.73
.679	320.19	432.07	655.55	767.34	879.14	1102.71	1326.29	1773.45	2220.61
.680	320.87	432.91	656.98	769.02	881.06	1105.13	1329.20	1777.34	2225.49
.681	321.55	433.84	658.41	770.69	882.98	1107.54	1332.11	1781.24	2230.38
.682	322.23	434.77	659.84	772.36	884.90	1109.96	1335.02	1785.14	2235.27
.683	322.92	435.70	661.23	774.04	886.82	1112.38	1337.93	1789.05	2240.17
.684	323.61	436.64	662.66	776.72	888.75	1114.80	1340.85	1792.96	2245.07
.685	324.30	437.58	664.13	777.40	890.68	1117.22	1343.77	1796.87	2249.97
.686	324.99	438.51	665.56	779.08	892.61	1119.64	1346.69	1800.78	2254.88
.687	325.68	439.44	666.99	780.76	894.54	1132.07	1349.61	1804.70	2259.79
.688	326.37	440.38	668.42	782.44	896.47	1134.50	1352.54	1808.62	2264.70
.689	327.06	441.32	669.86	784.12	898.40	1136.93	1355.47	1812.52	2269.61
.690	327.75	442.26	671.30	785.81	900.33	1129.36	1358.40	1816.46	2274.53
.691	328.44	443.20	672.73	787.50	902.26	1131.79	1361.33	1820.39	2279.45
.692	329.13	444.14	674.17	789.19	904.20	1134.23	1364.26	1824.32	2284.37
.693	329.82	445.08	675.61	790.88	906.14	1136.67	1367.19	1828.25	2289.30
.694	330.51	446.02	677.05	792.57	908.18	1139.11	1370.13	1832.19	2294.23
.695	331.20	446.96	678.49	794.26	910.02	1141.55	1373.07	1836.13	2299.18
.696	331.89	447.90	679.93	795.95	911.96	1143.99	1376.01	1840.07	2304.13
.697	332.58	448.86	681.37	797.64	913.90	1146.43	1378.95	1844.01	2309.08
.698	333.27	449.83	682.81	799.33	915.84	1148.87	1381.90	1847.96	2314.03
.699	333.96	450.80	684.26	801.02	917.79	1151.32	1384.85	1851.91	2318.98
.700	334.66	451.69	685.71	802.72	919.74	1153.77	1387.80	1855.86	2323.92
.701	335.35	452.63	687.15	804.42	921.69	1156.22	1390.75	1859.82	2328.88
.702	336.04	453.57	688.60	806.12	923.64	1158.67	1393.70	1863.78	2333.84
.703	336.74	454.51	690.05	807.82	925.59	1161.12	1396.66	1867.74	2338.81
.704	337.44	455.46	691.50	809.52	927.54	1163.58	1399.56	1871.70	2343.78
.705	338.14	456.41	692.95	811.22	929.49	1166.04	1402.58	1875.66	2348.75
.706	338.83	457.35	694.40	812.92	931.44	1168.50	1405.54	1879.73	2353.73
.707	339.52	458.30	695.85	814.62	933.40	1170.96	1408.50	1883.70	2358.71
.708	340.22	459.05	697.30	816.33	935.36	1173.42	1411.47	1887.67	2363.69
.709	340.92	460.00	698.76	818.04	937.32	1175.88	1414.44	1891.65	2368.67
.710	341.62	461.15	700.22	819.75	939.28	1178.34	1417.41	1895.53	2373.66
.711	342.32	462.10	701.67	821.46	941.24	1180.79	1420.38	1899.51	2378.66
.712	343.02	463.05	703.12	823.17	943.20	1183.24	1423.35	1903.50	2383.66
.713	343.72	464.00	704.58	824.88	945.16	1185.69	1426.32	1907.49	2388.66
.714	344.42	464.95	706.04	826.59	947.13	1188.14	1429.30	1911.48	2393.66
.715	345.12	465.91	707.50	828.30	949.10	1190.69	1432.28	1915.47	2398.66
.716	345.82	466.86	708.96	830.01	951.07	1193.16	1435.26	1919.47	2403.67
.717	346.52	467.81	710.42	831.72	953.04	1195.64	1438.24	1923.47	2408.68
.718	347.22	468.77	711.88	833.44	955.01	1198.12	1441.23	1927.47	2413.70
.719	347.92	469.73	713.35	835.16	956.98	1200.60	1444.22	1931.47	2418.72
.720	348.62	470.69	714.82	836.88	958.95	1203.08	1447.21	1935.48	2423.74
.721	349.32	471.64	716.30	838.60	960.92	1205.56	1450.20	1939.49	2428.77
.722	350.02	472.59	717.76	840.32	962.89	1208.04	1453.19	1943.50	2433.80
.723	350.72	473.55	719.23	842.04	964.87	1210.53	1456.19	1947.51	2438.83
.724	351.42	474.51	720.70	843.76	966.85	1213.02	1459.19	1951.53	2443.87
.725	352.13	475.47	722.15	845.49	968.83	1215.51	1462.19	1955.55	2448.91
.726	352.83	476.43	723.62	847.21	970.81	1218.00	1465.19	1959.57	2453.96
.727	353.53	477.39	725.09	848.94	972.79	1220.49	1468.19	1963.60	2459.01
.728	354.24	478.35	726.56	850.66	974.77	1222.98	1471.20	1967.63	2464.06

Depth on Weir.	LENGTH OF THE WEIR.								
Feet.	3 Feet.	4 Feet.	6 Feet.	7 Feet.	8 Feet.	10 Feet.	12 Feet.	16 Feet.	20 Feet.
.729	354.95	479.31	728.03	852.38	976.76	1225.48	1474.21	1971.66	2469.11
.730	355.66	480.28	729.51	854.13	978.75	1227.98	1477.22	1975.69	2474.16
.731	356.36	481.24	730.98	855.86	980.73	1230.48	1480.23	1979.73	2479.22
.732	357.06	482.20	732.45	857.59	982.72	1232.98	1483.24	1983.77	2484.29
.733	357.76	483.16	733.93	859.32	984.71	1235.48	1486.26	1987.81	2489.36
.734	358.46	484.12	735.31	861.05	986.70	1238.98	1489.28	1991.85	2494.43
.735	359.16	485.09	736.89	862.79	988.69	1240.49	1492.30	1995.90	2499.50
.736	359.87	486.05	738.37	864.53	990.68	1243.00	1495.32	1999.95	2504.58
.737	360.58	487.01	739.85	866.27	992.67	1245.51	1498.34	2004.00	2509.66
.738	361.30	487.98	741.33	868.01	994.67	1248.02	1501.36	2008.05	2514.74
.739	362.02	488.95	742.81	869.75	996.67	1250.53	1504.39	2012.11	2519.83
.740	362.74	489.92	744.30	871.49	998.67	1253.05	1507.42	2016.17	2524.92
.741	363.45	490.89	745.78	873.23	1000.67	1255.56	1510.45	2020.23	2530.02
.742	364.16	491.86	747.26	874.97	1002.67	1258.08	1513.48	2024.30	2535.12
.743	364.87	492.83	748.75	876.71	1004.67	1260.60	1516.47	2028.37	2540.22
.744	365.58	493.80	750.24	878.46	1006.67	1263.12	1519.52	2032.44	2545.32
.745	366.29	494.77	751.73	880.21	1008.68	1265.64	1522.60	2036.51	2550.42
.746	367.00	495.74	753.22	881.95	1010.69	1268.16	1525.64	2040.69	2555.53
.747	367.71	496.71	754.71	883.70	1012.70	1270.68	1528.68	2044.77	2560.65
.748	368.43	497.68	756.10	885.45	1014.72	1273.21	1531.72	2048.85	2565.77
.749	369.15	498.65	757.59	887.20	1016.73	1275.74	1534.77	2052.93	2570.89
.750	369.86	499.63	759.18	888.95	1018.73	1278.27	1537.82	2056.92	2576.01
.751	370.57	500.60	760.67	890.70	1020.74	1280.80	1540.87	2061.01	2581.14
.752	371.28	501.57	762.16	892.45	1022.75	1283.33	1543.92	2065.10	2586.27
.753	372.00	502.54	763.65	894.20	1024.76	1285.87	1546.97	2069.19	2591.40
.754	372.72	503.52	765.15	895.96	1026.78	1288.41	1550.03	2073.28	2596.54
.755	373.43	504.50	766.65	897.72	1028.80	1290.95	1553.09	2077.39	2601.68
.756	374.14	505.47	768.15	899.48	1030.82	1293.49	1556.15	2081.49	2606.83
.757	374.85	506.45	769.65	901.24	1032.84	1296.03	1559.21	2085.60	2611.98
.758	375.57	507.43	771.15	903.00	1034.86	1298.57	1562.28	2089.71	2617.13
.759	376.29	508.41	772.65	904.76	1036.88	1301.11	1565.35	2093.82	2622.28
.760	377.01	509.39	774.15	906.52	1038.90	1303.66	1568.42	2097.93	2627.44
.761	377.73	510.37	775.65	908.28	1040.92	1306.21	1571.49	2102.05	2632.60
.762	378.45	511.35	777.15	910.04	1042.95	1308.76	1574.06	2106.17	2637.77
.763	379.17	512.33	778.65	911.81	1044.98	1311.31	1577.63	2110.29	2642.94
.764	379.89	513.21	780.16	913.58	1047.01	1313.86	1580.70	2114.41	2648.11
.765	380.61	514.29	781.67	915.35	1049.04	1316.41	1583.78	2118.53	2653.28
.766	381.33	515.27	783.17	917.12	1051.07	1318.96	1586.86	2122.66	2558.46
.767	382.05	516.95	784.68	918.89	1053.10	1321.52	1589.94	2126.79	2563.64
.768	382.77	517.93	786.19	920.66	1055.13	1324.08	1593.02	2130.93	2568.82
.769	383.49	518.92	787.70	922.43	1057.16	1326.64	1596.11	2135.07	2574.01
.770	384.21	519.21	789.21	924.21	1059.20	1329.20	1599.20	2139.20	2679.20
.771	384.93	520.19	790.72	925.98	1061.24	1331.76	1602.29	2143.34	2684.40
.772	385.65	521.17	792.23	927.75	1063.28	1334.22	1605.38	2147.48	2689.60
.773	386.37	522.16	793.74	929.53	1065.32	1336.89	1608.47	2151.63	2694.80
.774	387.09	523.15	795.25	931.31	1067.36	1339.46	1611.57	2155.78	2700.00
.775	387.82	524.14	796.77	933.09	1069.40	1342.03	1614.67	2159.93	2705.20
.776	388.54	525.12	798.28	934.87	1071.44	1344.60	1617.77	2164.09	2710.41
.777	389.26	526.11	799.79	936.65	1073.48	1347.17	1620.87	2168.19	2715.62
.778	389.98	527.10	801.31	938.43	1075.53	1349.75	1623.97	2172.35	2720.84
.779	390.61	528.09	802.83	940.21	1077.58	1352.33	1627.07	2176.51	2726.06
.780	391.44	529.08	804.35	941.99	1079.63	1354.91	1630.18	2180.73	2731.28
.781	392.16	530.07	805.87	943.77	1081.68	1357.49	1633.29	2184.90	2736.51
.782	392.88	531.06	807.39	945.56	1083.73	1360.07	1636.40	2189.07	2741.74
.783	393.61	532.05	808.91	947.35	1085.78	1362.65	1639.51	2193.24	2746.97
.784	394.34	533.04	810.43	949.14	1087.83	1365.23	1642.62	2197.41	2752.21
.785	395.07	534.04	811.96	950.93	1089.89	1367.82	1645.74	2201.59	2757.45
.786	395.79	535.03	813.48	952.72	1091.94	1370.40	1648.86	2205.77	2762.69

Depth on Weir.	LENGTH OF THE WEIR.								
Feet.	3 Feet.	4 Feet.	6 Feet.	7 Feet.	8 Feet.	10 Feet.	12 Feet.	16 Feet.	20 Feet.
.787	396.51	536.02	815.00	954.51	1094.00	1372.99	1651.98	2209.95	2767.94
.788	397.24	537.01	816.53	956.30	1096.06	1375.58	1655.10	2214.14	2773.19
.789	397.97	538.01	818.06	958.09	1098.12	1378.17	1658.22	2218.33	2778.44
.790	398.71	539.01	819.59	959.88	1100.18	1380.76	1661.35	2222.52	2783.69
.791	399.44	540.00	821.12	961.67	1102.24	1383.35	1664.48	2226.71	2788.95
.792	400.17	540.99	822.65	963.47	1104.30	1385.95	1667.61	2230.91	2794.21
.793	400.90	541.00	824.18	965.27	1106.36	1388.55	1670.74	2235.11	2799.48
.794	401.63	542.00	825.71	967.07	1108.43	1391.15	1673.87	2239.31	2804.75
.795	402.36	543.99	827.24	968.87	1110.50	1393.75	1677.00	2243.51	2810.02
.796	403.09	544.99	828.77	970.67	1112.57	1396.35	1680.14	2247.72	2815.30
.797	403.82	545.99	829.30	972.47	1114.64	1398.95	1683.28	2251.93	2820.58
.798	404.55	546.99	830.84	974.27	1116.71	1401.56	1686.42	2256.14	2825.86
.799	405.28	547.99	832.38	976.07	1118.78	1404.17	1689.56	2260.35	2831.14
.800	406.02	548.99	834.92	977.88	1120.85	1406.78	1692.71	2264.57	2836.43
.801	406.75	549.99	836.45	979.68	1122.92	1409.39	1695.86	2268.79	2841.72
.802	407.48	550.99	837.99	981.49	1124.99	1412.00	1699.01	2273.01	2847.02
.803	408.22	551.99	839.53	983.30	1127.07	1414.61	1702.16	2277.23	2852.32
.804	408.96	552.99	841.07	985.11	1129.15	1417.22	1705.31	2281.46	2857.62
.805	409.69	554.00	842.61	986.92	1131.23	1419.84	1708.46	2285.69	2862.92
.806	410.42	554.00	844.15	988.73	1133.31	1422.46	1711.61	2289.92	2868.23
.807	411.15	555.00	845.69	990.34	1135.39	1425.08	1714.77	2294.15	2873.54
.808	411.89	556.00	847.23	992.35	1137.47	1427.70	1717.93	2298.39	2878.85
.809	412.63	557.01	848.78	994.16	1139.55	1430.32	1721.09	2302.63	2884.17
.810	413.37	559.02	850.33	995.98	1141.64	1432.95	1724.25	2306.87	2889.49
.811	414.10	559.02	851.87	997.79	1143.72	1435.57	1727.42	2311.12	2894.82
.812	414.83	560.03	853.42	999.61	1145.87	1438.20	1730.59	2315.37	2900.15
.813	415.57	561.04	854.97	1001.43	1147.96	1440.83	1733.76	2319.62	2905.48
.814	416.31	561.05	856.52	1003.25	1150.05	1443.86	1736.93	2323.87	2910.81
.815	417.05	564.06	858.07	1005.07	1152.08	1446.09	1740.10	2328.12	2916.14
.816	417.79	565.07	859.62	1006.89	1154.17	1448.72	1743.27	2332.38	2921.48
.817	418.53	566.08	861.17	1008.71	1156.26	1451.35	1746.45	2336.64	2926.82
.818	419.27	567.09	862.72	1010.53	1158.35	1453.99	1749.63	2340.90	2932.17
.819	420.01	568.10	864.27	1012.46	1160.45	1456.63	1750.81	2345.16	2937.52
.820	420.75	569.11	865.83	1014.19	1162.55	1459.27	1755.99	2349.43	2942.87
.821	421.59	570.12	867.38	1016.01	1164.65	1462.91	1759.17	2353.70	2948.23
.822	422.33	571.13	868.93	1017.84	1166.75	1465.55	1760.35	2357.97	2953.59
.823	423.07	572.14	870.49	1019.67	1168.85	1468.19	1763.54	2362.24	2958.95
.824	423.81	573.15	872.05	1021.50	1170.83	1470.83	1766.73	2366.52	2969.67
.825	424.45	574.17	873.61	1023.33	1173.05	1472.48	1771.92	2370.80	2969.67
.826	425.19	575.18	875.17	1025.16	1175.15	1475.13	1775.11	2375.08	2975.04
.827	425.93	576.19	876.73	1026.99	1177.25	1477.78	1778.30	2379.36	2980.42
.828	426.68	577.21	878.29	1028.82	1179.36	1480.43	1781.50	2381.65	2985.80
.829	427.43	578.24	881.41	1030.65	1181.47	1483.08	1784.70	2383.94	2991.18
.830	428.17	579.25	881.41	1032.49	1183.58	1485.74	1787.90	2392.23	2996.56
.831	428.91	580.26	882.97	1034.33	1185.69	1488.39	1791.10	2396.53	3001.95
.832	429.65	581.28	884.53	1036.17	1187.70	1491.05	1794.30	2400.83	3007.34
.833	430.40	582.30	886.10	1038.01	1189.81	1493.71	1797.51	2405.13	3012.73
.834	431.15	583.32	887.67	1039.85	1191.92	1496.37	1800.72	2409.43	3018.13
.835	431.89	584.34	889.24	1041.69	1194.13	1499.03	1803.93	2413.73	3023.53
.836	432.63	585.36	890.80	1043.53	1196.24	1501.69	1807.14	2418.04	3028.93
.837	433.37	586.38	892.37	1045.37	1198.35	1504.35	1810.35	2422.35	3034.34
.838	434.12	587.40	893.94	1047.21	1200.47	1507.02	1813.56	2426.66	3039.75
.839	434.87	588.42	895.51	1049.05	1202.59	1509.69	1816.78	2430.97	3045.16
.840	435.62	589.44	897.08	1050.90	1204.71	1512.36	1820.00	2435.29	3050.57
.841	436.36	590.46	898.65	1052.74	1206.84	1515.03	1823.22	2439.61	3055.99
.842	437.10	591.48	900.22	1054.59	1208.96	1517.70	1826.44	2443.93	3061.41
.843	437.85	592.50	901.79	1056.44	1211.08	1520.37	1829.66	2448.25	3066.83
.844	438.60	593.53	903.37	1058.29	1213.21	1523.05	1832.89	2452.58	3072.26

Depth on Weir.	LENGTH OF THE WEIR.								
Feet.	3 Feet.	4 Feet.	6 Feet.	7 Feet.	8 Feet.	10 Feet.	12 Feet.	16 Feet.	20 Feet.
.845	439.36	594.56	904.95	1060.14	1215.34	1525.73	1836.12	2456.91	3077.69
.846	440.11	595.58	906.52	1061.99	1217.47	1528.41	1839.35	2461.24	3083.13
.847	440.86	596.60	908.10	1063.84	1219.60	1531.09	1842.58	2465.57	3088.53
.848	441.61	597.62	909.68	1065.69	1221.73	1533.77	1845.81	2469.91	3093.97
.849	442.36	598.65	911.26	1067.54	1223.86	1536.45	1849.05	2474.25	3099.41
.850	443.11	599.68	912.84	1069.41	1225.99	1539.14	1852.29	2478.59	3104.89
.851	443.86	600.71	914.42	1071.26	1228.12	1541.82	1855.53	2482.93	3110.34
.852	444.61	601.74	916.00	1073.12	1230.25	1544.51	1858.77	2487.28	3115.79
.853	445.36	602.77	917.58	1074.98	1232.38	1547.20	1862.01	2491.63	3121.25
.854	446.11	603.80	919.16	1076.84	1234.52	1549.89	1865.25	2495.98	3126.71
.855	446.87	604.83	920.74	1078.70	1236.66	1552.58	1868.50	2500.33	3132.17
.856	447.62	605.85	922.32	1080.56	1238.80	1555.27	1871.75	2504.69	3137.64
.857	448.37	606.89	923.90	1082.32	1240.94	1557.96	1875.00	2509.05	3143.11
.858	449.12	607.92	925.49	1084.18	1243.08	1560.66	1878.25	2513.41	3148.58
.859	449.87	608.95	927.08	1086.05	1245.22	1563.36	1881.50	2517.77	3154.05
.860	450.63	609.98	928.67	1088.02	1247.37	1566.06	1884.75	2522.14	3159.53
.861	451.38	610.01	930.26	1089.88	1249.51	1568.76	1888.01	2526.51	3165.01
.862	452.13	611.01	932.85	1091.75	1251.65	1571.46	1891.27	2530.88	3170.49
.863	452.89	612.07	934.45	1093.62	1253.80	1574.16	1894.53	2535.25	3175.98
.864	453.65	613.11	936.04	1095.49	1255.85	1576.87	1897.79	2539.63	3181.47
.865	454.41	615.15	936.62	1097.36	1258.10	1579.58	1901.05	2544.01	3186.96
.866	455.16	616.18	938.21	1099.23	1260.25	1582.29	1904.32	2548.39	3192.46
.867	455.91	617.21	939.80	1101.10	1262.40	1585.00	1907.59	2552.77	3197.96
.868	456.67	618.25	941.39	1102.97	1264.55	1587.71	1910.86	2557.15	3203.46
.869	457.43	619.29	942.98	1104.85	1266.70	1590.42	1914.13	2561.54	3208.96
.870	458.19	620.33	944.59	1106.73	1268.86	1593.13	1917.40	2565.93	3214.47
.871	458.94	621.36	946.18	1108.60	1271.01	1595.84	1920.67	2570.32	3219.98
.872	459.71	622.40	947.78	1110.48	1273.17	1598.56	1923.95	2574.72	3225.50
.873	460.46	623.44	949.38	1112.36	1275.33	1601.28	1927.23	2579.12	3231.02
.874	461.22	624.48	950.98	1114.24	1277.49	1604.00	1930.51	2583.52	3236.54
.875	461.98	625.52	952.58	1116.12	1279.65	1606.72	1933.79	2587.92	3242.06
.876	462.74	626.56	954.18	1118.00	1281.81	1609.44	1937.07	2592.33	3247.59
.877	463.50	627.60	955.78	1119.88	1283.97	1612.16	1940.35	2596.34	3253.12
.878	464.26	628.64	957.38	1121.76	1285.13	1614.89	1943.64	2600.75	3258.65
.879	465.02	629.68	958.99	1123.64	1287.30	1617.62	1946.93	2605.16	3264.18
.880	465.78	630.72	960.60	1125.53	1290.47	1620.35	1950.22	2609.97	3269.72
.881	466.54	631.76	962.20	1127.38	1292.64	1623.08	1953.51	2614.39	3275.26
.882	467.30	632.80	963.80	1129.27	1294.81	1625.81	1956.81	2618.81	3280.81
.883	468.06	633.84	965.40	1131.16	1296.98	1628.54	1960.11	2623.23	3286.36
.884	468.82	634.89	967.00	1133.05	1299.15	1631.27	1963.41	2627.65	3291.91
.885	469.59	635.94	968.63	1134.97	1301.32	1634.01	1966.70	2632.08	3297.46
.886	470.35	636.98	970.23	1136.86	1303.49	1636.75	1970.00	2636.51	3303.02
.887	471.11	638.02	971.85	1138.75	1305.66	1639.49	1973.30	2640.94	3308.58
.888	471.87	639.07	972.46	1140.64	1307.83	1642.23	1976.60	2645.37	3314.14
.889	472.64	640.12	975.07	1142.54	1310.01	1644.97	1979.91	2649.81	3319.71
.890	473.41	641.17	976.68	1144.44	1312.19	1647.71	1983.22	2654.25	3325.28
.891	474.17	642.21	979.90	1146.33	1314.37	1650.45	1986.53	2658.76	3330.85
.892	474.93	643.25	971.51	1148.23	1316.55	1653.19	1989.84	2663.28	3336.45
.893	475.70	644.30	973.12	1150.13	1318.73	1655.93	1993.16	2667.80	3342.03
.894	476.47	645.35	974.14	1152.03	1320.91	1658.61	1996.48	2672.32	3347.61
.895	477.24	646.41	984.75	1153.93	1323.10	1661.44	1999.80	2676.48	3353.17
.896	478.00	647.46	986.37	1155.83	1325.28	1664.19	2003.12	2680.93	3358.76
.897	478.76	648.51	987.98	1157.73	1327.46	1666.94	2006.44	2685.39	3364.35
.898	479.54	649.56	989.61	1159.63	1329.65	1669.70	2009.76	2689.85	3369.94
.899	480.31	650.61	991.23	1161.53	1331.84	1672.46	2013.08	2694.31	3375.50
.900	481.07	651.66	992.85	1163.44	1334.03	1675.22	2016.40	2698.77	3381.14
.901	481.83	652.71	994.47	1165.34	1336.22	1677.98	2019.73	2703.24	3386.74
.902	482.69	653.76	996.09	1167.25	1338.41	1680.74	2023.06	2708.71	3392.35

Depth on Weir.				LENGTH OF THE WEIR.					
Feet.	3 Feet.	4 Feet.	6 Feet.	7 Feet.	8 Feet.	10 Feet.	12 Feet.	16 Feet.	20 Feet.
.903	483.37	654.81	997.71	1169.16	1340.60	1683.50	2026.39	2713.18	3397.96
.904	484.14	655.87	999.33	1171.07	1342.79	1686.26	2029.72	2717.65	3403.57
.905	484.91	656.93	1000.96	1172.98	1344.99	1689.02	2033.05	2721.12	3409.18
.906	485.68	657.98	1002.58	1174.89	1347.18	1691.79	2036.39	2725.60	3414.80
.907	486.45	659.03	1004.20	1176.80	1349.38	1694.56	2039.73	2730.08	3420.42
.908	487.22	660.09	1005.83	1178.71	1351.58	1697.33	2043.07	2734.56	3425.04
.909	487.99	661.15	1007.46	1180.62	1353.78	1700.10	2047.41	2739.04	3431.66
.910	488.76	662.21	1009.09	1182.54	1355.98	1702.87	2049.75	2743.53	3437.30
.911	489.53	663.26	1010.72	1184.45	1358.18	1705.64	2053.09	2748.02	3442.93
.912	490.30	664.32	1012.35	1186.36	1360.38	1708.41	2056.44	2752.51	3448.57
.913	491.07	665.38	1013.98	1188.28	1362.58	1711.19	2059.79	2757.00	3454.21
.914	491.84	666.44	1015.61	1190.28	1364.79	1713.97	2063.14	2761.49	3459.85
.915	492.62	667.50	1017.25	1192.12	1367.00	1716.75	2066.49	2765.99	3465.49
.916	493.39	668.56	1018.88	1194.04	1369.20	1719.53	2069.84	2770.49	3471.14
.917	494.16	669.62	1020.51	1195.96	1371.41	1722.31	2073.20	2774.99	3476.79
.918	494.93	670.68	1022.14	1197.88	1373.62	1725.09	2076.56	2779.50	3482.44
.919	495.71	671.74	1024.78	1199.80	1375.83	1727.87	2079.92	2784.01	3488.10
.920	496.49	672.80	1025.42	1201.73	1378.04	1730.66	2083.28	2788.52	3493.76
.921	497.26	673.86	1027.05	1203.65	1380.25	1733.45	2086.64	2793.03	3499.42
.922	498.03	674.92	1028.69	1205.57	1382.46	1736.24	2090.00	2797.55	3505.09
.923	498.80	675.98	1030.33	1207.50	1384.67	1739.03	2093.37	2802.07	3510.76
.924	499.58	677.05	1031.97	1209.43	1386.89	1741.82	2096.74	2806.59	3516.43
.925	500.36	678.12	1033.61	1211.36	1389.11	1744.61	2100.11	2811.11	3522.10
.926	501.13	679.18	1035.25	1213.29	1391.33	1747.40	2103.48	2815.63	3527.78
.927	501.91	680.24	1036.89	1215.22	1393.55	1751.20	2106.85	2820.16	3533.48
.928	502.69	681.30	1038.53	1217.15	1395.77	1754.00	2110.22	2824.69	3539.16
.929	503.47	681.37	1040.17	1219.08	1397.99	1756.80	2113.60	2829.23	3544.85
.930	504.25	683.44	1041.82	1221.02	1400.21	1758.60	2116.98	2833.75	3550.52
.931	505.02	684.50	1043.46	1222.95	1402.43	1761.40	2120.30	2838.29	3556.21
.932	505.80	685.57	1045.11	1224.88	1404.65	1764.20	2123.68	2842.83	3561.91
.933	506.58	686.64	1046.76	1226.82	1406.88	1767.00	2127.06	2847.37	3567.61
.934	507.36	687.73	1048.41	1228.76	1409.11	1769.81	2130.45	2851.91	3573.31
.935	508.14	688.78	1050.06	1230.70	1411.34	1772.62	2133.90	2856.45	3579.01
.936	508.92	689.85	1051.71	1232.64	1413.57	1775.43	2137.29	2861.00	3584.72
.937	509.70	690.92	1053.36	1234.58	1415.80	1778.24	2140.68	2865.55	3590.43
.938	510.48	691.99	1055.01	1236.52	1418.03	1781.05	2144.07	2870.10	3596.14
.939	511.26	693.06	1056.66	1238.46	1420.26	1783.86	2147.46	2874.66	3601.86
.940	512.04	694.13	1058.31	1240.40	1422.49	1786.67	2150.85	2879.22	3607.58
.941	512.82	695.20	1059.96	1242.34	1424.72	1789.48	2154.25	2883.78	3613.30
.942	513.60	696.27	1061.61	1244.38	1426.95	1792.30	2157.65	2888.34	3619.03
.943	514.38	697.34	1063.26	1246.23	1429.19	1795.12	2161.05	2892.90	3624.76
.944	515.16	698.41	1064.92	1248.18	1431.43	1797.14	2164.45	2897.46	3630.49
.945	515.95	699.49	1066.58	1250.13	1433.67	1800.76	2167.85	2902.03	3636.22
.946	516.73	700.56	1068.23	1252.18	1435.91	1803.58	2171.26	2906.60	3641.96
.947	517.51	701.63	1069.89	1254.13	1438.15	1806.40	2174.67	2911.17	3647.70
.948	518.29	702.71	1071.55	1256.08	1440.39	1809.23	2178.08	2915.75	3653.44
.949	519.07	703.79	1073.21	1258.03	1442.63	1812.06	2181.49	2920.33	3659.18
.950	519.86	704.87	1074.87	1259.88	1444.88	1814.89	2184.90	2924.91	3664.93
.951	520.64	705.94	1076.53	1261.13	1447.12	1817.72	2188.31	2929.49	3670.68
.952	521.42	707.01	1078.19	1263.08	1449.36	1820.55	2191.72	2934.07	3676.43
.953	522.20	708.09	1079.85	1265.03	1451.61	1823.38	2195.14	2938.65	3682.19
.954	522.99	709.17	1081.51	1266.99	1453.86	1826.21	2198.56	2943.24	3687.95
.955	523.78	710.25	1083.18	1269.65	1456.11	1829.05	2201.98	2947.85	3693.71
.956	524.56	711.33	1084.84	1271.61	1458.36	1831.88	2205.40	2952.44	3699.48
.957	525.34	712.41	1086.50	1273.57	1460.61	1834.72	2208.82	2957.04	3705.25
.958	526.13	713.49	1088.17	1275.53	1462.86	1837.56	2212.25	2961.64	3711.02
.959	526.92	714.57	1089.84	1277.49	1465.12	1840.40	2215.68	2966.24	3716.79
.960	527.71	715.65	1091.51	1279.45	1467.38	1843.24	2219.11	2970.84	3722.57

Depth on Weir.	LENGTH OF THE WEIR.								
Feet.	3 Feet.	4 Feet.	6 Feet.	7 Feet.	8 Feet.	10 Feet.	12 Feet.	16 Feet.	20 Feet.
.961	528.49	716.73	1093.18	1281.41	1469.63	1846.08	2222.54	2975.45	3728.35
.962	529.28	717.81	1094.85	1283.37	1471.89	1848.92	2225.97	2980.06	3734.13
.963	530.07	718.89	1096.52	1285.33	1474.15	1851.77	2229.40	2984.67	3739.92
.964	530.86	719.97	1098.20	1287.30	1476.41	1854.62	2232.84	2989.28	3745.71
.965	531.65	721.06	1099.86	1289.27	1478.67	1857.47	2236.28	2993.89	3751.50
.966	532.44	722.14	1101.53	1291.23	1480.93	1860.32	2239.72	2998.51	3757.30
.967	533.23	723.22	1102.60	1293.20	1483.19	1863.17	2243.16	3003.13	3763.10
.968	534.02	724.30	1104.27	1295.17	1485.45	1866.02	2246.60	3007.75	3768.90
.969	534.81	725.39	1105.95	1297.14	1487.71	1868.88	2250.04	3014.37	3774.70
.970	535.60	726.48	1108.23	1299.11	1489.98	1871.74	2253.49	3017.00	3780.50
.971	536.59	727.56	1109.90	1301.08	1492.25	1874.60	2256.94	3021.63	3786.31
.972	537.38	728.64	1111.58	1303.05	1494.52	1877.46	2260.39	3026.26	3792.12
.973	538.18	729.73	1113.18	1305.02	1496.79	1880.32	2263.84	3030.89	3797.94
.974	538.97	730.82	1114.86	1306.99	1499.06	1883.18	2267.29	3035.52	3803.76
.975	539.55	731.91	1116.62	1308.97	1501.33	1886.04	2270.74	3040.16	3809.58
.976	540.34	732.99	1118.30	1310.94	1503.60	1888.90	2274.20	3044.80	3815.41
.977	541.13	734.08	1119.99	1312.92	1505.87	1891.76	2277.66	3049.44	3821.24
.978	541.92	735.17	1121.66	1314.90	1508.14	1894.63	2281.12	3054.09	3827.07
.979	542.72	736.26	1123.31	1316.88	1510.42	1897.50	2284.58	3058.74	3832.90
.980	543.52	737.35	1125.02	1318.86	1512.70	1900.37	2288.04	3063.39	3838.73
.981	544.31	738.44	1126.70	1320.84	1514.97	1903.24	2291.50	3068.05	3844.57
.982	545.10	739.53	1128.38	1322.82	1517.25	1906.11	2294.97	3072.71	3850.41
.983	545.89	740.62	1130.07	1323.80	1519.53	1908.98	2298.44	3077.37	3856.25
.984	546.69	741.71	1131.76	1325.73	1521.81	1911.86	2301.91	3082.03	3862.09
.985	547.49	742.81	1133.45	1328.77	1524.09	1914.74	2305.38	3086.66	3867.95
.986	548.28	743.90	1135.13	1330.75	1526.37	1917.62	2308.85	3091.32	3873.80
.987	549.07	744.99	1136.82	1332.73	1528.65	1920.50	2312.33	3095.99	3879.66
.988	549.86	746.08	1138.51	1334.72	1530.94	1923.38	2315.81	3100.66	3885.22
.989	550.66	747.17	1140.20	1336.71	1533.23	1926.26	2319.29	3105.33	3891.38
.990	551.46	748.27	1141.89	1338.70	1535.52	1929.14	2322.76	3110.00	3897.24
.991	552.25	749.36	1143.58	1340.69	1537.80	1932.02	2326.24	3114.67	3903.11
.992	553.05	750.45	1145.27	1342.68	1540.09	1934.90	2329.72	3119.35	3908.98
.993	553.85	751.55	1146.96	1344.67	1542.38	1937.79	2333.20	3124.03	3914.85
.994	554.65	752.65	1148.66	1346.66	1544.67	1940.68	2336.69	3128.71	3920.72
.995	555.45	753.75	1150.36	1348.66	1546.96	1943.57	2340.18	3133.39	3926.60
.996	556.24	754.84	1152.05	1350.65	1549.25	1946.46	2343.67	3138.08	3932.48
.997	557.04	755.94	1153.74	1352.64	1551.54	1949.35	2347.16	3142.77	3938.37
.998	557.84	757.04	1155.44	1354.64	1553.84	1952.24	2350.65	3147.40	3944.26
.999	558.64	758.14	1157.14	1356.64	1556.14	1955.14	2354.14	3152.09	3950.15
1.000	559.44	759.24	1158.84	1358.64	1558.44	1958.04	2357.64	3156.84	3956.04
1.001	......	760.34	1160.54	1360.64	1560.74	1960.94	2361.14	3161.54	3961.94
1.002	......	761.44	1162.24	1362.64	1563.04	1963.84	2364.61	3166.24	3967.84
1.003	......	762.54	1163.94	1364.64	1565.34	1966.74	2368.14	3170.94	3973.74
1.004	......	763.64	1165.64	1366.64	1567.64	1969.64	2371.64	3175.64	3979.64
1.005	......	764.74	1167.34	1368.64	1569.94	1972.54	2375.14	3180.34	3985.55
1.006	......	765.84	1169.04	1370.64	1572.24	1975.44	2378.65	3185.05	3991.46
1.007	......	766.94	1170.74	1372.64	1574.54	1978.35	2382.16	3189.76	3997.37
1.008	......	768.04	1172.44	1374.64	1576.85	1981.26	2385.67	3194.47	4003.28
1.009	......	769.14	1174.19	1376.65	1578.16	1984.17	2389.18	3199.18	4009.20
1.010	......	770.25	1175.86	1378.66	1581.47	1987.08	2392.69	3203.90	4015.12
1.011	......	771.35	1177.56	1380.66	1583.78	1989.99	2396.20	3208.34	4021.05
1.012	......	772.45	1179.27	1382.67	1586.09	1992.90	2399.71	3213.06	4026.98
1.013	......	773.55	1180.98	1384.68	1588.40	1995.81	2403.23	3217.78	4032.91
1.014	......	774.66	1182.69	1386.69	1590.71	1998.73	2406.75	3222.51	4038.84
1.015	......	775.77	1184.40	1388.71	1593.02	2001.65	2410.27	3227.52	4044.77
1.016	......	776.87	1186.11	1390.72	1595.33	2004.57	2413.79	3232.25	4050.71
1.017	......	777.97	1187.82	1392.73	1597.64	2007.49	2417.31	3236.98	4056.65
1.018	......	779.08	1189.53	1394.74	1599.96	2010.41	2420.84	3241.71	4062.59

Depth on Weir.	LENGTH OF THE WEIR.							
Feet.	4 Feet.	6 Feet.	7 Feet.	8 Feet.	10 Feet.	12 Feet.	16 Feet.	20 Feet.
1.019	780.19	1191.24	1396.76	1602.28	2013.33	2424.37	3246.45	4068.54
1.020	781.30	1192.95	1398.78	1604.60	2016.25	2427.90	3251.19	4074.49
1.021	782.41	1194.66	1400.79	1606.92	2019.15	2431.43	3255.93	4080.44
1.022	783.52	1196.37	1402.81	1609.24	2022.06	2434.96	3260.67	4086.40
1.023	784.63	1198.09	1404.83	1611.56	2024.97	2438.49	3265.42	4092.36
1.024	785.74	1199.81	1406.85	1613.88	2027.88	2442.02	3270.17	4098.32
1.025	786.85	1201.53	1408.87	1616.21	2030.89	2445.56	3274.92	4104.28
1.026	787.96	1203.24	1410.89	1618.53	2033.82	2449.10	3279.67	4110.25
1.027	789.07	1204.96	1412.91	1620.85	2036.75	2452.64	3284.32	4116.22
1.028	790.18	1206.68	1414.93	1623.18	2039.88	2456.18	3289.18	4122.19
1.029	791.28	1208.40	1416.95	1625.51	2042.82	2459.72	3293.94	4128.16
1.030	792.40	1210.12	1418.98	1627.84	2045.56	2463.27	3298.70	4134.14
1.031	793.51	1211.84	1421.00	1630.17	2048.50	2466.82	3303.46	4140.12
1.032	794.62	1213.56	1423.03	1632.50	2051.44	2470.37	3308.23	4146.10
1.033	795.73	1215.28	1425.05	1634.83	2054.38	2473.92	3313.00	4152.09
1.034	796.84	1217.01	1427.07	1637.16	2057.32	2477.47	3317.77	4158.08
1.035	797.96	1218.74	1429.12	1639.50	2060.26	2481.02	3322.54	4164.07
1.036	799.07	1220.46	1431.15	1641.83	2063.20	2484.57	3327.32	4170.06
1.037	800.19	1222.18	1433.18	1644.16	2066.14	2488.13	3332.10	4176.06
1.038	801.31	1223.91	1435.21	1646.50	2069.09	2491.69	3336.88	4182.06
1.039	802.43	1225.64	1437.24	1648.84	2072.04	2495.25	3341.66	4188.06
1.040	803.55	1227.37	1439.27	1651.18	2074.99	2498.81	3346.44	4194.06
1.041	804.66	1229.10	1441.30	1653.52	2077.94	2502.37	3351.22	4200.07
1.042	805.78	1230.83	1443.33	1655.86	2080.89	2505.93	3356.01	4206.08
1.043	806.90	1232.56	1445.37	1658.20	2083.84	2509.50	3360.80	4212.09
1.044	808.02	1234.29	1447.41	1660.54	2086.80	2513.07	3365.59	4218.11
1.045	809.14	1236.02	1449.45	1662.89	2089.76	2516.64	3370.38	4224.13
1.046	810.15	1237.75	1451.49	1665.26	2092.72	2520.21	3375.18	4030.15
1.047	811.27	1239.48	1453.53	1667.57	2095.68	2523.78	3379.98	4036.18
1.048	812.39	1241.21	1455.57	1669.92	2098.64	2527.35	3384.78	4042.21
1.049	813.41	1242.94	1457.61	1672.27	2101.60	2530.93	3389.58	4048.24
1.050	814.73	1244.68	1459.65	1674.62	2104.56	2534.51	3394.39	4254.27
1.051	815.85	1246.41	1461.69	1676.97	2107.52	2538.09	3399.20	4260.31
1.052	816.97	1248.14	1463.73	1679.32	2110.49	2541.67	3404.01	4266.35
1.053	818.09	1249.88	1465.77	1681.67	2113.46	2545.25	3408.82	4272.39
1.054	819.21	1251.62	1467.82	1684.02	2116.43	2548.83	3413.63	4278.43
1.055	820.34	1253.36	1469.87	1686.38	2119.40	2552.41	3418.45	4284.48
1.056	821.46	1255.10	1471.92	1688.73	2122.37	2556.00	3423.27	4290.53
1.057	822.58	1256.84	1473.97	1691.09	2125.34	2559.59	3428.09	4296.58
1.058	823.71	1258.58	1476.02	1693.45	2128.31	2563.18	3432.91	4302.63
1.059	824.84	1260.32	1478.07	1695.81	2131.28	2566.77	3437.73	4308.69
1.060	825.97	1262.07	1480.12	1698.17	2134.26	2570.36	3442.56	4314.75
1.061	827.09	1263.81	1482.17	1700.53	2137.24	2573.95	3447.39	4320.82
1.062	828.21	1265.55	1484.22	1702.89	2140.22	2577.55	3452.22	4326.89
1.063	829.34	1267.29	1486.27	1705.25	2143.20	2581.15	3457.05	4332.96
1.064	830.47	1269.04	1488.32	1707.61	2146.18	2584.75	3461.89	4339.03
1.065	831.60	1270.79	1490.38	1709.98	2149.16	2588.35	3466.73	4345.10
1.066	832.72	1272.53	1492.43	1712.34	2152.14	2591.96	3471.57	4351.18
1.067	833.85	1274.28	1494.49	1714.70	2155.13	2595.55	3476.41	4357.26
1.068	834.98	1276.03	1496.55	1717.07	2158.12	2599.16	3481.25	4363.34
1.069	836.11	1277.78	1498.61	1719.44	2161.11	2602.71	3486.10	4369.42
1.070	837.24	1279.53	1500.67	1721.81	2164.10	2606.38	3490.95	4375.51
1.071	838.37	1281.28	1502.73	1724.18	2167.09	2609.99	3495.80	4381.60
1.072	839.50	1283.03	1504.79	1726.55	2170.08	2613.60	3500.65	4387.70
1.073	840.63	1284.78	1506.85	1728.92	2173.07	2617.21	3505.50	4393.80
1.074	841.76	1286.53	1508.91	1731.29	2176.06	2620.83	3510.36	4399.90
1.075	842.89	1288.28	1510.98	1733.67	2179.06	2624.45	3515.22	4406.00
1.076	844.02	1290.03	1513.04	1736.04	2182.06	2628.07	3520.08	4412.11

Depth on Weir.	LENGTH OF THE WEIR.							
Feet.	4 Feet.	6 Feet.	7 Feet.	8 Feet.	10 Feet.	12 Feet.	16 Feet.	20 Feet.
1.077	845.15	1291.78	1515.10	1738.42	2185.06	2631.69	3524.94	4418.22
1.078	846.28	1293.33	1517.17	1740.80	2188.06	2635.31	3529.81	4424.33
1.079	847.41	1295.28	1519.24	1743.18	2191.06	2638.93	3534.68	4430.44
1.080	848.55	1297.06	1521.31	1745.56	2194.06	2642.55	3539.55	4436.55
1.081	849.68	1298.81	1523.38	1747.94	2197.06	2646.18	3544.43	4442.67
1.082	850.81	1300.57	1525.45	1750.32	2200.06	2649.81	3549.30	4448.79
1.083	851.95	1302.33	1527.52	1752.70	2203.06	2653.44	3554.17	4454.91
1.084	853.09	1304.09	1529.59	1755.08	2206.07	2657.07	3559.04	4461.04
1.085	854.23	1305.85	1531.66	1757.47	2209.08	2660.70	3563.93	4467.17
1.086	855.36	1307.61	1533.73	1759.85	2212.09	2664.33	3568.81	4473.30
1.087	856.49	1309.37	1535.80	1762.23	2215.10	2667.97	3573.70	4479.13
1.088	857.63	1311.13	1537.87	1764.62	2218.11	2671.61	3578.59	4485.57
1.089	858.77	1312.89	1539.95	1767.01	2221.12	2675.25	3583.48	4491.71
1.090	859.91	1314.66	1542.03	1769.40	2224.14	2678.89	3588.37	4497.85
1.091	861.05	1316.42	1544.10	1771.79	2227.16	2682.53	3593.16	4504.00
1.092	862.19	1318.18	1546.17	1774.18	2230.18	2686.17	3598.06	4510.15
1.093	863.23	1319.95	1548.24	1776.57	2233.20	2689.81	3603.76	4516.30
1.094	864.37	1321.72	1550.32	1778.96	2236.22	2693.46	3608.66	4522.45
1.095	865.61	1323.49	1552.42	1781.36	2239.24	2697.11	3612.86	4528.61
1.096	866.75	1325.25	1554.50	1783.75	2242.26	2700.76	3617.76	4534.77
1.097	867.89	1327.60	1556.58	1786.15	2245.28	2704.41	3622.67	4540.93
1.098	869.03	1328.79	1558.66	1788.55	2248.30	2708.06	3627.58	4547.19
1.099	860.17	1330.56	1560.75	1790.95	2251.33	2711.71	3632.49	4553.36
1.100	871.31	1332.33	1562.84	1793.35	2254.36	2715.37	3637.40	4559.43
1.101	872.46	1334.10	1564.92	1795.75	2257.39	2719.03	3642.32	4565.60
1.102	873.61	1335.87	1567.00	1798.15	2260.42	2722.69	3647.24	4571.78
1.103	874.76	1337.64	1569.09	1800.55	2263.45	2726.35	3652.16	4577.96
1.104	875.91	1339.41	1571.18	1802.95	2266.48	2730.01	3657.08	4584.14
1.105	877.03	1341.19	1573.27	1805.36	2269.52	2733.68	3662.00	4590.32
1.106	878.17	1342.96	1575.36	1807.76	2272.55	2737.34	3666.93	4596.51
1.107	879.31	1344.73	1577.45	1810.16	2275.58	2741.01	3671.86	4602.70
1.108	880.45	1346.51	1579.54	1812.57	2278.62	2744.67	3676.79	4608.89
1.109	881.50	1348.29	1581.63	1814.98	2281.66	2748.34	3681.72	4615.08
1.110	882.75	1350.07	1583.73	1817.39	2284.70	2752.02	3686.65	4621.28
1.111	883.89	1351.85	1585.82	1819.80	2287.74	2755.69	3691.59	4627.48
1.112	885.04	1353.63	1587.91	1822.21	2290.78	2759.36	3696.53	4633.68
1.113	886.19	1355.41	1590.91	1824.62	2293.82	2763.04	3701.47	4639.89
1.114	887.34	1357.19	1592.11	1827.03	2296.87	2766.72	3706.41	4646.10
1.115	888.49	1358.97	1594.21	1829.45	2299.92	2770.40	3711.35	4652.31
1.116	889.64	1360.75	1596.31	1831.86	2302.97	2774.08	3716.30	4658.52
1.117	890.79	1362.53	1598.41	1834.27	2306.02	2777.76	3721.25	4664.74
1.118	891.94	1364.31	1600.51	1836.69	2309.07	2781.44	3726.20	4670.96
1.119	893.19	1366.10	1602.61	1839.11	2312.12	2785.13	3731.15	4677.18
1.120	894.24	1367.89	1604.71	1841.53	2315.17	2788.82	3736.11	4683.40
1.121	895.39	1369.67	1606.81	1843.95	2318.22	2792.51	3741.07	4689.63
1.122	896.54	1371.45	1608.91	1846.37	2321.28	2796.20	3746.03	4695.86
1.123	897.69	1373.24	1611.01	1848.79	2324.34	2799.89	3750.99	4702.09
1.124	898.84	1375.03	1613.12	1851.21	2327.40	2803.58	3755.95	4708.32
1.125	899.99	1376.82	1615.23	1853.64	2330.46	2807.28	3760.92	4714.56
1.126	901.14	1378.61	1617.33	1856.06	2333.52	2810.77	3765.89	4720.80
1.127	902.29	1380.40	1619.44	1858.48	2336.58	2814.27	3770.86	4727.04
1.128	903.44	1382.19	1621.55	1860.91	2339.64	2817.77	3775.83	4733.28
1.129	904.60	1383.98	1623.66	1863.34	2342.70	2821.27	3780.80	4739.53
1.130	905.76	1385.77	1625.77	1865.77	2345.77	2825.77	3785.78	4745.78
1.131	906.91	1387.56	1627.88	1868.20	2348.84	2829.47	3790.76	4752.03
1.132	908.06	1389.35	1629.99	1870.63	2351.91	2833.18	3795.74	4758.29
1.133	909.22	1391.14	1632.10	1873.06	2354.98	2836.89	3800.72	4765.55
1.134	910.38	1392.93	1634.21	1875.49	2358.05	2840.60	3805.70	4771.81

Depth on Weir.	LENGTH OF THE WEIR.							
Feet.	4 Feet.	6 Feet.	7 Feet.	8 Feet.	10 Feet.	12 Feet.	16 Feet.	20 Feet.
1.135	911.54	1394.73	1636.33	1877.92	2361.12	2844.31	3810.69	4777.07
1.136	912.69	1396.52	1638.44	1880.35	2364.19	2848.02	3815.68	4783.34
1.137	913.84	1398.32	1640.55	1882.79	2367.26	2851.73	3820.67	4789.61
1.138	915.00	1400.12	1642.67	1885.23	2370.33	2855.44	3825.66	4795.88
1.139	916.16	1401.92	1644.79	1887.67	2373.41	2859.16	3830.66	4802.15
1.140	917.32	1403.72	1646.91	1890.10	2376.49	2862.88	3835.66	4808.43
1.141	918.48	1405.52	1649.03	1892.54	2379.57	2866.60	3840.66	4814.71
1.142	919.64	1407.32	1651.15	1894.98	2382.65	2870.32	3845.66	4820.99
1.143	920.80	1409.12	1653.27	1897.42	2385.73	2874.04	3850.66	4827.27
1.144	921.96	1410.92	1654.39	1899.86	2388.81	2877.76	3855.66	4833.56
1.145	923.12	1412.72	1657.51	1902.31	2391.90	2881.49	3860.67	4839.85
1.146	924.28	1414.52	1659.63	1904.75	2394.98	2885.22	3865.68	4846.14
1.147	925.44	1416.32	1661.75	1907.19	2398.07	2888.95	3870.69	4852.44
1.148	926.60	1418.12	1663.87	1909.63	2401.16	2892.68	3875.70	4858.74
1.149	927.76	1419.92	1665.00	1912.08	2404.25	2896.41	3880.72	4865.04
1.150	928.93	1421.73	1668.13	1914.53	2407.34	2900.14	3885.74	4871.34
1.151	930.09	1423.53	1670.26	1916.98	2410.43	2903.87	3890.76	4877.65
1.152	931.25	1425.34	1672.39	1919.43	2413.52	2907.60	3895.78	4883.96
1.153	932.41	1427.15	1674.52	1921.88	2416.61	2911.35	3900.80	4890.27
1.154	933.57	1428.96	1676.65	1924.33	2419.71	2915.09	3905.82	4896.58
1.155	934.74	1430.77	1678.78	1926.79	2422.81	2918.83	3910.86	4902.90
1.156	935.90	1432.58	1680.90	1929.24	2425.91	2922.57	3915.89	4909.22
1.157	937.06	1434.39	1683.02	1931.69	2429.01	2926.31	3920.92	4915.56
1.158	938.23	1436.20	1685.15	1934.14	2432.11	2930.05	3925.96	4921.89
1.159	939.40	1438.01	1687.28	1936.60	2435.21	2933.80	3931.00	4928.19
1.160	940.57	1439.82	1689.44	1939.06	2438.31	2937.55	3936.04	4934.52
1.161	941.73	1441.63	1691.55	1941.52	2441.41	2941.30	3941.08	4940.85
1.162	942.90	1443.45	1693.69	1943.98	2444.51	2945.05	3946.12	4947.19
1.163	944.07	1445.27	1695.83	1946.44	2447.62	2948.80	3951.16	4953.53
1.164	945.24	1447.09	1697.97	1948.90	2450.73	2952.55	3956.21	4959.87
1.165	946.41	1448.89	1700.12	1951.36	2453.84	2956.31	3961.26	4966.21
1.166	947.57	1450.70	1702.26	1954.82	2456.95	2960.07	3966.31	4972.56
1.167	948.74	1452.51	1704.40	1957.28	2460.06	2963.83	3971.36	4978.91
1.168	949.91	1454.33	1706.54	1959.74	2463.17	2967.59	3976.41	4985.26
1.169	951.08	1456.15	1708.68	1962.21	2466.28	2971.35	3981.47	4991.61
1.170	952.25	1457.97	1710.83	1963.68	2469.40	2975.11	3986.54	4997.96
1.171	953.42	1459.89	1712.97	1966.15	2472.51	2978.87	3991.60	5004.32
1.172	954.59	1461.71	1715.11	1968.62	2475.63	2982.64	3996.66	5010.68
1.173	955.76	1463.53	1717.25	1971.09	2478.75	2986.41	4001.76	5017.04
1.174	956.93	1465.35	1719.40	1973.56	2481.87	2990.18	4006.83	5023.41
1.175	958.11	1467.07	1721.55	1976.03	2484.99	2993.95	4011.86	5029.78
1.176	959.28	1468.89	1723.70	1978.50	2488.11	2997.72	4016.93	5036.15
1.177	960.45	1470.71	1725.85	1980.97	2491.23	3001.49	4022.00	5042.52
1.178	961.62	1472.53	1728.00	1983.44	2494.35	3005.26	4027.08	5048.90
1.179	962.80	1474.36	1730.15	1985.92	2497.48	3009.04	4032.16	5055.20
1.180	963.98	1476.19	1732.30	1988.40	2500.61	3012.82	4037.24	5061.66
1.181	965.15	1478.01	1734.45	1990.84	2503.87	3016.60	4042.32	5068.05
1.182	966.32	1479.84	1736.60	1993.29	2506.87	3020.38	4047.40	5074.44
1.183	967.49	1481.67	1738.75	1995.74	2510.00	3024.16	4052.49	5080.83
1.184	968.67	1483.50	1740.90	1998.19	2513.13	3027.94	4057.58	5087.22
1.185	969.85	1485.33	1743.06	2000.80	2516.27	3031.73	4062.67	5093.61
1.186	971.02	1487.16	1745.21	2003.28	2519.40	3035.52	4067.76	5100.01
1.187	972.20	1488.99	1747.36	2005.76	2522.53	3039.31	4072.76	5106.41
1.188	973.38	1490.82	1749.52	2008.24	2525.67	3043.10	4077.86	5112.81
1.189	974.56	1492.65	1751.68	2010.72	2528.81	3046.89	4082.96	5119.22
1.190	975.74	1494.48	1753.84	2013.21	2531.95	3050.68	4088.15	5125.63
1.191	976.91	1496.31	1756.00	2015.69	2535.09	3054.47	4093.25	5132.04
1.192	978.09	1498.14	1758.16	2018.18	2538.23	3058.27	4098.36	5138.45

Depth on Weir.	LENGTH OF THE WEIR.							
Feet.	4 Feet.	6 Feet.	7 Feet.	8 Feet.	10 Feet.	12 Feet.	16 Feet.	20 Feet.
1.193	979.27	1499.97	1760.32	2020.67	2541.37	3062.07	4103.47	5144.86
1.194	980.45	1501.80	1762.58	2023.16	2544.51	3065.87	4108.58	5151.28
1.195	981.63	1503.64	1764.65	2025.65	2547.66	3069.67	4113.69	5157.70
1.196	982.81	1505.47	1766.81	2028.14	2550.80	3073.47	4118.80	5164.12
1.197	983.99	1507.31	1768.97	2030.63	2553.95	3077.27	4123.91	5170.55
1.198	985.17	1509.15	1771.13	2033.12	2557.10	3081.07	4129.03	5176.98
1.199	986.35	1510.99	1773.30	2035.62	2560.25	3084.88	4134.25	5183.41
1.200	987.54	1512.83	1775.47	2038.12	2563.40	3088.69	4139.27	5189.84
1.201	988.72	1514.67	1777.64	2040.61	2566.55	3092.50	4144.29	5196.28
1.202	989.90	1516.51	1779.81	2043.10	2569.70	3096.31	4149.41	5202.72
1.203	991.08	1518.35	1781.98	2045.60	2572.86	3100.12	4154.54	5209.16
1.204	992.26	1520.19	1784.15	2048.10	2576.02	3103.93	4159.67	5235.60
1.205	993.45	1522.03	1786.32	2050.60	2579.18	3107.75	4164.90	5222.05
1.206	994.63	1523.87	1788.49	2053.10	2582.34	3111.57	4170.03	5228.50
1.207	995.81	1525.71	1790.66	2055.60	2585.50	3115.39	4175.17	5234.95
1.208	996.99	1527.55	1792.83	2058.10	2588.66	3119.21	4180.31	5241.40
1.209	998.18	1529.40	1795.00	2060.60	2591.82	3123.03	4185.45	5247.86
1.210	999.37	1531.25	1797.18	2063.11	2594.98	3126.85	4190.59	5254.32
1.211	1000.56	1533.09	1799.35	2065.61	2598.14	3130.67	4195.73	5260.78
1.212	1001.75	1534.93	1801.52	2068.12	2601.31	3134.49	4200.87	5267.24
1.213	1002.94	1536.78	1803.70	2070.63	2604.48	3138.32	4206.21	5273.71
1.214	1004.13	1538.63	1805.88	2073.14	2607.65	3142.15	4211.17	5280.18
1.215	1005.31	1540.48	1808.06	2075.65	2610.82	3145.98	4216.32	5286.65
1.216	1006.49	1542.33	1810.24	2078.16	2613.99	3149.81	4221.47	5293.13
1.217	1007.68	1544.18	1812.44	2080.67	2617.16	3153.64	4226.62	5299.61
1.218	1008.87	1546.03	1814.62	2083.18	2620.33	3157.47	4231.78	5306.09
1.219	1010.06	1547.88	1816.80	2085.69	2623.50	3161.31	4236.94	5312.57
1.220	1011.25	1549.73	1818.97	2088.20	2626.68	3165.15	4242.10	5319.05
1.221	1012.44	1551.58	1821.15	2090.71	2629.85	3168.99	4247.26	5325.54
1.222	1013.63	1553.43	1823.33	2093.22	2633.03	3172.83	4252.43	5332.03
1.223	1014.82	1555.28	1825.52	2095.74	2636 21	3176.67	4257.60	5338.52
1.224	1016.01	1557.14	1827.71	2098.26	2639.38	3180.51	4262.77	5345.01
1.225	1017.20	1559.00	1829.89	2100.78	2642.57	3184.36	4267.94	5351.51
1.226	1018.39	1560.85	1832.07	2103.30	2645.75	3188.20	4273.11	5358.01
1.227	1019.58	1562.70	1834.26	2105.82	2648.93	3192.05	4278.28	5364.51
1.228	1020.77	1564.56	1836.45	2108.34	2652.12	3195.90	4283.46	5371.02
1.229	1021.96	1566.42	1838.64	2110.86	2655.31	3199.75	4288.64	5377.53
1.230	1023.16	1568.28	1840.83	2113.39	2658.50	3203.60	4293.82	5384.04
1.231	1024.35	1570.14	1843.02	2115.91	2661.69	3207.45	4299.00	5390.55
1.232	1025.54	1572.00	1845.21	2118.43	2664.88	3211.30	4304.19	5397.07
1.233	1026.74	1573.86	1847.40	2120.95	2668.07	3215.16	4309.38	5303.59
1.234	1027.94	1575.72	1849.59	2123.48	2671.26	3219.02	4314.58	5310.11
1.235	1029.14	1577.58	1851.79	2126.01	2674.45	3222.88	4319.76	5416.63
1.236	1030.33	1579.44	1853.98	2128.54	2677.64	3226.74	4324.75	5423.18
1.237	1031.52	1581.30	1856.18	2131.48	2680.83	3230.60	4329.95	5429.71
1.238	1032.72	1583.16	1858.38	2133.60	2684.03	3224.46	4335.15	5436.24
1.239	1033.92	1585.02	1860.58	2136.13	2687.23	3238.33	4340.35	5442.77
1.240	1035.12	1586.89	1862.78	2138.66	2690.43	3242.20	4345.74	5449.28
1.241	1036.31	1588.75	1864.98	2141.19	2693.63	3246.07	4350.94	5555.82
1.242	1037.51	1590.61	1867.18	2143.72	2696.83	3249.94	4356.14	5562.36
1.243	1038.71	1592.48	1869.38	2146.25	2700.03	3253.81	4361.35	5568.90
1.244	1039.91	1594.35	1871.58	2148.79	2703.23	3257.68	4366.56	5575.45
1.245	1041.11	1596.22	1873.78	2151.33	2706.44	3261.55	4371.77	5482.00
1.246	1042.31	1598.09	1875.98	2153.86	2709.64	3265.42	4376.98	5488.55
1.247	1043.51	1599.96	1878.18	2156.40	2712.85	3269.30	4382.20	5495.10
1.248	1044.71	1601.83	1880.38	2158.94	2716.06	3273.18	4387.42	5501.65
1.249	1045.91	1603.70	1882.59	2161.48	2719.27	3277.06	4392.64	5508.21
1.250	1047.11	1605.57	1884.80	2164.02	2722.48	3280.94	4397.86	5511.77

Depth on Weir.	LENGTH OF THE WEIR.							
Feet.	4 Feet.	6 Feet.	7 Feet.	8 Feet.	10 Feet.	12 Feet.	16 Feet.	20 Feet.
1.251	1048.31	1607.44	1887.00	2166.56	2725.69	3284.82	4403.08	5521.33
1.252	1049.51	1609.31	1889.20	2169.10	2728.90	3288.70	4408.30	5527.90
1.253	1050.71	1611.18	1891.41	2171.64	2732.11	3292.58	4413.53	5534.47
1.254	1051.91	1613.05	1893.62	2174.19	2735.33	3296.47	4418.76	5541.04
1.255	1053.11	1614.93	1895.83	2176.74	2738.55	3300.36	4423.99	5547.61
1.256	1054.31	1616.80	1898.04	2179.28	2741.77	3304.25	4429.22	5554.19
1.257	1055.51	1618.67	1900.25	2181.83	2744.99	3308.14	4434.55	5560.77
1.258	1056.71	1620.55	1902.46	2184.38	2748.21	3312.04	4439.69	5567.35
1.259	1057.91	1622.43	1904.67	2186.93	2751.43	3315.94	4445.93	5573.93
1.260	1059.12	1624.31	1906.89	2189.48	2754.65	3319.82	4450.17	5580.51
1.261	1060.32	1626.18	1909.10	2192.03	2757.87	3323.72	4455.41	5587.10
1.262	1061.52	1628.06	1911.31	2194.58	2761.09	3327.62	4460.65	5593.69
1.263	1062.72	1629.94	1913.53	2197.13	2764.32	3331.52	4465.90	5600.28
1.264	1063.93	1631.82	1915.75	2199.68	2767.55	3335.42	4471.15	5607.87
1.265	1065.14	1633.70	1917.97	2202.24	2770.78	3339.32	4476.40	5613.47
1.266	1066.62	1635.58	1920.19	2204.79	2774.01	3343.23	4481.65	5620.04
1.267	1067.83	1637.46	1922.41	2207.34	2777.24	3347.14	4487.90	5626.65
1.268	1069.04	1639.34	1924.63	2209.90	2780.47	3351.05	4493.16	5633.26
1.269	1070.25	1641.22	1926.85	2212.46	2783.70	3354.96	4498.42	5639.87
1.270	1071.19	1643.11	1929.07	2215.02	2786.94	3358.85	4502.68	5646.51
1.271	1072.40	1644.99	1931.29	2217.58	2790.17	3362.76	4507.94	5653.12
1.272	1073.61	1646.88	1933.51	2220.14	2793.41	3366.67	4513.20	5659.74
1.273	1074.82	1648.77	1935.73	2222.70	2796.65	3370.58	4518.47	5666.36
1.274	1076.03	1650.66	1937.95	2225.26	2799.89	3374.50	4523.74	5672.98
1.275	1077.24	1652.54	1940.18	2227.83	2803.13	3378.42	4529.01	5679.60
1.276	1078.45	1654.42	1942.40	2230.39	2806.37	3382.34	4534.28	5686.23
1.277	1079.66	1656.31	1944.63	2232.95	2809.61	3386.26	4539.55	5692.86
1.278	1080.87	1658.20	1946.86	2235.52	2812.85	3390.18	4544.83	5699.49
1.279	1082.08	1660.09	1949.09	2238.09	2816.09	3394.10	4550.11	5706.12
1.280	1083.29	1661.98	1951.32	2240.66	2819.34	3398.02	4555.39	5712.75
1.281	1084.50	1663.88	1953.55	2243.23	2822.34	3401.94	4560.67	5719.39
1.282	1085.71	1665.77	1955.78	2245.80	2825.84	3405.87	4565.93	5726.03
1.283	1086.91	1667.66	1958.01	2248.37	2829.09	3409.80	4571.22	5732.67
1.284	1087.12	1669.55	1960.24	2250.94	2832.34	3413.73	4576.51	5739.32
1.285	1089.35	1671.43	1962.47	2253.51	2835.59	3417.66	4581.82	5745.97
1.286	1090.56	1673.32	1964.70	2256.08	2838.84	3421.59	4587.21	5752.62
1.287	1091.77	1675.21	1966.93	2258.65	2842.09	3425.52	4592.50	5759.27
1.288	1092.99	1677.10	1969.16	2261.22	2845.34	3429.46	4597.79	5765.93
1.289	1094.21	1679.00	1971.40	2263.80	2848.60	3433.40	4603.09	5772.59
1.290	1095.43	1680.90	1973.64	2266.38	2851.86	3437.34	4608.29	5779.25
1.291	1096.64	1682.79	1975.85	2268.96	2855.12	3441.28	4613.59	5785.91
1.292	1097.85	1684.69	1978.11	2271.54	2858.38	3445.22	4618.89	5792.57
1.293	1099.07	1686.59	1980.35	2274.12	2861.64	3449.16	4624.20	5799.24
1.294	1100.29	1688.49	1982.59	2276.70	2864.90	3453.10	4629.51	5805.91
1.295	1101.51	1690.39	1984.83	2279.28	2868.16	3457.05	4634.82	5812.58
1.296	1102.72	1692.29	1987.07	2281.86	2871.42	3460.99	4640.13	5819.26
1.297	1103.93	1694.19	1989.31	2284.44	2874.68	3464.94	4645.44	5825.94
1.298	1105.15	1696.09	1991.55	2287.02	2877.95	3468.99	4650.75	5832.62
1.299	1106.37	1697.99	1993.80	2289.61	2881.22	3472.94	4656.07	5839.30
1.300	1107.59	1699.90	1996.05	2292.20	2884.49	3476.79	4661.39	5845.98
1.301	1108.81	1701.80	1998.29	2294.78	2887.76	3480.74	4666.71	5852.67
1.302	1110.03	1703.70	2000.53	2297.36	2891.03	3484.70	4672.03	5859.36
1.303	1111.25	1705.60	2002.77	2299.95	2894.30	3488.66	4677.35	5866.05
1.304	1112.47	1707.50	2005.02	2302.54	2897.57	3492.66	4682.68	5872.74
1.305	1113.69	1709.41	2007.27	2305.13	2900.85	3496.57	4688.01	5879.44
1.306	1114.91	1711.31	2009.52	2307.72	2904.12	3500.53	4693.34	5886.14
1.307	1116.13	1713.22	2011.77	2310.31	2907.40	3504.49	4698.67	5892.84
1.308	1117.35	1715.13	2014.02	2313.90	2910.68	3508.45	4704.00	5899.55

Depth on Weir.	LENGTH OF THE WEIR.							
Feet.	4 Feet.	6 Feet.	7 Feet.	8 Feet.	10 Feet.	12 Feet.	16 Feet.	20 Feet.
1.309	1118.52	1717.04	2016.27	2316.49	2913.96	3512.42	4709.34	5906.26
1.310	1119.80	1718.95	2018.52	2318.09	2917.24	3516.39	4714.68	5912.97
1.311	1121.62	1720.86	2020.77	2320.68	2920.52	3520.36	4720.02	5919.68
1.312	1122.24	1722.77	2023.02	2323.28	2923.80	3524.33	4725.36	5926.39
1.313	1123.47	1724.68	2025.27	2325.88	2927.08	3528.30	4730.70	5933.11
1.314	1124.70	1726.59	2027.53	2328.48	2930.37	3532.27	4736.04	5939.89
1.315	1125.92	1728.50	2029.79	2331.08	2933.66	3536.24	4741.40	5946.55
1.316	1127.36	1730.41	2032.04	2333.68	2936.94	3540.21	4746.75	5953.28
1.317	1128.59	1732.32	2034.29	2336.28	2940.23	3544.18	4752.10	5960.01
1.318	1129.81	1734.23	2036.55	2338.88	2943.52	3548.18	4757.45	5966.74
1.319	1131.04	1736.14	2038.81	2341.48	2946.81	3552.16	4762.80	5973.47
1.320	1132.04	1738.06	2041.07	2344.08	2950.10	3556.12	4768.16	5980.20
1.321	1133.26	1739.97	2043.33	2346.68	2953.39	3560.10	4773.52	5986.94
1.322	1134.48	1741.88	2045.59	2349.28	2956.68	3564.08	4778.88	5993.68
1.323	1135.71	1743.80	2047.85	2351.89	2959.97	3568.06	4784.24	6000.42
1.324	1136.94	1745.72	2050.11	2354.50	2963.07	3572.05	4789.60	6007.16
1.325	1138.17	1747.64	2052.38	2357.11	2966.57	3576.04	4794.97	6013.90
1.326	1139.39	1749.56	2054.64	2359.72	2969.87	3580.03	4800.36	6020.65
1.327	1140.62	1751.48	2056.90	2362.33	2973.17	3584.02	4805.73	6027.40
1.328	1141.85	1753.40	2059.16	2364.94	2976.47	3588.01	4811.10	6034.15
1.329	1143.08	1755.32	2061.43	2367.55	2979.77	3592.00	4816.47	6040.91
1.330	1144.31	1757.24	2063.70	2370.16	2983.07	3595.99	4821.83	6047.67
1.331	1145.54	1759.16	2065.96	2372.77	2986.37	3599.98	4827.21	6054.43
1.332	1146.77	1761.08	2068.23	2375.38	2989.68	3603.98	4832.59	6061.19
1.333	1148.00	1763.90	2070.50	2377.99	2992.99	3607.98	4837.97	6067.96
1.334	....	1764.92	2072.77	2380.61	2996.30	3611.98	4843.35	6074.73
1.335	....	1766.85	2075.04	2383.23	2999.61	3615.98	4848.74	6081.50
1.336	....	1768.77	2077.31	2385.84	3002.92	3619.98	4854.13	6088.27
1.337	....	1770.69	2079.58	2388.46	3006.23	3623.98	4859.52	6095.04
1.338	....	1772.60	2081.85	2391.08	3009.54	3627.99	4864.91	6101.82
1.339	....	1774.53	2084.12	2393.70	3012.85	3632.00	4870.20	6108.60
1.340	....	1776.47	2086.40	2396.32	3016.17	3636.01	4875.70	6115.38
1.341	....	1778.39	2088.67	2398.94	3019.48	3640.02	4881.11	6122.17
1.342	....	1780.32	2090.94	2401.56	3022.79	3644.03	4886.52	6128.96
1.343	....	1782.25	2093.21	2404.18	3026.11	3648.04	4891.93	6135.75
1.344	....	1784.18	2095.49	2406.80	3029.43	3652.05	4897.34	6142.54
1.345	....	1786.11	2097.77	2409.43	3032.75	3656.06	4902.70	6149.33
1.346	....	1788.04	2100.05	2412.05	3036.07	3660.08	4908.22	6156.13
1.347	....	1789.97	2102.33	2414.68	3039.39	3664.10	4913.65	6162.93
1.348	....	1791.90	2104.61	2417.31	3042.71	3668.12	4919.08	6169.73
1.349	....	1793.83	2106.89	2419.94	3046.03	3672.14	4924.51	6176.53
1.350	....	1795.77	2109.17	2422.57	3049.36	3676.16	4929.75	6183.34
1.351	....	1797.70	2111.45	2425.20	3052.68	3680.18	4935.17	6190.15
1.352	....	1799.63	2113.73	2427.83	3056.01	3684.20	4940.59	6196.96
1.353	....	1801.56	2116.01	2430.46	3059.34	3688.22	4946.01	6203.77
1.354	....	1803.50	2118.29	2433.09	3062.67	3692.25	4951.43	6210.59
1.355	....	1805.44	2120.58	2435.72	3066.00	3996.28	4956.85	6217.41
1.356	....	1807.37	2122.86	2438.35	3069.33	3700.31	4962.27	6224.23
1.357	....	1809.31	2125.14	2440.98	3072.66	3704.34	4967.70	6231.05
1.358	....	1811.25	2127.43	2443.62	3075.99	3708.38	4973.13	6237.88
1.359	....	1813.19	2129.72	2446.26	3079.33	3712.42	4978.56	6246.51
1.360	....	1815.13	2132.01	2448.90	3082.67	3716.45	4983.99	6251.54
1.361	....	1817.07	2134.30	2451.54	3086.01	3720.48	4989.43	6258.37
1.362	....	1819.01	2136.58	2454.18	3089.35	3724.52	4994.87	6265.21
1.363	....	1820.95	2138.87	2456.82	3092.69	3728.56	5000.31	6272.05
1.364	....	1822.89	2141.16	2459.46	3096.03	3732.60	5005.75	6278.89
1.365	....	1824.83	2143.46	2462.10	3099.37	3736.64	5011.19	6285.73
1.366	....	1826.77	2145.75	2464.74	3102.71	3740.68	5016.63	6292.58

Depth on Weir.	LENGTH OF THE WEIR.						
Feet.	6 Feet.	7 Feet.	8 Feet.	10 Feet.	12 Feet.	16 Feet.	20 Feet.
1.367	1828.71	2148.08	2467.38	3106.05	3744.72	5022.07	6299.43
1.368	1830.65	2150.33	2470.02	3109.40	3748.77	5027.52	6306.28
1.369	1832.59	2152.63	2472.66	3112.75	3752.82	5032.97	6313.13
1.370	1834.54	2154.93	2475.32	3116.10	3756.87	5038.42	6319.98
1.371	1836.48	2157.22	2477.96	3119.45	3760.92	5043.87	6326.84
1.372	1838.42	2159.52	2480.61	3122.80	3764.97	5049.33	6333.70
1.373	1840.37	2161.82	2483.26	3126.15	3769.02	5054.79	6340.56
1.374	1842.32	2164.12	2485.91	3129.50	3773.07	5060.25	6347.42
1.375	1844.27	2166.42	2488.56	3132.85	3777.13	5065.71	6354.28
1.376	1846.22	2168.72	2491.21	3136.20	3781.19	5071.17	6361.05
1.377	1848.17	2171.02	2493.86	3139.55	3785.25	5076.63	6368.02
1.378	1850.15	2173.32	2496.51	3142.91	3789.30	5082.10	6374.89
1.379	1852.10	2175.62	2499.16	3146.27	3793.36	5087.57	6381.77
1.380	1854.02	2177.92	2501.82	3149.63	3797.43	5093.04	6388.65
1.381	1855.97	2180.22	2504.47	3152.99	3801.49	5098.51	6395.53
1.382	1857.92	2182.52	2507.13	3156.35	3806.55	5103.98	6402.40
1.383	1859.87	2184.82	2509.79	3159.71	3810.62	5109.46	6409.30
1.384	1861.82	2187.13	2512.45	3163.07	3814.69	5114.94	6416.19
1.385	1863.78	2189.44	2515.11	3166.43	3817.76	5120.42	6423.08
1.386	1865.73	2191.74	2517.77	3169.79	3821.83	5125.90	6429.97
1.387	1867.68	2194.05	2520.43	3173.16	3825.90	5131.34	6436.86
1.388	1869.63	2196.36	2523.09	3176.53	3829.97	5136.82	6443.76
1.389	1871.69	2198.67	2525.75	3179.90	3834.05	5142.31	6450.66
1.390	1873.55	2200.98	2528.41	3183.27	3838.13	5147.84	6457.56
1.391	1875.50	2203.29	2531.07	3186.64	3842.12	5153.35	6464.47
1.392	1877.46	2205.60	2533.73	3190.01	3846.29	5158.86	6471.38
1.393	1879.42	2207.91	2536.40	3193.38	3850.37	5164.38	6478.29
1.394	1881.38	2210.22	2539.07	3196.75	3854.45	5169.90	6485.20
1.395	1883.34	2212.54	2541.74	3200.13	3858.53	5175.32	6492.11
1.396	1885.30	2214.85	2544.40	3203.50	3862.61	5180.82	6499.03
1.397	1887.26	2217.16	2547.07	3206.88	3866.69	5186.32	6505.95
1.398	1889.22	2219.47	2549.74	3210.26	3870.78	5191.82	6512.87
1.399	1891.18	2221.79	2552.41	3213.64	3875.87	5197.32	6519.85
1.400	1893.14	2224.11	2555.68	3217.02	3878.96	5202.83	6526.71
1.401	1895.10	2226.43	2557.75	3220.40	3883.05	5208.34	6533.64
1.402	1897.06	2228.75	2560.42	3223.78	3887.14	5213.85	6540.57
1.403	1899.02	2231.07	2563.09	3227.16	3891.23	5219.36	6547.50
1.404	1900.99	2233.39	2565.77	3230.55	3895.32	5224.88	6545.43
1.405	1902.96	2235.71	2568.45	3233.94	3899.42	5230.40	6561.37
1.406	1904.92	2038.03	2571.12	3237.32	3903.52	5235.92	6568.29
1.407	1906.88	2040.35	2573.79	3240.71	3907.62	5239.44	6575.21
1.408	1908.84	2042.67	2576.46	3244.10	3910.72	5244.96	6582.14
1.409	1910.81	2044.99	2579.13	3247.49	3913.82	5248.48	6589.17
1.410	1912.79	2247.32	2581.81	3250.88	3919.92	5254.01	6596.10
1.411	1914.76	2249.64	2584.94	3254.27	3924.02	5263.54	6603.05
1.412	1916.73	2251.96	2587.18	3257.66	3928.13	5267.07	6610.00
1.413	1918.70	2254.78	2589.88	3261.05	3932.24	5272.60	6616.95
1.414	1920.67	2256.61	2592.58	3264.45	3936.35	5278.13	6623.91
1.415	1922.64	2258.94	2595.25	3267.85	3940.46	5285.67	6630.87
1.416	1924.61	2261.27	2598.93	3271.65	3944.57	5289.22	6637.73
1.417	1926.58	2263.60	2601.61	3274.05	3948.68	5292.76	6644.60
1.418	1928.55	2265.93	2604.30	3279.45	3955.79	5296.21	6651.47
1.419	1930.52	2268.26	2606.99	3280.85	3959.90	5299.65	6658.34
1.420	1932.50	2270.59	2608.68	3284.85	3961.02	5313.37	6665.71
1.421	1934.47	2272.92	2611.37	3288.25	3965.14	5318.92	6672.69
1.422	1936.44	2275.25	2614.06	3291.65	3969.26	5324.47	6679.67
1.423	1938.42	2277.58	2616.75	3295.05	3973.38	5300.02	6686.65
1.424	1940.40	2279.91	2619.44	3298.46	3977.51	5335.57	6693.63

Depth on Weir.	LENGTH OF THE WEIR.						
Feet.	6 Feet.	7 Feet.	8 Feet.	10 Feet.	12 Feet.	16 Feet.	20 Feet.
1.425	1942.38	2282.25	2622.13	3301.87	3981.62	5341.12	6700.61
1.426	1944.35	2284.58	2624.83	3305.28	3985.75	5346.67	6707.63
1.427	1946.33	2286.91	2627.52	3308.69	3989.88	5352.23	6714.59
1.428	1948.31	2289.25	2630.21	3312.10	3994.01	5357.79	6721.58
1.429	1950.29	2291.59	2632.90	3315.51	3998.15	5363.35	6728.57
1.430	1952.27	2293.93	2635.60	3318.93	4002.25	5368.91	6735.57
1.431	1954.25	2296.27	2638.29	3322.34	4006.38	5374.47	6742.57
1.432	1956.23	2298.61	2640.99	3325.75	4010.51	5380.04	6749.57
1.433	1958.21	2300.95	2643.69	3329.17	4014.64	5385.61	6756.57
1.434	1960.19	2303.09	2646.39	3332.59	4018.78	5391.18	6763.57
1.435	1962.17	2305.63	2649.09	3336.01	4022.92	5396.75	6770.58
1.436	1964.15	2307.97	2651.79	3339.23	4027.06	5402.32	6777.59
1.437	1966.13	2310.31	2654.49	3342.45	4031.20	5407.90	6784.60
1.438	1968.11	2312.65	2657.19	3345.67	4035.34	5413.48	6791.61
1.439	1970.09	2315.00	2659.89	3348.89	4039.48	5419.06	6798.63
1.440	1972.09	2317.35	2662.60	3353.11	4043.62	5424.64	6805.65
1.441	1974.07	2319.69	2665.30	3356.53	4047.76	5430.22	6812.67
1.442	1976.05	2322.03	2668.00	3359.95	4051.90	5435.80	6819.70
1.443	1978.04	2324.38	2670.71	3363.38	4056.05	5441.39	6826.73
1.444	1980.03	2326.73	2673.42	3366.81	4060.20	5446.98	6833.76
1.445	1982.02	2329.08	2676.13	3370.24	4064.35	5452.57	6840.79
1.446	1984.01	2331.43	2678.84	3373.67	4068.50	5458.16	6847.80
1.447	1986.00	2333.78	2681.55	3377.10	4072.65	5463.75	6854.83
1.448	1987.99	2336.13	2684.26	3380.53	4076.80	5469.34	6861.87
1.449	1989.98	2338.48	2686.97	3383.96	4080.96	5474.94	6868.91
1.450	1991.97	2340.83	2689.69	3387.40	4085.12	5480.54	6875.97
1.451	1993.96	2343.18	2692.40	3390.83	4089.27	5486.14	6883.01
1.452	1995.95	2345.53	2695.11	3394.27	4093.43	5491.74	6890.06
1.453	1997.94	2347.88	2697.82	3397.71	4097.59	5497.34	6897.11
1.454	1999.93	2350.24	2700.54	3401.15	4101.75	5502.95	6904.16
1.455	2001.93	2352.60	2703.26	3404.59	4105.91	5508.56	6911.21
1.456	2203.92	2354.95	2705.97	3408.03	4110.07	5514.17	6918.27
1.457	2205.91	2357.30	2708.69	3411.47	4114.23	5519.78	6925.33
1.458	2207.91	2359.66	2711.41	3414.91	4118.40	5525.39	6932.39
1.459	2209.91	2362.06	2714.13	3418.35	4122.57	5531.01	6939.45
1.460	2011.91	2364.38	2716.85	3421.80	4126.74	5536.63	6946.52
1.461	2013.90	2366.74	2719.57	3425.24	4130.91	5542.25	6953.59
1.462	2015.90	2369.10	2722.29	3428.69	4135.08	5547.87	6960.66
1.463	2017.90	2371.46	2725.01	3432.14	4139.25	5553.49	6967.73
1.464	2019.90	2373.82	2727.74	3435.59	4143.43	5559.11	6974.80
1.465	2021.90	2376.18	2730.47	3439.04	4147.61	5564.74	6981.88
1.466	2023.90	2378.64	2733.19	3442.49	4151.78	5570.37	6988.96
1.467	2025.90	2380.90	2735.91	3445.94	4155.96	5576.00	6996.04
1.468	2027.90	2383.26	2738.64	3449.39	4160.14	5581.63	7003.12
1.469	2029.90	2385.63	2741.37	3452.84	4164.28	5587.26	7010.21
1.470	2031.90	2388.00	2744.10	3456.30	4168.50	5592.90	7017.30
1.471	2033.90	2390.36	2746.83	3459.75	4172.68	5598.54	7024.39
1.472	2035.90	2392.73	2749.56	3463.21	4176.86	5604.18	7031.48
1.473	2037.90	2395.10	2752.29	3466.67	4181.05	5609.82	7038.57
1.474	2039.91	2397.47	2755.02	3470.13	4185.24	5615.46	7045.67
1.475	2041.92	2399.84	2757.76	3473.59	4189.43	5621.10	7052.77
1.476	2043.92	2402.21	2760.49	3477.05	4193.62	5626.75	7059.87
1.477	2045.92	2404.58	2763.22	3480.51	4197.81	5632.41	7066.97
1.478	2047.93	2406.95	2765.95	3483.97	4201.90	5638.06	7074.08
1.479	2049.94	2409.32	2768.69	3487.44	4206.09	5643.72	7081.19
1.480	2051.95	2411.69	2771.43	3490.91	4210.39	5649.36	7088.30
1.481	2053.96	2414.06	2774.10	3494.38	4214.58	5655.04	7095.42
1.482	2055.97	2416.43	2776.77	3497.85	4218.78	5660.70	7102.54

Depth on Weir.	LENGTH OF THE WEIR.						
Feet.	6 Feet.	7 Feet.	8 Feet.	10 Feet.	12 Feet.	16 Feet.	20 Feet.
1.483	2057.96	2418.80	2779.44	3501.32	4222.98	5666.36	7109.66
1.484	2059.99	2421.18	2782.12	3504.79	4227.18	5672.02	7116.78
1.485	2062.00	2423.56	2785.13	3508.26	4231.38	5677.64	7123.90
1.486	2064.01	2425.93	2787.87	3511.73	4235.58	5683.30	7131.62
1.487	2066.02	2428.31	2790.61	3515.20	4239.78	5688.96	7138.15
1.488	2068.03	2430.69	2793.35	3518.67	4243.99	5694.63	7145.28
1.489	2070.04	2433.07	2796.09	3512.15	4248.20	5700.30	7152.41
1.490	2072.06	2435.45	2798.84	3525.63	4252.41	5705.97	7159.54
1.491	2074.07	2437.83	2801.58	3529.10	4256.62	5711.64	7166.68
1.492	2076.08	2440.21	2804.33	3532.58	4260.83	5717.31	7173.82
1.493	2078.09	2442.59	2807.08	3536.06	4265.04	5722.99	7180.96
1.494	2080.11	2444.97	2809.83	3539.54	4269.25	5728.67	7188.10
1.495	2082.13	2447.35	2812.58	3543.02	4273.47	5734.35	7195.24
1.496	2084.14	2449.73	2815.33	3546.50	4277.68	5740.03	7202.39
1.497	2086.16	2452.11	2818.08	3549.98	4281.90	5745.71	7209.54
1.498	2088.18	2454.50	2820.83	3553.46	4286.12	5751.40	7216.69
1.499	2090.20	2456.89	2823.58	3556.95	4290.34	5757.09	7223.84
1.500	2092.22	2459.28	2826.33	3560.44	4294.56	5762.78	7231.00
1.501	2094.24	2461.66	2829.08	3563.93	4298.78	5768.47	7238.16
1.502	2096.26	2464.04	2831.83	3567.42	4303.00	5774.16	7245.32
1.503	2098.28	2466.43	2834.59	3570.91	4307.22	5779.85	7252.48
1.504	2100.30	2468.82	2837.35	3574.40	4311.45	5785.55	7259.65
1.505	2102.32	2471.21	2840.11	3577.89	4315.68	5791.25	7266.82
1.506	2104.34	2473.60	2842.86	3581.38	4319.91	5796.95	7273.99
1.507	2106.36	2475.99	2845.62	3584.87	4324.14	5802.65	7281.16
1.508	2108.38	2478.38	2848.38	3588.36	4328.37	5808.35	7288.33
1.509	2110.41	2480.77	2851.14	3591.86	4332.60	5814.05	7295.51
1.510	2112.43	2483.17	2853.90	3595.36	4336.83	5819.76	7302.69
1.511	2114.45	2485.56	2856.66	3598.86	4341.06	5825.47	7309.87
1.512	2116.47	2487.95	2859.42	3602.36	4345.30	5831.18	7317.05
1.513	2118.50	2490.34	2862.18	3605.86	4349.54	5836.89	7324.24
1.514	2020.53	2492.74	2864.94	3609.36	4353.78	5842.60	7331.43
1.515	2122.56	2495.14	2867.71	3612.86	4358.02	5848.32	7338.62
1.516	2124.58	2497.53	2870.47	3616.36	4362.20	5854.04	7345.81
1.517	2126.61	2499.93	2873.24	3619.86	4366.44	5859.76	7353.01
1.518	2128.64	2502.33	2876.01	3623.37	4370.68	5865.48	7360.21
1.519	2130.67	2504.73	2878.78	3626.88	4374.92	5871.20	7367.41
1.520	2132.70	2507.13	2881.55	3630.39	4379.23	5876.92	7374.61
1.521	2134.73	2509.53	2884.32	3633.90	4383.48	5882.64	7381.81
1.522	2136.76	2511.93	2887.09	3637.41	4387.73	5888.37	7389.02
1.523	2138.79	2514.33	2889.86	3640.92	4391.98	5894.10	7396.23
1.524	2140.83	2516.73	2892.63	3644.43	4396.23	5899.83	7403.44
1.525	2142.86	2519.13	2895.40	3647.94	4400.48	5905.56	7410.65
1.526	2144.89	2521.53	2898.17	3651.45	4404.73	5911.29	7417.86
1.527	2146.92	2523.93	2900.94	3654.96	4408.98	5917.03	7425.08
1.528	2149.95	2526.33	2903.72	3658.48	4413.24	5922.77	7432.30
1.529	2151.99	2528.74	2906.50	3662.00	4417.50	5928.51	7439.52
1.530	2153.03	2531.15	2909.28	3665.52	4421.76	5934.25	7446.74
1.531	2155.06	2533.55	2912.05	3669.04	4426.02	5939.99	7453.97
1.532	2157.10	2535.96	2914.83	3672.56	4430.28	5945.74	7461.20
1.533	2159.14	2538.37	2917.61	3676.08	4434.54	5951.49	7468.43
1.534	2161.28	2540.78	2920.39	3679.60	4438.81	5957.24	7475.06
1.535	2163.21	2543.19	2923.17	3683.12	4443.08	5962.99	7482.90
1.536	2165.25	2545.60	2925.95	3686.64	4447.34	5968.74	7490.14
1.537	2167.29	2548.01	2928.73	3690.16	4451.61	5974.49	7497.38
1.538	2169.33	2550.42	2931.51	3693.69	4455.88	5980.25	7504.62
1.539	2171.37	2552.83	2934.29	3697.22	4460.15	5986.00	7511.86
1.540	2173.41	2555.24	2937.08	3700.75	4464.42	5991.77	7519.11

Depth on Weir.	LENGTH OF THE WEIR.						
Feet.	6 Feet.	7 Feet.	8 Feet.	10 Feet.	12 Feet.	16 Feet.	20 Feet.
1.541	2175.45	2557.65	2939.80	3704.28	4468.69	5997.53	7526.36
1.542	2177.49	2560.06	2942.64	3707.81	4472.96	6003.29	7533.61
1.543	2179.53	2562.48	2945.43	3711.34	4477.18	6009.05	7540.86
1.544	2181.57	2564.90	2948.22	3714.87	4481.38	6014.82	7548.11
1.545	2183.62	2567.32	2951.01	3718.41	4485.80	6020.59	7555.37
1.546	2185.66	2569.73	2953.80	3721.94	4490.08	6026.56	7562.63
1.547	2187.70	2572.14	2956.59	3725.47	4494.36	6032.33	7569.89
1.548	2189.75	2574.56	2959.38	3729.01	4498.64	6038.10	7577.15
1.549	2191.80	2576.98	2962.17	3732.55	4502.92	6043.87	7584.42
1.550	2193.85	2579.40	2964.96	3736.09	4507.21	6049.45	7591.69
1.551	2195.89	2581.82	2967.75	3739.63	4511.49	6055.23	7598.96
1.552	2197.93	2584.24	2970.54	3743.17	4515.78	6061.01	7606.23
1.553	2199.98	2586.66	2973.33	3746.71	4520.07	6066.79	7613.51
1.554	2202.03	2589.08	2976.13	3750.25	4524.36	6072.57	7620.79
1.555	2204.08	2591.51	2978.93	3753.79	4528.65	6078.36	7628.07
1.556	2206.13	2593.93	2981.72	3757.33	4532.94	6084.15	7635.35
1.557	2208.18	2596.35	2984.52	3760.87	4537.23	6089.94	7642.63
1.558	2210.23	2598.77	2987.32	3764.41	4541.52	6095.73	7649.92
1.559	2212.28	2601.20	2990.11	3767.95	4545.82	6101.52	7657.21
1.560	2214.33	2603.63	2992.92	3771.52	4550.12	6107.31	7664.50
1.561	2216.38	2606.05	2995.72	3775.57	4554.42	6113.10	7671.79
1.562	2218.43	2668.47	2998.52	3778.62	4558.72	6118.90	7679.09
1.563	2220.48	2670.90	3001.32	3782.17	4563.02	6124.70	7686.39
1.564	2222.54	2673.33	3004.12	3785.72	4567.02	6130.50	7693.69
1.565	2224.59	2615.76	3006.93	3789.28	4571.62	6136.30	7700.99
1.566	2226.64	2618.19	3009.73	3792.83	4575.92	6142.10	7708.29
1.567	2228.69	2620.62	3012.53	3796.38	4580.22	6147.91	7715.30
1.568	2230.75	2623.05	3015.34	3799.94	4584.53	6153.72	7722.61
1.569	2232.81	2625.48	3018.15	3803.50	4588.84	6159.53	7729.92
1.570	2234.87	2627.92	3020.96	3807.06	4593.15	6165.34	7737.53
1.571	2236.91	2630.35	3023.77	3810.62	4597.46	6171.15	7744.85
1.572	2238.95	2632.78	3026.58	3814.18	4601.77	6176.96	7752.17
1.573	2240.99	2635.21	3029.39	3817.74	4606.08	6182.78	7759.49
1.574	2243.04	2637.64	3032.20	3821.30	4610.40	6188.60	7766.81
1.575	2245.16	2640.08	3035.01	3824.86	4614.72	6194.42	7774.13
1.576	2247.22	2642.51	3037.82	3828.42	4619.03	6200.24	7781.46
1.577	2249.28	2644.95	3040.63	3831.98	4623.35	6206.06	7788.79
1.578	2251.34	2647.39	3043.44	3835.55	4627.67	6211.89	7796.12
1.579	2253.40	2649.83	3046.26	3838.12	4631.99	6217.72	7803.45
1.580	2255.46	2652.27	3049.08	3842.69	4636.31	6223.55	7810.78
1.581	2257.52	2654.71	3051.89	3846.25	4640.63	6229.38	7818.12
1.582	2259.58	2657.15	3054.70	3849.81	4644.95	6235.21	7825.46
1.583	2261.64	2659.59	3057.52	3853.37	4649.28	6241.04	7832.80
1.584	2263.71	2662.03	3060.34	3856.93	4653.60	6246.87	7840.14
1.585	2265.78	2664.47	3063.16	3860.55	4657.94	6252.71	7847.49
1.586	2267.84	2666.91	3065.98	3864.12	4662.27	6258.55	7854.84
1.587	2269.90	2669.35	3068.80	3867.69	4666.60	6264.39	7862.19
1.588	2271.97	2671.79	3071.62	3871.27	4670.93	6270.23	7869.54
1.589	2274.04	2674.24	3074.44	3874.85	4675.26	6276.07	7876.89
1.590	2276.11	2676.69	3077.27	3878.43	4679.60	6281.92	7884.25
1.591	2278.17	2679.13	3060.09	3882.01	4683.93	6287.77	7892.61
1.592	2280.24	2681.57	3082.91	3885.59	4688.26	6293.65	7899.97
1.593	2282.31	2684.02	3085.73	3889.17	4692.60	6299.55	7807.33
1.594	2284.38	2686.47	3088.56	3892.75	4696.94	6305.40	7814.70
1.595	2286.45	2688.92	3091.39	3896.34	4701.28	6311.18	7921.07
1.596	2288.51	2691.37	3094.22	3899.92	4705.62	6317.03	7928.44
1.597	2290.58	2693.82	3097.05	3903.50	4709.96	6322.89	7936.81
1.598	2292.65	2696.27	3099.88	3907.09	4714.30	6328.75	7944.18

Depth on Weir.	LENGTH OF THE WEIR.						
Feet.	6 Feet.	7 Feet.	8 Feet.	10 Feet.	12 Feet.	16 Feet.	20 Feet.
1.599	2294.72	2698.72	3102.71	3910.68	4718.65	6334.61	7951.56
1.600	2296.80	2701.17	3105.54	3914.27	4723.00	6340.47	7957.94
1.601	2298.87	2703.62	3108.37	3917.86	4727.35	6346.33	7965.32
1.602	2300.94	2706.07	3111.20	3921.45	4731.70	6352.20	7973.70
1.603	2303.02	2708.52	3114.03	3925.04	4736.05	6358.07	7981.08
1.604	2305.19	2710.97	3116.86	3928.63	4740.40	6363.94	7988.48
1.605	2307.17	2713.43	3119.70	3932.23	4744.75	6369.81	7994.86
1.606	2309.24	2715.88	3122.53	3935.82	4749.10	6375.68	8002.25
1.607	2311.31	2518.34	3125.36	3939.41	4753.46	6381.55	8009.64
1.608	2313.39	2520.80	3128.20	3943.01	4757.82	6388.43	8017.04
1.609	2315.47	2523.26	3131.04	3946.61	4762.18	6394.31	8024.44
1.610	2317.55	2725.72	3133.88	3950.21	4766.54	6399.19	8031.84
1.611	2319.63	2728.17	3136.72	3953.81	4770.90	6405.07	8039.24
1.612	2321.71	2730.63	3139.56	3957.41	4775.26	6410.95	8046.65
1.613	2323.89	2733.09	3142.44	3961.01	4779.62	6416.83	8054.06
1.614	2325.97	2735.55	3145.28	3964.61	4783.98	6422.72	8061.47
1.615	2327.95	2738.01	3148.08	3968.21	4788.34	6428.61	8068.88
1.616	2330.03	2740.47	3150.92	3971.81	4792.71	6434.50	8076.29
1.617	2332.11	2742.93	3153.76	3975.41	4797.08	6440.39	8083.71
1.618	2334.19	2745.40	3156.60	3979.02	4801.45	6446.28	8091.13
1.619	2336.28	2747.87	3159.45	3982.63	4806.82	6452.18	8098.55
1.620	2338.36	2750.33	3162.30	3986.24	4810.19	6458.08	8105.97
1.621	2340.44	2752.79	3165.14	3989.85	4814.56	6463.98	8113.39
1.622	2342.52	2755.25	3167.99	3993.45	4818.93	6469.88	8120.82
1.623	2344.60	2757.72	3170.84	3997.06	4823.30	6475.78	8128.25
1.624	2346.09	2760.19	3173.69	4000.67	4827.68	6481.68	8135.68
1.625	2348.78	2762.66	3176.54	4004.30	4832.06	6487.58	8143.11
1.626	2350.86	2765.12	3179.39	4007.91	4836.44	6493.49	8150.55
1.627	2352.94	2767.59	3182.24	4011.52	4840.82	6499.40	8157.99
1.628	2355.03	2770.06	3185.09	4015.14	4845.10	6505.31	8165.43
1.629	2357.12	2772.53	3187.94	4018.76	4849.48	6511.22	8172.87
1.630	2359.21	2775.00	3190.79	4022.38	4853.97	6517.13	8180.31
1.631	2361.29	2777.47	3193.64	4026.00	4858.35	6323.05	8181.76
1.632	2363.38	2779.94	3196.49	4029.62	4862.73	6328.07	8195.21
1.633	2365.47	2782.41	3199.35	4033.24	4867.12	6234.89	8202.66
1.634	2367.56	2784.88	3202.21	4036.86	4871.51	6240.81	8210.11
1.635	2369.65	2787.36	3205.07	4040.48	4875.90	6546.73	8217.56
1.636	2371.74	2789.83	3207.92	4044.10	4880.29	6552.65	8225.02
1.637	2373.83	2792.30	3210.78	4047.72	4884.68	6558.57	8232.48
1.638	2375.92	2794.78	3213.64	4051.35	4889.07	6564.50	8239.94
1.639	2378.01	2797.26	3216.50	4054.98	4893.46	6570.33	8247.40
1.640	2380.11	2799.74	3219.36	4058.61	4897.86	6576.36	8254.86
1.641	2382.20	2802.21	3222.22	4062.24	4902.26	6582.29	8262.33
1.642	2384.29	2804.69	3225.08	4065.87	4906.66	6588.22	8269.80
1.643	2386.38	2807.17	3227.94	4069.50	4911.06	6594.16	8277.27
1.644	2388.48	2809.65	3230.80	4073.13	4915.46	6600.10	8284.74
1.645	2390.58	2812.13	3233.67	4076.76	4919.86	6606.04	8292.22
1.646	2392.67	2814.61	3236.53	4080.39	4924.26	6611.98	8299.70
1.647	2394.77	2817.09	3239.40	4084.02	4928.66	6617.93	8307.18
1.648	2396.87	2819.57	3242.27	4087.66	4933.06	6623.87	8314.66
1.649	2398.97	2822.05	3245.14	4091.30	4937.47	6629.81	8322.14
1.650	2401.07	2824.54	3248.01	4094.94	4941.88	6635.75	8329.63
1.651	2403.16	2827.02	3250.87	4098.58	4946.29	6641.70	8337.12
1.652	2405.26	2829.50	3253.74	4102.22	4950.70	6647.65	8344.61
1.653	2407.36	2831.98	3256.61	4105.86	4955.11	6653.60	8352.10
1.654	2409.46	2834.47	3259.54	4109.58	4959.52	6659.55	8359.60
1.655	2411.56	2836.96	3262.35	4113.15	4963.94	6665.51	8367.10
1.656	2413.66	2839.44	3265.22	4116.79	4968.35	6671.47	8374.60

Depth on Weir.	LENGTH OF THE WEIR.						
Feet.	6 Feet.	7 Feet.	8 Feet.	10 Feet.	12 Feet.	16 Feet.	20 Feet.
1.657	2413.76	2841.93	3268.09	4120.43	4972.76	6677.43	8382.10
1.658	2417.86	2844.42	3270.96	4124.07	4977.18	6683.39	8389.60
1.659	2419.96	2846.91	3273.84	4127.71	4981.60	6689.35	8397.11
1.660	2422.07	2849.40	3276.72	4131.37	4986.02	6695.32	8404.62
1.661	2424.17	2851.89	3279.59	4135.02	4990.44	6701.28	8412.13
1.662	2426.27	2854.34	3282.47	4138.67	4994.86	6707.25	8419.64
1.663	2428.37	2856.79	3285.35	4142.32	4999.28	6713.22	8427.15
1.664	2430.48	2859.24	3288.23	4145.97	5004.70	6719.19	8434.67
1.665	2432.59	2861.85	3291.11	4149.62	5008.13	6725.16	8442.19
1.666	2434.69	2864.34	3293.99	4153.07	5012.56	6731.03	8449.71
1.667	2436.80	2866.83	3296.87	4156.52	5016.99	6736.90	8457.23
1.668	2438.91	2869.32	3299.75	4159.98	5021.42	6742.78	8464.75
1.669	2440.02	2871.82	3302.63	4163.44	5025.85	6748.66	8472.28
1.670	2443.13	2874.32	3305.51	4167.90	5030.28	6755.04	8479.81
1.671	2445.24	2876.81	3308.39	4171.56	5034.71	6761.02	8487.34
1.672	2447.35	2879.31	3311.27	4175.22	5039.14	6767.00	8494.87
1.673	2449.46	2881.81	3314.16	4178.88	5043.57	6772.99	8502.41
1.674	2451.57	2884.31	3317.05	4182.54	5048.01	6778.98	8509.95
1.675	2453.68	2886.81	3319.94	4186.20	5052.45	6784.97	8517.49
1.676	2455.79	2889.30	3322.82	4189.86	5056.89	6790.96	8525.03
1.677	2457.90	2891.80	3325.71	4193.52	5061.33	6796.95	8532.57
1.678	2460.01	2894.30	3328.60	4197.18	5065.77	6802.94	8540.12
1.679	2462.12	2896.80	3331.49	4200.85	5070.21	6808.94	8547.67
1.680	2464.24	2899.31	3334.38	4204.52	5074.66	6814.94	8555.22
1.681	2466.35	2901.81	3337.27	4208.19	5079.10	6820.94	8562.77
1.682	2468.46	2904.31	3340.16	4211.86	5083.34	6826.94	8570.33
1.683	2470.58	2906.81	3343.05	4215.53	5087.99	6832.94	8577.89
1.684	2472.70	2909.31	3345.94	4219.20	5092.44	6838.94	8585.45
1.685	2474.81	2911.82	3348.84	4222.87	5096.89	6844.95	8593.01
1.686	2476.92	2914.32	3351.73	4226.54	5101.34	6850.96	8600.57
1.687	2479.03	2916.83	3354.62	4230.21	5105.79	6856.97	8608.13
1.688	2481.15	2919.34	3357.52	4233.88	5110.24	6862.98	8615.70
1.689	2483.27	2921.85	3360.42	4237.55	5114.70	6868.99	8623.27
1.690	2485.39	2924.36	3363.32	4241.24	5119.16	6875.00	8630.84
1.691	2487.51	2926.86	3366.21	4244.91	5123.61	6881.01	8638.41
1.692	2489.63	2929.37	3369.11	4248.59	5128.07	6887.03	8645.99
1.693	2491.75	2931.88	3372.01	4252.27	5132.53	6893.05	8653.57
1.694	2493.87	2934.39	3374.91	4255.95	5136.99	6899.07	8661.15
1.695	2495.99	2936.90	3377.81	4259.63	5141.45	6905.09	8668.73
1.696	2498.11	2939.41	3380.71	4263.31	5145.91	6911.11	8676.31
1.697	2500.23	2941.92	3383.61	4266.99	5150.37	6917.13	8683.90
1.698	2502.35	2944.43	3386.51	4270.67	5154.88	6923.16	8691.49
1.699	2504.47	2946.94	3389.42	4274.36	5159.31	6929.18	8699.08
1.700	2506.60	2949.46	3392.33	4278.05	5163.78	6935.22	8706.67
1.701	2508.72	2951.97	3395.23	4281.74	5168.25	6941.25	8714.27
1.702	2510.84	2954.48	3398.13	4285.43	5172.72	6947.28	8721.87
1.703	2512.97	2957.00	3401.04	4289.12	5177.19	6953.32	8729.47
1.704	2515.10	2959.52	3403.95	4292.81	5181.66	6959.36	8737.07
1.705	2517.22	2962.04	3406.86	4296.50	5186.13	6965.40	8744.67
1.706	2519.34	2964.55	3409.77	4201.19	5190.60	6971.44	8752.28
1.707	2521.47	2967.07	3412.68	4204.88	5195.07	6977.48	8759.89
1.708	2523.60	2969.59	3415.59	4208.57	5199.55	6983.52	8767.50
1.709	2525.73	2972.11	3418.50	4212.26	5204.03	6989.57	8775.11
1.710	2527.86	2974.63	3421.41	4314.96	5208.51	6995.61	8782.72
1.711	2529.98	2977.15	3424.32	4318.65	5212.99	7001.66	8790.33
1.712	2532.10	2979.67	3427.23	4322.35	5217.47	7007.71	8797.94
1.713	2534.23	2982.19	3430.14	4326.05	5221.96	7013.76	8805.56
1.714	2536.36	2984.71	3433.06	4329.75	5226.45	7019.81	8813.18

Depth on Weir.	LENGTH OF THE WEIR.						
Feet.	6 Feet.	7 Feet.	8 Feet.	10 Feet.	12 Feet.	16 Feet.	20 Feet.
1.715	2538.50	2987.24	3435.98	4333.45	5230.94	7025.87	8820.82
1.716	2540.63	2989.76	3438.89	4337.15	5235.42	7031.93	8828.45
1.717	2542.76	2992.28	3441.80	4340.85	5239.90	7037.99	8836.08
1.718	2544.89	2994.80	3444.72	4344.55	5243.38	7044.05	8843.71
1.719	2547.03	2997.33	3447.64	4348.25	5247.87	7050.11	8851.34
1.720	2549.16	2999.86	3450.56	4351.96	5253.36	7056.17	8858.97
1.721	2551.29	3002.38	3453.48	4355.66	5257.85	7062.23	8866.61
1.722	2553.42	3004.91	3456.40	4359.37	5262.34	7066.30	8874.25
1.723	2555.55	3007.44	3459.32	4363.08	5266.84	7073.37	8881.89
1.724	2557.69	3009.97	3462.24	4366.79	5271.34	7079.43	8889.53
1.725	2559.83	3012.50	3465.17	4370.50	5275.84	7086.51	8897.17
1.726	2561.96	3015.03	3468.09	4374.21	5280.34	7092.58	8904.82
1.727	2563.10	3017.56	3471.01	4377.92	5284.84	7098.65	8912.47
1.728	2565.24	3020.09	3473.93	4381.63	5289.34	7104.72	8920.12
1.729	2567.38	3022.62	3476.86	4385.34	5293.84	7101.80	8927.77
1.730	2570.52	3025.15	3479.79	4389.06	5298.34	7116.88	8935.43
1.731	2572.65	3027.68	3482.71	4392.77	5302.84	7122.96	8943.09
1.732	2574.79	3030.21	3485.64	4396.49	5307.34	7129.04	8950.75
1.733	2576.93	3032.74	3488.57	4400.21	5311.85	7135.12	8958.41
1.734	2579.07	3035.28	3491.50	4403.93	5316.36	7141.21	8966.07
1.735	2581.21	3037.82	3494.43	4407.65	5320.87	7147.30	8973.74
1.736	2583.35	3040.35	3497.36	4411.36	5325.31	7153.39	8981.41
1.737	2585.49	3042.88	3500.29	4415.07	5329.84	7159.48	8989.08
1.738	2587.63	3045.42	3503.22	4418.78	5334.37	7165.57	8996.75
1.739	2589.78	3047.96	3506.15	4422.49	5338.90	7171.66	9004.42
1.740	2591.92	3050.50	3509.09	4426.20	5343.43	7177.76	9012.10
1.741	2594.06	3053.04	3512.02	4429.93	5347.94	7183.86	9019.78
1.742	2596.20	3055.58	3514.95	4433.67	5352.45	7189.96	9027.46
1.743	2598.34	3058.12	3517.88	4437.41	5356.97	7196.06	9035.14
1.744	2600.49	3060.66	3520.82	4441.15	5361.49	7202.16	9042.82
1.745	2602.64	3063.20	3523.76	4444.89	5366.01	7208.26	9050.51
1.746	2604.78	3065.74	3526.70	4448.62	5370.53	7214.36	9058.20
1.747	2606.92	3068.28	3529.64	4452.35	5375.05	7220.47	9065.89
1.748	2609.07	3070.82	3532.98	4456.08	5379.57	7226.58	9073.58
1.749	2611.22	3073.36	3535.52	4459.81	5384.10	7232.69	9081.28
1.750	2613.37	3075.91	3538.46	4463.55	5388.63	7238.80	9088.98
1.751	2615.51	3078.45	3541.40	4467.28	5393.76	7244.91	9096.68
1.752	2617.66	3080.99	3544.34	4471.01	5397.69	7251.02	9104.38
1.753	2619.81	3083.54	3547.28	4474.74	5402.22	7257.14	9112.08
1.754	2621.99	3086.09	3550.22	4478.47	5406.75	7263.26	9119.79
1.755	2624.11	3088.64	3553.17	4482.22	5411.28	7269.38	9127.50
1.756	2626.26	3091.18	3556.11	4485.96	5415.81	7275.51	9135.21
1.757	2628.41	3093.73	3559.05	4489.70	5420.34	7281.64	9142.92
1.758	2630.56	3096.28	3562.00	4493.44	5424.87	7287.77	9150.63
1.759	2632.71	3098.83	3564.95	4497.18	5429.41	7293.90	9158.38
1.760	2634.87	3101.38	3567.90	4500.92	5433.95	7300.01	9166.06
1.761	2637.01	3103.93	3570.84	4504.66	5438.39	7306.14	9173.78
1.762	2639.16	3106.48	3573.79	4508.40	5442.93	7312.27	9181.50
1.763	2641.31	3109.03	3576.74	4512.15	5447.47	7318.40	9189.22
1.764	2643.47	3111.58	3579.69	4515.90	5452.01	7324.53	9196.55
1.765	2645.64	3114.14	3582.64	4519.65	5456.66	7330.67	9204.68
1.766	2647.79	3116.69	3585.59	4523.40	5461.20	7336.81	9212.41
1.767	2649.94	3119.24	3588.54	4527.16	5465.76	7342.95	9220.14
1.768	2652.10	3121.79	3591.49	4530.90	5470.31	7349.05	9227.87
1.769	2654.26	3124.35	3594.44	4534.65	5474.86	7355.19	9235.61
1.770	2656.42	3126.91	3597.41	4538.40	5479.39	7361.37	9243.35
1.771	2658.57	3129.46	3600.36	4542.15	5484.94	7367.51	9251.09
1.772	2660.73	3132.02	3603.31	4545.90	5489.49	7373.66	9258.83

Depth on Weir.	LENGTH OF THE WEIR.						
Feet.	6 Feet.	7 Feet.	8 Feet.	10 Feet.	12 Feet.	16 Feet.	20 Feet.
1.773	2662.89	3134.58	3606.27	4549.65	5494.04	7379.81	9266.57
1.774	2665.05	3137.14	3609.23	4553.41	5498.59	7385.96	9274.32
1.775	2667.21	3139.70	3612.19	4557.17	5502.15	7392.11	9282.07
1.776	2669.37	3142.26	3615.15	4560.93	5506.70	7398.26	9289.82
1.777	2671.43	3144.82	3618.11	4564.69	5511.26	7404.42	9297.57
1.778	2673.59	3147.38	3621.07	4568.45	5515.82	7410.58	9305.33
1.779	2675.75	3149.94	3624.03	4572.21	5520.38	7416.74	9313.09
1.780	2678.01	3152.50	3626.99	4575.97	5524.94	7422.90	9320.85
1.781	2680.17	3155.06	3629.95	4579.73	5529.50	7429.06	9328.61
1.782	2682.33	3157.62	3632.91	4583.49	5534.06	7435.22	9336.37
1.783	2684.49	3160.18	3635.87	4587.25	5538.62	7441.38	9344.13
1.784	2686.66	3162.75	3638.83	4591.01	5543.19	7447.54	9351.90
1.785	2688.83	3165.32	3641.80	4594.78	5547.76	7453.71	9359.67
1.786	2690.99	3167.88	3644.76	4598.54	5552.33	7459.88	9367.44
1.787	2693.15	3170.44	3647.73	4602.31	5556.90	7466.05	9375.21
1.788	2695.31	3173.01	3650.70	4606.08	5561.47	7472.22	9382.99
1.789	2697.47	3175.58	3653.67	4609.85	5566.04	7478.40	9390.77
1.790	2699.64	3178.15	3656.64	4613.62	5570.61	7484.58	9398.55
1.791	2701.81	3180.71	3659.61	4617.39	5575.18	7490.76	9406.33
1.792	2703.98	3183.28	3662.58	4621.16	5579.75	7496.94	9414.11
1.793	2706.15	3185.85	3665.55	4624.93	5584.32	7503.12	9421.89
1.794	2708.32	3188.42	3668.52	4628.71	5588.90	7509.30	9429.68
1.795	2710.49	3190.99	3671.49	4632.49	5593.48	7515.48	9437.47
1.796	2712.66	3193.56	3674.46	4636.26	5598.06	7521.66	9445.26
1.797	2714.83	3196.13	3677.43	4640.03	5602.64	7527.85	9453.05
1.798	2717.00	3198.70	3680.40	4643.81	5607.22	7534.04	9460.85
1.799	2719.17	3201.27	3683.38	4647.59	5611.80	7540.23	9468.65
1.800	2721.34	3203.85	3686.36	4651.37	5616.39	7546.42	9476.45
1.801	2723.51	3206.42	3689.33	4655.15	5620.97	7552.61	9484.26
1.802	2725.68	3208.99	3692.30	4658.93	5625.55	7558.80	9492.07
1.803	2727.85	3211.57	3695.28	4662.61	5630.14	7565.00	9499.89
1.804	2730.02	3214.15	3698.26	4666.79	5634.73	7571.20	9507.71
1.805	2732.20	3216.73	3701.24	4670.28	5639.32	7577.40	9515.48
1.806	2734.37	3219.30	3704.22	4674.06	5643.91	7583.60	9523.29
1.807	2736.54	3221.87	3707.20	4677.85	5648.50	7589.86	9531.10
1.808	2738.72	3224.45	3710.18	4681.64	5653.09	7596.00	9538.92
1.809	2740.90	3228.03	3713.16	4685.43	5657.69	7602.21	9546.74
1.810	2743.08	3229.61	3716.15	4689.22	5662.29	7608.42	9554.56
1.811	2745.25	3232.19	3718.13	4693.01	5666.88	7614.63	9562.38
1.812	2747.43	3234.77	3721.11	4696.80	5671.47	7620.87	9570.20
1.813	2749.61	3237.35	3724.09	4700.59	5676.07	7627.07	9578.03
1.814	2751.79	3239.93	3727.08	4704.38	5680.67	7633.27	9585.86
1.815	2753.97	3242.52	3731.07	4708.17	5685.27	7639.48	9593.69
1.816	2756.15	3245.10	3734.05	4711.96	5689.87	7645.70	9601.52
1.817	2758.33	3247.68	3737.04	4715.75	5694.47	7651.92	9609.45
1.818	2760.51	3250.26	3740.03	4719.55	5699.07	7658.14	9617.39
1.819	2762.69	3253.84	3743.02	4723.35	5703.68	7664.36	9625.23
1.820	2764.87	3255.43	3746.01	4727.15	5708.29	7670.58	9632.87
1.821	2767.05	3258.01	3749.00	4730.95	5712.90	7676.81	9640.71
1.822	2769.23	3260.60	3751.99	4734.75	5717.51	7683.05	9648.55
1.823	2771.41	3263.19	3754.98	4738.55	5722.12	7689.29	9656.40
1.824	2773.59	3265.78	3757.97	4742.35	5726.73	7695.53	9664.25
1.825	2775.78	3268.37	3760.96	4746.15	5731.34	7701.72	9672.10
1.826	2777.96	3270.95	3763.95	4749.85	5735.95	7707.95	9679.95
1.827	2779.14	3273.54	3766.94	4753.65	5740.56	7704.18	9687.80
1.828	2781.32	3276.13	3769.93	4757.45	5745.18	7720.42	9695.66
1.829	2783.50	3278.72	3773.93	4761.26	5749.80	7726.66	9703.52
1.830	2786.69	3281.31	3775.93	4765.17	5754.42	7732.90	9711.38

Depth on Weir.	LENGTH OF THE WEIR.						
Feet.	6 Feet.	7 Feet.	8 Feet.	10 Feet.	12 Feet.	16 Feet.	20 Feet.
1.831	2788.87	3283.80	3778.92	4768.98	5759.04	7739.14	9719.24
1.832	2791.05	3286.39	3781.92	4772.80	5763.66	7745.38	9727.10
1.833	2793.24	3288.98	3784.92	4776.62	5768.28	7751.62	9734.97
1.834	2795.43	3291.57	3787.92	4780.44	5772.90	7757.86	9742.84
1.835	2797.62	3294.27	3790.92	4784.22	5777.52	7764.11	9750.71
1.836	2799.81	3296.86	3793.92	4788.03	5782.14	7770.36	9758.58
1.837	2802.00	3299.46	3796.92	4791.84	5786.76	7776.61	9766.45
1.838	2804.19	3302.06	3799.92	4795.65	5791.39	7782.86	9774.33
1.839	2806.38	3304.66	3702.92	4799.47	5796.02	7789.11	9782.21
1.840	2808.57	3307.25	3805.93	4803.29	5800.65	7795.37	9790.09
1.841	2810.76	3310.84	3808.93	4807.10	5805.28	7801.63	9797.97
1.842	2812.95	3313.44	3811.93	4810.92	5809.91	7807.89	9805.86
1.843	2815.04	3316.04	3814.93	4814.74	5814.54	7814.15	9813.75
1.844	2817.23	3318.64	3817.94	4818.56	5819.17	7820.41	9821.64
1.845	2819.52	3320.24	3820.95	4822.38	5823.81	7826.67	9829.53
1.846	2821.71	3322.84	3823.95	4826.20	5828.44	7832.93	9837.41
1.847	2823.90	3325.44	3826.96	4830.02	5833.07	7839.19	9845.30
1.848	2825.09	3328.04	3829.97	4833.84	5837.71	7845.46	9853.19
1.849	2827.29	3330.64	3832.98	4837.66	5842.35	7851.73	9861.09
1.850	2830.49	3333.24	3835.90	4841.49	5846.99	7858.00	9869.00
1.851	2832.68	3335.84	3838.00	4845.31	5851.63	7864.27	9876.90
1.852	2834.87	3338.44	3841.01	4849.14	5856.27	7870.54	9884.81
1.853	2837.09	3341.04	3844.02	4852.97	5860.91	7876.81	9892.72
1.854	2839.29	3343.65	3847.03	4856.80	5865.56	7883.09	9900.63
1.855	2841.47	3346.26	3851.05	4860.63	5870.21	7889.37	9908.54
1.856	2843.66	3348.86	3854.06	4864.46	5874.85	7895.65	9916.45
1.857	2845.85	3351.46	3857.07	4868.29	5879.50	7901.93	9924.36
1.858	2848.05	3354.01	3860.08	4872.12	5884.15	7908.21	9932.28
1.859	2850.25	3356.68	3863.10	4875.95	5888.80	7914.49	9940.20
1.860	2852.45	3359.29	3866.12	4879.79	5893.45	7920.78	9948.12
1.861	2854.65	3361.90	3869.13	4883.62	5897.10	7927.07	9956.04
1.862	2856.85	3364.51	3872.15	4887.46	5901.75	7933.36	9963.97
1.863	2859.05	3367.12	3875.17	4891.30	5906.41	7939.65	9971.90
1.864	2861.25	3369.73	3878.19	4895.14	5911.07	7945.94	9979.83
1.865	2863.46	3372.34	3881.21	4898.97	5916.73	7952.24	9987.75
1.866	2865.66	3374.95	3884.23	4902.81	5921.38	7958.53	9995.68
1.867	2867.86	3377.56	3887.25	4906.65	5926.04	7964.83	10003.61
1.868	2870.06	3380.17	3890.27	4910.49	5930.70	7971.13	10011.54
1.869	2872.26	3382.78	3893.30	4914.33	5935.36	7977.43	10019.47
1.870	2874.47	3385.39	3896.32	4918.17	5940.02	7983.73	10027.43
1.871	2876.67	3388.00	3899.34	4922.01	5944.68	7990.03	10035.37
1.872	2878.87	3390.61	3902.36	4925.85	5949.34	7996.33	10043.31
1.873	2881.08	3393.22	3905.38	4929.70	5954.01	8002.64	10051.26
1.874	2883.28	3395.86	3908.41	4933.55	5958.68	8008.95	10059.21
1.875	2885.49	3398.47	3911.44	4937.40	5963.35	8015.26	10067.16
1.876	2887.69	3301.08	3914.47	4941.24	5968.02	8021.57	10075.11
1.877	2889.89	3303.69	3917.50	4945.09	5972.69	8027.88	10082.06
1.878	2892.10	3306.31	3920.53	4948.94	5977.36	8034.19	10090.02
1.879	2894.31	3308.93	3923.56	4952.79	5982.03	8040.50	10097.98
1.880	2896.52	3411.55	3926.59	4956.64	5986.70	8046.82	10106.94
1.881	2898.73	3414.17	3929.62	4960.49	5991.37	8053.14	10114.90
1.882	2900.94	3416.79	3932.65	4964.34	5996.05	8059.46	10122.86
1.883	2903.15	3419.41	3935.68	4968.19	6000.73	8065.78	10130.83
1.884	2905.36	3422.03	3938.72	4972.05	6005.41	8072.00	10138.80
1.885	2907.57	3424.66	3941.75	4975.91	6010.09	8078.43	10146.77
1.886	2909.78	3427.68	3944.78	4979.77	6014.77	8084.75	10154.74
1.887	2911.99	3430.30	3947.81	4983.63	6018.45	8091.08	10162.71
1.888	2914.20	3432.92	3950.84	4987.49	6023.13	8097.41	10170.69

Depth on Weir.	LENGTH OF THE WEIR.						
Feet.	6 Feet.	7 Feet.	8 Feet.	10 Feet.	12 Feet.	16 Feet.	20 Feet.
-1.889	2916.41	3435.54	3953.88	4991.35	6028.81	8103.74	10178.67
1.890	2918.63	3437.77	3956.92	4995.21	6033.50	8110.07	10186.65
1.891	2920.84	3440.39	3959.95	4999.07	6038.18	8117.10	10194.63
1.892	2923.05	3443.01	3962.99	5002.93	6042.86	8123.44	10202.61
1.893	2925.27	3445.64	3966.03	5006.79	6047.55	8129.78	10210.60
1.894	2927.49	3448.27	3969.07	5010.65	6052.24	8136.12	10218.59
1.895	2929.70	3450.90	3972.11	5014.52	6056.93	8141.76	10226.58
1.896	2931.91	3453.53	3975.15	5018.38	6061.62	8148.10	10234.57
1.897	2934.12	3456.16	3978.19	5022.25	6066.31	8154.44	10242.56
1.898	2936.34	3458.79	3981.23	5026.12	6071.00	8160.78	10250.56
1.899	2938.56	3461.42	3984.27	5029.99	6075.70	8167.13	10258.56
1.900	2940.78	3464.05	3987.32	5033.86	6080.40	8173.48	10266.56
1.901	2942.99	3466.68	3990.36	5037.73	6085.09	8180.83	10274.56
1.902	2945.21	3469.31	3993.40	5041.60	6089.79	8187.18	10282.56
1.903	2947.43	3471.94	3996.44	5045.47	6094.49	8193.53	10290.57
1.904	2949.65	3474.57	3999.49	5049.34	6099.91	8199.88	10298.58
1.905	2951.87	3477.21	4002.54	5053.22	6103.89	8205.24	10306.59
1.906	2954.09	3479.84	4005.58	5057.09	6108.59	8211.60	10314.60
1.907	2956.31	3482.47	4008.63	5060.96	6113.29	8217.96	10322.61
1.908	2958.53	3485.10	4011.68	5064.84	6118.06	8224.32	10330.62
1.909	2960.75	3487.74	4014.73	5068.72	6122.77	8230.68	10338.64
1.910	2962.97	3490.38	4017.78	5072.60	6127.41	8237.04	10346.66
1.911	2965.19	3493.01	4020.83	5076.48	6132.12	8243.40	10354.68
1.912	2967.41	3495.64	4023.88	5080.36	6136.83	8249.76	10362.70
1.913	2969.63	3498.28	4026.93	5084.26	6141.54	8256.13	10370.73
1.914	2971.89	3500.92	4029.98	5088.14	6146.25	8262.50	10378.76
1.915	2974.08	3503.56	4033.04	5092.00	6150.96	8268.87	10386.79
1.916	2976.30	3506.20	4036.09	5095.88	6155.67	8275.24	10394.82
1.917	2978.52	3508.84	4039.14	5099.76	6160.38	8281.61	10402.85
1.918	2980.74	3511.48	4042.19	5103.64	6165.09	8287.99	10410.88
1.919	2982.97	3514.12	4045.25	5107.53	6169.81	8294.37	10418.92
1.920	2985.20	3516.76	4048.31	5111.42	6174.53	8300.75	10426.96
1.921	2987.42	3518.80	4051.36	5115.31	6179.25	8307.13	10435.00
1.922	2989.65	3521.44	4054.42	5119.20	6183.97	8313.51	10443.04
1.923	2991.88	3524.08	4057.48	5123.09	6188.69	8319.89	10451.08
1.924	2994.11	3526.72	4060.54	5126.98	6193.41	8326.27	10459.13
1.925	2996.34	3529.97	4063.60	5130.87	6198.13	8332.66	10467.18
1.926	2998.57	3532.61	4066.66	5134.76	6202.85	8339.05	10475.23
1.927	3000.80	3535.25	4069.72	5138.65	6207.57	8345.44	10483.28
1.928	3003.03	3537.90	4072.78	5142.54	6212.30	8351.83	10491.34
1.929	3005.26	3540.55	4075.84	5146.44	6217.03	8358.22	10499.40
1.930	3007.49	3543.20	4078.91	5150.34	6221.76	8364.61	10507.46
1.931	3009.72	3545.84	4081.97	5154.23	6226.49	8371.00	10515.52
1.932	3011.95	3548.49	4085.03	5158.13	6231.22	8377.40	10523.58
1.933	3014.18	3551.14	4088.09	5162.03	6235.95	8383.80	10531.64
1.934	3016.41	3553.79	4091.16	5165.93	6240.68	8390.20	10539.71
1.935	3018.64	3556.44	4094.23	5169.83	6245.42	8396.60	10547.78
1.936	3020.87	3559.09	4097.29	5173.73	6250.15	8403.00	10555.85
1.937	3023.10	3561.74	4100.36	5177.63	6254.88	8409.40	10563.92
1.938	3025.33	3564.39	4103.43	5181.53	6259.62	8415.80	10571.99
1.939	3027.57	3567.04	4036.50	5185.43	6264.36	8422.21	10580.07
1.940	3029.81	3569.69	4109.57	5189.34	6269.10	8428.62	10588.15
1.941	3032.04	3572.34	4112.64	5193.24	6273.84	8435.03	10596.23
1.942	3034.27	3574.99	4115.71	5197.14	6278.58	8441.44	10604.31
1.943	3036.51	3577.64	4118.78	5101.05	6283.32	8447.85	10612.39
1.944	3038.75	3580.30	4121.85	5104.96	6288.06	8454.27	10620.47
1.945	3040.99	3582.96	4124.93	5208.87	6292.81	8460.69	10628.56
1.946	3043.22	3585.61	4128.00	5212.78	6297.55	8467.11	10636.65

Depth on Weir.	LENGTH OF THE WEIR.						
Feet.	6 Feet.	7 Feet.	8 Feet.	10 Feet.	12 Feet.	16 Feet.	20 Feet.
1.947	3045.46	3588.26	4131.07	5216.69	6302.29	8473.53	10644.74
1.948	3047.70	3590.92	4134.14	5220.60	6307.04	8479.95	10652.83
1.949	3049.94	3593.58	4137.22	5224.51	6311.79	8486.37	10660.93
1.950	3052.18	3596.24	4140.30	5228.42	6316.54	8492.79	10669.03
1.951	3054.42	3598.90	4143.37	5232.33	6321.29	8499.21	10677.13
1.952	3056.66	3601.56	4146.45	5236.24	6326.04	8505.64	10685.23
1.953	3058.90	3604.22	4149.53	5240.16	6330.79	8512.07	10693.33
1.954	3061.14	3606.88	4152.61	5244.08	6335.55	8518.50	10701.44
1.955	3063.38	3609.54	4155.69	5248.00	6340.31	8524.93	10709.55
1.956	3065.62	3612.20	4158.77	5251.92	6345.06	8531.36	10717.66
1.957	3067.86	3614.86	4161.85	5255.84	6349.82	8537.79	10725.77
1.958	3070.10	3617.52	4164.93	5259.76	6354.58	8544.23	10733.88
1.959	3072.34	3620.18	4168.01	5263.68	6359.34	8550.67	10741.99
1.960	3074.59	3622.84	4171.09	5267.60	6364.10	8557.11	10750.11
1.961	3076.83	3625.50	4174.17	5271.52	6368.86	8563.55	10758.23
1.962	3079.07	3628.16	4177.25	5275.44	6373.52	8569.99	10766.35
1.963	3081.32	3630.83	4180.34	5279.36	6378.28	8576.43	10774.47
1.964	3083.57	3633.50	4183.43	5283.29	6383.05	8582.87	10782.59
1.965	3085.82	3636.17	4186.52	5287.22	6387.92	8589.32	10790.72
1.966	3088.06	3638.83	4189.60	5291.14	6392.68	8595.75	10798.85
1.967	3090.30	3641.49	4192.68	5295.07	6397.45	8602.20	10706.98
1.968	3092.55	3644.16	4195.77	5299.00	6402.22	8608.65	10715.11
1.969	3094.80	3646.83	4198.86	5302.93	6406.99	8615.10	10723.24
1.970	3097.05	3649.50	4201.95	5306.86	6411.76	8621.57	10631.38
1.971	3099.29	3652.17	4205.04	5310.79	6416.53	8629.02	10839.52
1.972	3100.54	3654.84	4208.13	5314.72	6421.30	8635.48	10847.66
1.973	3103.79	3657.51	4211.22	5318.65	6426.07	8641.94	10855.80
1.974	3106.04	3660.18	4214.31	5322.58	6430.85	8648.40	10863.95
1.975	3108.29	3662.85	4217.40	5326.52	6435.63	8653.86	10872.10
1.976	3110.54	3665.52	4020.49	5330.45	6440.41	8660.32	10880.25
1.977	3112.79	3668.19	4023.58	5334.38	6445.19	8666.78	10888.75
1.978	3115.04	3670.86	4026.68	5338.32	6449.97	8673.25	10896.90
1.979	3117.29	3673.53	4029.78	5342.26	6454.75	8679.72	10905.05
1.980	3119.54	3676.21	4232.88	5346.20	6459.53	8686.19	10912.85
1.981	3121.79	3678.88	4235.97	5350.14	6464.31	8692.66	10921.01
1.982	3124.04	3681.55	4239.06	5354.08	6469.09	8699.13	10929.17
1.983	3126.29	3684.22	4242.16	5358.02	6473.87	8705.60	10937.33
1.984	3128.55	3686.90	4245.26	5361.96	6478.66	8712.07	10945.49
1.985	3130.81	3689.58	4248.36	5365.91	6483.45	8718.55	10953.65
1.986	3133.06	3692.25	4251.46	5369.85	6488.24	8725.03	10961.82
1.987	3135.36	3694.93	4254.56	5373.79	6493.03	8731.51	10969.99
1.988	3137.57	3697.61	4257.66	5377.73	6497.82	8737.99	10978.16
1.989	3139.83	3700.29	4260.76	5381.68	6402.61	8744.47	10986.33
1.990	3142.09	3702.97	4263.86	5385.63	6507.41	8750.95	10994.50
1.991	3144.34	3705.59	4266.96	5389.58	6512.20	8757.43	10102.68
1.992	3146.59	3708.21	4270.06	5393.53	6516.99	8763.92	10112.68
1.993	3148.85	3710.83	4273.17	5397.48	6521.78	8770.41	10120.86
1.994	3151.11	3713.45	4276.28	5401.43	6526.57	8776.90	10129.04
1.995	3153.37	3716.37	4279.38	5405.38	6531.38	8783.39	11035.40
1.996	3155.63	3719.05	4282.48	5409.33	6536.18	8789.88	11043.59
1.997	3157.89	3721.73	4285.58	5413.28	6540.98	8796.37	11051.78
1.998	3160.15	3724.41	4288.69	5417.23	6545.78	8802.87	11059.97
1.999	3162.41	3727.10	4291.80	5421.19	6550.58	8808.37	11068.16
2.000	3164.67	3729.79	4294.91	5425.15	6555.39	8815.87	11076.35

TABLES

FOR FACILITATING THE COMPUTATION OF THE QUANTITY OF WATER FLOWING OVER WEIRS.

TABLE I.

To attain the greatest exactness, it is necessary to take account of the velocity of the water approaching the weir. The method adopted at Lowell for this purpose is to make a correction for it in the observed depth on the weir, by the formula

$$H' = \left[(H+h)^{\frac{3}{2}} - h^{\frac{3}{2}} \right]^{\frac{2}{3}} ;$$

in which

$H =$ the observed depth on the weir.

$h =$ the head due the mean velocity approaching the weir.

$H' =$ the corrected depth on the weir.

By developing into series and omitting the terms containing powers of $\frac{h}{H}$ above the first, h being always very small, relatively to H, this formula may, without sensible error, be put under the simpler form,

$$H' = H + h - \tfrac{2}{3}\sqrt{\tfrac{h^3}{H}}.$$

The mean velocity of the water approaching the weir is usually found, with sufficient exactness, by computing the discharge, approximately, from the observed depth on the weir, and dividing it by the section of the channel approaching the weir, the quotient being the velocity; the head due this velocity, or h, is found by Table I., which is computed by the formula,

$$h = \frac{V^2}{2\,g},$$

in which

$V =$ the mean velocity.

$g =$ the velocity acquired by a body at the end of the first second of its fall, in a vacuum; its value, for Lowell, being 32.1618.

TABLE II.

This is computed by the formula

$$Q = 3.33 \, (L - 0.1n \; H) \; H^{\frac{3}{2}},$$

in which

$Q =$ the quantity of water discharged, in cubic feet per second.

$L =$ the length of the weir in feet.

$H =$ the depth on the weir in feet, being the height of the surface of the water above the top of the weir, taken far enough from the weir to be unaffected by the curvature caused by the discharge, and corrected, if necessary, for the velocity of the water approaching the weir.

$n =$ the number of end contractions.

In computing the table, L is taken equal to 1, and n equal to 0.

The actual length of the weir being known, it is to be corrected for the end contractions, if any, by deducting from it one-tenth of the depth on the weir for each end contraction. If the length of the weir is the same as the width of the canal approaching it, there is no end contraction, and of course nothing to be deducted from the length of the weir. The discharge, as given by the table, multiplied by the length of the weir, corrected, if necessary, as above, gives the quantity of water discharged by the weir.

HEADS, IN FEET, DUE TO VELOCITIES FROM 0 TO 4.99 FEET PER SECOND.

Veloc'y	0	1	2	3	4	5	6	7	8	9
0.0	0.0000	0.0000	0.0000	0.0000	0.0000	0.0000	0.0001	0.0001	0.0001	0.0001
.1	0.0002	0.0002	0.0002	0.0003	0.0003	0.0003	0.0004	0.0004	0.0005	0.0006
.2	0.0006	0.0007	0.0008	0.0008	0.0009	0.0010	0.0011	0.0011	0.0012	0.0013
.3	0.0014	0.0015	0.0016	0.0017	0.0018	0.0019	0.0020	0.0021	0.0022	0.0024
.4	0.0025	0.0026	0.0027	0.0029	0.0030	0.0031	0.0033	0.0034	0.0036	0.0037
.5	0.0039	0.0040	0.0042	0.0044	0.0045	0.0047	0.0049	0.0051	0.0052	0.0054
.6	0.0056	0.0058	0.0060	0.0062	0.0064	0.0066	0.0068	0.0070	0.0072	0.0074
.7	0.0076	0.0078	0.0081	0.0083	0.0085	0.0087	0.0090	0.0092	0.0095	0.0097
.8	0.0099	0.0102	0.0105	0.0107	0.0110	0.0112	0.0115	0.0118	0.0120	0.0123
.9	0.0126	0.0129	0.0132	0.0134	0.0137	0.0140	0.0143	0.0146	0.0149	0.0152
1.0	0.0155	0.0159	0.0162	0.0165	0.0168	0.0171	0.0175	0.0178	0.0181	0.0185
.1	0.0188	0.0192	0.0195	0.0199	0.0202	0.0206	0.0209	0.0213	0.0216	0.0220
.2	0.0224	0.0228	0.0231	0.0235	0.0239	0.0243	0.0247	0.0251	0.0255	0.0259
.3	0.0263	0.0267	0.0271	0.0275	0.0279	0.0283	0.0288	0.0292	0.0296	0.0300
.4	0.0305	0.0309	0.0313	0.0318	0.0322	0.0327	0.0331	0.0336	0.0341	0.0345
.5	0.0350	0.0354	0.0359	0.0364	0.0369	0.0374	0.0378	0.0383	0.0388	0.0393
.6	0.0398	0.0403	0.0408	0.0413	0.0418	0.0423	0.0428	0.0434	0.0439	0.0444
.7	0.0449	0.0455	0.0460	0.0465	0.0471	0.0476	0.0482	0.0487	0.0493	0.0498
.8	0.0504	0.0509	0.0515	0.0521	0.0526	0.0532	0.0538	0.0544	0.0549	0.0555
.9	0.0561	0.0567	0.0573	0.0579	0.0585	0.0591	0.0597	0.0603	0.0609	0.0616
2.0	0.0622	0.0628	0.0634	0.0641	0.0647	0.0653	0.0660	0.0666	0.0673	0.0679
.1	0.0686	0.0692	0.0699	0.0705	0.0712	0.0719	0.0725	0.0732	0.0739	0.0746
.2	0.0752	0.0759	0.0766	0.0773	0.0780	0.0787	0.0794	0.0801	0.0808	0.0815
.3	0.0822	0.0830	0.0837	0.0844	0.0851	0.0859	0.0866	0.0873	0.0881	0.0888
.4	0.0895	0.0903	0.0910	0.0918	0.0926	0.0933	0.0941	0.0948	0.0956	0.0964
.5	0.0972	0.0979	0.0987	0.0995	0.1003	0.1011	0.1019	0.1027	0.1035	0.1043
.6	0.1051	0.1059	0.1067	0.1075	0.1084	0.1092	0.1100	0.1108	0.1117	0.1125
.7	0.1133	0.1142	0.1150	0.1159	0.1167	0.1176	0.1184	0.1193	0.1201	0.1210
.8	0.1219	0.1228	0.1236	0.1245	0.1254	0.1263	0.1272	0.1281	0.1289	0.1298
.9	0.1307	0.1316	0.1326	0.1335	0.1344	0.1353	0.1362	0.1371	0.1381	0.1390
3.0	0.1399	0.1409	0.1418	0.1427	0.1437	0.1446	0.1456	0.1465	0.1475	0.1484
.1	0.1494	0.1504	0.1513	0.1523	0.1533	0.1543	0.1552	0.1562	0.1572	0.1582
.2	0.1592	0.1602	0.1612	0.1622	0.1632	0.1642	0.1652	0.1662	0.1673	0.1683
.3	0.1693	0.1703	0.1714	0.1721	0.1734	0.1745	0.1755	0.1766	0.1776	0.1787
.4	0.1797	0.1808	0.1818	0.1829	0.1840	0.1850	0.1861	0.1872	0.1883	0.1894
.5	0.1904	0.1915	0.1926	0.1937	0.1948	0.1959	0.1970	0.1981	0.1992	0.2004
.6	0.2015	0.2026	0.2037	0.2049	0.2060	0.2071	0.2083	0.2094	0.2105	0.2117
.7	0.2128	0.2140	0.2151	0.2163	0.2175	0.2186	0.2198	0.2210	0.2221	0.2233
.8	0.2245	0.2257	0.2269	0.2280	0.2292	0.2304	0.2316	0.2328	0.2340	0.2352
.9	0.2365	0.2377	0.2389	0.2401	0.2413	0.2426	0.2438	0.2450	0.2463	0.2475
4.0	0.2487	0.2500	0.2512	0.2525	0.2537	0.2550	0.2563	0.2575	0.2588	0.2601
.1	0.2613	0.2626	0.2639	0.2652	0.2665	0.2677	0.2690	0.2703	0.2716	0.2729
.2	0.2742	0.2755	0.2769	0.2782	0.2795	0.2808	0.2821	0.2835	0.2848	0.2861
.3	0.2875	0.2888	0.2901	0.2915	0.2928	0.2942	0.2955	0.2969	0.2982	0.2996
.4	0.3010	0.3023	0.3037	0.3051	0.3065	0.3079	0.3092	0.3106	0.3120	0.3134
.5	0.3148	0.3162	0.3176	0.3190	0.3204	0.3218	0.3233	0.3247	0.3261	0.3275
.6	0.3290	0.3304	0.3318	0.3333	0.3347	0.3362	0.3376	0.3390	0.3405	0.3420
.7	0.3434	0.3449	0.3463	0.3478	0.3493	0.3508	0.3522	0.3537	0.3552	0.3567
.8	0.3582	0.3597	0.3612	0.3627	0.3642	0.3657	0.3672	0.3687	0.3702	0.3717
.9	0.3733	0.3748	0.3763	0.3779	0.3794	0.3809	0.3825	0.3840	0.3856	0.3871

DISCHARGE, IN CUBIC FEET PER SECOND, OF A WEIR ONE FOOT LONG, WITH-
OUT CONTRACTION AT THE ENDS; FOR DEPTHS FROM 0 TO 0.499 FEET.

Depth.	0	1	2	3	4	5	6	7	8	9
0.00	0.0000	0.0001	0.0003	0.0005	0.0008	0.0012	0.0015	0.0020	0.0024	0.0028
.01	0.0033	0.0038	0.0044	0.0049	0.0055	0.0061	0.0067	0.0074	0.0080	0.0087
.02	0.0094	0.0101	0.0109	0.0116	0.0124	0.0132	0.0140	0.0148	0.0156	0.0164
.03	0.0173	0.0182	0.0191	0.0200	0.0209	0.0218	0.0227	0.0237	0.0247	0.0256
.04	0.0266	0.0276	0.0287	0.0297	0.0307	0.0318	0.0329	0.0339	0.0350	0.0361
.05	0.0372	0.0384	0.0395	0.0406	0.0418	0.0430	0.0441	0.0453	0.0465	0.0477
.06	0.0489	0.0502	0.0514	0.0527	0.0539	0.0552	0.0565	0.0578	0.0590	0.0604
.07	0.0617	0.0630	0.0643	0.0657	0.0670	0.0684	0.0698	0.0712	0.0725	0.0739
.08	0.0753	0.0768	0.0782	0.0796	0.0811	0.0825	0.0840	0.0855	0.0869	0.0884
.09	0.0899	0.0914	0.0929	0.0944	0.0960	0.0975	0.0990	0.1006	0.1022	0.1037
0.10	0.1053	0.1069	0.1085	0.1101	0.1117	0.1133	0.1149	0.1166	0.1182	0.1198
.11	0.1215	0.1231	0.1248	0.1265	0.1282	0.1299	0.1316	0.1333	0.1350	0.1367
.12	0.1384	0.1402	0.1419	0.1436	0.1454	0.1472	0.1489	0.1507	0.1525	0.1543
.13	0.1561	0.1579	0.1597	0.1615	0.1633	0.1652	0.1670	0.1689	0.1707	0.1726
.14	0.1744	0.1763	0.1782	0.1801	0.1820	0.1839	0.1858	0.1877	0.1896	0.1915
.15	0.1935	0.1954	0.1973	0.1993	0.2012	0.2032	0.2052	0.2072	0.2091	0.2111
.16	0.2131	0.2151	0.2171	0.2191	0.2212	0.2232	0.2252	0.2273	0.2293	0.2314
.17	0.2334	0.2355	0.2375	0.2396	0.2417	0.2438	0.2459	0.2480	0.2501	0.2522
.18	0.2543	0.2564	0.2586	0.2607	0.2628	0.2650	0.2671	0.2693	0.2714	0.2736
.19	0.2758	0.2780	0.2802	0.2823	0.2845	0.2867	0.2890	0.2912	0.2934	0.2956
0.20	0.2978	0.3001	0.3023	0.3046	0.3068	0.3091	0.3113	0.3136	0.3159	0.3182
.21	0.3205	0.3228	0.3250	0.3274	0.3297	0.3320	0.3343	0.3366	0.3389	0.3413
.22	0.3436	0.3460	0.3483	0.3507	0.3530	0.3554	0.3578	0.3601	0.3625	0.3649
.23	0.3673	0.3697	0.3721	0.3745	0.3769	0.3794	0.3818	0.3842	0.3866	0.3891
.24	0.3915	0.3940	0.3964	0.3989	0.4014	0.4038	0.4063	0.4088	0.4113	0.4138
.25	0.4162	0.4187	0.4213	0.4238	0.4263	0.4288	0.4313	0.4339	0.4364	0.4389
.26	0.4415	0.4440	0.4466	0.4491	0.4517	0.4543	0.4568	0.4594	0.4620	0.4646
.27	0.4672	0.4698	0.4724	0.4750	0.4776	0.4802	0.4828	0.4855	0.4881	0.4907
.28	0.4934	0.4960	0.4987	0.5013	0.5040	0.5067	0.5093	0.5120	0.5147	0.5174
.29	0.5200	0.5227	0.5254	0.5281	0.5308	0.5336	0.5363	0.5390	0.5417	0.5444
0.30	0.5472	0.5499	0.5527	0.5554	0.5582	0.5609	0.5637	0.5664	0.5692	0.5720
.31	0.5748	0.5775	0.5803	0.5831	0.5859	0.5887	0.5915	0.5943	0.5972	0.6000
.32	0.6028	0.6056	0.6085	0.6113	0.6141	0.6170	0.6198	0.6227	0.6255	0.6284
.33	0.6313	0.6341	0.6370	0.6399	0.6428	0.6457	0.6486	0.6515	0.6544	0.6573
.34	0.6602	0.6631	0.6660	0.6689	0.6719	0.6748	0.6777	0.6807	0.6836	0.6866
.35	0.6895	0.6925	0.6954	0.6984	0.7014	0.7043	0.7073	0.7103	0.7133	0.7163
.36	0.7193	0.7223	0.7253	0.7283	0.7313	0.7343	0.7373	0.7404	0.7434	0.7464
.37	0.7495	0.7525	0.7555	0.7586	0.7616	0.7647	0.7678	0.7708	0.7739	0.7770
.38	0.7800	0.7831	0.7862	0.7893	0.7924	0.7955	0.7986	0.8017	0.8048	0.8079
.39	0.8110	0.8142	0.8173	0.8204	0.8235	0.8267	0.8298	0.8330	0.8361	0.8393
0.40	0.8424	0.8456	0.8488	0.8519	0.8551	0.8583	0.8615	0.8646	0.8678	0.8710
.41	0.8742	0.8774	0.8806	0.8838	0.8870	0.8903	0.8935	0.8967	0.8999	0.9032
.42	0.9064	0.9096	0.9129	0.9161	0.9194	0.9226	0.9259	0.9292	0.9324	0.9357
.43	0.9390	0.9422	0.9455	0.9488	0.9521	0.9554	0.9587	0.9620	0.9653	0.9686
.44	0.9719	0.9752	0.9785	0.9819	0.9852	0.9885	0.9919	0.9952	0.9985	1.0019
.45	1.0052	1.0086	1.0119	1.0153	1.0187	1.0220	1.0254	1.0288	1.0321	1.0355
.46	1.0389	1.0423	1.0457	1.0491	1.0525	1.0559	1.0593	1.0627	1.0661	1.0696
.47	1.0730	1.0764	1.0798	1.0833	1.0867	1.0901	1.0936	1.0970	1.1005	1.1039
.48	1.1074	1.1109	1.1143	1.1178	1.1213	1.1248	1.1282	1.1317	1.1352	1.1387
.49	1.1422	1.1457	1.1492	1.1527	1.1562	1.1597	1.1632	1.1668	1.1703	1.1738

DISCHARGE, IN CUBIC FEET PER SECOND, OF A WEIR ONE FOOT LONG, WITHOUT CONTRACTION AT THE ENDS; FOR DEPTHS FROM 0.500 TO 0.999 FEET.

Depth.	0	1	2	3	4	5	6	7	8	9
0.50	1.1773	1.1809	1.1844	1.1879	1.1915	1.1950	1.1986	1.2021	1.2057	1.2093
.51	1.2128	1.2164	1.2200	1.2235	1.2271	1.2307	1.2343	1.2379	1.2415	1.2451
.52	1.2487	1.2523	1.2559	1.2595	1.2631	1.2667	1.2703	1.2740	1.2776	1.2812
.53	1.2849	1.2885	1.2921	1.2958	1.2994	1.3031	1.3067	1.3104	1.3141	1.3177
.54	1.3214	1.3251	1.3287	1.3324	1.3361	1.3398	1.3435	1.3472	1.3509	1.3546
.55	1.3583	1.3620	1.3657	1.3694	1.3731	1.3768	1.3806	1.3843	1.3880	1.3918
.56	1.3955	1.3992	1.4030	1.4067	1.4105	1.4142	1.4180	1.4217	1.4255	1.4293
.57	1.4330	1.4368	1.4406	1.4444	1.4481	1.4519	1.4557	1.4595	1.4633	1.4671
.58	1.4709	1.4747	1.4785	1.4823	1.4862	1.4900	1.4938	1.4976	1.5014	1.5053
.59	1.5091	1.5130	1.5168	1.5206	1.5245	1.5283	1.5322	1.5361	1.5399	1.5438
0.60	1.5476	1.5515	1.5554	1.5593	1.5631	1.5670	1.5709	1.5748	1.5787	1.5826
.61	1.5865	1.5904	1.5943	1.5982	1.6021	1.6060	1.6100	1.6139	1.6178	1.6217
.62	1.6257	1.6296	1.6335	1.6375	1.6414	1.6454	1.6493	1.6533	1.6572	1.6612
.63	1.6652	1.6691	1.6731	1.6771	1.6810	1.6850	1.6890	1.6930	1.6970	1.7010
.64	1.7050	1.7090	1.7130	1.7170	1.7210	1.7250	1.7290	1.7330	1.7370	1.7410
.65	4.7451	1.7491	1.7531	1.7572	1.7612	1.7652	1.7693	1.7733	1.7774	1.7814
.66	1.7855	1.7896	1.7936	1.7977	1.8018	1.8058	1.8099	1.8140	1.8181	1.8221
.67	1.8262	1.8303	1.8344	1.8385	1.8426	1.8467	1.8508	1.8549	1.8590	1.8632
.68	1.8673	1.8714	1.8755	1.8796	1.8838	1.8879	1.8920	1.8962	1.9003	1.9045
.69	1.9086	1.9128	1.9169	1.9211	1.9252	1.9294	1.9336	1.9377	1.9419	1.9461
0.70	1.9503	1.9544	1.9586	1.9628	1.9670	1.9712	1.9754	1.9796	1.9838	1.9880
.71	1.9922	1.9964	2.0006	2.0048	2.0091	2.0133	2.0175	2.0217	2.0260	2.0302
.72	2.0344	2.0387	2.0429	2.0472	2.0514	2.0557	2.0599	2.0642	2.0684	2.0727
.73	2.0770	2.0812	2.0855	2.0898	2.0941	2.0983	2.1026	2.1069	2.1112	2.1155
74	2.1198	2.1241	2.1284	2.1327	2.1370	2.1413	2.1456	2.1499	2.1543	2.1586
.75	2.1629	2.1672	2.1716	2.1759	2.1802	2.1846	2.1889	2.1932	2.1976	2.2019
.76	2.2063	2.2107	2.2150	2.2194	2.2237	2.2281	2.2325	2.2369	2.2412	2.2456
.77	2.2500	2.2544	2.2588	2.2632	2.2675	2.2719	2.2763	2.2807	2.2851	2.2896
.78	2.2940	2.2984	5.3028	2.3072	2.3116	2.3161	2.3205	2.3249	2.3293	2.3338
.79	2.3382	2.3427	2.3471	2.3515	2.3560	2.3604	2.3649	2.3694	2.3738	2.3783
0.80	2.3828	2.3872	2.3917	2.3962	2.4006	2.4051	2.4096	2.4141	2.4186	2.4231
.81	2.4276	2.4321	2.4366	2.4411	2.4456	2.4501	2.4546	2.4591	2.4636	2.4681
.82	2.4727	2.4772	2.4817	2.4862	2.4908	2.4953	2.4999	2.5044	2.5089	2.5135
.83	2.5180	2.5226	2.5271	2.5317	2.5363	2.5408	2.5454	2.5500	2.5545	2.5591
.84	2.5637	2.5683	2.5728	2.5774	2.5820	2.5866	2.5912	2.5958	2.6004	2.6050
.85	2.6096	2.6142	3.6188	2.6234	2.6280	2.6327	2.6373	2.6419	2.6465	2.6511
.86	2.6558	2.6604	2.6650	2.6697	2.6743	2.6790	2.6836	2.6883	2.6929	2.6976
.87	2.7022	2.7069	2.7116	2.7162	2.7209	2.7256	2.7303	2.7349	2.7396	2.7443
.88	2.7490	2.7536	2.7583	2.7630	2.7677	2.7724	2.7771	2.7818	2.7865	2.7912
.89	2.7959	2.8007	2.8054	2.8101	2 8148	2.8195	2.8243	2.8290	2.8337	2.8385
0.90	2.8432	2.8479	2.8527	2.8574	2.8622	2.8669	2.8717	2.8764	2.8812	2.8860
.91	2.8907	2.8955	2.9003	2.9050	2.9098	2.9146	2.9194	2.9241	2.9289	2.9337
.92	2.9385	2.9433	2.9481	2.9529	2.9577	2.9625	2.9673	2.9721	2.9769	2.9817
.93	2.9865	2.9914	2.9962	3.0010	3.0058	3.0107	3.0155	3.0203	3.0252	3.0300
.94	3.0348	3.0397	3.0445	3.0494	3.0542	3.0591	3.0639	3.0688	3.0737	3.0785
.95	3.0834	3.0883	3.0931	3.0980	3.1029	3.1078	3.1127	3.1175	3.1224	3.1273
.96	3.1322	3.1371	3.1420	3.1469	3.1518	3.1567	3.1616	3.1665	3.1714	3.1764
.97	3.1813	3.1862	3.1911	3.1960	3.2010	3.2059	3.2108	3.2158	3.2207	3.2257
.98	3.2306	3.2355	3.2405	3.2454	3.2504	3.2554	3.2603	3.2653	3.2702	3.2752
.99	3.2802	3.2851	3.2901	3.2951	3.3001	3.3051	3.3100	3.3150	3.3200	3.3250

DISCHARGE, IN CUBIC FEET PER SECOND, OF A WEIR ONE FOOT LONG, WITH-
OUT CONTRACTION AT THE ENDS; FOR DEPTHS FROM 1.000 TO 1.499 FEET.

Depth.	0	1	2	3	4	5	6	7	8	9
1.00	3.3300	3.3350	3.3400	3.3450	3.3500	3.3550	3.3600	3.3650	3.3700	3.3751
.01	3.3801	3.3851	3.3901	3.3951	3.4002	3.4052	3.4102	3.4153	3.4203	3.4254
.02	3.4304	3.4354	3.4405	3.4455	3.4506	3.4557	3.4607	3.4658	3.4708	3.4759
.03	3.4810	3.4860	3.4911	3.4962	3.5013	3.5063	3.5114	3.5165	3.5216	3.5267
.04	3.5318	3.5369	3.5420	3.5471	3.5522	3.5573	3.5624	3.5675	3.5726	3.5777
.05	3.5828	3.5880	3.5931	3.5982	3.6033	3.6085	3.6136	3.6187	3.6239	3.6290
.06	3.6342	3.6393	3.6444	3.6496	3.6547	3.6599	3.6651	3.6702	3.6754	3.6805
.07	3.6857	3.6909	3.6960	3.7012	3.7064	3.7116	3.7167	3.7219	3.7271	3.7323
.08	3.7375	3.7427	3.7479	3.7531	3.7583	3.7635	3.7687	3.7739	3.7791	3.7843
.09	3.7895	3.7947	3.8000	3.8052	3.8104	3.8156	3.8209	3.8261	3.8313	3.8365
1.10	3.8418	3.8470	3.8523	3.8575	3.8628	3.8680	3.8733	3.8785	3.8838	3.8890
.11	3.8943	3.8996	3.9048	3.9101	3.9154	3.9206	3.9259	3.9312	3.9365	3.9418
.12	3.9470	3.9523	3.9576	3.9629	3.9682	3.9735	3.9788	3.9841	3.9894	3.9947
.13	4.0000	4.0053	4.0106	4.0160	4.0213	4.0266	4.0319	4.0372	4.0426	4.0479
.14	4.0532	4.0586	4.0639	4.0692	4.0746	4.0799	4.0853	4.0906	4.0960	4.1013
.15	4.1067	4.1120	4.1174	4.1228	4.1281	4.1335	4.1389	4.1442	4.1496	4.1550
.16	4.1604	4.1657	4.1711	4.1765	4.1819	4.1873	4.1927	4.1981	4.2035	4.2089
.17	4.2143	4.2197	4.2251	4.2305	4.2359	4.2413	4.2467	4.2522	4.2576	4.2630
.18	4.2684	4.2738	4.2793	4.2847	4.2901	4.2956	4.3010	4.3065	4.3119	4.3173
.19	4.3228	4.3282	4.3337	4.3392	4.3446	4.3501	4.3555	4.3610	4.3665	4.3719
1.20	4.3774	4.3829	4.3883	4.3938	4.3993	4.4048	4.4108	4.4158	4.4212	4.4267
.21	4.4322	4.4377	4.4432	4.4487	4.4542	4.4597	4.4652	4.4707	4.4763	4.4818
.22	4.4873	4.4928	4.4983	4.5038	4.5094	4.5149	4.5204	4.5260	4.5315	4.5370
.23	4.5426	4.5481	4.5537	4.5592	4.5647	4.5703	4.5759	4.5814	4.5870	4.5925
.24	4.5981	4.6036	4.6092	4.6148	4.6203	4.6259	4.6315	4.6371	4.6427	4.6482
.25	4.6538	4.6594	4.6650	4.6706	4.6762	4.6818	4.6874	4.6930	4.6986	4.7042
.26	4.7098	4.7154	4.7210	4.7266	4.7322	4.7378	4.7435	4.7491	4.7547	4.7603
.27	4.7660	4.7716	4.7772	4.7829	4.7885	4.7941	4.7998	4.8054	4.8111	4.8167
.28	4.8224	4.8280	4.8337	4.8393	4.8450	4.8506	4.8563	4.8620	4.8676	4.8733
.29	4.8790	4.8847	4.8903	4.8960	4.9017	4.9074	4.9131	4.9187	4.9244	4.9301
1.30	4.9358	4.9415	4.9472	4.9529	4.9586	4.9643	4.9700	4.9757	4.9814	4.9872
.31	4.9929	4.9986	5.0043	5.0100	5.0158	5.0215	5.0272	5.0330	5.0387	5.0444
.32	5.0502	5.0559	5.0616	5.0674	5.0731	5.0789	5.0846	5.0904	5.0961	5.1019
.33	5.1077	5.1134	5.1192	5.1249	5.1307	5.1365	5.1423	5.1480	5.1538	5.1596
.34	5.1654	5.1712	5.1769	5.1827	5.1885	5.1943	5.2001	5.2059	5.2117	5.2175
.35	5.2233	5.2291	5.2349	5.2407	5.2465	5.2523	5.2582	5.2640	5.2696	5.2756
.36	5.2814	5.2873	5.2931	5.2989	5.3048	5.3106	5.3164	5.3223	5.3281	5.3340
.37	5.3398	5.3456	5.3515	5.3573	5.3632	5.3691	5.3749	5.3808	5.3866	5.3925
.38	5.3984	5.4042	5.4101	5.4160	5.4219	5.4277	5.4336	5.4395	5.4454	5.4513
.39	5.4572	5.4630	5.4689	5.4748	5.4807	5.4866	5.4925	5.4984	5.5043	5.5102
1.40	5.5162	5.5221	5.5280	5.5339	5.5398	5.5457	5.5516	5.5576	5.5635	5.5694
.41	5.5754	5.5813	5.5872	5.5932	5.5991	5.6050	5.6110	5.6169	5.6229	5.6288
.42	5.6348	5.6407	5.6467	5.6526	5.6586	5.6646	5.6705	5.6765	5.6825	5.6884
.43	5.6944	5.7004	5.7064	5.7123	5.7183	5.7243	5.7303	5.7363	5.7423	5.7482
.44	5.7542	5.7602	5.7662	5.7722	5.7782	5.7842	5.7902	5.7962	5.8023	5.8082
.45	5.8143	5.8203	5.8263	5.8323	5.8384	5.8444	5.8504	5.8564	5.8625	5.8685
.46	5.8745	5.8806	5.8866	5.8926	5.8987	5.9047	5.9108	5.9168	5.9229	5.9289
.47	5.9350	5.9410	5.9471	5.9532	5.9592	5.9653	5.9714	5.9774	5.9835	5.9896
.48	5.9957	6.0017	6.0078	6.0139	6.0200	6.0261	6.0322	6.0382	6.0443	6.0504
.49	6.0565	6.0626	6.0687	6.0748	6.0809	6.0870	6.0931	6.0993	6.1054	6.1115

DISCHARGE, IN CUBIC FEET PER SECOND, OF A WEIR ONE FOOT LONG, WITH-
OUT CONTRACTION AT THE ENDS; FOR DEPTHS FROM 1.500 TO 1.999 FEET.

Depth.	0	1	2	3	4	5	6	7	8	9
1.50	6.1176	6.1237	6.1298	6.1360	6.1421	6.1482	6.1543	6.1605	6.1666	6.1727
.51	6.1789	6.1850	6.1912	6.1973	6.2034	6.2096	6.2157	6.2219	6.2230	6.2342
.52	6.2404	6.2465	6.2527	6.2588	6.2650	6.2712	6.2773	6.2835	6.2897	6.2959
.53	6.3020	6.3082	6.3144	6.3206	6.3268	6.3330	6.3391	6.3453	6.3515	6.3577
.54	6.3639	6.3701	6.3763	6.3825	6.3887	6.3949	6.4012	6.4074	6.4136	6.4198
.55	6.4260	6.4322	6.4385	6.4447	6.4509	6.4571	6.4634	6.4696	6.4758	6.4821
.56	6.4883	6.4945	6.5008	6.5070	6.5133	6.5195	6.5258	6.5320	6.5383	6.5445
.57	6.5508	6.5570	6.5633	6.5696	6.5758	6.5821	6.5884	6.5946	6.6009	6.6072
.58	6.6135	6.6198	6.6260	6.6323	6.6386	6.6449	6.6512	6.6575	6.6638	6.6701
.59	6.6764	6.6827	6.6890	6.6953	6.7016	6.7079	6.7142	6.7205	6.7268	6.7331
1.60	6.7394	6.7458	6.7521	6.7584	6.7647	6.7711	6.7774	6.7837	6.7901	6.7964
.61	6.8027	6.8091	6.8154	6.8217	6.8281	6.8344	6.8408	6.8471	6.8535	6.8598
.62	6.8662	6.8726	6.8769	6.8853	6.8916	6.8980	6.9044	6.9108	6.9171	6.9235
.63	6.9299	6.9363	6.9426	6.9490	6.9554	6.9618	6.9682	6.9746	6.9810	6.9874
.64	6.9937	7.0001	7.0065	7.0129	7.0193	7.0258	7.0322	7.0386	7.0450	7.0514
.65	7.0578	7.0642	7.0706	7.0771	7.0835	7.0899	7.0963	7.1028	7.1092	7.1156
.66	7.1221	7.1285	7.1349	7.1414	7.1478	7.1543	7.1607	7.1672	7.1736	7.1801
.67	7.1865	7.1930	7.1994	7.2059	7.2124	7.2188	7.2253	7.2318	7.2382	7.2447
·68	7.2512	7.2576	7.2641	7.2706	7.2771	7.2836	7.2901	7.2065	7.3080	7.3095
.69	7.3160	7.3225	7.3290	7.3355	7.3420	7.3485	7.3550	7.3615	7.3680	7.3745
1.70	7.3810	7.3876	7.3941	7.4006	7.4071	7.4136	7.4201	7.4267	7.4332	7.4397
.71	7.4463	7.4528	7.4593	7.4659	7.4724	7.4789	7.4855	7.4920	7.4986	7.5051
.72	7.5117	7.5182	7.5248	7.5313	7.5379	7.5445	7.5510	7.5576	7.5641	7.5707
.73	7.5773	7.5839	7.5904	7.5970	7.6036	7.6102	7.6167	7.6233	7.6299	7.6365
.74	7.6431	7.6497	7.6563	7.6628	7.6694	7.6760	7.6826	7.6892	7.6958	7.7024
.75	7.7091	7.7157	7.7223	7.7289	7.7355	7.7421	7.7487	7.7554	7.7620	7.7686
.76	7.7752	7.7819	7.7885	7.7951	7.8018	7.8084	7.8150	7.8217	7.8283	7.8349
.77	7.8416	7.8482	7.8549	7.8615	7.8682	7.8748	7.8815	7.8882	7.8948	7.9015
.78	7.9081	7.9148	7.9215	7.9281	7.9348	7.9415	7.9482	7.9548	7.9615	7.9682
.79	7.9749	7.9816	7.9882	7.9949	8.0016	8.0083	8.0150	8.0217	8.0284	8.0351
1.80	8.0418	8.0485	8.0552	8.0619	8.0686	8.0753	8.0820	8.0888	8.0955	8.1022
.81	8.1089	8.1156	8.1223	8.1291	8.1358	8.1425	8.1493	8.1560	8.1627	8.1695
.82	8.1762	8.1829	8.1897	8.1964	8.2032	8.2099	8.2167	8.2234	8.2302	8.2369
.83	8.2437	8.2504	8.2572	8.2640	8.2707	8.2775	8.2842	8.2910	8.2978	8.3046
.84	8.3113	8.3181	8.3249	8.3317	8.3385	8.3452	8.3520	8.3588	8.3656	8.3724
.85	8.3792	8.3860	8.3928	8.3996	8.4064	8.4132	8.4200	8.4268	8.4336	8.4404
.86	8.4472	8.4540	8.4608	8.4677	8.4745	8.4813	8.4881	8.4949	8.5018	8.5086
.87	8.5154	8.5223	8.5291	8.5359	8.5428	8.5496	8.5564	8.5633	8.5701	8.5770
.88	8.5838	8.5907	8.5975	8.6044	8.6112	8.6181	8.6250	8.6318	8.6387	8.6455
.89	8.6524	8.6593	8.6661	8.6730	8.6799	8.6868	8.6936	8.7005	8.7074	8.7143
1.90	8.7212	8.7281	8.7349	8.7418	8.7487	8.7556	8.7625	8.7694	8.7763	8.7832
.91	8.7901	8.7970	8.8039	8.8108	8.8177	8.8246	8.8316	8.8385	8.8454	8.8523
.92	8.8592	8.8662	8.8731	8.8800	8.8869	8.8939	8.9008	8.9077	8.9147	8.9216
.93	8.9285	8.9355	8.9424	8.9494	8.9563	8.9633	8.9702	8.9772	8.9841	8.9911
.94	8.9980	9.0050	9.0119	9.0189	9.0259	9.0328	9.0398	9.0468	9.0537	9.0607
.95	9.0677	9.0747	9.0816	9.0886	9.0956	9.1026	9.1096	9.1165	9.1235	9.1305
.96	9.1375	9.1445	9.1515	9.1585	9.1655	9.1725	9.1795	9.1865	9.1935	9.2005
.97	9.2075	9.2145	9.2216	9.2286	9.2356	9.2426	9.2496	9.2567	9.2637	9.2707
.98	9.2777	9.2848	9.2918	9.2988	9.3059	9.3129	9.3199	9.3270	9.3340	9.3411
.99	9.3481	9.3552	9.3622	9.3698	9.3763	9.3834	9.3904	9.3975	9.4045	9.4116

DISCHARGE, IN CUBIC FEET PER SECOND, OF A WEIR ONE FOOT LONG, WITHOUT CONTRACTION AT THE ENDS; FOR DEPTHS FROM 2.000 TO 2.499 FEET.

Depth	0	1	2	3	4	5	6	7	8	9
2.00	9.4187	9.4257	9.4328	9.4399	9.4469	9.4540	9.4611	9.4682	9.4752	9.4823
.01	9.4894	9.4965	9.5086	9.5106	9.5177	9.5248	9.5319	9.5390	9.5461	9.5532
.02	9.5603	9.5674	9.5745	9.5816	9.5887	9.5958	9.6029	9.6100	9.6171	9.6243
.03	9.6314	9.6385	9.6456	9.6527	9 6599	9.6670	9.6741	9.6812	9.6884	9 6955
.04	9.7026	9.7098	9.7169	9.7240	9.7312	9.7382	9.7455	9.7526	9.7598	9.7669
.05	9.7741	9.7812	9.7884	9.7955	9.8027	9.8098	9.8170	9.8242	9.8313	9.8385
.06	9.8457	9.8528	9.8600	9.8672	9.8744	9.8815	9 8887	9.8959	9.9031	9.9103
.07	9.9174	9 9246	9.9318	9.9390	9.9462	9.9534	9.9606	9.9678	9.9750	9.9822
.08	9.9894	9.9966	10.004	10.011	10.018	10.025	10.033	10.040	10.047	10.054
.09	10.062	10.069	10.076	10.083	10.090	10.098	10.105	10.112	10.119	10.127
2.10	10.134	10.141	10.148	10.156	10.163	10.170	10.177	10.185	10.192	10.199
.11	10.206	10.214	10.221	10.228	10.235	10.243	10.250	10.257	10.264	10.272
.12	10.279	10.286	10 293	10.301	10.308	10.315	10.323	10.330	10.337	10.344
.13	10.352	10 359	10.366	10.374	10.381	10.388	10.396	10.403	10.410	10.417
.14	10.425	10.432	10.439	10.447	10.454	10.461	10.469	10.476	10.483	10.491
.15	10.498	10.505	10.513	10.520	10.527	10.535	10 542	10.549	10.557	10.564
.16	10.571	10.579	10.586	10.593	10.601	10.608	10.615	10.623	10.630	10.637
.17	10.645	10.652	10.659	10.667	10.674	10.682	10.689	10.696	10.704	10.711
.18	10.718	10.726	10.783	10.741	10.748	10.755	10.763	10.770	10.777	10.785
.19	10.792	10.800	10.807	10.814	10.822	10.829	10.837	10.844	10.851	10.859
2.20	10.866	10.874	10.881	10.888	10.896	10.903	10 911	10.918	10.926	10.933
.21	10.940	10.948	10 955	10.963	10.970	10.978	10.985	10.992	11 000	11.007
.22	11.015	11.022	11.030	11.087	11.045	11.052	11.059	11.067	11.074	11.082
.23	11.089	11.097	11.104	11.112	11.119	11.127	11.134	11.141	11.149	11.156
.24	11.164	11.171	11.179	11.186	11.194	11.201	11.209	11.216	11 224	11.231
.25	11.239	11 246	11.254	11.261	11.269	11.276	11.284	11.291	11.299	11.306
.26	11.314	11.321	11.329	11.336	11.344	11.351	11.359	11.366	11.374	11.381
.27	11.389	11.396	11.404	11.412	11.419	11.427	11.434	11.442	11.449	11.457
.28	11.464	11.472	11.479	11.487	11.494	11.502	11.510	11.517	11.525	11.532
.29	11.540	11.547	11.555	11.562	11.570	11.578	11.585	11.593	11.600	11.608
2.30	11.615	11.623	11.631	11.638	11.646	11.653	11.661	11.669	11.676	11.684
.31	11.691	11.699	11.706	11.714	11.722	11.729	11.737	11.744	11.752	11.760
.32	11.767	11.775	11.783	11.790	11.798	11.805	11.813	11.821	11.828	11.836
.33	11.843	11.851	11.859	11.866	11.874	11.882	11.889	11.897	11.904	11.912
.34	11.920	11.927	11.935	11.943	11.950	11.958	11.966	11.973	11.981	11.989
.35	11.996	12.004	12.012	12.019	12.027	12.035	12.042	12.050	12.058	12.065
.36	12.073	12.081	12.088	12.096	12.104	12.111	12 119	12.127	12.134	12.142
.37	12.150	12.157	12.165	12.173	12.181	12.188	12.196	12.204	12.211	12.219
.38	12.227	12.234	12.242	12.250	12.258	12.265	12.273	12.281	12.288	12.296
.39	12.304	12.312	12.319	12.327	12.335	12.342	12.350	12.358	12.366	12.373
2.40	12.381	12.389	12.397	12.404	12.412	12.420	12.428	12.435	12.443	12.451
.41	12.459	12.466	12.474	12.482	12.490	12.497	12.505	12.513	12.521	12.528
.42	12.536	12.544	12.552	12.560	12.567	12.575	12.583	12.591	12.598	12.606
.43	12.614	12.622	12.630	12.637	12.645	12.653	12.661	12.669	12.676	12.684
.44	12.692	12.700	12.708	12.715	12.723	12.731	12.739	12.747	12.754	12.762
.45	12.770	12.778	12.786	12.794	12.801	12.809	12.817	12.825	12.833	12.840
.46	12.848	12.856	12.864	12.872	12.880	12.888	12.895	12.903	12.911	12.919
.47	12.927	12.935	12.942	12.950	12.958	12.966	12.974	12.982	12.990	12.997
.48	13.005	13.013	13.021	13.029	13.037	13.045	13.053	13.060	13.068	13.076
.49	13.084	13.092	13.100	13.108	13.116	13.124	13.131	13.139	13.147	13.155

DISCHARGE, IN CUBIC FEET PER SECOND, OF A WEIR ONE FOOT LONG, WITHOUT CONTRACTION AT THE ENDS; FOR DEPTHS FROM 2.500 TO 2.999 FEET.

Depth.	0	1	2	3	4	5	6	7	8	9
2.50	13.163	13.171	13.179	13.187	13.195	13.202	13.210	13.218	13.226	13.234
.51	13.242	13.250	13.258	13.266	13.274	13.282	13.290	13.297	13.305	13.313
.52	13.321	13.329	13.337	13.345	13.353	13.361	13.369	13.377	13.385	13.393
.53	13.401	13.409	13.417	13.424	13.432	13.440	13.448	13.456	13.464	13.472
.54	13.480	13.488	13.496	13.504	13.512	13.520	13.528	13.536	13.544	13.552
.55	13.560	13.568	13.576	13.584	13.592	13.600	13.608	13.616	13.624	13.632
.56	13.640	13.648	13.656	13.664	13.672	13.680	13.688	13.696	13.704	13.712
.57	13.720	13.728	13.736	13.744	13.752	13.760	13.768	13.776	13.784	13.792
.58	13.800	13.808	13.816	13.824	13.832	13.840	13.848	13.856	13.864	13.872
.59	13.880	13.888	13.896	13.904	13.912	13.920	13.928	13.936	13.944	13.953
2.60	13.961	13.969	13.977	13.985	13.993	14.001	14.009	14.017	14.025	14.033
.61	14.041	14.049	14.057	14.065	14.074	14.082	14.090	14.098	14.106	14.114
.62	14.122	14.130	14.138	14.146	14.154	14.162	14.171	14.179	14.187	14.195
.63	14.203	14.211	14.219	14.227	14.235	14.243	14.252	14.260	14.268	14.276
.64	14.284	14.292	14.300	14.308	14.316	14.325	14.333	14.341	14.349	14.357
.65	14.365	14.373	14.382	14.390	14.398	14.406	14.414	14.422	14.430	14.438
.66	14.447	14.455	14.463	14.471	14.479	14.487	14.496	14.504	14.512	14.520
.67	14.528	14.536	14.545	14.553	14.561	14.569	14.577	14.585	14.594	14.602
.68	14.610	14.618	14.626	14.634	14.643	14.651	14.659	14.667	14.675	14.684
.69	14.692	14.700	14.708	14.716	14.725	14.733	14.741	14.749	14.757	14.766
2.70	14.774	14.782	14.790	14.798	14.807	14.815	14.823	14.831	14.839	14.848
.71	14.856	14.864	14.872	14.881	14.889	14.897	14.905	14.913	14.922	14.930
.72	14.938	14.946	14.955	14.963	14.971	14.979	14.988	14.996	15.004	15.012
.73	15.021	15.029	15.037	15.045	15.054	15.062	15.070	15.078	15.087	15.095
.74	15.103	15.112	15.120	15.128	15.136	15.145	15.153	15.161	15.169	15.178
.75	15.186	15.194	15.203	15.211	15.219	15.227	15.236	15.244	15.252	15.261
.76	15.269	15.277	15.285	15.294	15.302	15.310	15.319	15.327	15.335	15.344
.77	15.352	15.360	15.369	15.377	15.385	15.394	15.402	15.410	15.419	15.427
.78	15.435	15.443	15.452	15.460	15.468	15.477	15.485	15.494	15.502	15.510
.79	15.519	15.527	15.535	15.544	15.552	15.560	15.569	15.577	15.585	15.594
2.80	15.602	15.610	15.619	15.627	15.635	15.644	15.652	15.661	15.669	15.677
.81	15.686	15.694	15.702	15.711	15.719	15.728	15.736	15.744	15.753	15.761
.82	15.769	15.778	15.786	15.795	15.803	15.811	15.820	15.828	15.837	15.845
.83	15.853	15.862	15.870	15.879	15.887	15.895	15.904	15.912	15.921	15.929
.84	15.938	15.946	15.954	15.963	15.971	15.980	15.988	15.997	16.005	16.013
.85	16.022	16.030	16.039	16.047	16.056	16.064	16.072	16.081	16.089	16.098
.86	16.106	16.115	16.123	16.132	16.140	16.148	16.157	16.165	16.174	16.182
.87	16.191	16.199	16.208	16.216	16.225	16.233	16.242	16.250	16.258	16.267
.88	16.275	16.284	16.292	16.301	16.309	16.318	16.326	16.335	16.343	16.352
.89	16.360	16.369	16.377	16.386	16.394	16.403	16.411	16.420	16.428	16.437
2.90	16.445	16.454	16.462	16.471	16.479	16.488	16.496	16.505	16.513	16.522
.91	16.530	16.539	16.547	16.556	16.565	16.573	16.582	16.590	16.599	16.607
.92	16.616	16.624	16.633	16.641	16.650	16.658	16.667	16.675	16.684	16.693
.93	16.701	16.710	16.718	16.727	16.735	16.744	16.752	16.761	16.770	16.778
.94	16.787	16.795	16.804	16.812	16.821	16.830	16.838	16.847	16.855	16.864
.95	16.872	16.881	16.890	16.898	16.907	16.915	16.924	16.932	16.941	16.950
.96	16.958	16.967	16.975	16.984	16.993	17.001	17.010	17.018	17.027	17.036
.97	17.044	17.053	17.062	17.070	17.079	17.087	17.096	17.105	17.113	17.122
.98	17.130	17.139	17.148	17.156	17.165	17.174	17.182	17.191	17.199	17.208
.99	17.217	17.226	17.234	17.243	17.251	17.260	17.269	17.277	17.286	17.295

J. B. Francis Tables.

VELOCITIES, IN FEET PER SECOND, DUE TO HEADS FROM 0 TO 4.99 FEET.

Head.	0	1	2	3	4	5	6	7	8	9
0.0	0.000	0.802	1.184	1.389	1.604	1.793	1.965	2.122	2.268	2.406
.1	2.536	2.660	2.778	2.892	3.001	3.106	3.208	3.307	3.403	3.496
.2	3.587	3.675	3.762	3.846	3.929	4.010	4.090	4.167	4.244	4.319
.3	4.393	4.465	4.537	4.607	4.677	4.745	4.812	4.878	4.944	5.009
.4	5.072	5.135	5.198	5.259	5.320	5.380	5.440	5.498	5.557	5.614
.5	5.671	5.728	5.783	5.893	5.894	5.948	6.002	6.055	6.108	6.160
.6	6.212	6.264	6.315	6.366	6.416	6.466	6.516	6.565	6.614	6.662
.7	6.710	6.758	6.805	6.852	6.899	6.946	6.992	7.038	7.083	7.129
.8	7.173	7.218	7.263	7.307	7.351	7.394	7.438	7.481	7.524	7.566
.9	7.609	7.651	7.693	7.734	7.776	7.817	7.858	7.899	7.940	7,980
1.0	8.020	8.060	8.100	8.140	8.179	8.218	8.257	8.296	8.335	8.373
.1	8.412	8.450	8.488	8.526	8.563	8.601	8.638	8.675	8.712	8.749
.2	8.786	8.822	8.859	8.895	8.931	8.967	9.003	9.038	9.074	9.109
.3	9.144	9.180	9.214	9.249	9.284	9.319	9.353	9.387	9.422	9.456
.4	9.490	9.523	9.557	9.591	9.624	9.658	9.691	9.724	9.757	9.790
.5	9.823	9.855	9.888	9.920	9.953	9.985	10.017	10.049	10.081	10.113
.6	10.145	10.176	10.208	10.240	10.271	10.302	10.333	10.364	10.395	10.426
.7	10.457	10.488	10.518	10.549	10.579	10.610	10.640	10.670	10.700	10.730
.8	10.760	10.790	10.820	10.850	10.879	10.909	10.938	10.967	10.997	11.026
.9	11.055	11.084	11.113	11.142	11.171	11.200	11.228	11.257	11.285	11.314
2.0	11.342	11.371	11.399	11.427	11.455	11.483	11.511	11.539	11.567	11.595
.1	11.622	11.650	11.678	11.705	11.733	11.760	11.787	11.814	11.842	11.869
.2	11.896	11.923	11.950	11.977	12.004	12.030	12.057	12.084	12.110	12.137
.3	12.163	12.190	12.216	12.242	12.269	12.295	12.321	12.347	12.373	12.399
.4	12.425	12.451	12.447	12.502	12.528	12.554	12.579	12.605	12.630	12.656
.5	12.681	12.706	12.732	12.757	12.782	12.807	12.832	12.857	12.882	12.907
.6	12.932	12.957	12.982	13.007	13.031	13.056	13.081	13.105	13.130	13.154
.7	13.179	13.203	13.227	13.252	13.276	13.300	13.324	13.348	13.372	13.396
.8	13.420	13.444	13.468	13.492	13.516	13.540	13.563	13.587	13.611	13.634
.9	13.658	13.681	13.705	13.728	13.752	13.775	13.798	13.822	13.845	13.868
3.0	13.891	13.915	13.938	13.961	13.984	14.007	14.030	14.053	14.075	14.098
.1	14.121	14.144	14.166	14.189	14.212	14.234	14.257	14.280	14.302	14.325
.2	14.347	14.369	14.392	14.414	14.436	14.459	14.481	14.503	14.525	14.547
.3	14.569	14.591	14.618	14.635	14.657	14.679	14.701	14.723	14.745	14.767
.4	14.789	14.810	14.832	14.854	14.875	14.897	14.918	14.940	14.961	14.983
.5	15.004	15.026	15.047	15.069	15.090	15.111	15.132	15.154	15.175	15.196
.6	15.217	15.238	15.259	15.281	15.302	15.322	15.344	15.364	15.385	15.406
.7	15.427	15.449	15.469	15.490	15.510	15.5 31	15.552	15.572	15.593	15 614
.8	15.634	15.655	15.675	15.696	15.716	15.737	15.757	15.778	15.798	15.818
.9	15.839	15.859	15.876	15.899	15.920	15.940	15.960	15.980	16.000	16.020
4.0	16.040	16.060	16.080	16.100	16.120	16 140	16.160	16.180	16.200	16.220
.1	16.240	16.259	16.279	16 299	16.319	16.338	16.358	16.878	16 397	16.417
.2	16.437	16.456	16.476	16.495	16.515	16.534	16.554	16 573	16.592	16.612
.3	16.631	16.650	16.670	16.689	16.708	16.727	16.747	16.766	16.785	16.804
.4	16.823	16.842	16.862	16.881	16.900	16.919	16.938	16.957	16.976	16.994
.5	17.013	17.032	17.051	17.070	17.089	17 108	17.126	17.145	17.164	17.183
.6	17.201	17.220	17.239	17.257	17.276	17.295	17.313	17.332	17.350	17.369
.7	17.387	17.406	17.424	17.443	17.461	17.480	17.498	17.516	17.535	17.553
.8	17.571	17.590	17.608	17.626	17.644	17.663	17.681	17.699	17.717	17.735
.9	17.753	17.772	17.790	17.808	17.826	17.844	17.862	17.880	17.898	17.916

VELOCITIES, IN FEET PER SECOND, DUE TO HEADS FROM 5 TO 9.99 FEET.

Head.	0	1	2	3	4	5	6	7	8	9
5.0	17.934	17.952	17.970	17.987	18.005	18.023	18.041	18.059	18.077	18.094
.1	18.112	18.130	18.148	18.165	18.183	18.201	18.218	18.236	18.254	18.271
.2	18.289	18.306	18.324	18.342	18.359	18.377	18.394	18.412	18.429	18.446
.3	18.464	18.481	18.499	18.516	18.533	18.551	18.568	18.585	18.603	18 620
.4	18.637	18.655	18.672	18.689	18.706	18.723	18.741	18 758	18 775	18.792
.5	18.809	18.826	18.843	18.860	18.877	18 894	18.911	18.928	18.945	18.962
.6	18.979	18.996	19.013	19.030	19.047	19.064	19.081	19.098	19.114	19.131
.7	19.148	19.165	19.182	19.198	19.215	19.232	19.248	19.265	19 282	19.299
.8	19.315	19.332	19.348	19.365	19.382	19.398	19.415	19.431	19.448	19.464
.9	19.481	19.497	19.514	19.530	19.547	19.563	19.580	19.596	19.613	19.629
6.0	19.645	19.662	19.678	19.694	19.711	19.727	19.743	19.760	19.776	19.792
.1	19.808	19.825	19.841	19.857	19.873	19.889	19.906	19 922	19.938	19.954
.2	19.970	19.986	20.002	20.018	20.034	20.050	20.067	20.083	20.099	20.115
.3	20.131	20.147	20.162	20.178	20.194	20.210	20 226	20.242	20.258	20.274
.4	20.290	20.306	20.321	20.337	20.353	20.369	20.385	20.400	20.416	20.432
.5	20.448	20.463	20.479	20.495	20.510	20.526	20.542	20.557	20.573	20.589
.6	20.604	20.620	20.635	20.651	20.667	20.682	20.698	20.713	20.729	20.744
.7	20.760	20.775	20.791	20.806	20.822	20.837	20.853	20.868	20.883	20.899
.8	20.914	20.929	20.945	20.960	20.976	20.991	21.006	21.021	21.037	21 052
.9	21.067	21.083	21.098	21.113	21.128	21.144	21.159	21.174	21.189	21.204
7.0	21.219	21.235	21.250	21.265	21.280	21.295	21.310	21.325	21.340	21.355
.1	21.370	21.386	21.401	21.416	21.431	21.446	21.461	21.476	21.491	21.506
.2	21.520	21.535	21.550	21.565	21.580	21.595	21.610	21.625	21.640	21.655
.3	21.669	21.684	21.699	21.714	21.729	21.743	21.758	21.773	21.78~	21.803
.4	21.817	21.832	21.847	21.861	21.876	21 891	21.906	21.920	21.935	21 950
.5	21.964	21.979	21.993	22.008	22.023	22.037	22.052	22.066	22.081	22.096
.6	22.110	22.125	22.139	22.154	22.168	22.183	22.197	22.212	22.226	22 241
.7	22.255	22.270	22.284	22.298	22.313	22.327	22.342	22.356	22.370	22.385
.8	22.399	22.414	22.428	22 442	22.457	22.471	22.485	22.499	22.514	22.528
.9	22 542	22.557	22.571	22.585	22.599	22.614	22.628	22.642	22.656	22.670
8.0	22.685	22.699	22.713	22.727	22.741	22.755	22.769	22.784	22.798	22.812
.1	22.826	22.840	22.854	22.868	22.882	22.896	22.910	22.924	22.938	22 952
.2	22.966	22.980	22.994	23.008	23.022	23 036	23.050	23.064	23.078	23 092
.3	23 106	23.120	23.134	23.148	23.162	23.175	23.189	23.203	23.217	23 231
.4	23.245	23.259	23.272	23.286	23.300	23.314	23 328	23.341	23 355	23.369
.5	23.383	23.396	23.410	23.424	23.438	23.451	23.465	23.479	23.492	23.506
.6	23.520	23.534	23.547	23.561	23.574	23.588	23 602	23.615	23.629	23 643
.7	23.656	23.670	23.683	23.697	43.711	23.724	23.738	23.751	23.765	23.778
.8	23.792	23.805	23.819	23.832	23.846	23.859	23.873	23.886	23.900	23 913
.9	23.927	23.940	23.953	23.967	23.980	23.994	24.007	24 020	24.034	24.047
9.0	24.061	24.074	24.087	24.101	24.114	24.127	24.141	24.154	24.167	24.181
.1	24.194	24.207	24.220	24.234	24.247	24.260	24.274	24.287	24.300	24 313
.2	24 326	24.340	24.353	24.366	24.379	24.392	24.406	24.419	24.432	24 445
.3	24 458	24.471	24.485	24.498	24.511	24.524	24.537	24.550	24.563	24.576
.4	24.589	24.603	24.616	24.629	24.642	24.655	24.668	24.681	24 694	24.707
.5	24.720	24.733	24.746	24.759	24.772	24.785	24.798	24.811	24 824	24.837
.6	24.850	24.863	24.876	24.888	24.901	24.914	24 927	24.940	24.935	24.966
.7	24.979	24.992	25.005	25.017	25.080	25.043	25.056	25.069	25.082	25.094
.8	25.107	25.120	25.133	25.146	25.158	25.171	25.184	25.197	25.209	25.222
.9	25.235	25.248	25.260	25.273	25.286	25.299	25.311	25.324	25.337	25.349

VELOCITIES, IN FEET PER SECOND, DUE TO HEADS FROM 10 TO 14.99 FEET.

Head.	0	1	2	3	4	5	6	7	8	9
10.0	25.362	25.375	25.387	25.400	25.413	25.425	25.438	25.451	25.463	25.476
.1	25.489	25.501	25.514	25.526	25.539	25.552	25.564	25.577	25.589	25.602
.2	25.614	25.627	25.640	25.652	25.665	25.677	25.690	25.702	25.715	25.728
.3	25.740	25.752	25.765	25.777	25.790	25.802	25.815	25.827	25.839	25.852
.4	25.864	25.877	25.889	25.902	25.914	25.926	25.939	25.951	25.964	25.976
.5	25.988	26.001	26.013	26.026	26.038	26.050	26.063	26.075	26.087	26.099
.6	26.112	26.124	26.136	26.149	26.161	26.173	26.186	26.198	26.210	26.222
.7	26.235	26.247	26.259	26.272	26.284	26.296	26.308	26.320	26.333	26.345
.8	26.357	26.369	26.381	26.394	26.406	26.418	26.430	26.442	26.454	26.467
.9	26.479	26.491	26.503	26.515	26.527	26.540	26.552	26.564	26.576	26.588
11.0	26.600	26.612	26.624	26.636	26.648	26.660	26.672	26.684	26.697	26.709
.1	26.721	26.733	26.745	26.757	26.769	26.781	26.793	26.805	26.817	26.829
.2	26.841	26.853	26.865	26.877	26.889	26.901	26.913	26.924	26.936	26.948
.3	26.960	26.972	26.984	26.996	27.008	27.020	27.032	27.044	27.056	27.067
.4	27.079	27.091	27.103	27.115	27.127	27.139	27.150	27.162	27.174	27.186
.5	27.198	27.210	27.221	27.233	27.245	27.257	27.269	27.280	27.292	27.304
.6	27.316	27.328	27.339	27.351	27.363	27.375	27.386	27.398	27.410	27.422
.7	27.433	27.445	27.457	27.468	27.480	27.492	27.504	27.515	27.527	27.539
.8	27.550	27.562	27.574	27.585	27.597	27.609	27.620	27.632	27.644	27.655
.9	27.667	27.678	27.690	27.702	27.713	27.725	27.736	27.748	27.760	27.771
12.0	27.783	27.794	27.806	27.817	27.829	27.841	27.852	27.864	27.875	27.887
.1	27.898	27.910	27.921	27.933	27.944	27.956	27.967	27.979	27.990	28.002
.2	28.013	28.025	28.036	28.048	28.059	28.071	28.082	28.094	28.105	28.117
.3	28.128	28.139	28.151	28.162	28.174	28.185	28.196	28.208	28.219	28.231
.4	28.242	28.253	28.265	28.276	28.288	28.299	28.310	28.322	28.333	28.344
.5	28.356	28.367	28.378	28.390	28.401	28.412	28.424	28.435	28.446	28.458
.6	28.469	28.480	28.491	28.503	28.514	28.525	28.537	28.548	28.559	28.570
.7	28.582	28.593	28.604	28.615	28.627	28.638	28.649	28.660	28.672	28.683
.8	28.694	28.705	28.716	28.727	28.739	28.750	28.761	28.772	28.783	28.795
.9	28.806	28.817	28.828	28.839	28.850	28.862	28.873	28.884	28.895	28.906
13.0	28.917	28.928	28.939	28.951	28.962	28.973	28.984	28.995	29.006	29.017
.1	29.028	29.039	29.050	29.061	29.073	29.084	29.095	29.106	29.117	29.128
.2	29.139	29.150	29.161	29.172	29.183	29.194	29.205	29.216	29.227	29.238
.3	29.249	29.260	29.271	29.282	29.293	29.304	29.315	29.326	29.337	29.348
.4	29.359	29.370	29.381	29.392	29.403	29.413	29.424	29.435	29.446	29.457
.5	29.468	29.479	29.490	29.501	29.512	29.523	29.533	29.544	29.555	29.566
.6	29.577	29.588	29.599	29.610	29.620	29.631	29.642	29.653	29.664	29.675
.7	29.686	29.696	29.707	29.718	29.729	29.740	29.751	29.761	29.772	29.783
.8	29.794	29.805	29.815	29.826	29.837	29.848	29.858	29.869	29.880	29.891
.9	29.901	29.912	29.923	29.934	29.944	29.955	29.966	29.977	29.987	29.998
14.0	30.009	30.020	30.030	30.041	30.052	30.062	30.073	30.084	30.094	30.105
.1	30.116	30.126	30.137	30.148	30.159	30.169	30.180	30.190	30.201	30.212
.2	30.222	30.233	30.244	30.254	30.265	30.276	30.286	30.297	30.307	30.318
.3	30.329	30.339	30.350	30.360	30.371	30.382	30.392	30.403	30.413	30.424
.4	30.435	30.445	30.456	30.466	30.477	30.487	30.498	30.508	30.519	30.529
.5	30.540	30.551	30.561	30.572	30.582	30.593	30.603	30.614	30.624	30.635
.6	30.645	30.656	30.666	30.677	30.687	30.698	30.708	30.719	30.729	30.739
.7	30.750	30.670	30.771	30.781	30.792	30.802	30.813	30.823	30.833	30.844
.8	30.854	30.865	30.875	30.886	30.896	30.906	30.917	30.927	30.938	30.948
.9	30.958	30.969	30.979	30.990	31.000	31.010	31.021	31.031	31.041	31.052

VELOCITIES, IN FEET PER SECOND, DUE TO HEADS FROM 15 TO 19.99 FEET.

Head.	0	1	2	3	4	5	6	7	8	9
15.0	31.062	31.072	31.083	31.093	31.103	31.114	31.124	31.134	31.145	31.155
.1	31.165	31.176	31.186	31.196	31.207	31.217	31.227	31.238	31.248	31.258
.2	31.268	31.279	31.289	31.299	31.310	31.320	31.330	31.340	31.351	31.361
.3	31.371	31.381	31.392	31.402	31.412	31.422	31.433	31.443	31.453	31.463
.4	31.474	31.484	31.494	31.504	31.514	31.525	31.535	31.545	31.555	31.565
.5	31.576	31.586	31.596	31.606	31.616	31.626	31.637	31.647	31.657	31.667
.6	31.677	31.687	31.698	31.708	31.718	31.728	31.738	31.748	31.758	31.768
.7	31.779	31.789	31.799	31.809	31.819	31.829	31.839	31.849	31.859	31.870
.8	31.880	31.890	31.900	31.910	31.920	31.930	31.940	31.950	31.960	31.970
.9	31.980	31.990	32.000	32.011	32.021	32.031	32.041	32.051	32.061	32.071
16.0	32.081	32.091	32.101	32.111	32.121	32.131	32.141	32.151	32.161	32.171
.1	32.181	32.191	32.201	32.211	32.221	32.231	32.241	22.251	32.261	32.271
.2	32.281	32.291	32.301	32.311	32.321	32.330	32.340	32.350	32.360	32.370
.3	32.380	32.390	32.400	32.410	32.420	32.4:0	32.440	32.450	32.460	32.470
.4	32.480	32.489	32.499	32.509	32.519	32.529	32.539	32.549	32.559	32.569
.5	32.579	32.588	32.598	32.608	32.618	32.628	32.637	32.647	32.657	32.667
.6	32.677	32.687	32.696	32.706	32.716	32.726	32.736	32.746	32.755	32.765
.7	32.775	32.785	32.795	32.804	32.814	32.824	32.834	32.844	32.854	32.863
.8	32.873	32.883	32.893	32.903	32.912	32.922	32.932	32.941	32.951	32.961
.9	32.971	32.980	32.990	33.000	33.010	33.019	33.029	33.039	33.049	33.058
17.0	33.068	33.078	33.088	33.097	33.107	33.117	33.126	33.136	33.146	33.156
.1	33.165	33.175	33.185	33.194	33.204	33.214	33.223	33.233	33.243	33.252
.2	33.262	33.272	33.281	33.291	33.301	33.310	33.320	33.330	33.339	33.349
.3	33.359	33.368	33.378	33.388	33.397	33.407	33.416	33.426	33.436	33.445
.4	33.455	33.465	33.474	33.484	33.493	33.503	33.513	33.522	33.532	33.541
.5	33.551	33.560	33.570	33.580	33.589	33.599	33.608	33.618	33.628	33.637
.6	33.647	33.656	33.666	33.675	33.685	33.694	33.704	33.713	33.723	33.733
.7	33.742	33.752	33.761	33.771	33.780	33.790	33.799	33.809	33.818	33.828
.8	33.837	33.847	33.856	33.866	33.875	33.885	33.894	33.904	33.913	33.923
.9	33.932	33.942	33.951	33.961	33.970	33.980	33.989	33.998	34.008	34.017
18.0	34.027	34.036	34.046	34.055	34.065	34.074	34.083	34.093	34.102	34.112
.1	34.121	34.131	34.140	34.149	34.159	34.168	34.178	34.187	34.197	34.206
.2	34.215	34.225	34.234	34.244	34.253	34.262	34.272	34.281	34.290	34.300
.3	34.309	34.319	34.328	34.337	34.347	34.356	34.365	34.375	34.384	34.393
.4	34.403	34.412	34.422	34.431	34.440	34.450	34.459	34.468	34.478	34.487
.5	34.496	34.505	34.515	34.524	34.533	34.543	34.552	34.561	34.571	34.580
.6	34.589	34.599	34.608	34.617	34.626	34.636	34.645	34.654	34.664	34.673
.7	34.682	34.691	34.701	34.710	34.719	34.728	34.738	34.747	34.756	34.766
.8	34.775	34.784	34.793	34.802	34.812	34.821	34.830	34.839	34.849	34.858
.9	34.867	34.876	34.886	34.895	34.904	34.913	34.922	34.932	34.941	34.950
19.0	34.959	34.968	34.978	34.987	34.996	35.005	35.014	35.024	35.033	35.042
.1	35.051	35.060	35.069	35.079	35.088	35.097	35.106	35.115	35.124	35.134
.2	35.143	35.152	35.161	35.170	35.179	35.188	35.198	35.207	35.216	35.225
.3	35.234	35.243	35.252	35.262	35.271	35.280	35.289	35.298	35.307	35.316
.4	35.325	35.334	35.344	35.353	35.362	35.371	35.380	35.389	35.398	35.407
.5	35.416	35.425	35.434	35.443	35.453	35.462	35.471	35.480	35.489	35.498
.6	35.507	35.516	35.525	35.534	35.543	35.552	35.561	35.570	35.579	35.588
.7	35.597	35.606	35.615	35.624	35.634	35.643	35.652	35.661	35.670	35.679
.8	35.688	35.697	35.706	35.715	35.724	35.733	35.742	35.751	35.760	35.769
.9	35.778	35.787	35.796	35.805	35.814	35.823	35.832	35.841	35.849	35.858

VELOCITIES, IN FEET PER SECOND, DUE TO HEADS FROM 20 TO 24.99 FEET.

Head.	0	1	2	3	4	5	6	7	8	9
20.0	35.867	35.876	35.885	35.894	35.903	35.912	35.921	35.930	35.939	35.948
.1	35.957	35.966	35.975	35.984	35.993	36.002	36.011	36.020	36.028	36.037
.2	36.046	36.055	36.064	36.073	36.082	36.091	36.100	36.109	36.118	36.127
.3	36.135	36.144	36.153	36.162	36.171	36.180	36.189	36.198	36.207	36.215
.4	36.2 4	96.233	36.242	36.251	36.260	36.269	36.278	36.286	36.295	36.304
.5	36.313	36.322	36.331	36.340	36.348	36.357	36.366	36.375	36.384	36.393
.6	36.401	36.410	36 419	36.428	36.437	36.446	36.454	36.463	36.472	36.481
.7	36.490	36.499	36.507	36.516	36.525	36.534	36.543	36.551	36.560	36.569
.8	36.578	36.587	36.595	36.604	36.613	36.622	36.630	36.639	36.648	36.657
.9	36.666	36.674	36.683	36.692	36.701	36.709	36.718	36.727	36.736	36.744
21.0	36.753	36.762	36.771	36.779	36.788	36.797	36.806	36.814	36.823	36.832
.1	36 841	36.849	36.858	36.867	36.875	36.884	36.893	36.902	36.910	36.919
.2	36.928	36.936	36.945	36.954	36.963	36.971	36.980	36.989	36.997	37.006
.3	37 015	37.023	37.032	37.041	37.049	37.058	37.067	37.076	37.084	37.093
.4	37.102	37.110	37.119	37.128	37.136	37.145	37.154	37.162	37.171	37.179
.5	37.188	37.197	37.205	37.214	37.223	37.231	37.240	37.249	37.257	37.266
.6	37.275	37.283	37.292	37.300	37.309	37.318	37.326	37.335	37.343	37.352
.7	37.361	37.369	37.378	37.387	37.395	37.404	37 412	37.421	37.430	37.438
.8	37.447	37.455	37.464	37.472	37.481	37.490	37.498	37.506	37.515	37.524
.9	37.532	37.541	37.550	37.558	37.567	37.575	37.584	37.592	37.601	37.610
22.0	37.618	37.627	37.635	37.644	37.652	37.661	37.669	37.678	37.686	37.695
.1	37.703	37.712	37.721	37.729	37.738	37.746	37.755	37.763	37.772	37.780
.2	37.789	37.797	37.806	37.814	37.823	37.832	37.840	37.848	37.857	37.865
.3	37 874	37.882	37.891	37.899	37.908	37.916	37.925	37.933	37.942	37.950
.4	37 959	37.967	37.975	37.984	38.001	38.001	38.009	38.018	38.026	38.035
.5	38.043	38.052	38.060	38.068	38.077	38.085	38.094	38.102	38.111	38.119
.6	38.128	38.136	38.144	38.153	38.161	38.170	38 178	38.187	38.195	38.203
.7	38.212	38.220	38.229	38.237	38.246	38.254	38.262	38.271	38.279	38.288
.8	38.296	38.304	38.313	38.321	38.330	38.338	38.346	38.355	38.363	38 371
.9	38.380	38.388	38 397	38.405	38.413	38.422	38.430	38.438	38.447	38.455
23.0	38.464	38.472	38.480	38.489	38.497	38.505	38.514	38.522	38.530	38 539
.1	38.547	38.555	38.564	38.572	38.580	38.589	38.597	38.605	38.614	38.622
.2	38.630	38.638	38.647	38.655	38.664	38.672	38.680	38.689	38 697	38.705
.3	38.714	38.722	38.730	38.738	38.747	38.755	38.763	38.772	38.780	38.788
.4	38.797	38.805	38.813	38.821	38.830	38.838	38.846	38.855	38.863	38.871
.5	38.879	38.888	38.896	38.904	38 912	38.921	38.929	38.937	38.945	38.954
.6	38.962	38.970	38.978	38.987	38.995	39.003	39 011	39.020	39.028	39.036
.7	39.044	39.053	39.061	39.069	39.077	39.086	39.094	39.102	39.110	39.119
.8	39.127	39.135	39.143	39.151	39.160	39.168	39.176	39.184	39.192	39.201
.9	39.209	39.217	39.225	39.233	39.242	39.250	39.258	39.266	39.274	39.283
24.0	39.291	39 299	39.307	39 315	39.324	39.332	39.340	39.348	39.356	39 364
.1	39 373	39.381	39.389	39.397	39.405	39.413	39.422	39.430	39.438	39.446
.2	39 454	39.462	39 470	39.479	39.487	39 495	39.503	39.511	39.519	39.527
.3	39.536	39.544	39.552	39.560	39.568	39.576	39.584	39.592	39.601	39 609
.4	39.617	39.625	39 633	39.641	39.649	39.657	39.666	39.674	39.682	39.690
.5	39.696	39.706	39.714	39.722	39.730	39.738	39.747	39.755	39.763	39.771
.6	39.779	39.787	39 795	39.803	39.811	39.819	39 827	39.835	39.844	39.852
.7	39.860	39.868	39.876	39 884	39.892	39.900	39.908	39.916	39.924	39.932
.8	39.940	39.948	39.956	39.964	39.972	39.981	39.989	39.997	40.005	40.013
.9	40.021	40.029	40.037	40 045	40.053	40.061	40.069	40.077	40 085	40.093

VELOCITIES, IN FEET PER SECOND, DUE TO HEADS FROM 25 TO 29.99 FEET.

Head.	0	1	2	3	4	5	6	7	8	9
25.0	40.101	40.109	40.117	40.125	40.133	40.141	40.149	40.157	40.165	40.173
.1	40.181	40.189	40.197	40.205	40.213	40.221	40.229	40.237	40.245	40.253
.2	40.261	40.269	40.277	40.285	40.293	40.301	40 309	40.317	40.325	40.333
.3	40.341	40.349	40.357	40.365	40.373	40.381	40.389	40.397	40.405	40.413
.4	40.421	40.428	40.436	40.444	40 452	40.460	4 1.468	40.476	40.484	40.492
.5	40.500	40.508	40.516	40 524	40.532	40.540	40 548	40.556	40.563	40.571
.6	40.579	40.587	40.595	40.603	40.611	40.619	40.627	40.635	40.643	40.651
.7	40.659	40.666	40.674	40.682	40.690	40.698	40.706	40.714	40.722	40.730
.8	40.738	40.745	40.753	40.761	40.769	40.777	40.785	40.793	40 801	40 809
.9	40.816	40.824	40.832	40.840	40.848	40.856	40.864	40 872	40.879	40.887
26.0	40.895	40.903	40.911	40.919	40.927	40.934	40.942	40.950	40.958	40.966
.1	40.974	40.982	40.989	40.997	41.005	41.013	41.021	41.029	41.036	41.044
.2	41.052	41.060	41.068	41.076	41.083	41.091	41.099	41.107	41.115	41.123
.3	41.130	41.138	41.146	41.154	41.162	41.169	41.177	41.185	41.193	41.201
.4	41.209	41.216	41.224	41.232	41.240	41.248	41.255	41.263	41.271	41.279
.5	41.287	41.294	41.302	41.310	41.318	41.325	41.333	41.341	41.349	41.357
.6	41.364	41.372	41.380	41.388	41.395	41.403	41.411	41.419	41.426	41.434
.7	41.442	41.450	41.458	41.465	41.473	41.481	41.489	41 496	41.504	41.512
.8	41.520	41.527	41.535	41.543	41.551	41.558	41.566	41.574	41.581	41.589
.9	41.597	41.605	41.612	41.620	41.628	41.636	41.643	41.651	41.659	41.666
27 0	41.674	41.682	41.690	41.697	41.705	41.713	41.720	41.728	41.736	41.744
.1	41.751	41.759	41.767	41.774	41.782	41.790	41.797	41.805	41.813	41.821
.2	41.828	41.836	41.844	41.851	41.859	41.867	41.874	41.882	41.890	41.897
.3	41.905	41.913	41.920	41.928	41.936	41.943	41.951	41.959	41.967	41.974
.4	41.982	41.989	41.997	42.005	42.012	42 020	42.028	42.035	42.043	42.051
.5	42.058	42.066	42.074	42.081	42.089	42.096	42.104	42.112	42.119	42.127
.6	42.135	42.142	42.150	42.158	42.165	42.173	42.180	42.188	42 196	42.203
.7	42.211	42.219	42.226	42.234	42.241	42.249	42 257	42.264	42.272	42.279
.8	42.287	42.295	42.302	42.310	42.317	42.325	42.333	42.340	42.348	42.355
.9	42.363	42.371	42.378	42.386	42.393	42.401	42.409	42.416	42.424	42.431
28.0	42.439	42.446	42.454	42.462	42.469	42 477	42.484	42.492	42.499	42.507
.1	42.515	42.522	42.530	42 537	42.545	42.552	42.560	42.568	42.575	42.583
.2	42.590	42.598	42.605	42.613	42.620	42.628	42.635	42.643	42.651	42.658
.3	42.666	42.673	42.681	42.688	42.696	42.703	42.711	42.718	42.726	42.733
.4	42.741	42.748	42.756	42.764	42.771	42.779	42.786	42.794	42.801	42.809
.5	42.816	42.824	42.831	42.839	42.846	42.854	42.861	42.869	42.876	42.884
.6	42 891	42.899	42.906	42.914	42.921	42 929	42.936	42.944	42.951	42.959
.7	42.966	42.974	42.981	42 989	42.996	43.004	43.011	43.018	43.026	43 033
.8	43.041	43.048	43.056	43.063	43.071	43.078	43 086	43.093	43 101	43.108
.9	43.116	43.123	43.130	43.138	43.145	43.153	43.160	43.168	43.175	43.183
29.0	43.190	43.198	43.205	43.212	43.220	43.227	43.235	43.243	43.250	43.257
.1	43.264	43.272	43 279	43.287	43.294	43.302	43.309	43 316	43.324	43 331
.2	43.339	43 346	43.354	43.361	43.368	43.376	43 383	43 391	43 398	43 405
.3	43 413	43.420	43 428	43.435	43.443	43 450	43.457	43.465	43.472	43.480
.4	43.487	43.494	43 502	43.509	43.517	43.524	43.531	43 539	43 546	43.553
.5	43 561	43.568	43 576	43.583	43.590	43.598	43 605	43.612	43.620	43 627
.6	43 635	43.642	43.649	43.657	43.664	43 671	43.679	43.686	43.694	43.701
.7	43.708	43.716	43 723	43.730	43.738	43.745	43.752	43.760	43.767	43.774
.8	43.782	43.789	43.796	43.804	43.811	43.818	43.826	43.833	43.840	43.848
.9	43.855	43.862	43.870	43.877	43.884	43.892	43 899	43.906	43.914	43.921

VELOCITIES, IN FEET PER SECOND, DUE TO HEADS FROM 30 TO
34.99 FEET.

Head.	0	1	2	3	4	5	6	7	8	9
30.0	43.928	43.936	43.943	43.950	43.958	43.965	43.972	43.980	43.987	43.994
.1	44.002	44.009	44.016	44.024	44.031	44.038	44.045	44.053	44.060	44.067
.2	44.075	44.082	44.089	44.097	44.104	44.111	44.118	44.126	44.133	44.140
.3	44.148	44.155	44.162	44.169	44.177	44.184	44.191	44.198	44.206	44.213
.4	44.220	44.228	44.235	44.242	44.249	44.257	44.264	44.271	44.278	44.286
.5	44.293	44.300	44.308	44.315	44.322	44.329	44.337	44.344	44.351	44.358
.6	44.366	44.373	44.380	44.387	44.395	44.402	44.409	44.416	44.423	44.431
.7	44.438	44.445	44.452	44.460	44.467	44.474	44.481	44.489	44.496	44.503
.8	44.510	44.518	44.525	44.532	44.539	44.546	44.554	44.561	44.568	44.575
.9	44.582	44.590	44.597	44.604	44.611	44.619	44.626	44.633	44.640	44.647
31.0	44.655	44.662	44.669	44.676	44.683	44.691	44.698	44.705	44.712	44.719
.1	44.727	44.734	44.741	44.748	44.755	44.762	44.770	44.777	44.784	44.791
.2	44.798	44.806	44.813	44.820	44.827	44.834	44.841	44.849	44.856	44.863
.3	44.870	44.877	44.884	44.892	44.899	44.906	44.913	44.920	44.927	44.935
.4	44.942	44.949	44.956	44.963	44.970	44.978	44.985	44.992	44.999	45.006
.5	45.013	45.020	45.028	45.035	45.042	45.049	45.056	45.063	45.070	45.078
.6	45.085	45.092	45.099	45.106	45.113	45.120	45.127	45.135	45.142	45.149
.7	45.156	45.163	45.170	45.177	45.184	45.192	45.199	45.206	45.213	45.220
.8	45.227	45.234	45.241	45.248	45.256	45.263	45.270	45.277	45.284	45.291
.9	45.298	45.305	45.312	45.319	45.327	45.334	45.341	45.348	45.355	45.362
32.0	45.369	45.376	45.383	45.390	45.397	45.405	45.412	45.419	45.426	45.433
.1	45.440	45.447	45.454	45.461	45.468	45.475	45.482	45.489	45.497	45.504
.2	45.511	45.518	45.525	45.532	45.539	45.546	45.553	45.560	45.567	45.574
.3	45.581	45.588	45.595	45.602	45.609	45.617	45.624	45.631	45.638	45.645
.4	45.652	45.659	45.666	45.673	45.680	45.687	45.694	45.701	45.708	45.715
.5	45.722	45.729	45.736	45.743	45.750	45.757	45.764	45.771	45.778	45.785
.6	45.792	45.799	45.807	45.814	45.821	45.828	45.835	45.842	45.849	45.856
.7	45.863	45.870	45.877	45.884	45.891	45.898	45.905	45.912	45.919	45.926
.8	45.933	45.940	45.947	45.954	45.961	45.968	45.975	45.982	45.989	45.996
.9	46.003	46.010	46.017	46.024	46.031	46.038	46.045	46.052	46.059	46.066
83.0	46.073	46.080	46.086	46.093	46.100	46.107	46.114	46.121	46.128	46.135
.1	46.142	46.149	46.156	46.163	46.170	46.177	46.184	46.191	46.198	46.205
.2	46.212	46.219	46.226	46.233	46.240	46.247	46.254	46.261	46.268	46.275
.3	46.281	46.288	46.295	46.302	46.309	46.316	46.323	46.330	46.337	46.344
.4	46.351	46.358	46.365	46.372	46.379	46.386	46.393	46.399	46.406	46.413
.5	46.420	46.427	46.434	46.441	46.448	46.455	46.462	46.469	46.476	46.483
.6	46.489	46.496	46.503	46.510	46.517	46.524	46.531	46.538	46.545	46.552
.7	46.559	46.566	46.572	46.579	46.586	46.593	46.600	46.607	46.614	46.621
.8	46.628	46.635	46.642	46.648	46.655	46.662	46.669	46.676	46.683	46.690
.9	46.697	46.703	46.710	46.717	46.724	46.731	46.739	46.745	46.752	46.759
34.0	46.765	46.772	46.779	46.786	46.793	46.800	46.807	46.814	46.820	46.827
.1	46.834	46.841	46.848	46.855	46.862	46.868	46.875	46.882	46.889	46.896
.2	46.903	46.910	46.916	46.923	46.930	46.937	46.944	46.951	46.958	46.964
.3	46.971	46.978	46.985	46.992	46.999	47.005	47.012	47.019	47.026	47.033
.4	47.040	47.047	47.053	47.060	47.067	47.074	47.081	47.088	47.094	47.101
.5	47.108	47.115	47.122	47.128	47.135	47.142	47.149	47.156	47.163	47.169
.6	47.176	47.183	47.190	47.197	47.203	47.210	47.217	47.224	47.231	47.238
.7	47.244	47.251	47.258	47.265	47.272	47.278	47.285	47.292	47.299	47.306
.8	47.312	47.319	47.326	47.333	47.340	47.346	47.353	47.360	47.367	47.374
.9	47.380	47.387	47.394	47.401	47.407	47.414	47.421	47.428	47.435	47.441

VELOCITIES, IN FEET PER SECOND, DUE TO HEADS FROM 35 TO 39.99 FEET.

Head.	0	1	2	3	4	5	6	7	8	9
35.0	47.448	47.455	47.462	47.469	47.475	47.482	47.489	47.496	47.502	47.509
.1	47.516	47.523	47.529	47.536	47.543	47.550	47.556	47.563	47.570	47.577
.2	47.584	47.590	47.597	47.604	47.611	47.617	47.624	47.631	47.638	47.644
.3	47.651	47.658	47.665	47.671	47.678	47.685	47.692	47.698	47.705	47.712
.4	44.719	47.725	47.732	47.739	47.745	47.752	47.750	47.766	47.772	47.779
.5	47.786	47.793	47.799	47.806	47.813	47.819	47.826	47.833	47.840	47.846
.6	47.853	47.860	47.867	47.873	47.880	47.887	47.893	47.900	47.907	47.914
.7	47.920	47.927	47.934	47.940	47.947	47.954	47.961	47.967	47.974	47.981
.8	47.987	47.994	48.001	48.007	48.014	48.021	48.028	48.034	48.041	48.048
.9	48.054	48.061	48.068	48.074	48.081	48.088	48.094	48.101	48.108	48.115
36.0	48.121	48.128	48.134	48.141	48.148	48.155	48.161	48.168	48.175	48.181
.1	48.188	48.195	48.201	48.208	48.215	48.221	48.228	48.235	48.241	48.248
.2	48.255	48.261	48.268	48.275	48.281	48.288	48.295	48.302	48.308	48.315
.3	48.321	48.328	48.335	48.341	48.348	48.355	48.361	48.368	48.375	48.381
.4	48.388	48.394	48.401	48.408	48.414	48.421	48.428	48 434	48.441	48.448
.5	48.454	48.461	48.467	48.474	48.481	48.487	48.494	48.501	48.507	48.514
.6	48.521	48.527	48.534	48.540	48.547	48.554	48.560	48.567	48.574	48.580
.7	48.587	48.593	48.600	48.607	48.613	48.620	48.626	48.633	48.640	48.646
.8	48.653	48.660	48.666	48.673	48.679	48.686	48.693	48.699	48.706	48.712
.9	48.719	48.726	48.732	48.739	48.745	48.752	48.759	48.765	48.771	48.778
37.0	48.785	48.792	48.798	48.805	48.811	48.818	48.824	48.831	48.838	48.844
.1	48.851	48.857	48.864	48.871	48.877	48.884	48.890	48.897	48.903	48.910
.2	48.917	48.923	48.930	48.936	48.943	48.950	48.956	48.963	48.969	48.976
.3	48.982	48.989	48.995	49 002	49.009	49.015	49.022	49.028	49.035	49.041
.4	49.048	49.055	49.061	49.068	49.074	49.081	49.087	49.094	49.100	49.107
.5	49.113	49.120	49.127	49.133	49.140	49.146	49.153	49.159	49.166	49.172
.6	49.179	49.185	49.192	49.199	49.205	49.212	49.218	49.225	49.231	49.238
.7	49.244	49.251	49.257	49.264	49.270	49.277	49.283	49.290	49.297	49 303
.8	49.310	49.316	49.323	49.329	49.336	49.342	49.349	49.355	49 362	49.368
.9	49.375	49.381	49.388	49.394	49.401	49.407	49 414	49.420	49.427	49.433
38.0	49.440	49.446	49.453	49.459	49.466	49.472	49 479	49.485	49.492	49.498
.1	49.505	49.511	49 518	49.524	49.531	49.537	49.544	49.550	49.557	49.563
.2	49.570	49.576	49 583	49.589	49.596	49.602	49.609	49.615	49.622	49.628
.3	49.635	49.641	49.648	49.654	49.661	49.667	49.673	49.680	49.686	49.693
.4	49.699	49.706	49 712	49.719	49.725	49.732	49.738	49.745	49.751	49.758
.5	49.764	49.770	49.777	49.783	49.790	49.796	49.803	49.809	49.816	49.822
.6	49 829	49.835	49.842	49.848	49.854	49 861	49.867	49.874	49.880	49.887
.7	49.893	49.900	49.906	49.912	49.919	49.925	49.932	49.938	49.945	49.951
.8	49.958	49.964	49.970	49.977	49.983	49.990	49.996	50.003	50.009	50.015
.9	50.022	50.028	50.035	50.041	50.048	50 054	50.060	50.067	50.073	50.080
39.0	50.086	50.093	50.099	50.105	50.112	50.118	50.125	50.131	50.137	50.144
.1	50.150	50.157	50.163	50.170	50.176	50.182	50.189	50.195	50.202	50.208
.2	50.214	50.221	50.227	50.234	50.240	50.246	50.253	50.259	50 266	50.272
.3	50.278	50.285	50.291	50.298	50.304	50.310	50.317	50.323	50.330	50.336
.4	50.342	50.349	50.355	50.362	50 368	50.374	50.381	50.387	50.393	50.400
.5	50.406	50.413	50.419	50.425	50.432	50.438	50 444	50.451	50.457	50.464
.6	50.470	50.476	50.483	50.489	50.495	50.502	50.508	50.515	50.521	50.527
.7	50 534	50.540	50.546	50 553	50.559	50.565	50.572	50.578	50.585	50.591
.8	50.597	50.604	50.610	50.616	50.623	50.629	50.635	50.642	50.648	50.654
.9	50.661	50.667	50.673	50.680	50 686	50.692	50.699	50.706	50.712	50.718

VELOCITIES, IN FEET PER SECOND, DUE TO HEADS FROM 40 TO 44.99 FEET.

Head.	0	1	2	3	4	5	6	7	8	9
40.0	50.724	50.731	50.737	50.743	50.750	50.756	50.762	50.769	50.775	50.781
.1	50.788	50.794	50.800	50.807	50.813	50.819	50.826	50.832	50.838	50.845
.2	50.851	50.857	50.863	50.870	50.876	50.882	50.889	50.895	50.901	50 908
.3	50.914	50.920	50.927	50.933	50.939	50.946	50.952	50.958	50.965	50.971
.4	50.977	50.983	50.990	50.996	51.002	51.009	51.015	51.021	51.028	51.034
.5	51.040	51.047	51.053	51.059	51.065	51.072	51.078	51.084	51.091	51.097
.6	51.103	51.110	51.116	51.122	51.128	51.135	51.141	51.147	51.154	51.160
.7	51.166	51.172	51.179	51.185	51.191	51.198	51.204	51.210	51.216	51.223
.8	51.229	51.235	51.241	51.248	51.254	51.260	51.267	51.273	51.279	51.285
.9	51.292	51.298	51.304	51.310	51.317	51.323	51.329	51.336	51.342	51.348
41.0	51.354	51.361	51.367	51.373	51.379	51.386	51.392	51.398	51.404	51.411
.1	51.417	51.423	51.429	51.436	51.442	51.448	51.454	51.461	51.467	51.473
.2	51.479	51.486	51.492	51.498	51.504	51.511	51.517	51.523	51.529	51.536
.3	51.542	51.548	51.554	51.561	51.567	51.573	51.579	51.586	51.592	51.598
.4	51.604	51.610	51.617	51.623	51.629	51.635	51.642	51.648	51.654	51.660
.5	51.667	51.673	51.679	51.685	51.691	51.698	51.704	51.710	51.716	51.723
.6	51.729	51.735	51.741	51.747	51.754	51.760	51.766	51.772	51.778	51.785
.7	51.791	51.797	51.803	51.809	51.816	51.822	51.828	51.834	51.841	51.847
.8	51.853	51.859	51.865	51.872	51.878	51.884	51.890	51.896	51.903	51.909
.9	51.915	51.921	51.927	51.934	51.940	51.946	51.952	51.958	51.964	51.971
42.0	51.977	51.983	51.989	51.995	52.002	52.008	52.014	52.020	52.026	52.032
.1	52.039	52.045	52.051	52.057	52.063	52.070	52.076	52.082	52.088	52.094
.2	52.100	52.107	52.113	52.119	52.125	52.131	52.137	52.144	52.150	52.156
.3	52.162	52.168	52.174	52.181	52.187	52.193	52.199	52.205	52.211	52.218
.4	52.224	52.230	52.236	52.242	52.248	52.255	52.261	52.267	52.273	52.279
.5	52.285	52.291	52.298	52.304	52.310	52.316	52.322	52.328	52.334	52.341
.6	52.347	52.353	52.359	52.365	52.371	52.377	52.384	52.390	52.396	52.402
.7	52.408	52.414	52.420	52.427	52.433	52.439	52.445	52.451	52.457	52.463
.8	52.470	52.476	52.482	52.488	52.494	52.500	52.506	52.512	52.519	52.525
.9	52.531	52.537	52.543	52.549	52.555	52.561	52.567	52.574	52.580	52.586
43.0	52.592	52.598	52.604	52.610	52.616	52.623	52.629	52.635	52.641	52.647
.1	52.653	52.659	52.665	52.671	52.678	52.684	52.690	52.696	52.702	52.708
.2	52.714	52.720	52.726	52.732	52.738	52.745	52.751	52.757	52.763	52 769
.3	52.775	52.781	52.787	52.793	52 799	52.806	52.812	52.818	52.824	52.830
.4	52 836	52.842	52.848	52.854	52.860	52.866	52.873	52.879	52.885	52.891
.5	52.897	52.903	52.909	52.915	52.921	52.927	52.933	52.939	52.945	52.952
.6	52.958	52.964	52.970	52.976	52.982	52.988	52.994	53.000	53.006	53.012
.7	53.018	53.024	53.030	53.037	53.043	53.049	53 055	53.061	53.067	53.073
.8	53.079	53.085	53.091	53.097	53.103	53.109	53.115	53.121	53.127	53.133
.9	53.139	53.146	53.152	53.158	53.164	53 170	53.176	53.182	53.188	53.194
44.0	53.200	53.206	53.212	53.218	53.224	53.230	53.236	53.242	53.248	53.254
.1	53.260	53 266	53.272	53.279	53.285	53.291	53.297	53.303	53.309	53.315
.2	53.321	53.327	53 333	53.339	53.345	53.351	53.357	53.363	53.369	53.375
.3	53.381	53.397	53.393	53.399	53.405	53 411	53 417	53.423	53.429	53.435
.4	53.441	53.447	53.453	53.459	53.465	53.471	53.477	53.483	53.489	53.495
.5	53.501	53.507	53.513	53.519	53.525	53 531	53.537	53.543	53.549	53.555
.6	53.561	53.567	53.573	53.579	53.586	53.592	53.598	53.604	53.610	53.616
.7	53.621	53.627	53.633	53.639	53.645	53.651	53.657	53.663	53.669	53.675
.8	53.681	53.687	53.693	53.699	53 705	53.711	53.717	53.723	53.729	53.735
.9	53.741	53.747	53.753	53.759	53.765	53.771	53.777	53.783	53.789	53.795

VELOCITIES, IN FEET PER SECOND, DUE TO HEADS FROM 45 TO 49.99 FEET.

Head.	0	1	2	3	4	5	6	7	8	9
45.0	53.801	53.807	53.813	53.819	53.825	53.831	53.837	53.843	53.849	53.855
.1	53.861	53.867	53.873	53.879	53.885	53.891	53.897	53.903	53.909	53.915
.2	53.921	53.927	53.932	53.938	53.944	53.950	53.956	53.962	53.968	53.974
.3	53.980	53.986	53.992	53.998	54.004	54.010	54.016	54.022	54.028	54.034
.4	54.040	54.046	54.052	54.058	54.064	54.069	54.075	54.081	54.087	54.093
.5	54.099	54.105	54.111	54.117	54.123	54.129	54.135	54.141	54.147	54.153
.6	54.159	54.165	54.170	54.176	54.182	54.188	54.194	54.200	54.206	54.212
.7	54.218	54.224	54.230	54.236	54.242	54.248	54.254	54.259	54.265	54.271
.8	54.277	54.283	54.289	54.295	54.301	54.307	54.313	54.319	54.325	54.331
.9	54.336	54.342	54.348	54.354	54.360	54.366	54.372	54.378	54.384	54.390
46.0	54.396	54.402	54.407	54.413	54.419	54.425	54.431	54.437	54.443	54.449
.1	54.455	54.461	54.467	54.472	54.478	54.484	54.490	54.496	54.502	54.508
.2	54.514	54.520	54.526	54.531	54.537	54.543	54.549	54.555	54.561	54.567
.3	54.573	54.579	54.585	54.590	54.596	54.602	54.608	54.614	54.620	54.626
.4	54.632	54.638	54.643	54.649	54.655	54.661	54.667	54.673	54.679	54.685
.5	54.690	54.696	54.702	54.708	54.714	54.720	54.726	54.732	54.737	54.743
.6	54.749	54.755	54.761	54.767	54.773	54.779	54.784	54.790	54.796	54.802
.7	54.808	54.814	54.820	54.826	54.831	54.837	54.843	54.849	54.855	54.861
.8	54.867	54.872	54.878	54.884	54.890	54.896	54.902	54.908	54.913	54.919
.9	54.925	54.931	54.937	54.943	54.949	54.954	54.960	54.966	54.972	54.978
47.0	54.984	54.990	54.995	55.001	55.007	55.013	55.019	55.025	55.030	55.036
.1	55.042	55.048	55.054	55.060	55.066	55.071	55.077	55.083	55.089	55.095
.2	55.101	55.106	55.112	55.118	55.124	55.130	55.136	55.141	55.147	55.153
.3	55.159	55.165	55.171	55.176	55.182	55.188	55.194	55.200	55.206	55.211
.4	55.217	55.223	55.229	55.235	55.240	55.246	55.252	55.258	55.264	55.270
.5	55.275	55.281	55.287	55.293	55.299	55.304	55.310	55.316	55.322	55.328
.6	55.334	55.339	55.345	55.351	55.357	55.363	55.368	55.374	55.380	55.386
.7	55.392	55.397	55.403	55.409	55.415	55.421	55.426	55.432	55.438	55.444
.8	55.450	55.455	55.461	55.467	55.473	55.479	55.484	55.490	55.496	55.502
.9	55.508	55.513	55.519	55.525	55.531	55.537	55.542	55.548	55.554	55.560
48.0	55.566	55.571	55.577	55.583	55.589	55.595	55.600	55.606	55.612	55.618
.1	55.623	55.629	55.635	55.641	55.647	55.652	55.658	55.664	55.670	55.675
.2	55.681	55.687	55.693	55.699	55.704	55.710	55.716	55.722	55.727	55.733
.3	55.739	55.745	55.750	55.756	55.762	55.768	55.774	55.779	55.785	55.791
.4	55.797	55.802	55.808	55.814	55.820	55.825	55.831	55.837	55.843	55.848
.5	55.854	55.860	55.866	55.872	55.877	55.883	55.889	55.895	55.900	55.906
.6	55.912	55.918	55.923	55.929	55.935	55.941	55.946	55.952	55.958	55.964
.7	55.969	55.975	55.981	55.987	55.992	55.998	56.004	56.009	56.015	56.021
.8	56.027	56.032	56.038	56.044	56.050	56.055	56.061	56.067	56.073	56.078
.9	56.084	56.090	56.096	56.101	56.107	56.113	56.118	56.124	56.130	56.136
49.0	56.141	56.147	56.153	56.159	56.164	56.170	56.176	56.181	56.187	56.193
.1	56.199	56.204	56.210	56.216	56.222	56.227	56.233	56.239	56.244	56.250
.2	56.256	56.262	56.267	56.273	56.279	56.284	56.290	56.296	56.302	56.307
.3	56.313	56.319	56.324	56.330	56.336	56.342	56.347	56.353	56.359	56.364
.4	56.370	56.376	56.381	56.387	56.393	56.399	56.404	56.410	56.416	56.421
.5	56.427	56.433	56.439	56.444	56.450	56.456	56.461	56.467	56.473	56.478
.6	56.484	56.490	56.495	56.501	56.507	56.513	56.518	56.524	56.530	56.535
.7	56.541	56.547	56.552	56.558	56.564	56.569	56.575	56.581	56.586	56.592
.8	56.598	56.604	56.609	56.615	56.621	56.626	56.632	56.638	56.643	56.649
.9	56.655	56.660	56.666	56.672	56.677	56.683	56.689	56.694	56.700	56.706

Swain Turbine Co. Tables.

Head in Feet	Velocity due Head	Cubic Feet per Second to one H. P.	Section of Stream in Square Feet to 1 H. P.	Section of Stream in Square Inches to 1 H. P.	Head in Feet	Velocity due Head	Cubic Feet per Second to one H. P.	Section of Stream in Square Feet to 1 H. P.	Section of Stream in Square Inches to 1 H. P.
1	8.02	8.8186	1.0995	158.3280	51	57.27	.1729	.003019	.4347
2	11.34	4.4093	.3888	55.9872	52	57.84	.1695	.002930	.4219
3	13.89	2.9395	.2116	30.4704	53	58.39	.1663	.002848	.4101
4	16.04	2.2046	.1374	19.7856	54	58.93	.1633	.002771	.3990
5	17.92	1.7637	.0984	14.1696	55	59.48	.1603	.002695	.3880
6	19.65	1.4697	.0747	10.7568	56	60.01	.1574	.002622	.3775
7	21.22	1.2598	.0593	8.5392	57	60.56	.1547	.002554	.3677
8	22.68	1.1023	.0486	6.9884	58	61.08	.1520	.002488	.3582
9	24.06	.9798	.0407	5.8608	59	61.61	.1494	.002424	.3490
10	25.36	.8818	.0347	4.9968	60	62.12	.1469	.002364	.3404
11	26.60	.8016	.0301	4.3344	61	62.71	.1445	.002304	.3317
12	27.78	.7348	.0264	3.8016	62	63.15	.1422	.002251	.3241
13	28.92	.6783	.0234	3.3696	63	63.66	.1399	.002197	.3163
14	30.01	.6299	.0209	3.0096	64	64.16	.1377	.002146	.3090
15	31.06	.5879	.0189	2.7216	65	64.66	.1356	.002097	.3019
16	32.08	.5511	.0171	2.4624	66	65.16	.1336	.002050	.2952
17	33.07	.5187	.0156	2.2464	67	65.65	.1316	.002004	.2885
18	34.03	.4899	.0143	2.0592	68	66.14	.1296	.001959	.2820
19	34.96	.4641	.0132	1.9008	69	66.62	.1278	.001918	.2761
20	35.87	.4409	.0122	1.7568	70	67.11	.1259	.001876	.2701
21	36.75	.4199	.0114	1.6416	71	67.58	.1242	.001837	.2645
22	37.61	.4008	.0106	1.5264	72	68.06	.1224	.001796	.2589
23	38.46	.3834	.0099	1.4256	73	68.53	.1208	.001762	.2537
24	39.29	.3674	.0093	1.3392	74	69.00	.1191	.001726	.2485
25	40.10	.3527	.0087	1.2528	75	69.46	.1175	.001691	.2435
26	40.89	.3391	.0082	1.1808	76	69.92	.1160	.001659	.2388
27	41.67	.3266	.0078	1.1232	77	70.38	.1145	.001626	.2341
28	42.44	.3149	.0074	1.0656	78	70.84	.1130	.001595	.2296
29	43.19	.3040	.0070	1.0080	79	71.29	.1116	.001565	.2253
30	43.93	.2939	.0066	.9504	80	71.74	.1102	.001536	.2211
31	44.65	.2844	.00636	.9158	81	72.19	.1088	.001507	.2170
32	45.37	.2755	.00607	.8740	82	72.63	.1075	.001480	.2131
33	46.07	.2672	.00579	.8337	83	73.07	.1062	.001453	.2092
34	46.77	.2593	.00554	.7977	84	73.51	.1049	.001425	.2052
35	47.45	.2519	.00530	.7632	85	73.95	.1037	.001402	.2018
36	48.12	.2449	.00509	.7329	86	74.38	.1025	.001379	.1985
37	48.78	.2383	.00488	.7027	87	74.81	.1013	.001354	.1949
38	49.44	.2320	.00469	.6753	88	75.24	.1002	.001331	.1916
39	50.09	.2261	.00451	.6494	89	75.67	.0990	.001308	.1883
40	50.72	.2204	.00434	.6249	90	76.09	.0979	.001286	.1851
41	51.35	.2150	.00418	.6019	91	76.51	.0969	.001266	.1823
42	51.98	.2099	.00403	.5803	92	76.93	.0958	.001245	.1792
43	52.59	.2050	.00389	.5601	93	77.35	.0948	.001225	.1764
44	53.20	.2004	.00376	.5414	94	77.76	.0938	.001206	.1736
45	53.80	.1959	.00364	.5241	95	78.18	.0928	.001188	.1710
46	54.40	.1917	.00352	.5068	96	78.59	.0918	.001168	.1681
47	54.99	.1876	.00341	.4910	97	79.00	.0909	.001150	.1656
48	55.57	.1837	.00330	.4752	98	79.40	.0899	.001132	.1630
49	56.14	.1799	.00320	.4608	99	79.81	.0890	.001115	.1605
50	56.71	.1763	.00310	.4475	100	80.22	.0881	.001096	.1581

Head in Feet.	Velocity due Head.	Cubic Feet per Second to one H. P.	Section of Stream in Square Feet to 1 H. P.	Section of Stream in Square Inches to 1 H. P.	Head in Feet.	Velocity due Head.	Cubic Feet per Second to one H. P.	Section of Stream in Square Feet to 1 H. P.	Section of Stream in Square Inches to 1 H. P.
101	80.61	.0873	.001082	.1558	151	98.56	.0584	.0005925	.08533
102	81.01	.0864	.001066	.1535	152	98.89	.0580	.0005865	.08445
103	81.40	.0856	.001051	.1513	153	99.21	.0576	.0005805	.08359
104	81.80	.0847	.001035	.1490	154	99.54	.0572	.0005746	.08274
105	82.19	.0839	.001020	.1468	155	99.86	.0568	.0005687	.08189
106	82.58	.0831	.001006	.1448	156	100.18	.0565	.0005640	.08121
107	82.97	.0824	.000993	.1429	157	100.50	.0561	.0005582	.08038
108	83.35	.0816	.000979	.1409	158	100.82	.0558	.0005534	.07968
109	83.74	.0809	.000966	.1391	159	101.14	.0554	.0005477	.07886
110	84.12	.0801	.000952	.1370	160	101.46	.0551	.0005430	.07819
111	84.50	.0794	.000939	.1352	161	101.77	.0547	.0005374	.07738
112	84.88	.0787	.000927	.1334	162	102.09	.0544	.0005328	.07672
113	85.26	.0780	.000914	.1316	163	102.40	.0541	.0005283	.07307
114	85.64	.0773	.000902	.1298	164	102.72	.0537	.0005227	.07536
115	86.01	.0766	.000890	.1281	165	103.03	.0534	.0005182	.07462
116	86.39	.0760	.000879	.1265	166	103.34	.0531	.0005138	.07398
117	86.76	.0753	.000867	.1248	167	103.65	.0528	.0005094	.07335
118	87.13	.0747	.000857	.1234	168	103.96	.0524	.0005040	.07257
119	87.50	.0741	.000848	.1221	169	104.27	.0521	.0004996	.07194
120	87.86	.0734	.000835	.1202	170	104.58	.0518	.0004953	.07132
121	88.23	.0728	.000825	.1188	171	104.89	.0515	.0004911	.07071
122	88.59	.0722	.000814	.1172	172	105.19	.0512	.0004867	.07008
123	88.95	.0716	.000804	.1157	173	105.50	.0509	.0004824	.06946
124	89.31	.0711	.000796	.1146	174	105.80	.0506	.0004782	.06886
125	89.67	.0705	.000786	.1131	175	106.11	.0503	.0004740	.06825
126	90.03	.0699	.000776	.1117	176	106.41	.0501	.0004708	.06779
127	90.39	.0694	.000767	.1104	177	106.71	.0498	.0004666	.06719
128	90.74	.0688	.000758	.1091	178	107.01	.0495	.0004625	.06660
129	91.10	.0683	.000749	.1078	179	107.31	.0492	.0004584	.06600
130	91.45	.0678	.0007413	.10674	180	107.61	.0489	.0004544	.06543
131	91.80	.0673	.0007331	.10556	181	107.91	.0487	.0004513	.06498
132	92.15	.0668	.0007249	.10438	182	108.21	.0484	.0004472	.06439
133	92.50	.0663	.0007167	.10320	183	108.50	.0481	.0004433	.06383
134	92.85	.0658	.0007086	.10203	184	108.80	.0479	.0004402	.06338
135	93.19	.0653	.0007007	.10090	185	109.10	.0476	.0004362	.06281
136	93.54	.0648	.0006926	.09973	186	109.39	.0474	.0004332	.06238
137	93.88	.0643	.0006849	.09862	187	109.68	.0471	.0004294	.06183
138	94.22	.0639	.0006780	.09763	188	109.98	.0469	.0004264	.06140
139	94.56	.0634	.0006704	.09653	189	110.27	.0466	.0004225	.06084
140	94.90	.0629	.0006626	.09541	190	110.56	.0464	.0004106	.06042
141	95.24	.0625	.0006582	.09449	191	110.85	.0461	.0004158	.05987
142	95.58	.0621	.0006497	.09356	192	111.14	.0459	.0004129	.05945
143	95.91	.0616	.0006422	.09247	193	111.43	.0456	.0004092	.05892
144	96.25	.0612	.0006358	.09155	194	111.72	.0454	.0004063	.05850
145	96.58	.0608	.0006295	.09064	195	112.01	.0452	.0004035	.05810
146	96.92	.0604	.0006231	.08972	196	112.29	.0449	.0003998	.05757
147	97.25	.0599	.0006159	.08868	197	112.58	.0447	.0003970	.05716
148	97.58	.0595	.0006097	.08779	198	112.86	.0445	.0003942	.05676
149	97.91	.0591	.0006036	.08691	199	113.15	.0443	.0003915	.05637
150	98.23	.0587	.0005975	.08604	200	113.43	.0440	.0003879	.05585

Head in Feet.	Velocity due Head.	Cubic Feet per Second to one H.P.	Section of Stream in Square Feet to 1 H.P.	Section of Stream in Square Inches to 1 H.P.	Head in Feet.	Velocity due Head.	Cubic Feet per Second to one H.P.	Section of Stream in Square Feet to 1 H.P.	Section of Stream in Square Inches to 1 H.P.
201	113.72	.0438	.00038515	.055461	251	127.08	.03513	.00027644	.039607
202	114.00	.0436	.00038245	.055072	252	127.33	.03499	.00027478	.039568
203	114.28	.0434	.00037976	.054685	253	127.58	.03485	.00027316	.039335
204	114.56	.0432	.00037709	.054300	254	127.83	.03471	.00027153	.039100
205	114.84	.0430	.00037443	.053917	255	128.08	.03458	.00026998	.038877
206	115.12	.0428	.00037178	.053536	256	128.33	.03444	.00026837	.038645
207	115.40	.0426	.00036915	.053157	257	128.58	.03431	.00026683	.038423
208	115.68	.0423	.00036566	.052655	258	128.83	.03418	.00026531	.038204
209	115.96	.0421	.00036305	.052279	259	129.08	.03404	.00026371	.037974
210	116.23	.0419	.00036049	.051910	260	129.33	.03391	.00026219	.037755
211	116.51	.0417	.00035790	.051537	261	129.58	.03378	.00026061	.037527
212	116.79	.0415	.00035533	.051167	262	129.83	.03365	.00025918	.037321
213	117.06	.0413	.00035281	.050804	263	130.08	.03353	.00025776	.037117
214	117.34	.0412	.00035111	.050559	264	130.32	.03340	.00025629	.036905
215	117.61	.0410	.00034852	.050186	265	130.57	.03327	.00025480	.036691
216	117.88	.0408	.00034611	.049839	266	130.82	.03315	.00025340	.036489
217	118.15	.0406	.00034363	.049482	267	131.06	.03302	.00025194	.036279
218	118.43	.0404	.00034112	.049121	268	131.31	.03290	.00025055	.036079
219	118.70	.0402	.00033886	.048767	269	131.55	.03278	.00024918	.035881
220	118.97	.0400	.00033621	.048414	270	131.80	.03266	.00024779	.035681
221	119.24	.0399	.00033461	.048183	271	132.04	.03254	.00024644	.035487
222	119.51	.0397	.00033218	.047833	272	132.28	.03242	.00024508	.035291
223	119.78	.0395	.00032977	.047486	273	132.53	.03230	.00024371	.035094
224	120.05	.0393	.00032736	.047139	274	132.77	.03218	.00024237	.034901
225	120.31	.0391	.00032499	.046798	275	133.01	.03206	.00024103	.034708
226	120.58	.0390	.00032343	.046573	276	133.25	.03195	.00023977	.034526
227	120.85	.0388	.00032105	.046231	277	133.49	.03183	.00023822	.034303
228	121.11	.0386	.00031871	.045894	278	133.74	.03172	.00023717	.034152
229	121.38	.0385	.00031710	.045662	279	133.98	.03160	.00023585	.033962
230	121.64	.0383	.00031486	.045339	280	134.22	.03149	.00023461	.033783
231	121.91	.0381	.00031252	.045002	281	134.46	.03138	.00023337	.033605
232	122.17	.0380	.00031104	.044789	282	134.69	.03127	.00023223	.033441
233	122.43	.0378	.00030874	.044458	283	134.93	.03116	.00023093	.033253
234	122.70	.0376	.00030643	.044125	284	135.17	.03105	.00022971	.033078
235	122.96	.0375	.00030497	.043915	285	135.41	.03094	.00022849	.032902
236	123.22	.0373	.00030271	.043590	286	135.65	.03083	.00022727	.032726
237	123.48	.0372	.00030126	.043381	287	135.88	.03072	.00022608	.032555
238	123.74	.0370	.00029901	.043057	288	136.12	.03062	.00022494	.032391
239	124.00	.0368	.00029677	.042734	289	136.36	.03051	.00022374	.032218
240	124.26	.0367	.00029534	.042528	290	136.59	.03040	.00022257	.032050
241	124.52	.0365	.00029312	.042209	291	136.83	.03030	.00022144	.031887
242	124.78	.0364	.00029171	.042006	292	137.06	.03020	.00022034	.031728
243	125.03	.0362	.00028953	.041692	293	137.30	.03009	.00021906	.031547
244	125.29	.0361	.00028813	.041490	294	137.53	.02999	.00021807	.031402
245	125.55	.0359	.00028594	.041175	295	137.76	.02989	.00021697	.031243
246	125.80	.0358	.00028457	.040978	296	138.00	.02979	.00021586	.031083
247	126.06	.0357	.00028319	.040779	297	138.23	.02969	.00021478	.030916
248	126.31	.0355	.00028105	.040471	298	138.46	.02959	.00021370	.030772
249	126.57	.0354	.00027968	.040273	299	138.69	.02949	.00021263	.030616
250	126.82	.0352	.00027811	.040047	300	138.92	.02939	.00021154	.030461

Head in Feet.	Velocity due Head.	Cubic Feet per Second to one H. P.	Section of Stream in Square Feet to 1 H. P.	Section of Stream in Square Inches to 1 H. P.	Head in Feet.	Velocity due Head.	Cubic Feet per Second to one H. P.	Section of Stream in Square Feet to 1 H. P.	Section of Stream in Square Inches to 1 H. P.
301	139.16	.029297	.00021047	.030307	351	150.27	.025124	.00016719	.024075
302	139.39	.029200	.00020948	.030165	352	150.49	.025052	.00016646	.023970
303	139.62	.029104	.00020845	.030016	353	150.70	.024981	.00016570	.023860
304	139.85	.029008	.00020742	.029658	354	150.91	.024911	.00016506	.023768
305	140.08	.028913	.00020640	.020721	355	151.13	.024841	.00016436	.023667
306	140.31	.028819	.00020539	.029576	356	151.34	.024771	.00016367	.023568
307	140.54	.028725	.00020439	.029432	357	151.55	.024702	.00016299	.023470
308	140.77	.028631	.00020338	.029286	358	151.76	.024633	.00016231	.023372
309	141.00	.028539	.00020240	.029145	359	151.98	.024564	.00016162	.023273
310	141.22	.028447	.00020143	.029035	360	152.19	.024496	.00016095	.023176
311	141.45	.028355	.00020046	.028866	361	152.40	.024428	.00016028	.023083
312	141.68	.028264	.00019949	.028726	362	152.61	.024360	.00015962	.022985
313	141.90	.028174	.00019854	.028589	363	152.82	.024293	.00015896	.022890
314	142.13	.028084	.00019759	.028452	364	153.03	.024226	.00015830	.022795
315	142.36	.027995	.00019665	.028317	365	153.24	.024100	.00015766	.022703
316	142.58	.027907	.00019572	.028183	366	153.45	.024094	.00015701	.022609
317	142.81	.027819	.00019479	.028049	367	153.66	.024028	.00015637	.022517
318	143.03	.027731	.00019388	.027918	368	153.87	.023963	.00015578	.022425
319	143.26	.027644	.00019296	.027786	369	154.08	.023898	.00015510	.022334
320	143.48	.027558	.00019206	.027656	370	154.29	.023834	.00015447	.022243
321	143.71	.027472	.00019106	.027516	371	154.49	.023769	.00015385	.022154
322	143.93	.027387	.00019027	.027398	372	154.70	.023705	.00015323	.022065
323	144.15	.027302	.00018939	.027272	373	154.91	.023642	.00015261	.021975
324	144.38	.027217	.0001F850	.027144	374	155.12	.023579	.00015200	.021888
325	144.60	.027134	.00018764	.027020	375	155.33	.023516	.00015139	.021800
326	144.82	.027050	.00018678	.026896	376	155.53	.023453	.00015079	.021713
327	145.04	.026968	.00018593	.026773	377	155.74	.023391	.00015019	.021627
328	145.27	.026886	.00018507	.026650	378	155.95	.023329	.00014959	.021540
329	145.49	.026804	.00018423	.026529	379	156.15	.023268	.00014901	.021457
330	145.71	.026723	.00018339	.026408	380	156.36	.023206	.00014841	.021371
331	145.93	.026642	.00018256	.026288	381	156.56	.023145	.00014783	.021287
332	146.15	.026562	.00018174	.026170	382	156.77	.023085	.00014725	.021204
333	146.37	.026482	.00018092	.026052	383	156.97	.023025	.00014668	.021121
334	146.59	.026403	.00018011	.025935	384	157.18	.022965	.00014610	.021038
335	146.81	.026324	.00017931	.025820	385	157.38	.022905	.00014552	.020956
336	147.03	.026245	.00017843	.025693	386	157.59	.022846	.00014497	.020875
337	147.25	.026168	.00017771	.025590	387	157.79	.022787	.00014441	.020795
338	147.46	.026090	.00017692	.025476	388	157.99	.022728	.00014385	.020714
339	147.68	.026018	.00017614	.025364	389	158.20	.022669	.00014329	.020633
340	147.90	.025937	.00017536	.025251	390	158.40	.022611	.00014274	.020554
341	148.12	.025861	.00017459	.025140	391	158.60	.022554	.00014220	.020476
342	148.33	.025781	.00017383	.025032	392	158.81	.022496	.00014165	.020397
343	148.55	.025710	.00017307	.024922	393	159.01	.022430	.00014111	.020319
344	148.77	.025635	.00017231	.024812	394	159.21	.022382	.00014058	.020243
345	148.98	.025561	.00017157	.024706	395	159.41	.022325	.00014004	.020165
346	149.20	.025487	.00017082	.024598	396	159.62	.022269	.00013951	.020089
347	149.41	.025413	.00017008	.024491	397	159.82	.022213	.00013896	.020013
348	149.63	.025340	.00016928	.024376	398	160.02	.022157	.00013846	.019933
349	149.85	.025268	.00016869	.024281	399	160.22	.022101	.00013794	.019863
350	150.06	.025196	.00016790	.024177	400	160.42	.022046	.00013742	.019788

Head in Feet.	Velocity due Head.	Cubic Feet per Second to one H. P.	Section of Stream in Square Feet to 1 H. P.	Section of Stream in Square Inches to 1 H. P.	Head in Feet.	Velocity due Head.	Cubic Feet per Second to one H. P.	Section of Stream in Square Feet to 1 H. P.	Section of Stream in Square Inches to 1 H. P.
401	160.62	.021991	.00013691	.019715	451	170.34	.019553	.00011478	.016528
402	160.82	.021936	.00013640	.019641	452	170.53	.019510	.00011440	.016473
403	161.02	.021882	.00013589	.019568	453	170.72	.019467	.00011402	.016418
404	161.22	.021828	.00013539	.019496	454	170.91	.019424	.00011365	.016365
405	161.42	.021774	.00013489	.019424	455	171.09	.019381	.00011327	.016310
406	161.62	.021720	.00013438	.019350	456	171.28	.019339	.00011290	.016257
407	161.82	.021667	.00013389	.019280	457	171.47	.019296	.00011253	.016204
408	162.02	.021614	.00013340	.019209	458	171.66	.019254	.00011216	.016151
409	162.21	.021561	.00013291	.019139	459	171.84	.019212	.00011180	.016099
410	162.41	.021508	.00013243	.019069	460	172.03	.019170	.00011143	.016045
411	162.61	.021456	.00013194	.018999	461	172.22	.019129	.00011107	.015994
412	162.81	.021404	.00013146	.018930	462	172.40	.019087	.00011071	.015942
413	163.01	.021352	.00013098	.018861	463	172.59	.019046	.00011035	.015890
414	163.20	.021301	.00013052	.018794	464	172.78	.019005	.00010999	.015838
415	163.40	.021249	.00013004	.018725	465	172.96	.018964	.00010964	.015788
416	163.60	.021198	.00012957	.018658	466	173.15	.018924	.00010929	.015737
417	163.79	.021147	.00012911	.018591	467	173.34	.018883	.00010893	.015685
418	163.99	.021097	.00012864	.018524	468	173.52	.018843	.00010859	.015636
419	164.19	.021046	.00012818	.018457	469	173.71	.018803	.00010824	.015586
420	164.38	.020996	.00012772	.018391	470	173.89	.018763	.00010790	.015537
421	164.58	.020946	.00012726	.018325	471	174.08	.018723	.00010755	.015487
422	164.77	.020897	.00012682	.018262	472	174.26	.018683	.00010721	.015438
423	164.97	.020847	.00012633	.018195	473	174.45	.018644	.00010687	.015389
424	165.16	.020798	.00012592	.018132	474	174.63	.018604	.00010653	.015340
425	165.36	.020749	.00012547	.018067	475	174.81	.018565	.00010620	.015292
426	165.55	.020700	.00012503	.018004	476	175.00	.018526	.00010586	.015243
427	165.75	.020652	.00012459	.017940	477	175.18	.018487	.00010553	.015196
428	165.94	.020604	.00012416	.017879	478	175.36	.018449	.00010520	.015148
429	166.13	.020556	.00012373	.017817	479	175.55	.018410	.00010487	.015101
430	166.33	.020508	.00012329	.017753	480	175.73	.018372	.00010454	.015053
431	166.52	.020460	.00012286	.017691	481	175.91	.018333	.00010421	.015006
432	166.71	.020413	.00012244	.017631	482	176.10	.018295	.00010388	.014958
433	166.91	.020366	.00012201	.017569	483	176.28	.018258	.00010357	.014914
434	167.10	.020319	.00012159	.017508	484	176.46	.018220	.00010325	.014868
435	167.29	.020272	.00012117	.017448	485	176.64	.018182	.00010293	.014821
436	167.48	.020226	.00012076	.017389	486	176.83	.018145	.00010261	.014775
437	167.67	.020179	.00012034	.017328	487	177.01	.018108	.00010229	.014729
438	167.87	.020133	.00011993	.017269	488	177.19	.018070	.00010198	.014685
439	168.06	.020087	.00011952	.017210	489	177.37	.018033	.00010166	.014639
440	168.25	.020042	.00011912	.017153	490	177.55	.017996	.00010135	.014594
441	168.44	.019996	.00011871	.017094	491	177.73	.017960	.00010105	.014551
442	168.63	.019951	.00011831	.017036	492	177.91	.017924	.00010074	.014506
443	168.82	.019906	.00011791	.016979	493	178.10	.017987	.00010043	.014461
444	169.01	.019861	.00011751	.016921	494	178.28	.017851	.00010012	.014417
445	169.20	.019817	.00011712	.016865	495	178.46	.017815	.00009982	.014374
446	169.39	.019772	.00011672	.016807	496	178.64	.017779	.00009952	.014330
447	169.58	.019728	.00011633	.016751	497	178.82	.017743	.00009922	.014287
448	169.77	.019684	.00011594	.016695	498	179.00	.017708	.00009892	.014244
449	169.96	.019640	.00011555	.016639	499	179.18	.017672	.00009862	.014201
450	170.15	.019596	.00011516	.016583	500	179.35	.017637	.00009833	.014159

Dia. in inch.	Circ'm in ft. in.	Area in square inch.	Dia. in inch.	Circ'm in ft. in.	Area in sq. inch.	Dia. in inch.	Circ'm in ft. in.	Area in Square feet.
1/16	.196	.0030	5	1 3 5/8	19.635	10 7/8	2 10 1/8	.6499
1/8	.392	.0122	5 1/8	1 4 1/8	20.629	11	2 10 1/2	.6652
3/16	.589	.0276	5 1/4	1 4 1/2	21.647	11 1/8	2 10 7/8	.6804
1/4	.785	.0490	5 3/8	1 4 7/8	22.690	11 1/4	2 11 1/4	.6958
5/16	.981	.0767	5 1/2	1 5 1/4	23.758	11 3/8	2 11 3/4	.7143
3/8	1.178	.1104	5 5/8	1 5 5/8	24.850	11 1/2	3 0 1/8	.7270
7/16	1.374	.1503	5 3/4	1 6	25.967	11 5/8	3 0 1/2	.7429
1/2	1.570	.1963	5 7/8	1 6 3/8	27.108	11 3/4	3 0 7/8	.7590
9/16	1.767	.2485	6	1 6 3/4	28.274	11 7/8	3 1 1/4	.7752
5/8	1.963	.3068	6 1/8	1 7 1/4	29.464	12	3 1 5/8	.7916
11/16	2.159	.3712	6 1/4	1 7 5/8	30.679	12 1/8	3 2	.8082
3/4	2.356	.4417	6 3/8	1 8	31.919	12 1/4	3 2 1/2	.8250
13/16	2.552	.5185	6 1/2	1 8 3/8	33.183	12 3/8	3 2 7/8	.8419
7/8	2.748	.6013	6 5/8	1 8 3/4	34.471	12 1/2	3 3 1/4	.8590
15/16	2.945	.6903	6 3/4	1 9 1/8	35.784	12 5/8	3 3 5/8	.8762
1	3 1/8	.7854	6 7/8	1 9 1/2	37.122	12 3/4	3 4	.8937
1 1/8	3 1/2	.9940	7	1 10	38.484	12 7/8	3 4 3/8	.9113
1 1/4	3 7/8	1.227	7 1/8	1 10 3/8	39.871	13	3 4 3/4	.9291
1 3/8	4 1/4	1.484	7 1/4	1 10 3/4	41.282	13 1/8	3 5 1/4	.9470
1 1/2	4 5/8	1.767	7 3/8	1 11 1/8	42.718	13 1/4	3 5 5/8	.9642
1 5/8	5 1/8	2.073	7 1/2	1 11 1/2	44.178	13 3/8	3 6	.9835
1 3/4	5 1/2	2.405	7 5/8	1 11 7/8	45.663	13 1/2	3 6 3/8	1.0019
1 7/8	5 7/8	2.761	7 3/4	2 0 3/8	47.173	13 5/8	3 6 3/4	1.0206
2	6 1/4	3.141	7 7/8	2 0 3/4	48.707	13 3/4	3 7 1/8	1.0294
2 1/8	6 5/8	3.546	8	2 1 1/8	50.265	13 7/8	3 7 1/2	1.0584
2 1/4	7	3.976	8 1/8	2 1 1/2	51.848	14	3 7 7/8	1.0775
2 3/8	7 3/8	4.430	8 1/4	2 1 7/8	53.456	14 1/8	3 8 3/8	1.0968
2 1/2	7 3/4	4.908	8 3/8	2 2 1/4	55.088	14 1/4	3 8 3/4	1.1193
2 5/8	8 1/4	5.412	8 1/2	2 2 5/8	56.745	14 5/8	3 9 1/8	1.1360
2 3/4	8 5/8	5.939	8 5/8	2 3	58.426	14 1/2	3 9 1/2	1.1569
2 7/8	9	6.491	8 3/4	2 3 3/8	60.132	14 5/8	3 9 7/8	1.1749
3	9 3/8	7.068	8 7/8	2 3 7/8	61.862	14 3/4	3 10 1/4	1.1961
3 1/8	9 3/4	7.669	9	2 4 1/4	63.617	14 7/8	3 10 5/8	1.2164
3 1/4	10 1/4	8.295	9 1/8	2 4 5/8	65.396	1 5	3 11 1/8	1.2370
3 3/8	10 5/8	8.946	9 1/4	2 5	67.200	15 1/8	3 11 1/2	1.2577
3 1/2	11	9.621	9 3/8	2 5 3/8	69.029	15 1/4	3 11 7/8	1.2785
3 5/8	11 3/8	10.320	9 1/2	2 5 3/4	70.882	15 3/8	4 0 1/4	1.2996
3 3/4	11 3/4	11.044	9 5/8	2 6 1/4	72.759	15 1/2	4 0 5/8	1.3208
3 7/8	12 1/8	11.793	9 3/4	2 6 5/8	74.662	15 5/8	4 1	1.3422
4	1 0 1/4	12.566	9 7/8	2 7	76.588	15 3/4	4 1 1/2	1.3637
4 1/8	1 0 7/8	13.364	10	2 7 3/8	78.540	15 7/8	4 1 7/8	1.3855
4 1/4	1 1 3/8	14.186	10 1/8	2 7 3/4	80.515	16	4 2 1/4	1.4074
4 3/8	1 1 3/4	15.033	10 1/4	2 8 1/8	82.516	16 1/8	4 2 5/8	1.4295
4 1/2	1 2 1/4	15.904	10 3/8	2 8 1/2	84.540	16 1/4	4 3	1.4517
4 5/8	1 2 1/2	16.800	10 1/2	2 8 7/8	86.590	16 3/8	4 3 3/8	1.4741
4 3/4	1 2 7/8	17.720	10 5/8	2 9 1/4	88.664	16 1/2	4 3 3/4	1.4967
4 7/8	1 3 1/4	18.665	10 3/4	2 9 3/4	90.762	16 5/8	4 4 1/4	1.5195

Diam. in inch.	Circ'm in ft. in.	Area in Square feet.	Diam. in ft. in.	Circ'm in ft. in.	Area in Square feet.	Diam. in ft. in.	Circ'm. in ft. in.	Area in Square feet.
16¾	4 4⅝	1.5424	22⅝	5 11	2.7980	2 9	8 7⅝	5.9398
16⅞	4 5	1.5655	22¾	5 11½	2.8054	2 9¼	8 8½	6.0291
17	4 5⅜	1.5888	22⅞	5 11⅞	2.8658	2 9½	8 9¼	6.1201
17⅛	4 5¾	1.6123	23	6 0¼	2.8903	2 9¾	8 10	6.2129
17¼	4 6⅛	1.6359	23⅛	6 0⅝	2.9100	2 10	8 10¾	6.3051
17⅜	4 6½	1.6597	23¼	6 1	2.9518	2 10¼	8 11½	6.3981
17½	4 6⅞	1.6836	23⅜	6 1⅜	2.9937	2 10½	9 0⅜	6.4911
17⅝	4 7⅜	1.7078	23½	6 1¾	3.0129	2 10¾	9 1⅛	6.5863
17¾	4 7¾	1.7321	23⅝	6 2¼	3.0261	2 11	9 1⅞	6.6815
17⅞	4 8⅛	1.7566	23¾	6 2⅝	3.0722	2 11¼	9 2¾	6.7772
18	4 8½	1.7812	23⅞	6 3	3.1081	2 11½	9 3½	6.8738
18⅛	4 8⅞	1.8061	2 0	6 3⅜	3.1418	2 11¾	9 4¼	6.9701
18¼	4 9¼	1.8311	2 0¼	6 4⅛	3 2075	3 0	9 5	7.0688
18⅜	4 9¾	1.8562	2 0½	6 4⅞	3.2731	3 0¼	9 5⅞	7.1671
18½	4 10⅛	1.8816	2 0¾	6 5¾	3.3410	3 0½	9 6⅝	7.2664
18⅝	4 10½	1.9071	2 1	6 6½	3.4081	3 0¾	9 7½	7.3662
18¾	4 10⅞	1.9328	2 1¼	6 7¼	3.4775	3 1	9 8¼	7.4661
18⅞	4 11¼	1.9586	2 1½	6 8⅛	3.5468	3 1¼	9 9	7.5671
19	4 11⅝	1.9847	2 1¾	6 8⅞	3.6101	3 1½	9 9⅞	7.6691
19⅛	5 0	1.9941	2 2	6 9⅝	3.6870	3 1¾	9 10½	7.7791
19¼	5 0½	2.0371	2 2¼	6 10½	3.7583	3 2	9 11⅜	7.8681
19⅜	5 0⅞	2.0637	2 2½	6 11¼	3.8302	3 2¼	10 0⅛	7.9791
19½	5 1¼	2.0904	2 2¾	7 0	3.9042	3 2½	10 0⅞	8.0846
19⅝	5 1⅝	2.1172	2 3	7 0¾	3.9761	3 2¾	10 1¾	8.1891
19¾	5 2	2.1443	2 3¼	7 1⅝	4.0500	3 3	10 2½	8.2951
19⅞	5 2⅜	2.1716	2 3½	7 2⅜	4.1241	3 3¼	10 3¼	8.4026
20	5 2⅞	2.1990	2 3¾	7 3⅛	4.2000	3 3½	10 4	8.5091
20⅛	5 3¼	2.2265	2 4	7 3⅞	4.2760	3 3¾	10 4⅞	8.6171
20¼	5 3⅝	2.2543	2 4¼	7 4¾	4.3521	3 4	10 5⅝	8.7269
20⅜	5 4	2.2822	2 4½	7 5½	4.4302	3 4¼	10 6⅜	8.8361
20½	5 4⅜	2.3103	2 4¾	7 6¼	4.5083	3 4½	10 7¼	8.9462
20⅝	5 4¾	2.3386	2 5	7 7	4 5861	3 4¾	10 8	9.0561
20¾	5 5⅛	2.3670	2 5¼	7 7⅞	4.6665	3 5	10 8¾	9.1686
20⅞	5 5½	2.3956	2 5½	7 8⅝	4.7467	3 5¼	10 9½	9.2112
21	5 5⅞	2.4244	2 5¾	7 9½	4.8274	3 5½	10 10⅜	9.3936
21⅛	5 6⅜	2.4533	2 6	7 10¼	4.9081	3 5¾	10 11⅛	9.5061
21¼	5 6¾	2.4824	2 6¼	7 11	4.9901	3 6	11 11⅞	9.6212
21⅜	5 7⅛	2.5117	2 6½	7 11¾	5.0731	3 6¼	11 0¾	9.7364
21½	5 7½	2.5412	2 6¾	8 0⅝	5.1573	3 6½	11 1½	9.8518
21⅝	5 7⅞	2.5708	2 7	8 1⅜	5.2278	3 6¾	11 2¼	9.9671
21¾	5 8¼	2.6007	2 7¼	8 2⅛	5.3264	3 7	11 3	10.084
21⅞	5 8¾	2.6306	2 7½	8 2⅞	5.4112	3 7¼	11 3⅞	10.202
22	5 9⅛	2.6608	2 7¾	8 3¾	5.4982	3 7½	11 4⅝	10.320
22⅛	5 9½	2.6691	2 8	8 4½	5.5850	3 7¾	11 5⅜	10.439
22¼	5 9⅞	2.7016	2 8¼	8 5⅜	5.6729	3 8	11 6¼	10.559
22⅜	5 10¼	2.7224	2 8½	8 6⅛	5.7601	3 8¼	11 7	10.679
22½	5 10⅝	2.7632	2 8¾	8 6⅞	5.8491	3 8½	11 7¾	10.800

Diam. in ft. in.	Circ'm in ft. in.	Area in Square feet	Diam. in ft. in.	Circ'm. in ft. in.	Area in Square feet.	Diam in ft. in.	Circ'm in ft. in.	Area in Square feet.
3 8¾	11 8½	10.922	4 8½	14 9½	17.411	5 8¼	17 10⅜	25.405
3 9	11 9⅜	11.044	4 8¾	14 10¼	17.565	5 8½	17 11⅛	25.592
3 9¼	11 10⅛	11.167	4 9	14 11	17.720	5 8¾	17 11⅞	25.779
3 9½	11 10⅞	11.291	4 9¼	14 11⅞	17.876	5 9	18 0¾	25 964
3 9¾	11 11¾	11.415	4 9½	15 0⅝	18.033	5 9¼	18 1½	26.155
3 10	12 0½	11.534	4 9¾	15 1⅜	18.189	5 9½	18 2¼	26.344
3 10¼	12 1¼	11.666	4 10	15 2¼	18.347	5 9¾	18 3⅛	26.534
3 10½	12 2	11.793	4 10¼	15 2⅞	18.506	5 10	18 3⅞	26.725
3 10¾	12 2⅞	11.920	4 10½	15 3¾	18.665	5 10¼	18 4⅝	26.916
3 11	12 3⅝	12.048	4 10¾	15 4½	18.825	5 10½	18 5½	27.108
3 11¼	12 4⅜	12.176	4 11	15 5¼	18.985	5 10¾	18 6¼	27.301
3 11½	12 5¼	12.305	4 11¼	15 6⅛	19.147	5 11	18 7	27.494
3 11¾	12 6	12.435	4 11½	15 6⅞	19.309	5 11¼	18 7¾	27.688
4 0	12 6¾	12.566	4 11¾	15 7¾	19.471	5 11½	18 8⅝	27.883
4 0¼	12 7½	12.697	5 0	15 8½	19.635	5 11¾	18 9⅜	28.078
4 0½	12 8⅜	12.829	5 0¼	15 9¼	19.798	6 0	18 10⅛	28.274
4 0¾	12 9⅛	12.962	5 0½	15 10	19.963	6 0¼	18 10⅞	28.471
4 1	12 9⅞	13.095	5 0¾	15 10¾	20.128	6 0½	18 11¾	28.663
4 1¼	12 10¾	13.229	5 1	15 11⅝	20.294	6 0¾	19 0½	28.866
4 1½	12 11½	13.364	5 1¼	16 0⅜	20.461	6 1	19 1¼	29.065
4 1¾	13 0¼	13.499	5 1½	16 1¼	20.629	6 1¼	19 2⅛	29.264
4 2	13 1	13.635	5 1¾	16 1⅞	20.797	6 1½	19 2⅞	29.466
4 2¼	13 1⅞	13.772	5 2	16 2¾	20.965	6 1¾	19 3⅝	29.665
4 2½	13 2⅝	13.909	5 2¼	16 3½	21.135	6 2	19 4½	29.867
4 2¾	13 3⅜	14.047	5 2½	16 4¼	21.305	6 2¼	19 5¼	30.069
4 3	13 4¼	14.186	5 2¾	16 5⅛	21.476	6 2½	19 6	30.271
4 3¼	13 5	14.325	5 3	16 5⅞	21.647	6 2¾	19 6¾	30.475
4 3½	13 5¾	14.465	5 3¼	16 6¼	21.819	6 3	19 7⅝	30.679
4 3¾	13 6½	14.606	5 3½	16 7½	21.992	6 3¼	19 8⅜	30.884
4 4	13 7⅜	14.748	5 3¾	16 8¼	22.166	6 3½	19 9⅛	31.090
4 4¼	13 8⅛	14.890	5 4	16 9	22.333	6 3¾	19 9⅞	31.296
4 4½	13 8⅞	15.033	5 4¼	16 9¾	22.515	6 4	19 10¾	31 503
4 4¾	13 9¾	15.176	5 4½	16 10⅝	22.621	6 4¼	19 11½	31.710
4 5	13 10½	15.320	5 4¾	16 11⅜	22.866	6 4½	20 0¼	31.919
4 5¼	13 11¼	15.465	5 5	17 0⅛	23.043	6 4¾	20 1⅛	32.114
4 5½	14 0	15.611	5 5¼	17 0⅞	23.221	6 5	20 1⅞	32.337
4 5¾	14 0⅞	15.757	5 5½	17 1¾	23.330	6 5¼	20 2⅝	32.548
4 6	14 1⅝	15.904	5 5¾	17 2½	23.578	6 5½	20 3½	32.759
4 6¼	14 2⅜	16.051	5 6	17 3⅜	23.758	6 5¾	20 4¼	32.970
4 6½	14 3¼	16.200	5 6¼	17 4⅛	23.938	6 6	20 5	33.183
4 6¾	14 4	16.349	5 6½	17 4⅞	24.119	6 6¼	20 5¾	33.396
4 7	14 4¾	16.498	5 6¾	17 5⅝	24.301	6 6½	20 6½	33.619
4 7¼	14 5½	16.649	5 7	17 6½	24.483	6 6¾	20 7⅜	33.824
4 7½	14 6⅜	16.800	5 7¼	17 7¼	24.666	6 7	20 8⅛	34.039
4 7¾	14 7⅛	16.951	5 7½	17 8	24.850	6 7¼	20 8⅞	34.255
4 8	14 7⅞	17.104	5 7¾	17 8¾	25.034	6 7½	20 9¾	34.471
4 8¼	14 8⅝	17.257	5 8	17 9⅝	25.220	6 7¾	20 10½	34.688

Diam. in ft. in.	Circ'm in ft. in.	Area in square feet.	Diam. in ft. in.	Circ'm. in ft in.	Area in square feet.	Diam in ft. in.	Circ'm in ft. in.	Area in square feet.
6 8	20 11¼	34.906	9 7	30 1¼	72.1309	13 6	42 4⅞	143.1391
6 8¼	21 0⅛	35.125	8	30 4⅜	73.3910	7	42 8	144.9111
6 8½	21 0⅞	35.344	9	30 7½	74.6620	8	42 11⅛	146.6949
6 8¾	21 1⅝	35.564	10	30 11⅝	75.9433	9	43 2¼	148.4896
6 9	21 2⅜	35.784	11	31 1¾	77.2362	10	43 5½	150.2943
6 9¼	21 3¼	36.006	10 0	31 5	78.5400	11	43 8⅝	152.1109
6 9½	21 4	36.227	1	31 8⅛	79.8540	14 0	43 11¾	153.9384
6 9¾	21 4¾	36.450	2	31 11¼	81.1795	1	44 2⅞	155.7758
6 10	21 5½	36.674	3	32 2⅜	82.5160	2	44 6	157.6250
6 10¼	21 6⅜	36.897	4	32 5½	83.8627	3	44 9⅛	159.4852
6 10½	21 7⅛	37.122	5	32 8⅝	85.200	4	45 0¼	161.3553
6 10¾	21 7⅞	37.347	6	32 11¾	86.588	5	45 3½	163.2373
6 11	21 8¾	37.573	7	33 2⅞	87.9697	6	45 6⅝	165.1303
6 11¼	21 9½	37.700	8	33 6⅛	89.3608	7	45 9¾	167.0331
6 11½	21 10¼	38.027	9	33 9¼	90.7627	8	46 0⅞	168.9479
6 11¾	21 11	38.256	10	34 0⅜	92.1749	9	46 4	170.8735
7 0	21 11⅞	38.4846	11	34 3½	93.5986	10	46 7⅛	172.8091
1	22 3	39.4060	11 0	34 6⅝	95.0334	11	46 11¼	174.7565
2	22 6⅛	40.3388	1	34 9¾	96.4783	15 0	47 1½	176.7150
3	22 9¼	41.2825	2	35 0⅞	97.9347	1	47 4⅝	178.6832
4	23 0⅜	42.2367	3	35 4⅛	99.4021	2	47 7¾	180.6634
5	23 2⅛	43.2022	4	35 7¼	100.8797	3	47 10⅞	182.6545
6	23 6¾	44.1787	5	35 10⅝	102.3689	4	48 2½	184.6555
7	23 11	45.1656	6	36 1½	103.8691	5	48 5⅛	186.6684
8	24 1⅛	46.1638	7	36 4½	105.3794	6	48 8¼	188.6923
9	24 4⅛	47.1730	8	36 7¾	106.9013	7	48 11⅜	190.7260
10	24 7¼	48.1926	9	36 10⅞	108.4342	8	49 2⅝	192.7716
11	24 10⅜	49.2236	10	37 2¾	109.9772	9	49 5¾	194.8282
8 0	25 1½	50 2656	11	37 5¼	111.5319	10	49 8⅞	196.8946
1	25 4⅝	51.3178	12 0	37 8⅜	113.0976	11	50 0	198.9730
2	25 7⅞	52.3816	1	37 11½	114.6732	16 0	50 3⅛	201.0624
3	25 11	53.4562	2	38 2⅝	116.2607	1	50 6¼	203.1615
4	26 2⅛	54.5412	3	38 5¾	117 8590	2	50 9⅜	205.2726
5	26 5¼	55.6377	4	38 8⅞	119.4674	3	51 0½	207.3946
6	26 8⅜	56.7451	5	39 0	121.0876	4	51 3¾	209.5264
7	26 11½	57.8628	6	39 3¼	122.7187	5	51 6½	211.6703
8	27 2¾	58.9920	7	39 6⅜	124.3598	6	51 10	213.8251
9	27 5¾	60.1321	8	39 9½	126.0127	7	52 1⅛	215.9896
10	27 9	61.2826	9	40 0⅝	127.6765	8	52 4¼	218.1662
11	28 0⅛	62.4445	10	40 3¾	129.3504	9	52 7⅜	220.3537
9 0	28 3¼	63.6174	11	40 6⅞	131.0360	10	52 10½	222.5510
1	28 6⅜	64.8006	13 0	40 10	132.7326	11	53 1⅝	224.7603
2	28 9½	65.9951	1	41 1⅛	134.4391	17 0	53 4⅞	226.9806
3	29 0⅝	67.2007	2	41 4⅜	136.1574	1	53 8	229.2105
4	29 3¾	68.4166	3	41 7½	137.8867	2	53 11⅛	231.4625
5	29 7	69.6440	4	41 10⅝	139.6260	3	54 2⅛	233.7055
6	29 10⅛	70.8823	5	42 1⅝	141.3771	4	54 5⅜	235.9682

CHARLA.

MISS CHARLA A. ADAMS.

A Green Mountain girl, receiving three months' schooling in the summer and occasional spells in the winter. At thirteen away to the Lowell mills, graduating from there at nineteen as mathematician of my testing work, and as I had never owned a schoolbook until buying them for my children, it will readily be conceived that we were not handicapped by the Massachusetts school system.

Without exception Charla was the most expeditious mathematician and best adapted for the purpose of any one I have ever known engaged in the work.

SPIRITUALISM.

A Living Religion, of Demonstration, Personal Responsibility, and Consolation.

This belief has been latent in the human heart since the dawn of recorded intelligence down to the present time, and is now openly accepted by the most intelligent as the truth, yet sneered at by the dollar stamped clergy from self interest.

The cause has had a terrible load to carry in carrying the vagaries of its professed friends, and had it not been based upon eternal truth it would have been annihilated long ago.

Its mediums, mere mortals of very ordinary clay, instead of being encouraged and aided to seek the truth, have too often been surrounded by ruffianly bands of bigoted ignorance, and in frequent cases female mediums have been married by lazy loafers of the male species, solely as a means of obtaining a living without labor, and the wife has often been compelled to do what, if properly cherished, she would never have thought of doing.

Then again, as its expounders in many cases have belonged to, to say the least, not the most learned, the vagaries published are not always well established, to say nothing of the long words required to express the profound depths of the writer's ideas.

Then the *smellers* that seek for fraud, the self appointed witch finders of the Gagool type described by Rider Haggard in "King Solomon's Mines," who through monumental conceit and ignorantly conceived notions of spirit etiquette assume the office of censor of spiritual management, may retard but can never stop the onward march of its grand and humane truths.

From infancy I have ever desired to know the why of any mystery. My first visit to a haunted house was in my eighth year. Of course the Rochester knockings interested me, but a wandering life of ten years' previous experience in strange lands had knocked many of childhood's conceits from my mind and broadened the horizon of my ideas; personal experience also had caused consideration. It was not uncommon for me at that time to suddenly become unconscious and begin to repeat lines of poetry that would be seemingly printed upon the wall of the room in front of me. As the last word was repeated, there would be exactly such a change in appearance as takes place in a kaleidoscope and more lines would come in view. As this was about a year before the advent of the knockings my declamations were considered uncanny. A vivid impression of the fact was always left upon my mind but the lines could never be remembered.

Then followed a phase of gradually rousing from sleep to a consciousness of two or three voices near by arguing a case, so real that it would cause me to turn and try various methods to ascertain whether it was a dream; suddenly all would cease but the impression would remain for days, yet the subject could never be recalled though perfectly understood the moment before it ceased.

To this followed visions of beautiful landscapes, rarely persons or animal life, but the colors of mosses, leaves, stones, and the thousand details so perfect that at times I would get up and walk across the room to make sure of being awake. For years these were believed to be optical illusions, but I know better now and deeply regret that such gifts were not more thankfully received. Another phase followed and to some extent is still with me, namely, impressions, often as palpable as spoken words. These usually come when receiving or reading a letter, message, or communication, in one case causing me to pitch a letter containing a check for $150 into the waste basket, for doing which the sender at times attempts to be sarcastic.

For years I took but little interest in Spiritualism, but as its adherents increased it became a power, and I took the Christian's ideal of good, the dollar, as a standard of its popularity.

At the time I was publishing a quarterly paper, EMERSON'S TURBINE REPORTER, five thousand copies each issue to fill contract with advertisers. It had paid expenses less postage up to that time. I announced that after four more issues the paper would be discontinued and a book take its place; then commenced a series of articles on Spiritualism herewith republished in their order. The first issue containing the article paid all expense, the next $25 above, the third over $100, and the last over $200, a supplement being required for advertising space.

SPIRITUALISM.

The wonderful stories of spiritual manifestations going the rounds of the press have caused a desire for more light relative thereto ; such manifestations, under various phases, have been common since the dawn of history ; in ancient times the leaders of the people made them useful, now those that would be leaders are careful to ignore interest in them. Editors that are loudest in screeching, " See how independent we are !" dare not publish an article upon the subject without launching it from the top of the fence that it may be fitted for either side, by the ever convenient, "I told you so !" Why this unmanly hedging? A little inquiry will satisfy any one that the world is ready for the truth. It is true that there is a feeble " tweet, tweet, tweet" going out from the pulpit, as there doubtless was nineteen centuries since, but the time now, as then, is unfavorable to pulpits ; intelligence plays the deuce with such places ; there is little consistency in talking about the Bible being a guide, while building structures in which to worship the son of a carpenter, so very nice that one of that class has little chance of ever seeing the inside after taking his tools out ; five to twenty thousand dollar salaries have little in common with the veritable Jesus of Nazareth, though in full accordance with the pulpit article. The time for such is passing away, the sneered at manifestations have had much to do with the change and Church creeds are kept in the background as being too illiberal for the times. Nearly every book of note now issued is spiced with the belief ; our conversation is mixed with its phrases ; if one doubts the general infusion let him get into quiet conversation with the first person met, and the chances are ten to one that some wonderful experience having a bearing upon the subject will be related. Some of the best known manufacturers with whom I am acquainted are deeply interested as investigators. Such, invariably, are thinkers, and usually successful in their business, some of them very remarkably so. A large portion of Turbine builders are open believers in Spiritualism, and it is but fair to state that, in not one single instance has one of that belief misrepresented results obtained from a test of wheel, while the contrary has often been the case with builders ever ready to sneer at the Spiritualist. It is true that Spiritualism has been " exposed " almost daily for the last twenty-five years, yet it will not down. Would it not be wiser to meet the case fairly and learn what right it has to consideration ? It does not matter what this or that professor has to say upon the subject, unless said after fair examination ; the prefix adds nothing to the individual's power of discernment ; besides, such persons are usually specialists, and have some hobby upon the brain. Professor Univalve spends twenty years in ascertaining the exact number of wrinkles that a mussel of respectable habits should have in his shell at maturity. Prof. Thimble does not believe in spirits, and, like a cow, has no interest in a Hereafter. Our educational professors are so deeply engaged in searching for the roots of words, that the useless abominations in spelling of those words, against which nature through every child learning to read, is constantly protesting, are unnoticed by them, and the stone at one end of the bag to balance the grist is constantly carried, and is likely to be, unless the " heathen Japs " relieve us of the useless weight. It is useless to expect such minds to investigate anything aside from their own narrow world, and perhaps it is better that it is so, for the few have done the thinking for the many too long already. What a turning over of things there would be if prejudice could be annihilated and questions be decided upon merit ! A sort of moral undertow compels general progression now ; froth rises to the top and becomes the most conspicuous ; shallow minds, without investigation, pronounce anything humbug that is new and beyond their comprehension. Could such control events La Place's statement that " What we know is little, what we don't know, immense," would ever remain true. The *cui bono* of the truckling editor, while pandering to popular prejudice, is simply a tribute paid to such minds, and is doubly shallow when written within sight of a score of steeples all claiming to point the way to the spirit land, and upon exactly the same evidence as the sneered at manifestations, the latter witnessed by ourselves, friends, and neighbors, the former by——well, whom ? It is a matter of little consequence whether Prof. Thimble is interested in the matter or not, the world has been, is, and ever will be, interested ; for myself, all other gain would be as nothing compared with the knowledge that life here is but the beginning of eternal conscious progress, that separation from our loved ones is but temporary. If the manifestations are of spiritual origin as claimed they offer the only tangible evidence of a Hereafter. If not of spiritual, but of earthly, origin, may they not be the harbingers of knowledge of boundless importance to humanity? If neither of spiritual nor earthly origin in a proper sense, but the result of mere trickery, then they have a fearful bearing upon evidence. I have seen a table rise upon two legs and walk out of the dining-room into the parlor and return, with no visible person touching it. I have seen two heavy men try in vain to hold a table to the floor ; this in Mechanics Hall, Lowell, Mass., and before an audience of four hundred persons ;. no one pretended to doubt the fact. I have taken a common accordion in my hand, holding it by the molding around the valve ; the instrument extended at arm's length from my side ; the key end of the instrument immediately rose to a level in line with my arm, but extended from me, and then commenced to play a very lively tune ; the sun was shining full upon the instrument. I have taken a slate in my hand, or one end of it, the other being held by the medium ; a bit of pencil was placed upon the slate, which was then held beneath the table, not up against it, but at least a foot below, and in plain sight. The pencil commenced to write immediately ; several messages were produced in less time than I could have written one ; one of the messages was as follows : " There is a large band of us around you ; if you will sit at home we will show you things that are wonderful." I have had the Eddys at my house, also several other well-known mediums ; have had to do with nearly all the best known public mediums, and many not generally known to be such. I have seen the " exposers " such as Carbonell, have spent hours with them at a time in private, and witnessed their *modus operandi*, have seen excellent imitations, as I have also of greenbacks, but an expert can readily see and explain the difference. Have often had such mediums as Foster and Read try to play tricks upon me, at the same time have seen things that trickery could not accomplish. I have witnessed the most of the various kinds of manifestations described by R. D. Owen, and others he has not described ; mind-reading will account for Mans field's letter answering, and some other mysteries, but there is something deeper and beyond. It is singular that a people so boastful of intelligence should be so shy of investigations outside of Congress. The following letter to the *N. Y. Graphic* displays more true manhood than is generally to be met with in regard to the subject.

ELMIRA, N. Y., November 11, 1874.

GENTLEMEN : Your circular indicates a most reasonable request. It is indeed a burning shame that men called scientific and investigators should be so hopelessly materialistic that they will not look towards the only windows through which the twilight of a great discovery is now shining.

Thirty years ago I would have sacrificed everything to undertake, without encouragement, the work to which you now invite me and others. But, as matters now stand, I have not the time or strength to do the work ; and had I both, my standing is not such among men of science that discoveries made by me, however important, could even arrest attention, much less command respect.

Profoundly distrustful of much that honest but untrained men tell us as to spiritual manifestations, it yet, remains that where there is so much smoke of notoriety there must be some fire of fact. How much let him declare whom you succeed in pressing into your service as investigator and reporter. Sincerely regretting that I cannot be the man, I remain very truly yours, THOS. K. BEECHER.

If people in general were candid thinkers, like Mr. Beecher, we might hope for a speedy solution of the matter, but, unfortunately, the majority take their opinion second-handed, while the balance divide into two parties, seemingly running in opposite directions, but in seeming only. The one believe everything, the other nothing ; the leaders of the first, with heads shaped like a pineapple cheese, or perhaps more on the shed roof style, the slope being such that one is left in doubt whether the forehead extends to the crown of the head, or the top of the head reaches down to the eyes ; these swear by the *Banner of Light* ; their followers are expected to swallow mountains or mites ; mediums by such are spoken of as " too sensitive for ordinary treatment," " heaven borned," " of the angels," etc., etc. (while in fact, as a general thing, public mediums are lazy sensualists, generally acting the part of Harold Skimpole, and never forgetting to take the " Fypunnote "). and a score of that ilk are cancerous excretions of the cause. The other party simply panders to popular prejudice, and naturally gravitates toward the *Scientific American*, a fair offset to the *Banner of Light*, the one certainly knowing as much of spirits as the other does of science. The writers of this party are generally nicely bespectacled young men with weak eyes, knees, and heads, and considerable alphabet tailed on to their address, with a strong flavor of the apothecary apprentice about them. The organ of this party has just been handed to me, and in it the announcement is gravely made that the manifestations called materializations were invented by one Gordon, of New York, about two years since (don't state whether he patented them through that agency or not). The materializations were common ten years since, and it was in answer to a request that he would witness them, that the following letter was written.

BOSTON, November 28, 1865.

JAMES EMERSON, LOWELL, MASS. :

Dear Sir :—I hope I shall find time sooner or later to attend some of the best managed so-called "spiritual" *seances*, but just now I am too much occupied to do anything more than listen to the wonderful stories you are told about them.

Yours, in haste,

O. W. HOLMES.

It is often asserted that if one commences to investigate the so-called manifestations, he soon becomes infatuated, and a believer. Well, suppose the discovery of a gold mine to be announced, do experts ever delve in a "salted" mine twenty-five years ? If the assertion is true it would rather seem to favor the idea that there is something to become infatuated with, but persons are often credited with being what they are not, as will be seen by this letter.

20 MORNINGTON RD., LONDON, N. W.,
Aug. 19, 1872.

JAMES EMERSON :

Dear Sir :—Long traveling about on business has prevented me from previously acknowledging the receipt of your most interesting letter giving an account of some phenomena you have witnessed in the presence of Dr. Slade. After the very extraordinary things you have seen, I am particularly struck with what you state your opinion to be —viz. : that the " Spirit World " has nothing to do with them, but that the phenomena belong to our physical bodies. If you could explain what you have stated to me, and could give me the reasons which cause you to think that the exertion of force (not that of the medium physically) and the writing of messages by a piece of pencil not held in a human hand, are connected with our physical body and not with invisible, independent, intelligent beings, I should be very pleased. The latter opinion is the one most generally held by those who have studied the phenomena here. For myself, I confess I do not go as far as some, and until I can get good proofs of identity I prefer to keep to the " force " only, for there I am safe. With many thanks for your polite attention, believe me very sincerely yours, WILLIAM CROOKS.

It would be impossible for me to explain fully, why I believe the manifestations to be of physical origin, but such ever has been, and continues to be, my opinion ; there is a lack of connection as well as an earthiness, that seems to locate them with ourselves, but for all that, there is ground for the spiritual claim ; the water of a river partakes of the soil through which it flows, but remains water for all that. The manifestations partake of their earthly surroundings. P. H. Vander Wyede, through *Scientific American*, says the manifestations are silly ; one has but to read one of his articles to see why he finds them so. Dr. Hammond published an article in which he pronounced them to be the result of trickery ; his career while Surgeon-General will perhaps account for his belief, but enough of such. The weak minded are credited with being the most interested in such matters, but in all the *seances* with which I have had to do, either public or private, there has never been any trouble in filling the house with the best mechanics known, mill agents, school superintendents and teachers, doctors, lawyers, ministers, members of Congress, etc., etc. The belief of the better class of spiritualists is substantially that taught by Jesus of Nazareth, and it is singular that a belief so sensible and beautiful has not produced a literature to correspond. That such is not the case is probably owing to the fact that the best minds tinged with that belief feel that more good can be done through the liberal religious movement, which may be the case, but it leaves the cause of Spiritualism in the care of those who have done it little credit, and at the close of a quarter of a century there is not a paper published in that interest that a gentleman would care to be seen reading in car or hotel. Watching the falling of an apple, the rattling tea-kettle cover, or flying a kite, were perhaps not the most dignified of employments, but the results have revolutionized the world. The "Spiritual Manifestations" may, or may not be, of equal importance, but believing them to be of God, or nature, as the reader chooses, and that they may be made useful, I at least shall do what is in my power to ascertain their cause.

POPULAR SCIENCE.

During our war of rebellion the idea became prevalent that our flunkyism relative to English opinion would be cured ; and such might have been the case had it not been for a great change in the management of our leading papers. Previous to the war, writers of age, talent, and experience were employed thereon ; now, through motives of economy, boys take the place of such. The former never quoted the *Scientific American* as authority, in fact, never quoted it at all. The boy writers swallow its wonderful statements unquestioned ; while our local editor, with his three hundred subscribers, made up of those who advertise "pull-backs," codfish, tin-ware and skillets, pulls off his hat in reverence, as he catches sight of a "New Discovery," by Prof. Tyndall, or "The Mystery Solved," by Prof. Carpenter ; though were he a reader and thinker, he would readily recognize the fact that both discovery and solution were old a hundred years ago. Look at the following fresh from the press, and which fairly represents Mr. Tyndall as a scientist.

Fresh Discovery and Practical Suggestions.

PROF. TYNDALL ON HEAT.

Having caused a ball of lead to fall from the roof of a theater on to a stone, he drew the ball up again and let it down gently with a string and pulley. The heat generated by the collision in the first instance was the exact equivalent of the heat produced in his finger and thumb, and in the string in the second instance. The outlay of muscular force expended in drawing up the ball was made obvious by causing the ball to be drawn up again by a small engine worked by compressed air. The exact equivalent of the heat evolved by a quantity of coal, completely consumed by consumption with oxygen, sufficient to lift a weight of 50 tons to a height of 100 feet above the earth, would be produced by the collision of that mass with the earth when allowed to fall. Given the velocity of a body, the heat generated by the destruction of that velocity could be easily calculated, and some time ago he was led to the conclusion that the stoppage of a rifle bullet would produce sufficient heat to fuse the metal. This conclusion was proved in the Franco-German war, when bullets which had been stopped by contact with a bone showed, on being extracted, undoubted marks, in many cases, of fusion. The same thing had also been illustrated incidentally in the experiments with gun-cotton at Stowmarket.

This "Fresh Discovery" was a part of the stock in trade of a gassy lecturer, named Boynton, who traveled the country some thirty years since. He elaborated it, however, by adding that the "average laborer consumes fourteen ounces of carbon per day, and fourteen ounces of carbon, consumed by a man or a steam engine, will lift the same weight of brick to a given height." The statement was repeated by myself to an old physician, then of Worcester, Mass. "Humph !" was his rejoinder, "heard that in lectures at college when I was a boy." Mr. Tyndall seems to be a sort of Rip Van Winkle, and to have waked from a nap of a few centuries. A few years since he announced that he had discovered that heat moves in waves. That fact was a theme for angry discussion among stove builders a half century since ; a portion favoring the use of sheet iron because its "flexibility caused it to throw off heat in more rapid waves than could be possible with its more rigid competitor, cast iron." That heat moves in waves is a fact that has been perceptible since hot surfaces existed. This discovery by Mr. Tyndall was soon followed by the announcement that he had also discovered that *motion* moves in waves, which could hardly seem new to any one who ever saw the ocean, felt the waves of an earthquake, or who, as a boy, ever gave the end of a long rope a flip, thus causing a wave to run its whole length. The Indian, however, who has watched the flight of an arrow or lance, may have his doubts as to the invariable applicability of the rule. Not long since, the editor of the *Scientific American* urged the substitution of death by electricity for that of hanging ; innocently stating that Prof. Tyndall, while experimenting, was knocked senseless by a shock, and on recovery announced the fact that it didn't hurt, thus adding another to his characteristic discoveries. Mr. Tyndall is probably more generally known through his "Prayer Gauge" proposition, than in any other way; but in this he retained his consistency. It would be difficult to find a boy of ten who has not heard very positive doubts expressed as to the efficacy of prayer; and such doubts have been expressed by writers for more than two thousand years. "Can the Ethiopian change his skin or the leopard his spots," is plain enough. Franklin was equally plain when he suggested that it would save time and answer the same purpose to ask a blessing over the food in the lump, when it was housed in the fall, as to do it at each meal daily. Paine in his "Age of Reason," Allen in his "Oracles of Reason," and many other writers have done the same. Yet it is hardly likely that any observant person has doubted the benefit of prayer to the petitioner, but merely that the Creator is unlikely to change his laws at the solicitation of individuals. A wish is a prayer. To "cry" is to pray. The new born child utters its first prayer with its first breath, and probably with about the same consciousness of its real needs as have those who make the most show of praying. Plato, or one of his friends, once remarked: "It would be well to hesitate before praying, as the gods might answer the prayer." We may readily conceive that things would become somewhat tangled, if the prayers, even of a single Sunday, were all granted Prayer, or striving with a matter, brings reconciliation with the existing conditions. Moulton showed himself to be a close observer, when he concluded to let "Theodore write himself out," before trying to stop his proceedings. Every woman feels better after she has had her "good cry." We all pray; quite likely Brother Seventhly would not consider our prayers orthodox, but that is not important. What is needed is to be more real, more self-dependent. Superficial characters like Tyndall are soon forgotten. Look back twenty-five years, and learn how quickly noted individuals, who have no real claim upon humanity, pass from memory. Twenty-five years ago there was a very popular man, named Edward Everett, who went toodling round the country, very much in the style of Tyndall; that is, with many words and but few ideas. Scarce ten years have passed since his death, yet he is nearly forgotten, and is sure to be entirely so when the generation in which he lived has passed away. Twenty-five years ago the names of John Brown and Abraham Lincoln were far less familiar than they are likely to be centuries hence. Twenty-five years ago the *Tribune* was edited by a MAN, and though issued from an unnoticeable, dingy, old building, every one was asking: "What does the *Tribune*, or what does Greeley say?" Now, edited by a sort of Tyndall, and advertised by its towering steeple, that rises from a base as narrow and as fiery as a Calvinist's creed, there are none so weak as to ask or care what is said by it or its editor. There is hardly a person in the country, of ordinary intelligence, who would be at a loss for a reply, if asked to give a reason why the memory of Franklin is still fresh and respected; yet not one in ten thousand of the persons who would be influenced thereby could give any reason why the opinions of Profs. Tyndall or Carpenter should have any weight in this country. It is said these two persons court the society of Mrs. Lewes, which is likely to be the case, for these gentlemen are very anxious to shine, even if they have to do so by the borrowed light from a woman. And it has recently been in order for flunkydom, to glorify the authoress of "Daniel Deronda"; but if any mortal can tell why, I, for one, would be glad to learn. I have *worked* my way through the book twice, but the opinion still continues with me, that it is a mess of garrulous twaddle, and deserves to sink as it has into oblivion. Gwendolen, like other prostitutes, sells herself for a consideration; then is too shallow either to accept the situation or to fight it out. Daniel Deronda, though young, has the wisdom of a Solomon, and is as passionless as was old David in his dotage. Faugh! What a world this would be if filled with Daniel Derondas! There is one point, however, in which the work should be useful to us, namely: If the most intelligent classes of England are so far back in barbarism in relation to the standing of woman, as indicated by that work and Reade's "Woman Hater," then this country certainly has no call to go there for information upon any subject whatever, or to be tickled by the second hand clap-trap that is published in the *Science Monthly* over the signatures of such scientists as Tyndall and Carpenter.

Midnight Musings.

BY WASHINGTON IRVING.

I am now alone in my chamber. The family have long since retired. I have heard their steps die away, and the doors clap to after them. The murmur of voices and the peal of remote laughter no longer reach the ear. The clock from the church, in which so many of the former inhabitants of this house lie buried, has chimed the awful hour of midnight.

I have sat by the window, and mused upon the dusky landscape, watching the lights disappearing one by one from the distant village ; and the moon, rising in her silent majesty, and leading up all the silver pomp of heaven. As I have gazed upon these quiet groves and shadowing lawns, silvered over and imperfectly lighted by streaks of dewy moonshine, my mind has been crowded by "thick coming fancies" concerning those spiritual beings which

> " Walk the earth
> Unseen both when we wake and when we sleep."

Are there, indeed, such beings ? Is this space between us and the Deity filled up by innumerable orders of spiritual beings, forming the same gradations between the human soul and divine perfection that we see prevailing from humanity down to the meanest insect ? It is a sublime and beautiful doctrine inculcated by the early fathers, that there are guardian angels appointed to watch over cities and nations, to take care of good men, and to guard and guide the steps of helpless infancy. Even the doctrine of departed spirits returning to visit the scenes and beings which were dear to them during the bodies' existence, though it has been debased by the absurd superstitions of the vulgar, in itself is awfully solemn and sublime.

However lightly it may be ridiculed, yet the attention involuntarily yielded to it whenever it is made the subject of serious discussion, and its prevalence in all ages and countries, even among newly discovered nations that have had no previous interchange of thought with other parts of the world, prove it to be one of those mysterious and instinctive beliefs, to which, if left to ourselves, we should naturally incline.

In spite of all the pride of reason and philosophy, a vague doubt will still lurk in the mind, and perhaps will never be eradicated, as it is a matter that does not admit of positive demonstration. Who yet has been able to comprehend and describe the nature of the soul ; its mysterious connection with the body ; or in what part of the frame it is situated ? We know merely that it does exist ; but whence it came, and when it entered into us, and how it is retained, and where it is seated, and how it operates, are all matters of mere speculation, and contradictory theories. If, then, we are thus ignorant of this spiritual essence, even while it forms a part of ourselves, and is continually present to our consciousness, how can we pretend to ascertain or deny its powers and operations, when released from its fleshly prison-house?

Everything connected with our spiritual nature is full of doubt and difficulty. "We are fearfully and wonderfully made;" we are surrounded by mysteries, and we are mysteries even to ourselves. It is more the manner in which this superstition has been degraded, than its intrinsic absurdity, that has brought it into contempt. Raise it above the frivolous purposes to which it has been applied, strip it of the gloom and horror with which it has been enveloped, and there is none, in the whole circle of visionary creeds, that could more delightfully elevate the imagination, or more tenderly affect the heart. It would become a sovereign comfort at the bed of death, soothing the bitter tear wrung from us by the agony of mortal separation.

What could be more consoling than the idea that the souls of those we once loved were permitted to return and watch over our welfare ?—that affectionate and guardian spirits sat by our pillows when we slept, keeping a vigil over our most helpless hours ?—that beauty and innocence, which had languished into the tomb, yet smiled unseen around us, revealing themselves in those blest dreams wherein we live over again the hours of past endearments ? A belief of this kind would, I should think, be a new incentive to virtue, rendering us circumspect, even in our most secret moments, from the idea that those we once loved and honored were invisible witnesses of all our actions.

It would take away, too, from that loneliness and destitution which we are apt to feel more and more as we get on in our pilgrimage through the wilderness of this world and find that those who set forward with us lovingly and cheerily on the journey have one by one dropped away from our side. Place the superstition in this light, and I confess I should like to be a believer in it. I see nothing in it that is incompatible with the tender and merciful nature of our religion, or revolting to the wishes and affections of the heart.

There are departed beings that I have loved as I never again shall love in this world ; that have loved me as I never again shall be loved. If such beings do even retain in their blessed spheres the attachments which they felt on earth ; if they take an interest in the poor concerns of transient mortality, and are permitted to hold communion with those whom they have loved on earth, I feel as if now, at this deep hour of night, in this silence and solitude, I could receive their visitation with the most solemn but unalloyed delight.

In truth, such visitations would be too happy for this world ; they would take away from the bounds and barriers that hem us in and keep us from each other. Our existence is doomed to be made up of transient embraces and long separations. The most intimate friendship—for what brief and scattered portions of time does it exist ! We take each other by the hand ; and we exchange a few words and looks of kindness ; and we rejoice together for a few short moments ; and then days, months, years intervene, and we have no intercourse with each other. Or, if we dwell together for a season, the grave soon closes its gates, and cuts off all further communion ; and our spirits must remain in separation and widowhood, until they meet again in that more perfect state of being, where soul shall dwell with soul, and there shall be no such thing as death, or absence, or any other interruption of our union.

The foregoing is taken from one of our school books that has continued in use for more than fifty years, which would seem to warrant its popularity. It expresses my own views so perfectly, that it is republished as an introductory to remarks upon the modern phase of the same subject. It is now generally admitted by the intelligent, that whether the belief in spirit communion is or is not well founded, at least there are strange phenomena connected therewith that demand investigation. At the same time there is a shallow, ignorant, loud-mouthed class that derides every attempt to solve the mystery. The press pander to this class in order to become popular therewith, or through natural stupidity. The first is well represented in the Springfield *Republican*, which is racy, full of gossip, but every article seems written in a style to render it applicable at any time to the side then the most popular. The influence gained by such a course seems to be made plain in the fact, that at the determination of any public matter that paper, almost invariably, stands on the losing side. Its neighbor, the *Union*, seems to fill the other position. Servile as a partisan, dumb with astonishment at the announcement of any " wonderful discovery " at a distance ; but implacably hostile to anything near by that is out of the beaten track, though it may be readily verified by personal observation. Perhaps a " little story " will best illustrate. In my young days, a neighbor of my father had a ram of such combative propensities that he was kept in a small enclosure surrounded by a granite wall. It was soon understood that, before making a charge, he took aim, then closed his eyes and went it blind ; so that it was fun to drop inside, make a few " Masonic passes," then look out for the rush that was sure to follow, when prudence dictated a flank movement and the ram would bring up against the wall, the contact having as little tendency to demolish the granite as to enlighten the ram. But the strong points of the editors of such papers are yearly described in the stock reports of our cattle shows, and it is useless to waste space upon them here. From my earliest childhood I have had an intense desire to learn the *why* of any seeming mystery, and I believe that it is not only the right, but it is the positive duty of every human being to take every possible opportunity to do so. I have never had any desire to invent " perpetual motion," or seek buried treasures ; but my wanderings and investigating habits have made me slow to limit the possibilities. " Table tippings " seem contrary to the laws of gravitation, but when certified to by so many they deserve consideration, because they have a bearing upon evidence in general. Millions of lives have been sworn away upon the tithe of evidence that can be produced in proof of the verity of spirit communion. "It is electricity!" shouts Mr. Shallow. Very likely, but what then? What is electricity? Suppose some traveler, out of breath, should rush into the study of Prof. Snoodinks, who has calmly settled down upon this electricity hypothesis, shouting: "Sir, sir! I have been traveling in the East for five years to find out about the marks that were placed upon the ancient structures, and have discovered all about them." "Glorious," answered Snoodinks, "let us hear, quick!" "Why, they are letters or words," says our discoverer. Imagine Snoodinks' look of disgust, as he exclaims: "Why, you infernal donkey, have you been traveling five years to find out what everybody else knew? It is not what they are, but what they mean, that is wanted." So of the phenomena connected with spiritualism. I have seen tables walk up and down stairs, around the house, give communications, etc., etc. "O, you were mesmerized." Possibly, but if mesmerized in this, why not in other matters? What value is there in evidence? This matter has a very important bearing in the every-day affairs of this life, and the judge or juror who fails to improve every opportunity to gain information upon the phase of our system that may have such an important influence, in my opinion, is criminally negligent; and a doctor who neglects to inform himself upon the matter may well turn back to Hippocrates for information, and it will depend more upon luck than his skill if seventeen out of forty-two of his patients recover, as was the case with Hippocrates. My study of the subject has had more to do with its physical than spiritual bearing, still I have studied the latter sufficiently to know that it offers the best evidence extant, that this life is but a prelude to another. It seems strange to me that Brother Nehemiah cannot see that in denouncing spiritualism he is only injuring his own cause, and is only hastening the time when his hearers will become confirmed materialists. Only his conceited blindness prevents him from seeing that the lady who is so attentive, while he is sniveling and declaiming in his weak way, is only looking at some other lady's "pull-back," with the intention of copying or criticising it; she neither knows nor cares anything about what he is saying. She goes to meeting from habit, and to show her own or to see how others are dressed.

Let her lose her loved ones, then his twaddle becomes husks, and she seeks more tangible evidence of an hereafter where she shall meet them again. Were he of even average intellect he would respect the sorrows of such; his devil theory denotes his caliber, and is just suitable for grannies in breeches. After twenty years and more of investigation, I cannot accept the spiritual theory as a solution of the mystery, though it may prove much that is claimed by the spiritualists, and I think it does, but it is a broader matter, it covers our life here. If it is electricity, it is time to try and find out what electricity is. It has happened that for more than a year past I have had this power in my own family, and have had a chance to study it at leisure, not in the dark particularly but in any of the twenty-four hours of the day. To me it seems to be our life that flows through our body operating it as a river operates a mill. The mill or the body may decay but this power or the river flows on forever. We have abundance of communications which are quite as likely to purport to come from those who prove to be living as from those who have "gone before." We are not mediums, nor do we exhibit this power for money or to the merely curious, but whenever at leisure we are always happy to have intelligent seekers call for the purpose of witnessing its effect and operation.

TABLE TIPPINGS.

In the last issue of the Reporter the fact was mentioned that for months past we have had what are termed *Table Tippings* in my family. The statement attracted more attention than was expected, and many who laughed at the matter a few years since have expressed a desire to know more of my experience. Great indignation is often expressed by the believers in Spiritualism, because scientists do not investigate the manifestations, but that is not so easy to do as may at first appear; peculiar conditions are required; then there are few public mediums willing to be thoroughly investigated: beyond this, real scientists, like Franklin, are scarce. He, silly man, believed investigation should precede decision; but the popular scientists of to-day are so wise that anything new is at once condemned. If facts prove them to be in error, they *damn the facts;* a plan that saves trouble, but one unlikely to lead to discoveries of importance. Much has been said about Agassiz's refusal to investigate the subject, but Mr. Agassiz was simply a specialist, puffed up with conceit through our adulation. That he was a weak-minded man is evident from the following extract taken from his own statement :—

EXPERIENCE OF PROF. AGASSIZ, GIVEN BY
HIMSELF TO REV. C. H. TOWNSHEND.

" Desirous of knowing what to think of animal magnetism, I for a long time sought an opportunity of making some experiments in regard to it upon myself, so as to avoid the doubts which might arise on the nature of the sensations which we have heard described by magnetized persons. M. Desor, yesterday, in a visit which he made to Berne, invited Mr. Townshend, who had previously magnetized him, to accompany him to Neuchatel and try to magnetize me. These gentlemen arrived here with the evening courier, and informed me of their arrival. At eight o'clock I went to them. We continued at supper till half-past nine o'clock, and about ten Mr. Townshend commenced operating on me. While we sat opposite to one another, he in the first place only took hold of my hands and looked at me fixedly. I was firmly resolved to arrive at a knowledge of the truth, whatever it might be ; and therefore the moment I saw him endeavoring to exert an action upon me I silently addressed the Author of all things, beseeching him to give me the power to resist the influence.

"AGASSIZ."

Think of a grown-up man praying that he may be able to resist the proof of a fact ; it puts one in mind of the tramp seeking work, and praying to God that he may not find it. We hear too much of men who have gained popularity through the puffing of those who wish to make themselves known thereby. We know that the scientific men of England proved the impossibility of tunnels like that of the Thames, of railroads, telegraphs ; in fact the impracticability of anything new. England owes her greatness to her mechanics, and would hardly miss them if her whole clique of popular scientists should emigrate. What do we know of the abilities of such men as Huxley, Tyndall, and Carpenter, or care what they say ? We see millions of foreigners, and as a mass know them to be much lower, intellectually, than our own people ; is it likely that countries that produce so much ignorance, produce the greatest thinkers ? See what an Englishman says : —

" Not only in oratory is the American the superior of the Englishman. You excel us in oysters, in corn bread, in sweet potatoes, in canvas-back ducks, and, I venture to say, in kindliness and hospitality. In intellect, I take it, we are about level ; but doubt whether you give yours full play. If you did, you would depend upon yourselves."—*B. L. Farjeon's New York Speech.*

And why do we not depend upon ourselves ? We are taxed heavily for schools in which to give all an education. Are those schools a failure ? If so, is it not time that the howl of the insatiate teacher for more pay should cease ? Many of our papers assume the rôle of teacher, but their writers are usually mere machines that run in well worn ruts ; one of these in the Springfield *Republican* writes substantially as follows : " Herbert Spencer, probably the greatest thinker of the age, expresses the opinion that the marriage relation of to-day is not likely to be considered desirable in the not distant future." This stale idea that was common with Lycurgus, still later with Plato, and has been entertained by hundreds of communistic societies, the theme of innumerable lectures and the practice of the Oneida community for forty years, is given as proof of originality. The *Republican* gushes with adulation. The " Great Dr. Hammond " is one of its superior idols. Will it inform its readers whether the said Doctor as Surgeon General was ignominiously expelled from the army ; if so, is his assertion that Spiritualism is a humbug, and its so-called manifestations the result of trickery, of any account when placed against that of so many quite as intelligent as himself who believe to the contrary ? It is easy for a noisy person to find followers, and a single rowdy will make more noise than is made by a thousand intelligent persons ; consequently, it is no proof that Spiritualism is unpopular, because a few ignorant persons shout humbug. The one witness in court that swears positively to have seen a crime committed would have more weight than a thousand who should swear that they did not see it, yet it is the ignorant and prejudiced who have not seen, that are the most strenuous in shouting humbug in relation to the spiritual manifestations. Fifteen years ago the professional exposer drew full houses ; now he soon has to pawn his traps in order to get away from his last place of exhibition. One fact that is open to all should attract the attention of the intelligent; we know that such men as Sumner, Beecher, Agassiz and others have spent months in preparing a lecture that is given a hundred times, yet Cora L. V. Hatch, *that was*, who is certainly not remarkably talented, will take the same subject given to her as she rises to speak, and give as polished and profound lecture as those who have taken months to prepare it. It would not be desirable to have any one believe simply because others do so, but when men like Abraham Lincoln, William H. Seward, and others of the same abilities accept Spiritualism as a fact, it certainly cannot be derogatory to those who think less to consider the subject fairly. My attention was called to what were termed " table tippings " soon after the Fox sisters made their debut, but it was not my lot to meet with anything of the kind for a number of years that caused me to look upon the subject with favor. " Table tipping " violated the law of gravitation, and my faith in that law was positive. In 1865, Horatio, William, and Mary Eddy were at my house in Lowell, Mass., five days, each evening giving public séances to large audiences in Mechanics' Hall. At those exhibitions the laws of gravitation and cohesion seemed of little account. The mediums were ironed by the police, but it made no difference; hundreds of feet of cordage were used in tying each medium separately, then together, to staples in their cabinet. They were literally wound over as a woman winds a rag in a ball of yarn, but their coats would be taken off from under all of this cordage, or put on in the same way in fifteen seconds after being shut into their cabinet. Sewing the knots made no difference, for the cords and knots were invariably the same throughout the séance as when first tied. I have had much experience in handling cordage at sea, and in other business, and have tied many mediums, but so far have never succeeded in tying one so but what the cords would come off at request. I have had to do with nearly all of the mediums of note known in the Eastern States, and as a general thing have not had cause through the acquaintance to respect them, and have often wondered why such remarkable gifts are given to such low characters; but the beautiful pond-lily springs from the slimy depths of the frog-pond. I have spent hours in private with professional exposers, have seen excellent imitations, but the observer who has seen the real and imitation and cannot see the difference must be dull indeed. There would be no lack of exposers if the real mediums could explain the *modus operandi*, for there are few of the noted ones, in my opinion, who would not for a consideration readily act as such. I have witnessed nearly all of the various manifestations that have been described, and shall briefly mention a few. Sitting with Slade in New York, the slate was not held up against the table but a foot below. *I saw the writing as it was done, each letter and line*, but no hand or other means of operating the pencil could be seen, though at request a hand was twice shown above the table, seemingly an Indian hand; it was noon and the sun shining on the table at the time. While the writing was being done there was such a strain downwards that it surprised me that the frame was not stripped from the slate. Watkins, the slate writer, probably as little of a man and as much of a medium as has yet been developed, was at my home a week; he placed a bit of pencil upon a slate and then turned another slate of the same size upon the first; each of us held an end of the slates together; in a moment the pencil was heard to move as though writing; soon, three light taps were heard, then the slates were pushed toward me, Watkins not even looking at them; on opening them the following message, plainly written, was found: "*My dear friend, I come to you to let you know that I live. Ansel Cain.*" Mr. Cain was not an intimate friend of mine, though we had conversed upon the subject of Spiritualism, and he had given me the impression that he doubted a future existence, though he evidently desired such. The communication was copied at the time, as were the following which were given immediately afterwards: " *My dear brother, I am glad to see you here this morning, and hope you will believe that this is me. Moses W. E.*" "*My dear papa, I will come to you again some day. I am happy, so is mother. God bless you all. Your loving daughter, Hattie.*" Of the source of the communications others may judge. That they came as stated, I know. Numerous communications of a similar nature were received by myself and others through Mr. Watkins while he was at my house. He got them anywhere that he made the attempt, out on the door steps, in the bushes. I saw him get one in a smoking-car on the Boston and Albany Railroad. The communications were not always of a spiritual nature, but such as they were, any one that would pay could have them, and considering the way they were given hardly any one mentally higher than an idiot could have been tricked thereby. Mrs. Huntoon (Mary Eddy) was invited to my house for the gratification of my own family and special friends. Numerous hands and faces were shown, instruments were played upon, then passed out to the audience. One woman, or form of a woman, came out into the room, showed her night-cap and dress of ancient days, then voices, shouts, and a pistol shot. ".Oh, so low!" exclaims the high toned. Certainly, they have always been so ; think of the frogs, vermin, turning rods into snakes, water into wine, etc. Yes, but why not do them in the light ?. Sure enough, why was the earth created in darkness ; why did God require a bush as a cabinet when he appeared to Moses, or a cloudy pillar at the door of the Tabernacle ? Why did the angels come to Lot in the evening, or release the Apostles in darkness ? The Christian fabric rests upon dreams and darkness ; the veil was rent and saints rose from their graves in the dark ; the Ascension was in a cloud ; a kernel of grain, or the roots of a tree, require darkness from which to produce manifestations of growth and life ; the body commences and obtains its form in darkness, receives the spirit or life in darkness. Is it strange then that certain phases of the manifestations require darkness ? Only the shallow minded will be surprised at the fact. After our séance I happened into the kitchen where I found Mrs. Huntoon looking around that part of the room where the cabinet had stood and saying to herself, " I do wish I could find where the bullet goes to," which caused me to ask if a ball cartridge was discharged from the pistol the previous evening. " Yes, we always use regular cartridges," was her reply, which seemed decidedly interesting. Her pistol was called for and cleaned. Then from her supply of cartridges I loaded its seven chambers, placed it in a small empty closet, put a guitar, bell, and tambourine with it, then hung a curtain at the door, after which Mrs. Huntoon's hands were tied behind her and as secure as I could tie them. My assistant "Charla " sewed the knots firmly with thread. Four chairs were placed in front of the curtain for the family, then Mrs. Huntoon took a seat in the closet, and in less than ten seconds, hands and a face were shown through the curtain, all of the instruments were played upon, then bang, bang, went the pistol, and a third time at my request. Immediately after the third discharge the medium stepped out to the light, tied exactly as when she entered ; not a sign of a bullet mark could be found. I took the pistol and discharged another cartridge at the floor of the closet ; the bullet from that is plain enough to be seen. The medium was then asked to step into the closet and have the spirits untie her, which was done while I was taking my watch from my pocket in order to time the untying. It certainly was not one second in being done. As no mention is made of the fact that the discharged bullets cannot be found, it can hardly be considered a trick. Never bother, however, to tie a medium ; trust to the production ; if the medium is tied, note the time required for any manifestation, and whether there has been an effort in the production ; the real medium keeps cool, the exposer is often covered with perspiration through his struggles. Suppose a letter is written to a spirit friend to be answered by Mansfield, write as follows : " My dear friend, give me some test by which I may know that I am in communication with you." Do this mechanically, keeping your mind upon other matters, and be sure to have no thought of what the test is to be ; if this is done, sealing the letter is of no account, and the writer will be more fortunate than myself if anything satisfactory is received. In a dark circle where hands are felt, observe closely whether the movements are like those of a person groping in the dark, or every attempt is accomplished without blundering. I have tried hard to study the manifestations carefully and candidly, but to do it advantageously requires the regular attendance at stated hours of several persons, and it is not easy to find such. It is generally supposed that an intermingling of the sexes is necessary, but that is not certain. I have often entertained theories about the matter that have as often been dispelled; whatever the power, if an appointment is made it is kept without fail, even if forgotten by the earthly party interested ; one moment we have what seems absolute proof of spirit communion, the next something is given that makes the matter doubtful. We have abundance of communications, often two try to communicate at the same time, mixing the letters as would be done by two telegraph wires getting twisted together. We are now using our sixth table, five having been destroyed. Table tipping but poorly expresses the movements with us, and no person with a particle of the true scientist about him could fail to be interested in the ever-changing movements. My wife, her sister, and myself constitute the sitters; we simply place our hands upon the table without any attempt to control its movements It travels through the house, up stairs or down, swings upon my head and shoulders and rushes me backwards, and in darkness through rooms and doors without touching a casing, though the table is nearly as wide as the doors, or perhaps it will bear down until it crushes me to the floor. I think it can press down three hundred pounds. Sometimes while I am sitting in a chair it will swing on to my back, hook its legs to my chair and turn me around or drag me along, or perhaps tip me over, then drag me on the carpet. A recent freak was to tip itself over, then pick up the chairs on its legs, call for the alphabet, spell out "confusion," then disengage itself from the chairs, set them upright in place. Any movement is made just as well in the blackest darkness as in the light. It will move quickly to the window and tap the glass rapidly without injury, though it is so dark that nothing can be seen. Its communications are as varied as is our conversation. My boy was asleep; the question was asked, "Do you know where Jimmie is?" "*Yes, his body is up stairs, his mind is wandering through immensity.*" To the question, "Why do we get so many unreliable communications?" was answered by those purporting to be special friends, "*When we withdraw our control it leaves you open to the influence of elements of which you know not.*" I spanked my boy one evening because he was raising the d——l generally. On returning to the table where another person and myself had been sitting it gave me a hearty thump, knocking me against the wall and handled me very roughly, which caused me to laugh, as its force could be calculated. My laughter seemed objectionable for it immediately whirled itself into the hall and dashed its corners into the walls —the marks still remain. Being pitched back into the kitchen it tipped on end, called for the alphabet and spelled out as follows : "*Learn patience and discretion with your child or you will be the sufferer.*"

One evening a gentleman was anxious to get the full name of one whose initials had been given ; he had urged for some time, when the alphabet was called for, and what purported to be the spirit of another person spelled out, "*She is gone away.*" "Who is she?" was our inquiry. "*The one whose name is desired,*" was the reply. "Well, can't you give it?" was then asked. "*No.*" "Why, don't the spirits all know each other?" we asked. The alphabet was called for, seemingly impatiently, and it spelled out, "*Do you know all that come to the telegraph office?*" The table calls for the alphabet by two peculiar upward movements, but how those and other peculiarities were understood by us is not positively known, but I think through impressions. We have hundreds of communications, each characteristic of its purported source, all of which can be reconciled with the spiritualistic claim ; if the spirit life is but a continuation of this, the only change being separation from the body, which has been used as a cabinet or cage in this life, and in the same way, the strange and unreliable communications are readily accounted for. If a business man should put up a speaking tube from his place of business to a distant city, leaving the distant end open to the public, the *gamins* would be likely to send him queer messages occasionally. Much has been said in derision about Frank J. Baxter and the "Abe Bunter" matter, but that is not an uncommon phase though it adds to the mystery. I will give a case in my own experience almost identical, and for which there is abundance of evidence to substantiate the fact if necessary. I shall give the particulars literally, that the case may be clearly understood. I was experimenting, asking questions, which were answered by a planchette, purporting to be controlled by my mother ; many questions had been answered, but in such a set way that they were unsatisfactory ; finally I asked, "Mother, do you know where Mr. Buck is now?" "*Yes, he is here.*" "Oh! no, no, mother, that won't do, Mr. Buck is not dead." "*Yes, he is, he died four months ago.*" I did not believe it, but wrote the next morning to my daughter at Lebanon, N. H., requesting her to ascertain Mr. Buck's whereabouts, giving no intimation of my reason for desiring her to do so. In a day or two her reply came and was as follows : "*Cousin Isa was at Newport about a month ago, and while there news came that Mr. Buck was dead, and had been dead three months,"* certainly seeming good proof of spirit communion. Yet, Mr. Buck was living at the time, and is yet, I believe. If placed upon a jury to decide the question of spirit communion my verdict would be *"Not proven"*; still proof of the fact that seems almost positive may be obtained in abundance, but that *almost* invariably stands in the way. That the subject is of more importance than that of any discovery which has been made for thousand years is my firm belief; in my opinion, it is *our life*, and offers the key to life and health; the force that tips the table moves our limbs and bodies, operating our movements as a river operates a mill, continuing with us from the birth of the spirit through eternity; our brains are simply instruments through which we receive ideas as tunes are rendered by a piano, the average mind receiving ideas as water flows into a hole to the general level, the thinker *pumps* his higher and becomes the advanced leader. The infidel Paine, of 1776, was but the Unitarian of 1876. The "Autocrat of the Breakfast Table" radical twenty years since is accepted by the multitude to-day. The manifestations, however, seem more the reflections of the past, than representations of spirit life of the future. We can readily decide whether we see the reflection of an object in a perfect mirror, or the object itself through plate glass, though exactly the same view may be presented in either case; yet it might not be easy to explain the difference. It is a practice of writers to lay out a general plan of a work, then to smooth up and fill in the details as it is written out. The completed "Edwin Drood" of Dickens by the spiritual medium, I believe to be the rough sketch of Dickens in this life. That spirits of murdered persons do not return and expose their murderers is strong presumptive proof that such return is impossible; for, notwithstanding all that can be said about the spirits not believing in hanging, etc., it is universally conceded that prevention is better than cure, and if the fact were once established that exposure was probable through the spirit's return it would act as the strongest preventative. The following extract from a lecture on the "Law of Influence" seems deserving of consideration : —

"May not that energy known as electricity be the universal medium for the application of the creative and reproductive force or influence to matter? It not only conveys the signs of thought through the telegraph and telephone; it also transmits our thought-force with our thought-touch through nerve and muscle to our hands and feet. On the same principle the thought-force with the touch of the Creator through this electric hand may extend constantly to each world and to every atom of material organism."

That there is a force that produces strange manifestations is a fact too well established to allow of its being ignored, and the proper course would seem to be to grapple with it and solve its nature : to its spiritual bearing I have given but little attention, though the following lines express my own feelings upon that point : —

Oh shades of loved ones gone before!
Do you still exist on some unknown shore?
In a brighter land and advanced state,
Where souls from earth with angels mate,
Where free from pain and earthly strife,
The soul aspires for a higher life,
Where a purer love to each is given,
Surrounding all with the joys of heaven?

And are the joys of that unknown shore
So complete that earth attracts no more?
Hath earthly ties nor kindred's tears
No responsive throb in those brighter spheres?
Or do you in the spirit form
Remain with earthly friends to roam,
To fill our hearts with gentle love
And lead us on to that home above?

We loved you here, we love you still;
You have gone before, 'twas our Father's will.
Though we still remain in our earthly homes,
Our hearts oft turn to our loved ones gone.
Yes, gone before at the Father's will,
But in memory cherished at the old homes still;
At the table, the fireside, in each sunny spot, yet
Your influence is felt, ah! we shall never forget!

No, never forget this side the " dark river."
May He influence our lives, the all bounteous Giver;
Guide us o'er its silent waters to that unknown shore
That we may meet with loved ones to part never more,
And through the seeming love of those gone before
We've the most tangible proof of that unknown shore;
That we shall meet again with our dearest friends
In an advanced life, when the present ends.

SCIENCE AND RELIGION.

The time seems approaching with giant strides when any religion irrecon-
cilable with reason will have no place except with the ignorant or venal.
The morals of to-day, compared with those inculcated two thousand years
ago, show little in favor of what during the last fifteen centuries has passed
for religion ; we know that during the time when that religion held unlimited sway, the period is known as that of the " Dark Ages." Our great improvements and inventions are the work of an age of *infidelity*, and, if experience is of any value, it is evident that our elevation and salvation depend
upon our own exertions guided by reason, and that religious dogmas only
benefit those who teach them. A venal priesthood has hedged religion with
superstition until it has seemed impossible to elucidate the matter by natural means ; besides, it has been said that the bigotry of science is only second to that of religion ; but science means knowledge, and knowledge has
nothing to do with bigotry, either in religion or science. We do not dispute
about the sum of two and two, or as to whether the sunlight is greater than
that of a tallow candle ; ignorance causes the dissensions, and the greater
the ignorance the more tenacious the opinion. Superstition away, science
will readily prove all seeming miracles to be either delusive or the effect of
natural causes. The purpose of this article, however, is not to meddle directly with religion, but to cause a scientific consideration of the claim of
" special inspiration " of the *Bible*, and the subject of " election " or " predestination." An opinion upon inspiration to be of any value must be based
upon evidence, and such evidence can only be obtained from observation.
First, science readily demonstrates the fact that, physically considered, man
is but a complicated machine, each organ being fitted for certain duties, and
as a whole, by the consumption of a given quantity of carbon, he or a steam
engine will raise the same weight to a given height. Such being the case, is
it unreasonable to suppose the mental organs are also mechanical, and that
the brain transmits ideas, as the larynx does that of tones, or a violin tunes ?
There is abundance of evidence to prove that intelligence comes from a
fountain outside of ourselves, open to all, but to each individual in accordance with the quality of that individual's brain or instrument of transmission. How often we read accusations of plagiarisms between authors when
in fact neither had ever seen the writings of the other ; how common it is
for two persons to commence at the same time to speak of the same matter.
Every inventor realizes how liable he is to be anticipated if he delays the
completion of a device. Persons of the lowest intelligence, like " Blind
Tom," will perform wonders without consciousness of how it is done. Ignorant " mediums " will deliver off hand the most profound lectures. Ministers and authors in a state of somnambulism have written articles of a superior character to what they could write in their normal condition ; problems
have been solved in the same way. Can we suppose that the ideas of a lifetime are stowed away in a person's head ? We may divide the head of a
man or a fiddle into minute pieces without finding either an idea or a tune ;
then is it not reasonable to believe the brain, like the fiddle, to be a mere
instrument of transmission, and that some new intelligence is operating it
when things are done in our sleep or unconsciousness that are impossibilities in our waking hours ? When inspiration is fully understood we may
rest assured that, like other discoveries, we shall find it very simple, and
that we have looked too far away for the solution. Predestination ! Who
believes or even thinks of an idea so obsolete ? asks the reader. More than
generally supposed, my friends, though under various names. Those who
believe a large portion of our race doomed to hell ; the Adventist, who believes in the annihilation of the wicked ; last, but not least, the Materialist,
under which name may be found the shallow-minded of every station of

life, from the shoveler in the bog to the pseudo-scientist who believes himself to have exhausted the source of knowledge ; the sleek priest, sanctimonious preacher, and pretentious professor, all preach or teach something, but at heart believe in "nothing"; the professor in science at the expense of consistency, for a fundamental principle of science is that to exist at all is to exist forever. Predestination, religiously considered, is a hazy matter, but treated rationally becomes very clear, as do election and annihilation, and seems the proper termination of a large portion of the human family, as may readily be made to appear. I have before me a tool, called by its inventor, "*the imp*"; it is but one of many of a similar character, that is, a combination of old devices, thus forming something new. "Twelve useful tools in one," says the inventor, a screw-driver, rule, hammer, carpet-stretcher, tile, saw, etc., etc., made in two pieces, which by a peculiar joint are readily united, then becoming wrench and pincers, thus creating "the Imp"; disunited, "the Imp" is annihilated, dissolved into the commonest of tools ; so of the average human mind. Pat, the shoveler, dissolved, leaves a residuum of ideas relative to pipes, tobacco, dogs, pigs, jokes, absolution, wakes, etc., etc.; the pseudo-scientist, of big words in which old ideas running in still older ruts have been dressed. The effect of the disintegration of such mountebanks as Justin D. Fulton, Talmage, and others, may be witnessed where some one is skimming the scum or froth from a cauldron. Observe how an air bubble explodes here, another there; soon it has vanished, there is nothing but a little common dirt left. Take preachers like our weak but amiable Doctor Adams, those who will neither learn themselves, nor, so far as they can hinder, let others ; who search their Bible through for evidence that spirit communion with man was once very common, in order to prove thereby its impossibility. At the disintegration of such reverend delusions what can there be left but a little sediment of John the Baptist, intolerance of John Calvin, superstition of grandmothers and puling of babies ? Is there one particle of originality in such persons that can give hope, rationally considered, from which an individuality can be constructed for a continued existence ? If not, is not annihilation the predestined end of all who fail to work out an individuality for themselves; while *election* as naturally follows for those who do ?

"SPIRITS, OR WHAT?"

Under the above heading, the Boston *Herald* of February 28th ultimo gave a very circumstantial account of what were claimed to be materialized spirit forms witnessed at Rochester, N. H.; we can hardly take up a paper without finding something of the kind described, and unless desirous of passing down to posterity as a superstitious set of materialistic idiots it is time that some attempt should be made to elucidate the cause of such appearances. Is the question, however, logical in connection with the account ? Spirit is *immaterial* intelligent being. The account describes material forms that must have been those of ordinary human beings, or reincarnations of persons once known in this life, neither offering any proof of life beyond the grave ; but such materializations are likely to gain the attention of the multitude sooner than those of a more intellectual character; they are evidently of this life and have to do with our well being here, emanating from the same cause or force as that which causes "table tippings, spirit raps," etc., and I believe the same as that which produces all of our physical movements ; and this force seems traceable back for centuries. Recall the monks of Luther's time, mere animals with only animal desires, their religion a formula, denying the right of thought ; forbidden to marry, but, unless sadly belied, the fathers of many children. Think of the stern old sectarian with his coarse animal nature and belief in woman's subjection, thinking it a sin to smile, but scriptural to gratify his passions ; it was the rule for such Christians, from John Rogers to Lyman Beecher, to have many children. Turn from those running in ruts to those beginning to *think*. The astounding "spiritual manifestations" in the family of Samuel Wesley prove the mediumistic temperament of the children. Those old enough will readily recall the ecstatic shouts and convulsive ways and worship of the early Methodist. Can any one remember such with large families of children ? Yet they were not credited with a disposition to mortify the flesh. Come now to the spiritual medium (male) ; look the list through, see how few of them are fathers, yet many of them have a very *corn-fed* look, and they are generally noted for liberal views, and it would seem that their *life* force is expended in the production of their so-called physical manifestations. Go back centuries and it will be found that wherever these manifestations have appeared in families they have almost invariably done so through the children. See the Rev. Joseph Glanvill's account of the "disturbances in the Mompesson family, 1661 to 1663"; also, Adam Clark's account of those in the Wesley family ; so of modern times. Half a dozen children seated at a table soon get *table tippings* or *raps;* the same number of octogenarians might sit until doomsday without doing so, which would seem to indicate that they are the product of a surplus of the life force. Solomon in his prime could undoubtedly have caused the heaviest extension table to dance a hornpipe, but after getting to the *vanity and vexation* stage, would have found a teapoy too heavy. Ignorance sneers at the treatment of old David, as described in the 1st Book and Chap. of Kings ; but in my opinion, a profound depth of knowledge of the life force is indicated therein, that is not thought of by the medical fraternity of to-day. See how readily women, babies, and dogs take to rosy, robust men; then see the same dog with hanging head and tail describe the segment of a circle as he passes the lank, saturnine specimen of humanity. Robust men usually mate with fragile women ; animal propensity would seem to demand an equally robust mate, but it is evident that nature guides ; the one has a surplus, the other lacks the life force, and each attracts the other. Married couples are seldom effective as *table tippers*, though each carry their proportion of force mixed with others. Why is the invalid strengthened by taking iron into the system unless because of its being a good conductor of electricity or this life force? Singing or music has the same effect upon the *manifestations* in all their phases as upon human beings. The foregoing suggestions are offered for the consideration of observers ; they relate to the physical bearings of the phenomena ; but there are other phases that offer strong proof that the spirit germ from the great ocean of intelligence takes possession of the body in order to gain an individuality, the body itself like a vegetable starting from seed, drawing sustenance from the earth and returning to the same at maturity ; then how important that life in the body should be natural. *Suffer little children to come unto me and forbid them not, for of such is the kingdom of heaven.* In the face of such a command, how dares a being so ignorant as a Moody, attempt to warp the mind of a child into harmony with the superstitions of his own perverted nature ? The question answers itself ; it is only through ignorance that he so dares. The mind of a child is a study for the profound. How natural it is, and how its simple inquiries confound the ghastly theories of a Calvin. Suppose a forest to be cultivated by cutting the tops from a portion of the trees, leaving unsightly stubs, the branches from others, leaving bare poles, all the branches from one side of others, and so on, would not such work be considered that of barbarians ? From the depths of my soul I believe it to be a greater sin to teach a child any other motive for doing right than for right's sake, than it was for Fagin to teach Oliver Twist and his companions to steal. So long as a mercenary priesthood can live on the credulity of the ignorant, so long will such as Moody be encouraged to peddle out superstition, that the educated clergy would be ashamed to mention ; but as a matter of policy it would seem better to live here by sawing wood, then return to God a full fledged individual soul ready to commence a higher life, than to live at ease preaching platitudes, then to "melt back into the universe" with the spirit germ so shrunken that it will naturally gravitate to the body of some lower animal in which to make a new effort for a higher life. With Moodys, there will be Ingersolls,* for the two are *cause* and *effect*, ignorant fanaticism and cupidity. No well-read thinking person can well doubt that Christianity has put humanity back a thousand years

* Bob Ingersoll, as the ready champion of star-route thieves and other praying rascals, wiping his modest brow and posing before an audience as a model man and instructor of the world, invariably recalls to my mind Mr. R. Riderhood, so well described by Dickens, in " Our Mutual Friend." The feature however to me seemingly the most to be regretted is the fact that superstition has caused such ignorance that an audience can be found willing to listen to ideas that were musty with age a thousand years since, and which have been reiterated a thousand times by far abler and more disinterested men than Mr. Ingersoll has ever shown himself to be.

(Jesus of Nazareth would find little sympathy in a Christian church to-day) ; or that the Bible has been perverted through selfishness ; but the strange phenomena, known as spiritual manifestations, are likely to furnish proof that its leading ideas are correct, and that its seeming miracles were the effect of natural causes. These have interested me for many years, but it is only since their appearance in my own family that I have been able to study them with any satisfaction ; as we are not *mediums* their appearance with us is supposed to be the result of an earnest desire and cultivation of the means to bring them. Either my wife, her sister, "Charla," or myself can get *table tippings* or *raps*, sitting with almost any other person, but such communications as were published in last issue of Reporter are only obtained when "Charla" is one of the sitters ; those were obtained by calling the alphabet, the table moving at the proper letter ; they also come through her mind by seeming inspiration. I mention her as "Charla," because as such she is known to engineers, turbine builders, and manufacturers in more than half of the states of the Union, as the young lady assistant in my testing business; quiet, and of a mathematical turn, but certainly not a poetess, yet in answer to wished-for information, communications like the following come through her mind like a flash of light :

False, with the true, you'll one day find,
Is but the crossing of your mind
With our dispatches as they are sent
From our Summer Land to your Continent.

If the truth you wish to find,
You must study your own mind,
Learn its workings, its relation
With the great unknown creation;
A gem we promise you shall find,
In a knowledge of the human mind.
Spirit friends around you gather,
Wishing much to help their brother,
Seeking earnestly to find
This gem of truth in the human mind.

Spirits from the other shore
Oft come tapping at your door,
Wishing to inspire your mind
With happier thoughts of their design.

Sleep is a rest for the weary mind,
Which, as it wanders free,
Oft catches inspiration
And brings it back to thee.

Ofttimes when the mind doth wander,
While the body is at rest,
Strange elements of earth
This freed mind doth impress.

Many times when the mind doth wander,
Ideas which to earth are grand,
Are in his sleeping hours
Stamped on the mind of man.

Then when he doth awake,
And reason and thought control,
These ideas are developed
And given to the world.

Your lives are tangled in with ours ;
We are not far away,
We join men and women in their work,
And children in their play.

Much has been said about the twaddle that purports to come from the spirits of noted persons, but if persons will pander, twist and be all things to all men for the sake of becoming noted, there is no good reason for supposing the spirit of such will retain a very positive individuality after separation from the body. Separate the parts of a twenty-four bladed jack knife, and the corkscrew would have the same right as any other piece to call it-

self the many bladed knife. We have had many communications the past
two years, but none of a lower character than such as I have published ; and
I do not believe a low or silly communication was ever given in what is
called a *circle*, unless there was a mind in that circle to match. The communications are often oracular and difficult of application, simply because
they are answers to ideas conversed about hours, perhaps days, before. The
following is one of the kind and was given through the table : *"None realize
for how great an object they live."* Our minds and conversation affect the
manifestations but do not control them. Our ideas are opposed quite as
freely by this force or intelligence as by persons in the flesh. Communications purporting to come from persons who prove to be living are very common, but that depends somewhat upon who the *sitters* are. I have sat with
persons and obtained communications as fast as they could be spelled out,
and found them to be nothing but the passing thoughts of the sitter's mind.
I find, also, that any idea thoroughly established in my own mind is pretty
sure to crop out in the communications. An earnest wish or desire, though
it may not be gratified, is very likely to receive notice, so that my faith is becoming strong that there *is* efficacy in prayer ; not through any change of
God's laws but in accordance therewith. Franklin drew lightning from the
clouds, so I believe that one or many persons praying earnestly for a given
purpose might produce an effect ; we little know yet the power of mind
upon mind or mind upon matter. These manifestations have been offered
for man's study since the dawn of history ; I believe they offer a key to a
knowledge of our life, health, and surroundings that can be obtained in no
other way. Through them I believe it will be made clear that crime is a
disease, and that there is something more than a moral influence in the
contact of individuals. I find in sitting with certain persons that my
strength or life force is taken from me to a very disagreeable extent, while
the contrary is the case with others. There are persons who seem to leave
a part of themselves with us for weeks, so that if we sit for the manifestations by ourselves, we have what purport to be their matters to attend to.
One case has interested me much ; it was what purported to be the spirit
sister of a person from a distant state. She gave a communication as a test
for her brother, which was sent and by him disowned ; when she came
again she was rated soundly for her deception, and requested to keep away
unless she could be truthful. Nothing more was heard of her for six
months when her brother came to our house for perhaps five minutes ; that
evening his sister put in an appearance. "How happens it that you have
not been here for so long a time," was my inquiry. "*You scolded,*" was the
reply. "Ah, Annie, you told fibs, you remember." "*No, I didn't.*" "Ah,
yes, that test to your brother." "*I didn't give any.*" "Who did then ?"
was asked. "*The one that broke the window.*" Nothing had been said about
a broken window at the time, but a month previous our table had been
pitched into the window. Singing the "Braes of Balquither" has been as
effective in bringing Annie to us as "rubbing the lamp" proved in bringing
Aladdin's genie. We have had what purported to be the spirits of many persons come to us, some of them very noted ones; the latter have almost invariably been followed by imitators. Enthusiastic spiritualists exult in the thought
that these manifestations are breaking up the pulpit influence, but that is
solely because pulpits are occupied by materialists at heart, who preach for
those who pay best, without faith in their own teachings: and consequently,
who are the first to laugh at the idea that any of their dogmas may be sustained by tangible evidence : yet these so-called spiritual manifestations do
furnish a plausibility for many of them. The *materializations*, if real, of which
I have no doubt, are reincarnations, and give ground for the belief in the
resurrection of the body for judgment, and the resurrection of Jesus of Nazareth. A careful study of the forces that produce the materialization will
at least cause the observer to hesitate before rejecting as impossible the
idea of such conception as that claimed for him ; not of course through any
miraculous process, but through a concentration of sexual force, as a concentration of the elements under certain conditions produce earthquakes,
tornadoes, whirlwinds, etc. We have much to learn yet, and until sure that
we are quite as wise as our Creator it is not worth while to ascribe to miracles or the devil, what may well take place through natural causes, though
we may not understand the why. The shallow may sneer at these manifestations, but the thinker who has studied them carefully under favorable

conditions will feel more inclined to bow in humility and thankfulness before his Creator and to earnestly ask for more light. In conclusion, I would say that from my own experience during many years of unprejudiced investigation I believe the matter to be susceptible of practical solution.

SCIENTISTS AND PROFESSORS.

In earlier times, when learning, as it was termed, was confined to a favored few, the masses looked up to those as leaders in knowledge, and the most competent to direct ; habit continues the practice with too little consideration of who really are scientists, or what has any professor ever done to cause the prefix to denote ability ; it is used for the most frivolous callings outside, such as horse tamers, balloonists, dancers, etc., etc.; in college it denotes a sort of conduit for the transmission of stereotyped knowledge from books to pupils. Would a smart, practical person desire the prefix, or be willing to settle down in the conduit business? It is very common in case of some terrible bridge or dam disaster for the papers to teem with articles from such professors, explaining the faults and how it should have been done, etc., etc. A year or two since a Professor Tansey, or some such name, came out in a profound report as to how car heating by steam from the locomotive should be done, causing a road to fit up a train with his device. His ignorance was only equaled by his conceit. His plan is hardly likely to be adopted ; the influence of the prefix is justly passing away. What a field there is open for the pruning of our language of its useless synonyms and outrageous spelling, yet if done it is sure to be done by some one not anxious for the prefix of professor to his name.

Of scientists who are they? Certainly not of the Huxley and Tyndall type who are forgotten as soon as dead. While such names as Franklin, Morse, Fulton, Watt, Howe, and others, become more familiar as time passes by. We have too much adulation for heroes before taking any pains to ascertain whether such credit is deserved.

How often Herbert Spencer is referred to as a great thinker, etc. Like Athenæus, Rabelais, and others, he is a great reader, compiles, credits, and publishes ; but if he has suggested anything remarkable for originality it has not met my view. So of Edison, the most slopped over of inventors of the time ; can any one name what his inventions are except, perhaps, his talking doll? He receives a large income from some invention, and undoubtedly is deserving of commendation, but the courts do not mention his name in connection with the telephone nor any other exclusively original invention ; such flunkyism is not creditable to our people.

In a lecture on scientific training, delivered recently in England, Dr. Siemens, who has been one of the most distinguished practical inventors of the age, struck out on a line of thought which is not often traversed. He said that, at this period in the development of science, it was easily possible to give to a young man a too thorough training in science. This, he maintained, was the defect in the great technical schools in Germany. They turned out men who made admirable supervisors of existing systems ; but they were at the same time men who understood existing systems with such completeness that their improvement by new contrivances or inventions seemed to them practically hopeless. The inventor was a man who had far less respect, and possibly much less knowledge, of the methods already in use ; but, for this reason, his mind was free, and he was not hampered by old traditions. In corroboration of this, it might be said that our greatest inventors have rarely been men who have received in their youth a thorough scientific training. By analogy, it might be said of them that their very ignorance led them to rush in where the wise did not dare to tread.

SPIRITUAL MANIFESTATIONS.

It is nearly a score of years since the foregoing articles were commenced and thirty-five since commencing to study the phenomena. This has brought me in contact with ministers, congressmen, doctors, lawyers, curiosity seekers, and mechanics—the latter by far the most intelligent observers from the nature of their make up, for they judge understandingly of the space and time necessary for effect. Indeed, I do not believe any one will ever arrive at the highest standard in any calling if destitute of the mechanic's creative and organizing faculty. The claims of the mesmerist have been familiar to me for a half century, and seem so blended with those of the spiritualist that I am unable to separate them.

Hypnotism seems a subterfuge for retreating from a position impossible to maintain, and a claim to share the honor after braver hearts have won the battle.

Theosophy, seemingly the "old clo'" of Spiritualism, offers little that is new, if A. P. Sinnett is to be considered the exponent. His "Karma," in itself to me interesting, is made up of old, old ideas and stories. Rabelais says, Alexander the Great—of course re-incarnated—is making a poor living mending old stockings. Cyrus is a cowherd, Themistocles is a glass maker, Cicero a fire kindler, Ulysses mows hay, etc., etc. The story of the warning and fall of ceiling upon the bed is better told in "The Error" by G. P. R. James. "Esoteric Buddhism" seems to be made up from the maunderings of the Apocalypse and maudlin gush of some weak minded evangelist.

The Foreign Missionary society, that great maelstrom of cupidity, gullibility, and credulity, claims to seek for heathen where the theosophist seeks for wonders, but there is a large field for the best efforts of both at home.

The "Rochester Knockings" offered nothing new, but the time was ripe for a demonstrable belief. Those knockings presented evidence to the masses that the opening of the gate depended upon the merit of the applicant and not upon the favor of the priest. Back to the dark may readily be traced the gushing forth of the spiritual application for recognition, too often met by the priestly devils with fire and sword.

The Cock Lane Ghost of 1760 answered questions by raps as is now done. Joseph Glanvil's Demon of Tedworth, 1661, is of the same kind. Peter Piquet, case Civil Court of Tours; the Holy Maid of Kent, beheaded by the butcher king, 1534; Joan of Arc, burned at the stake, 1431; then the thousands upon thousands murdered as witches,—but enough. As we trace the gory trail from the Egyptian priest down through Torquemada, Loyola, Luther, Calvin, Cotton Mather, Jonathan Edwards, the Andover school of theology, and efforts to force the closing of the Columbia ı Exhibition Sunday, we find fire and blood lavishly shed when possible, then threats of hell fire in more enlightened times to keep the masses in subjection to this hierarchic control for selfish interest. The question, "Do devils die?" is one of terrible interest to mankind, for it is only a question of power if such continue to exist, whether the Smithfield fires and horrors of the Inquisition shall not again be revived.

A half century since belief in Spiritualism was general in an undefined way ; all writers treated it as such. G. P. R. James's works abound in it, Scott, Marryat, Bulwer, Ainsworth, Burney, Jane Porter, Charlotte Bronte, all in fact accepted the belief. "Midnight Musings," by Washington Irving, is taken from the "American First Class Reader," published in 1831, and popular in our schools for a half century.

> " Ye spirits of Washington, Warren, Montgomery,
> Look down from above with bright aspect serene ;
> Come, soldiers, a tear and a toast to their memory,
> Rejoicing they'll see us as they once have been."

If that is not Spiritualism, what is it ? Yet it commenced one of our most popular songs early in the century. Have our people become better for rejecting such belief now ? A belief in spirit communion is the oldest, most encouraging, sensible, and progressive of any, but to be properly appreciated superstition and materialistic conventional ideas thereof must be abandoned. How shall I study the matter intelligently ? was my inquiry when sitting for manifestations one Sunday morning.

"Study your own mind, you may find a gem," was the instant reply. "Who are you ?" was asked. "Ballou," was answered. The evening after seemingly from another source came the following :—

> If the truth you wish to find, you must study your own mind,
> Learn its workings, its relation with the great unknown creation ;
> A gem we promise you shall find, in a knowledge of the human mind.
> Spirit friends around you gather, wishing much to help their brother,
> Seeking earnestly to find this gem of truth in the human mind.

False communications were constantly coming when outsiders sat with us, which caused me to impatiently inquire the cause. This reply followed :—

> False, with the true, you'll one day find,
> Is but the crossing of your mind
> With our dispatches as they are sent
> From our Summer Land to your Continent.

One day while waiting for an assistant to return, Charla commenced to converse about unreliable communications that came the night previous, then remarked, "There is so much that is totally unreliable that I don't believe there is any spirit life at all," which caused me to commence a remonstrance which was cut short by "Hush! hush!" from her, "I hear." Then after a short pause she repeated the following :—

> Your lives are tangled in with ours;
> We are not far away,
> We join men and women in their work,
> And children in their play.

Soon after, this followed :—

> If you will but be faithful,
> We will sometime prove to you,
> That spirit friends surround you,
> Who can and will be true.

> Harmony is heaven's own law;
> And to get the truth you ask,
> Conditions must be perfect,
> And your sittings not a task.

> We want you all to be of good cheer,
> And help each other while you are here;
> For in the life to come your riches consist
> Of the good you do to others in this.

My housekeeper, made up after the Mrs. Jellyby pattern, was inclined to be away much of the week, then pick up Sunday. I had remonstrated until tired, then let the matter pass with indifference. We made a practice of having a sitting Sunday morning. A racket in the laundry could be heard in the library, and one Sunday morning while reading I noticed that washing was being done. It continued for a short time then Charla and her sister came in to sit at the table, which was done for perhaps five minutes, then the table started for the door, then through the hall and kitchen to the back door, out through that down to the basement door; through that to the set tubs, there it immediately swung upon the sister's head and pulled that down and bumped it upon the edge of the tub, started again up the cellar stairs, through the hall and into the library, there floated up so that the top of the table hung upon the projecting cornice of the book shelves, hung there perhaps a minute, floated off and down in between the chairs, where we commenced, called for the alphabet and spelled out the following :—

> Give the seventh day to rest,
> To thought, culture, and to us.

The following Sunday the table started back over the same course previously named, but instead of entering the basement it continued on down towards a small water power. When partly down the hill the women looked up to the windows of a neighbor where several persons stood looking at our table performance. My assistants fled for the house, leaving me alone with the table, which closed the performance; but the next day's mail brought me notice that an Iowa court had appointed me chairman of a commission to settle a case reported in the first part of this book.

(See page 95.) Of course it is a matter of conjecture as to whether there was any connection between the table journey and the appointment. I think there was.

In answer to a pertinent question the following was the reply :—

> Sleep is a rest for the weary mind,
> Which, as it wanders free,
> Oft catches inspiration,
> And brings it back to thee.
>
> Ofttimes when the mind doth wander,
> While the body is at rest,
> Strange elements of earth
> This freed mind doth impress.
>
> Many times when the mind doth wander,
> Ideas which to earth are grand
> Are, in his sleeping hours,
> Stamped on the mind of man.
>
> Then, when he doth awake,
> And reason and thought control,
> These ideas are developed
> And given to the world.

Almost identically the same ideas were published in one of the Boston papers the same week, credited as coming from the Concord school of philosophy in explanation of " *The Whichness of Which*" as editorially explained. It was common to receive a communication purporting to come from Ballou, then, as accidentally would happen, to take up a *Banner of Light* and find the same subject treated in a lecture by Mrs. Richmond in the same way. With Charla alone, if she felt interested, answers to questions would be given that to me seem to the point ; but too often she was indifferent and the communications were the same. There was a persistent assertion that if I would persevere there was a band of spirits around me that in time would find a medium through whom reliable communications would be sent me. A niece came to visit us, a believer in the Advent doctrine, and that the spiritual manifestations were from the devil. Out of mere curiosity and bravado, perhaps, she consented to try the table with me and in five minutes was entranced and the series of communications that follow commenced.

This niece was subject to catalepsy and rarely was with us more than two or three days at a time. A heavy table would start from the side of the room and go to her ; or standing between her and myself that table would turn somersaults between us. In her cataleptic conditions in the light, her boots would be taken off and thrown across the room ; in that condition two persons could not raise her from the lounge. Sitting in darkness, her hands firmly clasped by others, her boots and stockings would be taken off and concealed in some out of the way place. Often she would come screaming from her room saying that some form had appeared to her, her description of which rendered recognition easy, in short, she seemed capable of producing every phase of manifestations known ; raps with her meant such as could be heard all over the house, she personated the spirits of those who had died in asylums, and described their cruel treatment, etc., etc.

I would like to study the materializations more, but want no medium that requires tying or test conditions.

Oct. 25, 1878.

First reliable telegram : May I come in ? Certainly, and welcome. My name is Julius N. Ives. I died September 15. I was seventy-six years of age, or should be now. I lived in Cromwell, Ct. A letter to the postmaster brought the following reply :—

CROMWELL, Oct. 29, 1878.

MR. EMERSON. *Dear Sir :*—Yours of the 25th of October is before me. Would say that Julius N. Ives came from Middletown, 18th of January, 1878, and made it his home in Cromwell with his brother, till he died, September 12, 1878, aged seventy-five.

A few questions : Was your niece a medium? Was she ever acquainted with Mr. Ives? Did she or any one see his death in a paper, etc.? I ask these questions because some have said that it might be the case.

Respectfully yours, JOHN STEVENS, P.M.

WILLIMANSETT, MASS.

Telegrams from——W-e-l-l where? Dec. 9, 1878. While sitting at the table one evening, there was a call for the alphabet, and as it was called, a message as follows was immediately spelled out :—

There is an old man here trying to get control of the medium.

All right, was our reply, go on.

Do you allow strangers to come in here? was asked. Certainly, you are very welcome!

Well, I didn't know as you would, but I was looking round and would kinder like to look in. I am from Saco, Maine. I was a blacksmith there many years. How old? Why, about seventy, but cannot tell exactly, for I have hardly recovered consciousness. There was a blank for a while, but I died about four months ago or early in the fall. My name was John Gains.

Can I give the name of some one there to write to? well, I guess I can, w-e-l-l, let me see ; why, write to S. S. Mitchell, Druggist, Main Street. He and I were old friends. Tell him that I would like to take one of them sly drinks from the barrel. Ah, it wasn't every one that could get a drink there, but I could, notwithstanding the Maine law.

Had you no family? was asked.

W-e-l-l my family—was kinder scattered. Oh, yes! I had a son, named Albert, he is in Washington, yes, and I had a darter, her name is—Sarah—Sarah, oh, Sarah Elizabeth, she is married, no, she is a widder, her name is—well I can't think of that chap's name. Oh, if you write, ask Mitchell about Horace Watterhouse. Poor fellow ! he worked for me thirty or forty years and at times would go upon a spree and I used to take care of him, but now, poor fellow, I don't know how he gets along. I am looking round and will come again soon.

He came the next night and was told that a letter had been sent to inquire about those "sly drinks."

There now, did you write about them? Certainly I did! W-a-l-l there now, I hadn't orter said so, but I allers was saying such things! Had Mr. Gains visited us in the body he could have appeared no more real ; he remained with us some time, gave many particulars that made him very welcome as a visitor. The communication was immediately sent to Mr. Mitchell, who seems to have employed a lawyer to look the matter up as may be seen.

SACO, ME., Dec. 12, 1878.

MR. POSTMASTER:—Will you please be so kind as to inform me if there is a man now residing in Willimansett by the name of James Emerson? If so, about how old is he? What is his occupation? Is he a man of good standing in the community? To what religious denomination does he belong, if any? How long has he resided in your place and where did he come from when he came to your place?

These questions and information are not asked for the purpose of injuring, in any way, Mr. Emerson or any other person, but from the best of motives, and I can satisfy you of my reliability if necessary.

Please give the information and greatly oblige,

Very respectfully, your most obedient servant,
F. W. GUPTILL,
Counselor and Attorney at Law, 99 Main Street.

A reply by return mail is desirable.

Mr. Guptill was furnished with the required information, then Mr. Mitchell made his reply, but it will be seen that he does not plead to the sly drinks. As I refused to suppress the communication Mrs. Emmons was called in as shown.

SACO, Dec. 16, 1878.

MR. JAMES EMERSON. *Dear Sir:*—In reply to your letters of inquiry about the late Mr. Gains I have to say:—

1. John Gains, a well known citizen of this place, a blacksmith, and a man of considerable property, died here last September.

He left several children, all of whom lived with him except his only son, Albert, who has resided in Washington many years. Of his daughters, one is a widow and her name is Sarah Elizabeth.

2. A man by the name of Horace Watterhouse worked for Mr. Gains many years and was always carefully looked after by him when the poor fellow (as the communication calls him) had yielded too much to his passion for drink—he still continues the blacksmith business under the direction of the administratrix.

Mr. Gains was one of the best friends I had in Saco and I don't wish to have the communication published, neither do I think his family would; still I should like to hear further from you in regard to this matter, although not a "believer." Yours truly, S. S. MITCHELL.

SACO, Dec. 23, 1878.

MR. EMERSON:—I have read your letters to Mr. S. S. Mitchell with a good deal of interest.

I think the communications that you have received are certainly remarkable, although very unsatisfactory.

Among my most valued friends are some of your belief, so that I have *seen something* and *heard a great* deal of spiritual manifestations without, however, having my views at all affected by it.

When our friends depart this life, I hope and believe it is for a better and happier existence—hence their burden of earthly care and trouble must be left behind. And because the infirmity of "poor Horace" was a trouble to my father during his lifetime seems to me to be the *very* reason why he should be relieved from it now. And if he is able to communicate with his friends here, there are matters (mysterious to them, but clear as the noonday to him) that would claim his attention. And I do not understand why one member of his family should be remembered while another is forgotten.

I shall be interested in any further developments that may occur, and trust that you will sacrifice your desire to publish this matter at least for the present, to the wishes of the friends and family of the late John Gains.

Most respectfully yours, MRS. S. E. EMMONS,
Box 117. Saco, Me.

October 15, 1878.

The next was: My name is Charlotte Wooster. I lived in Litchfield, Conn., and died September 12. I was twenty—no, I cannot remember my age. Tell my friends not to mourn for me. I am happy, and do not wish to come back.

Reply :—

Nov. 6, 1878.

JAMES EMERSON. *Dear Sir:*—Charlotte Wooster, a daughter of Joseph Wooster of this village, died here on the 7th day of September last past, in her thirty-third year.

Very truly,   L. W. WAPELLS, P. M.

The next came as follows : Anna S. Cookson, Coopers Mills, Maine. I died—no, I cannot remember when, but recently. Tell my friends not to grieve for me.

COOPERS MILLS, MAINE, NOV. 1, 1878.

JAMES EMERSON, ESQ. *Dear Sir:*—In reply to your letter, would state that Anna S. Cookson died the 20th of October. Was twenty years and six months old. She was sick but five or six days. Cause of her sickness and death, to the public unknown. Should be happy to hear from you again on the matter, and would like to know if the spirit told the cause of her death.   Yours truly,   GEORGE W. GREENE,

Assistant Postmaster, Coopers Mills, Me.

She died Tuesday, the communication came the following Friday evening, or there was an interval of three days between death and communication.

We were again informed that a stranger desired to get control of the medium. On doing so the name of Hazen Kimball of Hopkinton, N. H., was given, and his age as seventy-six. Reply :—

HOPKINTON, N. H., Nov. 13, 1878.

Dear Sir:—Hazen Kimball died March 28, 1877, aged seventy-six years and seven months. Lost a relative in Chelsea, Mass., myself. Should be pleased to hear from the party.

With respect,   DAVID L. GAGE, P. M.

The next was : My name was Stephen Sibley, of Chelsea, Mass. I died the 9th of June. I was sixty-four years of age. Reply :—

No date.

JAMES EMERSON. *Sir :*—Stephen Sibley, a resident of this city for more than forty years past, and one of its principal business men, died on the 9th day of June last, aged sixty-four years, three months, and sixteen days.

SAMUEL BASSETT, City Clerk.

WILLIMANSETT, MASS., Nov. 18, 1878.

Dear Sir :—It has happened recently that I have had seven communications from those purporting to be in the spirit world. I have written to each place where these spirits claim to have lived while in this life. Six of the seven have been answered and confirmed in every essential particular, the only difference being a day or two in date of death. The last communication was as follows :—

My name is Cyrus Alden, of Leeds, Maine, ninety-three years old, a soldier of 1812.

Will you inform me whether such a person has resided there within your knowledge, and oblige,

Yours truly,   JAMES EMERSON.

Sir :—There was such a man as Cyrus Alden, died March, 1877. I think some one is fooling you by getting these dates, and pretending that they came from spirits.

This answer was from the one to whom my letter was addressed, the postmaster of the place named; one of the profound kind that knows it all.   J. E.

Nov. 3, 1878.
Another: I wish to control Alice. I died in Paterson, N. J. Well, I can't remember the date, but in the early part of the summer. I was seventy-five years of age.

Write to the Grant Locomotive Company, Paterson, N. J., for information.   WILLARD W. FAIRBANKS.

PATERSON, N. J , Nov. 5, 1878.
JAMES EMERSON, ESQ., Willimansett, Mass.
Dear Sir :—Willard W. Fairbanks was formerly superintendent of these works; he died last May or June, aged seventy-five years. I have sent your letter to his family.   Very respectfully,
D. B. GRANT,
General Manager.

WILLIMANSETT, MASS., May 22, 1882.
MR. POSTMASTER, Franklin Falls, New Hampshire.
Dear Sir :—On Saturday evening last, while sitting in conversation with a niece, she suddenly became seemingly unconscious; then, shortly, in a very feeble voice, exclaimed: "My name was Benson (Samuel Benson), of Franklin Falls, New Hampshire. I was eighty years of age. I died four or five months ago, or in January last, of heart disease." Now, neither my niece or myself had ever heard of Franklin Falls, though some thirty-five years ago I resided a short time in Warner, also in Concord, and knew of Franklin through what were at that time termed the Akin boys, or the Akins, who were considered inventors of various devices, an awl haft for one. I would be obliged to you if you will be so kind as to inform me if there was such a person as Mr. Benson who died there in accordance with what I have written.
Yours truly,
JAMES EMERSON.

FRANKLIN FALLS, May 23, 1882.
Samuel Benson died about the time mentioned. Was about eighty years old.   Yours truly,
P. M.,
Franklin Falls, N. H.

Another answer:

FRANKLIN, N. H., July 10, 1882.
JAMES EMERSON, ESQ. *Dear Sir:*—I think you left a few words out of your first question. I understand its import to be this :—
1st. Was there such a resident of Franklin Falls as Samuel Benson ?
Answer: There was.
2d. When did he die?
Answer: January 21, 1882.
3d. What was his age at the date of death?
Answer: His physician gave me his age as eighty-two years, seven months, and four days. But the *Merrimack Journal*, published here January 27, 1882, third page, second column, says he was nearly eighty. His daughter is away on a journey. As soon as I learn her address, I will try to remove the doubt.
4th. Was there any supposed cause of his death?
Answer: His physician says it was a disease of the heart called *angina pectoris*.   Very respectfully,
J. L. THOMPSON.

⌐ or years past my investigations have mostly been at my home ; the same care has been observed as in mechanical, hydraulic, dynamic, and caloric trials. Witnessing the developing of cause and effect has amply repaid the time expended.

During the past two years my boy, now nearly thirteen years of age, and myself have formed our *circle*, at times others have joined, but such usually come with preconceived ideas, generally destitute of desire or ability to judge of force, time, or space necessary for effect, consequently it is time lost.

One, a doctress, came, prodded the boy with a pin, then began to orate about " reflex action "; she possibly had some idea of what she meant, I had not. The most of my investigations with my boy are in the dark, not all. An ordinary dressing table is used.

Often while sitting with this table between us the boy is thrown upon the bed, the round in the feet of the table placed across my knees, the top of the table resting against my forehead, the boy's chair is placed on top of the table, where there is barely room for the chair to stand, the boy is then placed standing in the chair, where he begins to declaim.

Then the voice of a child takes the place of his. This voice pronounces the longest words just as well when the boy is gagged as when his mouth is free. Untying feats have been performed by the boy, but it has seemed too brutal for me to care to experiment in that way.

He has shown feats of strength of the *Lulu Hurst* order that would be impossible in his normal condition.

The mesmeric influence is from the spirit side, at least not from me. A simple word, Minnie, would cause him to drop while crossing the room.

During the French trial some two years since, reported in our leading papers, I invited Judge Bond to witness the influence that might be brought to bear against another. He declined to do so, and often since the query has arisen in my mind as to whether the bench has a tendency to expand the mind. The practice of the law cannot be productive of the best thoughts. I have at times witnessed materializations but should prefer to study such at my home. I have seen some I believed real, others that I did not. At times the medium's person or hands are used direct, but where the conditions are right the forms or hands are drawn from the medium or those present. Where voices are heard the organs of those present are used to produce them. The spirit of some one may be present but the organs of the physical being must be used to make that spirit's presence perceptible.

MIND READING

is now so generally accepted that I do not care to go into particulars in proof, though it would be easy to furnish positive evidence of the fact, and that it begins with the very young even before the child can talk.

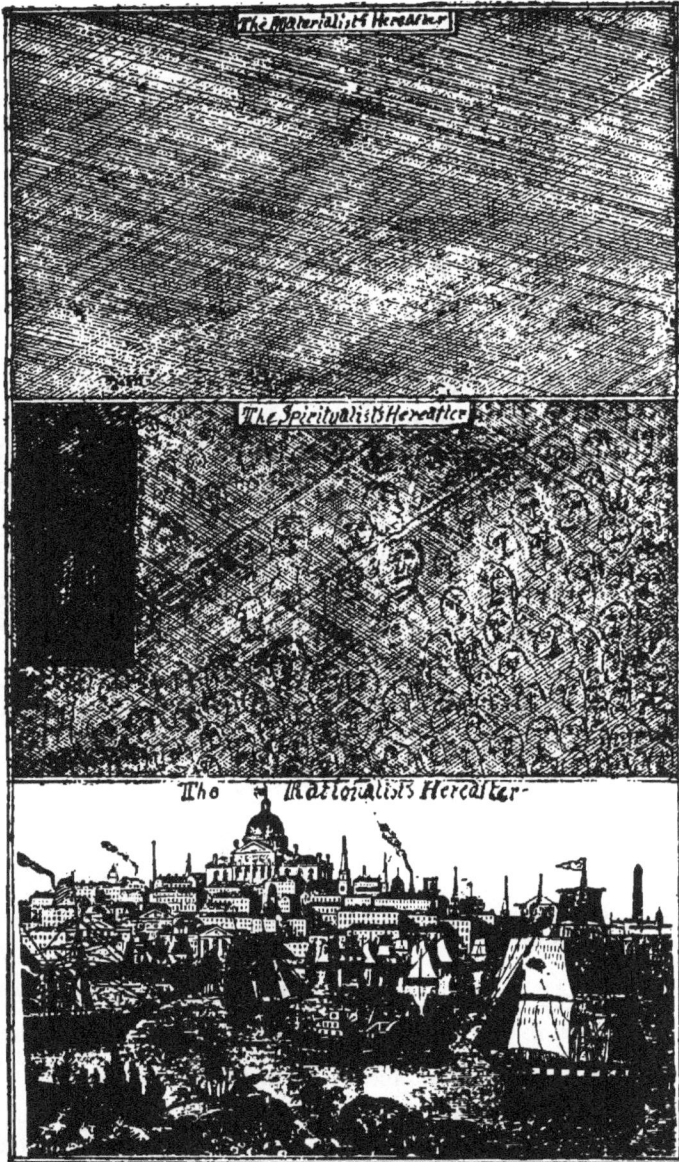

The Materialists Hereafter

The Spiritualists Hereafter

The Rationalists Hereafter

PROPHESYING FUTURE EVENTS.

Of the possibility of foreseeing events abundance of evidence has been furnished me, but two cases will only be given here.

Sitting with Charla and her sister, our attention was called, then:—

Five years from this day, one of you three
Folded in the bosom of mother earth will be.
April 21, 1878.

The communication caused us often to think of it. Three years or so passed by, then a communication from some medium came to us saying that we three were not meant, but the three of the family, Charla, her brother, and sister; the remaining members of their family. Some months afterwards news came of the sudden death of the brother in the far west.

ANOTHER CASE MORE DECISIVE.

At my home three of us were sitting at the table talking of the death of a little child of one of the sitters; the other sitter was entranced and said, "Another little child will soon come over here." I said, "I hope not from the one who has just lost the one spoken of." "There is a star over her head, which signifies peace. It is not hers," was the reply. Then she exclaimed, "Within four months from this day and within sixty rods from this house another little child will come here."

The time passed on; about two months after a little child sickened and died, then it was said that the prophecy had been fulfilled, but to me it seemed not, for she had no connection with the sitters.

The fourth month had well advanced when the youngest child of the prophetess sickened and died but a few days previous to the expiration of the time. These statements may be depended upon to the letter.

A COINCIDENCE?

"Not a sparrow shall fall without the Father's notice."

Every effect has a cause, even a coincidence.

During an investigation of materialization in my library a chair was suddenly jerked from me, at the same instant a streak of moonlight pierced through the blinds upon the medium, her arms were folded and the face of the dead could not have appeared more serene than was hers.

By opening the doors of a cabinet quickly after arms have appeared from the aperture, a sort of halo of those arms may be traced back to the shoulders of the mediums though the real arms are firmly bound down behind their bodies. I have seen arms that must have been projected eight feet from the bodies of the mediums.

It is time the brutal tying and test conditions were done away with and the investigation should be done by kind but cautious observers. If the ruffianly can only be convinced through brutality let them go unconvinced. A ruffian is none the less a ruffian because well dressed.

To investigate intelligently one must expect to meet spirit friends as they left the body.

Seeking spirits is done too much upon the plan of searching the scriptures.

The Christian that exacts twelve per cent. interest does not look for the passage that condemns usury but that which says, Render unto Cæsar the things that are Cæsar's, the other fellow to render, he is likely to find what he seeks.

A sponge surrounded by fluid takes in sediment according to the fineness of its fiber. To make my meaning clear, take, say, the authoress of the Little Pilgrim. Communications through her should be sweet, charitable, but impracticable except in Bellamy's Utopia a thousand years hence. The authoress of Beyond the Gates should produce communications very proper, slightly progressive, not too much so, for her patrons have weak digestive powers, an overdose would cause the grip, which would have a disastrous effect upon the dollar product.

The authoress of Is this Your Son, My Lord? should produce bright, brainy, intelligent, progressive, practical, womanly ideas, clear in style as the tone of a silver bell. Would there were more like her. Mentally I shake hands with her and say, Go on.

Unless under favorable conditions the investigator is as likely to obtain reflections of his own mind as information from the spirit world. The physical man is but an engine operated by spirit power and becomes dead as that power is withdrawn, as does the steam or electric engine, but, unlike those, once withdrawn it cannot again enter.

Surrounded by intelligence the ordinary brain receives and guides it in ordinary channels. The seeker finds more.

The question is daily asked, Do you think spirits would come back and tip tables ? Certainly, if that is the way they can best make their presence known. They do worse things than tip tables while in the body, why not after leaving that ? The murderer evoluting at once to a reserved seat along side the Father is not an encouraging idea; the real effect of ending the career of a murderer on the gallows is more likely to be like that of having a sewer empty its contents into a thickly settled neighborhood.

Spirits returning show all of the passions common in this life. I have seen a table dash at a man with all the fierceness possible in the physical man. So much force was used that the two legs caught by the defendant were splintered in a moment, and the contest was continued until every joint of the table was separated.

Where is the heaven located that Christians talk so much of and seem so much to dread starting for ?

More than fifty years ago at a prayer meeting, a *brother* kindly informed me that I was liable to be sent to hell that night. "And just think," he exclaimed, "if this earth was made up of fine sand and a bird should carry away a grain once in a million of years, in time it would all be gone, and your punishment would be no nearer ending than at its commencement; while all of that time I hope to be singing the praises of the Creator, not through any merit of my own but through the atoning blood of Christ."

There is no reason to doubt but what the soul of such a saint would be small, yes, very small; but even the soul of any one willing to be saved through the sufferings of another may be as large as a fine grain of sand. As there are thousands of such souls freed from the body daily, it can be comparatively but a short time before the bulk of souls will exceed that of the earth, and as the other planets should be in the same condition what is to become of such souls? But first, where are such souls to come from? If you constantly check out without depositing, your checks will not be cashed. So of souls. The Spiritualist's hobby is no different; give us sense instead of old fads. How common it is for those who have advocated the belief in Spiritualism to cool and say there is no advance, often becoming as materialistic as a Christian.

It should not be so. The phenomena are grand and instructive if studied with consideration. Materialized forms invariably represent this life and I have been unable to obtain a shadow of evidence that there is any progression in the spirit life; reason does not teach any such thing. Progression is in this life, the spirit clings to this life. The body and everything we know of physically dies, disintegrates, and comes again in the same or some other form. So of the spirit, in time like the body it disintegrates and comes again. The bosh of the Spiritualist crank talking about ancient spirits, is bosh indeed. By constant renewal the child or man who dies prematurely again has his chance, and in this way man will evolute to a higher being. Bulwer has tried to describe a man six thousand years of age and a poor stick he made of him. Rider Haggard in weak imitation tried to delineate a woman of two thousand years and had her terminate in appearance as a monkey. There is no data for such. What would Archimedes have done had he tried to describe the locomotive of to-day? The gods grind slowly but fine. "I know that my Redeemer liveth and that I shall see God," I believe to be correct, the Godhood in man being the redeemer, and in time man will have so improved as to be capable of seeing the Creator. Man, being of God, yet progresses by degrees, and there is no reason to suppose that the Creator could have made a perfect world at the start.

The ideas of the Spiritualist are but the old fads of Christianity.

"Where two or three are gathered together in my name there will I be in the midst of them," is quite as much to the point with the Spiritualist as it once was with Christians.

"Organization" is the constant whine of a certain grade of Spiritualists.

Organization of an army means concentration of brute force under one mind, individuality spreads information.

Wesley was familiar with spirit manifestations, his preaching gushed with spiritual fervor.

Universalism was aggressive, argumentative, and full of push.

Unitarianism represented by Theodore Parker was a power that moved the world.

All, now organized and immensely respectable a–n–d frozen to death.

Could their houses be utilized as ice houses and summer hotels, there would be no burning through spontaneous combustion nor from the friction of high insurance.

It has happened several times to my knowledge that noted inspirational speakers of the Spiritualists have been taken up by the Unitarians to be educated. The influence seems to be death to inspiration, for such speakers are heard of no more.

Inspiration, I believe to be as much in force to-day as at any time since man's existence; it may be from good or bad influence.

There is not one article in this book relative to Spiritualism or religion presented as intended when commenced.

Don't organize, and better still don't try.

Communications purporting to come from the spirit world have long been familiar to us, mostly of a personal character, very interesting to us but unlikely to interest others. At first such communications were looked upon as of too sacred a source to be treated as an ordinary matter, but experience soon taught us that such came from minds like our every day associates and are to be treated the same ; the long-faced saint is not the prevailing intelligence there any more than here.

To the question, what are dreams, the reply was :—

> " Dreams are little devils
>   That play tunes upon your brain,
> When you awake they escape
>   And at night return again."

The following was given to an unbelieving sitter of the puritanic faith :—

> " All the time that is not well spent in this life while you are here,
>   Is borrowed from the early part of your life in the other sphere.
>   Look up higher, look up higher,
>   Spirit friends will come the nigher,
>   Whispering to you all the time
>   Of a life that is sublime.
>   Sublime because it is so broad,
>   That every soul knows there's a God ;
>   A God who doeth all things well,
>   Who has never made a fiery hell."

INSPIRATION.

Through priestly selfishness we have been educated to think that inspiration comes from an incomprehensible source located in an impossible place, instead of being an every day affair right at home.

If nature abhors a vacuum it no less abhors the useless. A heaven or hell of the Andover type could benefit neither God nor his children, while we have ages of evidence to prove that a belief in such myths produces the meanest minds.

To think, opens the channels for inspiration. That there is a spiritual body, one more substantial than generally supposed, I believe susceptible of tangible proof. The conventional cold blast, attendant upon ghostly visit, is not a matter of imagination, but a reality. The first symptom of approaching trance is cold extremities, at least with the neophyte. Excitement leads to ecstasy, often to inspiration, which may come to the individual or through a medium sitting with others, when the character of the communication may be dominated by the minds of those sitters, or the communication may come from the minds of those or from a disembodied mind near by, or, like sinking a well for water, a stream may be struck, the source of which is far away. These communications, like the ancient oracles, may have a very different meaning from that seeming at the time. For instance, a story is told that the duke of Buckingham, being very popular, aspired to the throne of Henry the VIII. An aspirant for favor consulted a medium and was told that the head of the duke would soon be the highest in the land, which was taken to mean that he would soon be king, instead of which he was beheaded and his head stuck upon a pole above the gates of the city. Communications may be answers to questions asked mentally by the sitter previous to sitting in a *circle*. Such a jumble is made of the matter that but little dependence can be placed in communications, but the fault is with us.

As the news of President Lincoln's assassination spread over the country, gloom and excitement prevailed, business men sent to their minister for mottoes to be placed over their doors, ministers preached and choristers chanted of it. As I was preparing to attend a meeting that would discuss the event, the following lines came to me which are here given as proof of a fact for which I can vouch:—

A world's true friend is dead,
Oh God, and is it thus
The sins of ourselves and sires
Should fall upon the just?
Could not the four past years
With their hosts of bloody dead
Satiate the slaver's cruel heart,
That he strikes our nation's head?

Could not Fort Pillow's blood scenes
Fill e'en a fiend's fiercest dreams,
That they shoot and starve our men
In their filthy prison dens?
That slavery may be understood,
Must we give our great and good?
Must the highest in the land
Fall, by the assassin's cowardly hand?

Oh God, we know thou art just,
Fill our mourning hearts with trust,
Lead our newly appointed head
In the footsteps of the dead.
For ourselves we give our prayers,
For the slain, our heartfelt tears.
On slavery be the stain,
Let us hope for freedom's gain.

www.ingramcontent.com/pod-product-compliance
Lightning Source LLC
Chambersburg PA
CBHW020901210326
41598CB00018B/1743